NONLINEAR THEORY OF PSEUDODIFFERENTIAL
EQUATIONS ON A HALF-LINE

NORTH-HOLLAND MATHEMATICS STUDIES 194

(Continuation of the Notas de Matemática)

Editor: Jan van Mill
Faculteit der Exacte Wetenschappen
Amsterdam, The Netherlands

ELSEVIER

2004

Amsterdam – Boston – Heidelberg – London – New York – Oxford
Paris – San Diego – San Francisco – Singapore – Sydney – Tokyo

NONLINEAR THEORY OF PSEUDODIFFERENTIAL EQUATIONS ON A HALF-LINE

Nakao HAYASHI
Osaka University
Osaka, Japan

and

Elena Kaikina
Instituto Tecnológico de Morelia
Morelia, México

ELSEVIER

2004

Amsterdam – Boston – Heidelberg – London – New York – Oxford
Paris – San Diego – San Francisco – Singapore – Sydney – Tokyo

ELSEVIER B.V.
Sara Burgerhartstraat 25
P.O. Box 211, 1000 AE Amsterdam
The Netherlands

ELSEVIER Inc.
525 B Street, Suite 1900
San Diego, CA 92101-4495
USA

ELSEVIER Ltd
The Boulevard, Langford Lane
Kidlington, Oxford OX5 1GB
UK

ELSEVIER Ltd
84 Theobalds Road
London WC1X 8RR
UK

First edition 2004

Library of Congress Cataloging in Publication Data
A catalog record is available from the Library of Congress.

British Library Cataloguing in Publication Data
A catalogue record is available from the British Library.

ISBN: 0-444-51569-0
ISSN: 0304-0208

⊗ The paper used in this publication meets the requirements of ANSI/NISO Z39.48-1992 (Permanence of Paper).
Printed in Hungary.

Contents

Preface

This book is the first attempt to systematically develop a general theory of the initial-boundary value problems for nonlinear evolution equations

$$u_t + \mathbb{N}\left(u\right) + \mathbb{K}u = f,$$

with pseudodifferential operators $\mathbb{K}u$ on a half-line or on a segment, where the nonlinear term $\mathbb{N}\left(u\right)$ depends on the unknown function $u(x,t)$ and its derivatives. We study traditionally important problems of a theory of nonlinear partial differential equations, such as local and global existence of classical and generalized solutions to the initial-boundary value problem, properties of smoothing of discontinuous initial perturbations and the asymptotic behavior of solutions for large time. Up to now the theory of nonlinear initial-boundary value problems with a general pseudodifferential operator $\mathbb{K}$ involving dispersive or dissipative equations, such as Korteweg-de Vries and nonlinear Schrödinger equations, has not been well developed due to its difficulty. There are many open natural questions which we need to study. Firstly how many boundary data should we pose on the initial-boundary value problems with a pseudodifferential operator $\mathbb{K}$ for its correct solvability? There are some results in the case of nonlinear differential equations. However, as far as we know there are few results in the case of nonlinear nonlocal equations. The methods developed in this book are applicable to a wide class of dispersive and dissipative nonlinear equations, both local and nonlocal. One of the main purposes of this book is to develop general methods for studying the large time asymptotics of solutions to initial-boundary value problems for nonlinear nonlocal equations. To that purpose we apply a minimal set of estimates of solutions which always include the $\mathbf{L}^{\infty}$-norm, so we use a concept of the semiclassical solution.

We would like to thank Professors T. Ozawa, F. Ruiz Paredes, L. Guardado Zavala and K. Kato, who have read the manuscript and made many useful suggestions and comments. Especially we would like to thank Professors P. Naumkin and I. Shishmarev for many fruitful discussions on the subject of this book.

CHAPTER 1

Introduction

The theory of nonlinear evolutionary equations plays an important role in the contemporary mathematical physics. Such equations serve as a basis for mathematical models to describe the different phenomena in modern Physics, Biology, Technology and others fields of science. As examples, we mention famous equations of Burgers, Schrödinger, Korteweg-de Vries, Klein-Gordon (see [131], [24]), which are picked out from complicated natural mathematical models via the separation of the basic mechanism responsible for interesting nonlinear effects. These equations are deduced from the fundamental conservation laws (of energy, mass, impulse, and so on.), so they have wide applications in describing the distinct processes of wave propagation, which are dissimilar at first sight. It is natural to study the principles of evolution first basing on simple nonlinear evolution equations, which enable us to fix principle directions in the development of the analytic theory of nonlinear evolutionary equations.

Tremendous progress in the study of nonlinear evolutionary equations is related to the appearance of powerful supercomputers allowing us to carry out complicated numerical experiments. A correctly organized numerical experiment permits us to discover new nonlinear laws and effects that are difficult to predict analytically, in addition gives us some hypothesis for further analytical study (see [37]). Exactly in this way the Inverse Scattering Transform (IST) Method was discovered in 1967 (see [16]). Now IST is used to find explicit solutions of a particular class of nonlinear evolutionary equations and to study its surprising properties. In many cases it is possible to find asymptotic behavior of solutions by the IST method (see [1] and references cited therein). Unfortunately we do not know any effective criteria of the applicability of the IST method, and its application is limited today. On the other hand it is possible to obtain many properties of solutions for nonlinear evolutionary equations by virtue of general functional analytic methods. Thus the traditional questions such as existence and uniqueness of classical and generalized solutions, global in time existence, and differentiability properties of solutions were solved for a wide class of nonlinear evolutionary equations. Various nonlinear effects were studied, for example blow up of waves in finite time, existence and stability of self-similar solutions, the existence of the symmetry groups, solitons, and so on.

Sometimes the results of a nonlinear theory can be obtained by linearizing the problem and applying the perturbation theory with respect to a small parameter.

It is clear that this approach does not work, when the nonlinear effects are essential in the problem: hence the behavior of solutions does not resemble the corresponding linear case. Therefore it is very important to develop the nonlinear theory. Unfortunately there is no general approach for solving nonlinear equations: every equation has its intrinsic individual structure and requires a set of special analytical method of investigation. On the other hand many of methods developed for particular equations can be generalized for some classes of nonlinear evolutionary equations. For example, existence of generalized solutions is a well developed field (see [91]). However, generalized solutions do not permit us to ascertain of the behavior of solutions; therefore, classical (or semi classical) solutions should be considered, for example, when studying the asymptotic behavior of solutions.

Asymptotic methods play an important role in the theory of nonlinear evolutionary equations. Only a few nonlinear equations can be solved explicitly, so it is very important to have an approximate analytical representation of solutions and to estimate an error. Asymptotic formulas allows us to know such basic properties of solutions as how solutions grow or decay in different regions, where solutions are monotonous and where they oscillate, which information about initial data is preserved in the asymptotic representation of the solution after large time, and so on. It is interesting to study the influence of the nonlinear term in the asymptotic behavior of solutions. For example, compared with the corresponding linear case, solutions can obtain rapid oscillations, can converge to a self-similar profile, can grow or decay further, and so on. It is very difficult to obtain this information via numerical experiments, since computational errors increment for large time values and put into doubt the validity of the result. Thus asymptotic methods are important not only from the theoretical point of view, but also they are widely used in practice as a complement to numerical methods. Note that in applications "large time" does not mean necessarily large value. Actually this is a finite time, which is sufficient for finishing all transient processes related with the initial perturbation.

The theory of asymptotic methods for nonlinear evolutionary equations is relatively young and traditional questions of general theory are far from their conclusion. A description of the large time asymptotic behavior of solutions of nonlinear evolution equations requires principally new approaches and the reorientation of points of view in the asymptotic methods. For example, in the linear theory usually it is sufficient and acceptable to suppose that functions are infinitely differentiable and have a compact support; however, in the nonlinear theory these requirements are too strong.

Asymptotic theory is difficult even in the case of linear evolutionary equations (see [25], [128]). Moreover in the case of nonlinear equations it is necessary to prove global existence of classical solutions and to obtain some additional estimates to clarify the asymptotic expansions. Every type of nonlinearity should be studied individually, especially in the case of large initial data. In spite of the importance and actuality of asymptotic methods there are relatively few results for nonlinear evolutionary equations - not only for the initial-boundary value problems, but also in the case of the Cauchy problem. Nevertheless a great number of publications have dealt with asymptotic representations of solutions to the Cauchy problem for nonlinear equations in the last twenty years. While not attempting to provide a complete review of this publications, we do list some known results [5], [7], [8], [9], [11], [12], [26], [27], [31], [32], [41], [42], [46], [55], [69], [80], [81], [104], [103], [106], [117], [118], [123], [138], [139], [140], where there were obtained optimal time decay estimates and asymptotic formulas of solutions to different nonlinear local and nonlocal equations in the case of the Cauchy problem. However, there are few results concerning large time asymptotic behavior of solutions for the initial-boundary value problems.

This book is the first attempt to develop systematically a general theory of the initial-boundary value problems for nonlinear evolution equations

$$
(0.1) \qquad u_t + \mathbb{N}(u) + \mathbb{K}u = f,
$$

with pseudodifferential operators $\mathbb{K}u$ on a half-line or on a segment, where the nonlinear term $\mathbb{N}(u)$ depends on the unknown function $u(x,t)$ and its derivatives. We define a general class $\mathcal{K}^\alpha$ of symbols $K(p)$ as a set of complex functions $K(p)$ such that $K(p)$ is analytic for all $\operatorname{Re} p \geq \sigma \geq 0$ and

$$
K(p) = \sum_j C_j p^{\alpha_j} + O(p^\beta) \text{ as } p \to \infty, \ \operatorname{Re} p \geq \sigma
$$

where $\alpha_j \geq 1$ and $\beta < 1$. We use the notation of the symbol O as $\phi = O(\psi)$ means that there exists a constant C such that $|\phi| \leq C |\psi|$.

We introduce the pseudodifferential operator $\mathbb{K}$ on a half-line by virtue of the inverse Laplace transformation

$$
(0.2) \quad \mathbb{K}u \ = \ \frac{1}{2\pi i} \sum_j C_j \int_{-i\infty+\sigma}^{i\infty+\sigma} e^{px} p^{\alpha_j} \left(\widehat{u}(p) - \sum_{l=1}^{[\alpha_k]} \partial_x^{l-1} u(0) p^{-l} \right) dp
$$

$$
+ \frac{1}{2\pi i} \int_{-i\infty+\sigma}^{i\infty+\sigma} e^{px} \left(K(p) - \sum_j C_j p^{\alpha_j} \right) \widehat{u}(p) dp,
$$

where $\widehat{u}(p,t) = \int_0^{+\infty} e^{-px} u(x,t) dx$ is the Laplace transform of the function $u(x,t)$ with respect to the space variable $x \geq 0$. Note that the pseudodifferential operator $\mathbb{K}$ converts into a purely differential operator, when the symbol $K(p)$ is an integer power of p.

Similarly we define the pseudodifferential operator $\mathbb{K}u$ on a segment $[0, a]$

$$\mathbb{K}u \;=\; \frac{(1 - \theta(x - a))}{2\pi i} \sum_j C_j \int_{-i\infty}^{i\infty} e^{px} p^{\alpha_j}$$

$$\times \left(\hat{u}(p,t) - \sum_{l=1}^{[\alpha_j]} \frac{\partial_x^{l-1} u(0,t) - e^{-ap} \partial_x^{l-1} u(a,t)}{p^l} \right) dp$$

(0.3) $$+ \frac{(1 - \theta(x - a))}{2\pi i} \int_{-i\infty+\sigma}^{i\infty+\sigma} e^{px} \left(K(p) - \sum_j C_j p^{\alpha_j} \right) \hat{u}(p) dp,$$

here $\hat{u}(p) = \int_0^a e^{-px} u(x) dx$ and $\theta(x) = 0$ for $x \le 0$ and $\theta(x) = 1$ for $x > 0$.

Note that the inverse Laplace transform gives us the function which is equal to 0 for all $x < 0$, so that in the multiplication by the function $(1 - \theta(x - a))$ the operator $\mathbb{K}u$ vanishes outside of the interval $(0, a)$. Thus the solution $u(x, t)$ can be considered for all $x \in \mathbf{R}$ prolonged by zero outside of the segment $[0, a]$.

There are many physical problems, which are described in a unified way by the nonlinear nonlocal equation (0.1). We recall famous classical examples. The nonlinear nonlocal Whitham equation

(0.4) $$u_t + u u_x + \mathbb{K}u = 0,$$

represents, as particular cases, many equations that are of great interest for physical applications, for example, the Korteveg-de Vries equation $u_t + u u_x + \alpha u_{xxx} = 0$ (if $K(p) = -i\alpha p^3$) and Korteveg-de Vries-Burgers equation $u_t + u u_x + \alpha u_{xxx} + \beta u_{xx} = 0$ (if $K(p) = \alpha p^3 - \beta p^2$); both are well-known in the theory of surface waves in water and in nonlinear acoustics for fluids with gas bubbles.

Ott, Sudan and Ostrovskiy [**109**], [**108**] proposed the following generalization of the Korteveg-de Vries equation (here $K(p) = -i\alpha p^3 + 1 + \sqrt{p}$)

$$u_t + u u_x + \alpha u_{xxx} + u + \frac{1}{\pi} \int_0^x \frac{u_s(s,t)}{\sqrt{x-s}} ds = 0,$$

for studying the ion-acoustic waves in plasma with the Landau damping.

All these equations contain the nonlinear term $u u_x$ of the shallow water theory. It is reasonable to start from equations with other kinds of nonlinearities that are typical for problems of modern mathematical physics. Thus we arrive at the following model nonlinear nonlocal Shrödinger equation

$$u_t + |u|^2 u + \mathbb{K}u = 0.$$

A particular case of this type of equation is the generalized Landau-Ginzburg equation ([**125**])

$$u_t + i|u|^2 u - u_{xx} + \alpha u = 0,$$

where $\alpha > 0$. Another example is the equation ([**137**])

$$u_t + |u|^2 u + \int_0^{+\infty} q(x-s) u(s,t) ds = 0,$$

encountered in the theory of the Langmuir waves in plasma. Finally we mention the Kuramoto-Sivashinskiy ([**89**], [**121**]) equation as an example of another type of nonlinearity

$$u_t + \frac{1}{2}\left(u_x\right)^2 + u_{xx} + \alpha u_{xxxx} = 0,$$

which is applied, for instance, in the theory of combustion to model a flame front and also in the study of two dimensional turbulence. As we can see all the above-mentioned physical examples are described in a unified way by equation (0.1).

Thus, the study of the nonlinear nonlocal equations enables us to proceed from the analysis of individual equations to the investigation of wide classes of nonlinear equations that are of great interest for a physical application and their rigorous mathematical study. Therefore we now focus our attention on the development of a theory of nonlinear nonlocal equations.

The case of the Cauchy problem is more developed and a number of publications have been devoted to classical models of nonlinear nonlocal equations. The existence, uniqueness and some qualitative properties of the solutions to the Cauchy problems for some classes of nonlinear nonlocal dissipative equations were studied in [**67**] through [**103**]. Large time asymptotic behavior of solutions to the Cauchy problem for dissipative and dispersive nonlinear nonlocal equations was investigated in [**76**] through [**103**].

Up to now the theory of nonlinear initial-boundary value problems is not developed as stated in the Preface. There are many natural open questions which we need to study. First we consider the following question: how many boundary data we should pose on problem (0.1) with pseudodifferential operator $\mathbb{K}$ for its correct solvability? For example, do we need to impose any boundary condition for the Kobelev-Ostrovskii equation ([**86**])

$$u_t + uu_x + \int_0^x \frac{u_y\left(y, t\right)}{\sqrt{x - y}}dy = 0, \ x > 0$$

if $\mathbb{K}u = \int_0^x \frac{u_y(y,t)}{\sqrt{x-y}}dy$? There are some results in the case of nonlinear differential equations with the usual differential operator $\mathbb{K}$. However, as far as we know there are no results in the case of nonlocal operator $\mathbb{K}$.

In the present book we study traditionally important problems of a theory of nonlinear partial differential equations, such as local and global in time existence of classical and generalized solutions to the initial-boundary value problem, properties of smoothing of discontinuous initial perturbations and the asymptotic behavior of solutions for large time. The methods developed in this book are applicable to a wide class of dispersive and dissipative nonlinear equations, both local and nonlocal.

One of the main purposes of this book is to develop general methods for studying large time asymptotics of solutions to the initial-boundary value problem for nonlinear nonlocal equations. For that purpose we apply a minimal set of estimates of solutions which always include the $\mathbf{L}^\infty$-norm, so we use a concept of the semiclassical solution (see Chapter 3, Subsection 2.1 below).

We now outline the sketch of this monograph. In the first chapter we give only some principal results about Sobolev spaces and Laplace transformation, which will be extensively used in the book.

In Chapter 3 we introduce pseudodifferential operators $\mathbb{K}$ on a half-line by virtue of the inverse Laplace transform (see (0.2)) and then in Section 2.2 we study the linear initial-value problem for equation

$$u_t + \mathbb{K}u = f(x,t), (t,x) \in \mathbf{R}^+ \times \mathbf{R}^+.$$

To obtain an explicit form of the Green function we use an approach based on the Laplace transformation with respect to the spatial variable contrary to the standard application of the Laplace transformation with respect to the time variable. We prove that the amount of boundary data which we need to put in the problem for its well posedness is equal to the integer part of $\left[\frac{\alpha}{2}\right]$, where α is the order of the operator $\mathbb{K}$, which is not equal to an odd integer (in the case of odd integer order of operator $\mathbb{K}$ the amount of the boundary data depends also on the sign of the highest derivative, see Chapter 6). As we see from (0.2) the boundary values $\partial_x^{j-1}u(0,t)$, $j = 1, ..., [\alpha]$ are involved in the definition of the pseudodifferential operator $\mathbb{K}$; hence, the remainder $[\alpha] - \left[\frac{\alpha}{2}\right]$ boundary values should be obtained from the problem. In order to do this we develop a method based on the properties of symbol $K(p)$. Finally we give a general sketch of proofs of the existence of local and global solutions for the nonlinear nonlocal equation (0.1). Then we apply developed theory to the well-known particular nonlinear model equations choosing appropriately functional spaces and time decay estimates. The choice of these particular nonlinear model equations is interesting not only from the physical but also from the mathematical point of view since they include all difficulties arising in the nonlinear theory of evolution equations, such as critical and sub critical nonlinearities, large initial data, dispersive and dissipative operators, and so on.

Chapter 4 is devoted to the study of the initial-boundary value problem for the nonlinear nonlocal Schrödinger dissipative equation: $u_t + ia(t)|u|^\rho u + \mathbb{K}u = 0$ in the case of the symbol $K(p) = Ep^\alpha$.

In Chapter 5 we investigate the nonlinear nonlocal Whitham equation (0.4). Here we have nonlinearity of the shallow water type: uu_x. This nonlinearity contains the derivative of the unknown function and so represents a difficulty of the so-called derivative loss. Therefore in addition to the methods presented in previous chapters we must use the smoothing properties of the strongly dissipative operator $\mathbb{K}u$. Also we adopt here an approach based on the estimates of the Green function.

In Chapter 6 we consider the initial-boundary value problem on a half-line for the Korteweg-de Vries-Burgers equation: $u_t + uu_x - u_{xx} \pm u_{xxx} = 0$. In this case we have the operator $\mathbb{K}u = -u_{xx} \pm u_{xxx}$, i.e. the symbol $K_\pm(p) = -p^2 \pm p^3$ is not homogeneous and also the order of p in the second term is an odd integer. Thus we consider the critical case in the sense that the number of the boundary data also depends on the sign of the highest derivative in the equation. Note that in the case of the Cauchy problem for the Korteweg-de Vries-Burgers equation the nonlinearity uu_x of the shallow water type is critical from the point of view of the large time asymptotic behavior of solutions since the nonlinearity in the equation has the same decay rate as the linear terms. In the case of the initial-boundary value problem, due to the homogeneous boundary data, the solution obtains an additional decay rate and as a result the nonlinear term in the boundary-value problem appears to be super critical contrary to the corresponding Cauchy problem. The main difficulty in the boundary value problem is in the evaluation of the contribution of the boundary data to the large time asymptotic formulas of the solutions. Our approach here is based on the detailed $\mathbf{L}^p$ estimates of the Green function.

The aim of Chapter 7 is to study the initial-boundary value problem without a smallness condition on the data, taking the Korteweg-de Vries-Burgers equation as an example. As we know (see Chapter 6) the solutions of the initial-boundary value problem have more rapid time decay in comparison with solutions to the Cauchy problem due to the zero boundary value. Therefore the use of the symmetry property of the nonlinearity allows us to estimate the $\mathbf{L}^2$ - norm or $\mathbf{H}^{1,0}$-norm of solutions without any assumption on the size of the initial data.

In Chapter 8 we consider the case of the nonlinearity of the type u_x^2 and study as an example the initial-boundary value problem for the nonlinear KdVB equation $u_t + u_x^2 - u_{xx} + u_{xxx} = 0$.

The above-mentioned results are concerned with dissipative nonlinear equations, i.e. when the operator $\mathbb{K}u$ is strongly dissipative: $\operatorname{Re} K(p) > 0$ for all $\operatorname{Re} p = 0$.

In Chapter 9 we consider the case of a dispersive equation and study the Dirichlet problem for the Korteweg-de Vries (KdV) equation $u_t + uu_x + u_{xxx} = 0$. As we know in the case of the Cauchy problem the quadratic nonlinearity of the shallow water type uu_x is sub critical for large time, since the nonlinear term decays more slowly than the linear part of the equation. On the contrary the nonlinear term uu_x appears to be super critical for the boundary-value problem on the positive line since the solution obtains an additional time-decay. In this chapter we will obtain large time asymptotic formulas of solutions. Local and global existence of solutions in Sobolev spaces for the initial-boundary value problem

$$u_t + uu_x + u_{xxx} + u_x = f$$

on the positive line were studied in papers [20], [13]. Their results are also valid for equations without term u_x. In [36] an integral formula was used to obtain the formal long time asymptotics of solutions, and in [13] the integral formula which is different from that used in [36] was employed to derive various smoothing properties of solutions. Our approach here is based on the estimates in the weighted Sobolev spaces via the integral equation. The integral formula is obtained by using the Laplace transform with respect to space variable. The Laplace transform involves the boundary values $u(0,t)$, $u_x(0,t)$, $u_{xx}(0,t)$, one of which $u(0,t)$ is given as the Dirichlet boundary data in the problem, and the rest of the values $u_x(0,t)$ and $u_{xx}(0,t)$ must be determined in the construction of the solution in terms of the given data. In order to achieve this we use the method developed in the previous chapters.

Chapter 10 is devoted to the Neumann boundary value problem for the KdV equation. In the case of the Neumann boundary-value problem the time decay rate of the solutions is slower as compared to the Dirichlet boundary problem so that the nonlinear term is now critical.

In Chapter 11 we consider the initial-boundary value problem for the nonlinear Landau-Ginzburg type equations with Dirichlet boundary conditions. Here we do not need to use our theory of pseudodifferential operators on a half-line since the Green function can be constructed by the standard method of odd prolongation. The main goal of this chapter is the study of the large time behavior of solutions in the critical and sub critical cases. Our result shows that solutions of the initial-boundary value problem decay faster than those of the corresponding Cauchy problem due to the homogeneous Dirichlet boundary conditions. Therefore the critical power for the initial-boundary value problem for the nonlinear Landau-Ginzburg type equations is different from that of the case of the Cauchy problem.

In Chapter 12 we use the methods of the boundary value problems to solve the Burgers equation with pumping on the whole line with different possible types of the boundary data at infinity: rarefaction wave, shock wave and zero boundary conditions.

In Chapter 13 we study the global existence and large time asymptotic behavior of solutions to the initial-boundary value problem for the Korteweg-de Vries-Burgers equation on a segment $[0, 1]$. Here we do not assume the smallness condition for the initial data. In the case of large initial data it is more difficult to obtain exact large time asymptotic representation of solutions, so there are few results (see, for example [**76**]). Another difficulty in the study of the boundary value problem for the Korteweg-de Vries-Burgers equation on a segment $[0, 1]$ is that the linear operator $-\partial_x^2 + \partial_x^3$ is not self-adjoint, and we cannot apply the standard Fourier method. To avoid this difficulty we apply the Laplace transformation with respect to space variable to derive the Green function of the resulting equation. To obtain $\mathbf{L}^p$-estimates of the Green function we use the method of Chapter 6.

In Chapter 14 we present for the first time a general theory of the initial-boundary value problem for nonlinear evolution equations with pseudodifferential operators $\mathbb{K}$ on a segment (see Formula (0.3)) on the example of the nonlinear nonlocal Schrödinger equation. There are many natural open questions which we need to study in this respect. First is how many boundary data do we have to impose in the initial-boundary value problem with pseudodifferential operator $\mathbb{K}$ for its correct solvability? The methods proposed in this chapter can be applied to a wide class of local equations with non self-adjoint differential operators, where due to the lack of completeness it is difficult to apply the usual Fourier method, if we want to take boundary data into account.

The aim of Chapter 15 is to find the large time asymptotic representation of solutions to the periodic problem. We intend to find the main term of the asymptotics and to give an estimate of the remainder in the uniform norm. The study of the periodic problem is in many respects easier than the boundary value problems and the Cauchy problem, and, typically, the periodic results are exponential, whereas the problem on the line appears more delicate and often possesses algebraic decay rates. We remove the requirement of smallness of the initial data taking into account some additional symmetry of the nonlinearity in the model evolution equation under study. In comparison with the boundary value problems and the Cauchy problem, we can consider not only decaying large time asymptotics but we can also treat the cases where the solutions exponentially decay with time, oscillate or grow exponentially with time, depending on the linear part of the equation and the structure of the nonlinearity.

A more detailed description of the contents is given in the introduction to each chapter. We have strived to keep the presentation in each chapter as independent as possible, with some unavoidable repetitions as consequence.

In conclusion, we note that above-mentioned equations contain nonlinearities of the following types uu_x, u_x^2 and $i|u|^\rho u, \rho > 1$ and the symbols $K(p)$ of the forms $K(p) = p^\alpha$ or $K(p) = p^n + p^m$, with n, m natural and α real. We believe that the methods developed in this book could be applied to many other nonlinear nonlocal equations with different types of nonlinearities and symbols $K(p)$ and with more general non homogeneous and mixed types of the boundary.

CHAPTER 2

Preliminaries

Sobolev space and Laplace transform are very important tools in the theory of the initial-boundary value problems. In this chapter we give only the main results, which will be used in this book. More details about Sobolev spaces can be found, for example, in [**92**], [**134**]. For the Laplace transform we refer the reader to [**129**].

1. Laplace transform

1.1.   Definition and main properties. We say that the function $f(t)$ belongs to a class of functions $\mathcal{T}_a$ if

1) $f(t) \equiv 0$ for $t < 0$

2) the integral $\int_0^{+\infty} e^{-p_0 t} f(t) dt$ is convergent for $\operatorname{Re} p_0 = a$.

DEFINITION 1. *The Laplace transform of the function $f(t)$ is defined by the improper integral*

$$(1.1) \qquad\qquad \widehat{f}(p) = \int_0^{+\infty} e^{-pt} f(t) dt.$$

THEOREM 1. *If $f(t) \in \mathcal{T}_a$ integral (1.1) is convergent in domain $\operatorname{Re} p > a$. Moreover for any $x_0 > a$ integral (1.1) converges uniformly with respect to $\operatorname{Re} p \geq x_0 > a$.*

PROOF. Let $\phi(t) = e^{-p_0 t} f(t)$ and $F(t) = -\int_t^{+\infty} \phi(\tau) d\tau$. Note that $F'(t) = \phi(t)$. Since $\int_0^{+\infty} e^{-p_0 t} f(t) dt < \infty$, then for any $\varepsilon > 0$ there exists some T_0, such that

$$|F(t)| < \varepsilon$$

for $t \geq T_0$. Now let us consider the integral

$$\int_{T_1}^{T_2} e^{-pt} f(t) dt,$$

where $T_2 > T_1$. We have

$$\int_{T_1}^{T_2} e^{-pt} f(t) dt = \int_{T_1}^{T_2} e^{-(p-p_0)t} \phi(t) dt = \int_{T_1}^{T_2} e^{-(p-p_0)t} F'(t) dt.$$

1

Integrating by parts in the last integral we get

$$\int_{T_1}^{T_2} e^{-(p-p_0)t} F'(t)dt = e^{-(p-p_0)T_2} F(T_2) - e^{-(p-p_0)T_1} F(T_1)$$

$$+(p-p_0)\int_{T_1}^{T_2} e^{-(p-p_0)t} F(t)dt.$$

Therefore we obtain

$$\left| \int_{T_1}^{T_2} e^{-pt} f(t)dt \right| \le \left(e^{-\operatorname{Re}(p-p_0)T_2} + e^{-\operatorname{Re}(p-p_0)T_1} \right) \varepsilon$$

$$+\varepsilon \frac{|p-p_0|}{\operatorname{Re}(p-p_0)} \left(-e^{-\operatorname{Re}(p-p_0)T_2} + e^{-(p-p_0)T_1} \right)$$

$$\le \left(2 + \frac{|p-p_0|}{\operatorname{Re}(p-p_0)} \right) e^{-\operatorname{Re}(p-p_0)T_0} \varepsilon$$

for T_1, $T_2 > T_0$ and $\operatorname{Re}(p-p_0) > 0$. Hence integral $\int_0^{+\infty} e^{-pt} f(t)dt$ is convergent for $\operatorname{Re} p > \operatorname{Re} p_0 = a$ via the Cauchy criteria. In the same way we can prove that this integral is convergent uniformly in $\operatorname{Re} p \ge \operatorname{Re} p_1 > \operatorname{Re} p_0$. Theorem 1 is proved. $\quad\square$

Thus we see that the Laplace transform $\widehat{f}(p)$ is defined in domain $\operatorname{Re} p > a$. Moreover we obtain

$$\left| \widehat{f}(p) \right| \longrightarrow 0, \quad |p| \longrightarrow \infty.$$

THEOREM 2. *If $f(t) \in \mathcal{T}_a$ then the Laplace transformation $\widehat{f}(p)$ is an analytic function for $\operatorname{Re} p > a$.*

PROOF. Via Theorem 1 the integral

$$\int_0^{+\infty} e^{-pt} f(t)dt$$

is convergent in $\operatorname{Re} p > a$. We represent this integral as

$$\widehat{f}(p) = \int_0^{+\infty} e^{-pt} f(t)dt = \sum_{n=0}^{\infty} \int_{t_n}^{t_{n+1}} e^{-pt} f(t)dt = \sum_{n=0}^{\infty} f_n(p),$$

where $t_0 < t_1 ... < t_i < t_{i+1} ... < +\infty$. Taking into account that

$$\sum_{n=N}^{\infty} f_n(p) = \int_{t_{N+1}}^{+\infty} e^{-pt} f(t)dt$$

we see that the series $\sum_{n=0}^{\infty} f_n(p)$ converges uniformly in domain $\operatorname{Re} p \ge x_0 > a$. All functions $f_n(p) = \int_{t_n}^{t_{n+1}} e^{-pt} f(t)dt$ are analytic; hence via the Weierstrass Theorem (see [68], [124]) the function $\widehat{f}(p)$ is analytic in domain $\operatorname{Re} p > a$. Theorem 2 is proved. $\quad\square$

Now we formulate the main property of the Laplace transformation.

THEOREM 3. *If $f(t) \in \mathcal{T}_a$, and $D^n f(t) \in \mathcal{T}_a$, then*

$$\widehat{D^n f} = p^n \left(\widehat{f}(p) - \sum_{j=1}^{n} D^{j-1} f(0) p^{-j} \right)$$

for $\operatorname{Re} p > a$.

1.2. Inverse Laplace Transform.

THEOREM 4. *If $f(t) \in \mathcal{T}_a$ and*

$$\widehat{f}(p) = \int_0^{+\infty} e^{-pt} f(t) dt,$$

then for $\sigma > a$

(1.2)
$$f(t) = \frac{1}{2\pi i} \int_{\sigma - i\infty}^{\sigma + i\infty} e^{pt} \widehat{f}(p) dp.$$

Integral (1.2) is called by the inverse Laplace transform or the Mellin's integral.

PROOF. Let us denote

$$\phi(t) = e^{-xt} f(t), \quad x > a.$$

This function can be represented as a Fourier transform

$$\phi(t) = \frac{1}{2\pi} \int_{-\infty}^{+\infty} d\xi \int_{-\infty}^{+\infty} \phi(\theta) e^{i\xi(t-\theta)} d\theta.$$

Since $f(\theta) = 0$, $\theta < 0$ we get

$$e^{-xt} f(t) = \frac{1}{2\pi} \int_{-\infty}^{+\infty} e^{i\xi t} d\xi \int_0^{+\infty} e^{-(x+i\xi)\theta} f(\theta) d\theta.$$

Taking $p = x + i\xi$ we obtain

$$f(t) = \frac{1}{2\pi i} \int_{x - i\infty}^{x + i\infty} e^{pt} \widehat{f}(p) dp.$$

Theorem 4 is proved. □

Now we prove the necessary and sufficient conditions for the existence of the inverse Laplace transform for the function $F(p)$. First we define a class of functions $\mathcal{H}_a(p)$.

DEFINITION 2. *Function $F(p) \in \mathcal{H}_a(p)$ if*
1) $F(p)$ is analytic in domain $\operatorname{Re} p > a$,
2) for any $\varepsilon > 0$ and $\sigma_0 > a$ there exist constants $C_\varepsilon(\sigma_0) \geq 0$ and $m = m(\sigma_0) \geq 0$ such that

$$|F(p)| \leqslant C_\varepsilon(\sigma_0) e^{\varepsilon\sigma} (1 + |p|^m), \sigma > \sigma_0$$

3) the norm $\|F(\sigma + i\cdot)\|_{\mathbf{L}^1} < +\infty$ for any $\sigma > a$.

THEOREM 5. *In order that function $f(t) \in \mathcal{T}_a$ it is necessary and sufficient that the Laplace transform $F(p) \in \mathcal{H}_a(p)$. Moreover the original $f(t)$ can be reconstructed via $F(p)$ by the Mellin integral*

$$(1.3) \qquad f(t) = \frac{1}{2\pi i} \int_{\sigma - i\infty}^{\sigma + i\infty} e^{pt} F(p) dp.$$

PROOF. Let $f(t) \in \mathcal{T}_a$ then $F(p)$ is analytic for $\operatorname{Re} p > a$ and for any $\sigma > \sigma_0 > a$ we have $p = \sigma + iw$

$$|F(p)| = \left| \int_0^{+\infty} e^{-pt} f(t) dt \right| < C.$$

Thus we get that $F(p) \in \mathcal{H}_a(p)$.

Sufficiency. First we consider the case, where

$$(1.4) \qquad |F(p)| \leqslant \frac{C_\varepsilon(\sigma_0) e^{\varepsilon\sigma}}{|p-a|^\alpha}, \alpha > 1.$$

Then the integral

$$(1.5) \qquad f(t) = \frac{1}{2\pi i} \int_{\sigma - i\infty}^{\sigma + i\infty} e^{pt} F(p) dp, (\sigma > a)$$

is convergent uniformly with respect to t and $f(t)$ is continuous with respect to $t \in (-\infty, +\infty)$. Let us prove that $f(t)$ does not depend on $\sigma > a$. We denote $\sigma_2 > \sigma_1 > a$. Via the Cauchy theorem we have

$$(1.6) \qquad \int_{\Gamma(d)} e^{pt} F(p) dp = 0,$$

where the contour

$$\Gamma(d) = \{ p \in \mathbb{C}, p \in [\sigma_1 - id, \sigma_2 - id) \cap [\sigma_2 - id, \sigma_2 + id)$$
$$\cap [\sigma_2 + id, \sigma_1 + id) \cap [\sigma_1 + id, \sigma_1 - id] \}.$$

By (1.4) we have

$$\left| \int_{\sigma_1 + id}^{\sigma_2 + id} e^{pt} F(p) dp \right|$$
$$\leq \int_{\sigma_1}^{\sigma_2} e^{\sigma t} F(\sigma + id) d\sigma \leq C_\varepsilon(\sigma_1) \int_{\sigma_1}^{\sigma_2} \frac{e^{\sigma(\varepsilon + t)} d\sigma}{|(\sigma - a)^2 + d^2|^{\frac{\alpha}{2}}} \to 0,$$

then taking a limit $d \to +\infty$ in (1.6) we obtain

$$\int_{\sigma_1 - i\infty}^{\sigma_1 + i\infty} e^{pt} F(p) dp = \int_{\sigma_2 - i\infty}^{\sigma_2 + i\infty} e^{pt} F(p) dp.$$

We rewrite (1.5) as

$$(1.7) \qquad f(t) = \frac{e^{\sigma t}}{2\pi} \int_{-\infty}^{+\infty} e^{iwt} F(\sigma + iw) dw, (\sigma > a).$$

Using (1.4) and choosing $\sigma > \sigma_0 > a$ we get

$$
\begin{aligned}
|f(t)| &\leq \frac{e^{\sigma t}}{2\pi} \int_{-\infty}^{+\infty} |F(\sigma + iw)| \, dw \\
&\leq \frac{C_\varepsilon(\sigma_0)e^{\sigma(t+\varepsilon)}}{2\pi} \int_{-\infty}^{+\infty} \frac{dw}{|(\sigma - a)^2 + w^2|^{\frac{\alpha}{2}}} \\
&\leq C_\varepsilon(\sigma_0)e^{\sigma(\varepsilon + t)}.
\end{aligned}
$$

Let $t < -\varepsilon$, $\varepsilon > 0$. Taking a limit $\sigma \to +\infty$ we obtain $f(t) = 0$, and since $\varepsilon > 0$ is arbitrary we obtain $f(t) = 0$ for all $t < 0$. From (1.7) it follows that

$$
F(p) = \int_0^{+\infty} e^{-pt} f(t) dt.
$$

Therefore $f(t) \in \mathcal{T}_a$ and $f \leftrightarrow F(p)$, $\sigma > a$. Now we consider a general case. We denote

$$
F_1(p) = \frac{F(p)}{|p - b|^k}
$$

for $b \leq a, \sigma_0 > a$ and integer $k > m(\sigma_0) + 1$. Note that $F_1(p)$ is analytic in $\operatorname{Re} p = \sigma > a$ and

$$
\begin{aligned}
F_1(p) &\leq \frac{C_\varepsilon(\sigma_0)e^{\varepsilon\sigma}(1 + |p|^m)}{|p - a|^k} \\
&\leq \frac{C_\varepsilon(\sigma_0)e^{\varepsilon\sigma}(1 + |p|^m)}{|p - a|^{k-m} |p - a|^m} < \frac{C_\varepsilon(\sigma_0)e^{\varepsilon\sigma}}{|p - a|^{k-m}}
\end{aligned}
$$

for any $\sigma > \sigma_0$ if $k - m > 1$. We proved that there exists a function $f_1(t) \in \mathcal{T}_a$ such that

$$
(1.8) \qquad f_1(t) = \frac{1}{2\pi i} \int_{\sigma - i\infty}^{\sigma + i\infty} e^{pt} F_1(p) dp \longleftrightarrow F_1(p), \sigma > \sigma_0.
$$

Since $\int_{\sigma - i\infty}^{\sigma + i\infty} |F(\sigma + iw)| \, dw < +\infty, \sigma > a$ we have

$$
\begin{aligned}
f(t) &= (\frac{d}{dt} - b)^k f_1(t) \\
&= \frac{1}{2\pi i} (\frac{d}{dt} - b)^k \int_{\sigma - i\infty}^{\sigma + i\infty} e^{pt} F_1(p) dp \\
&= \frac{1}{2\pi i} \int_{\sigma - i\infty}^{\sigma + i\infty} (\frac{d}{dt} - b)^k e^{pt} F_1(p) dp \\
&= \frac{1}{2\pi i} \int_{\sigma - i\infty}^{\sigma + i\infty} (p - b)^k e^{pt} F_1(p) dp \\
&= \frac{1}{2\pi i} \int_{\sigma - i\infty}^{\sigma + i\infty} e^{pt} F(p) dp.
\end{aligned}
$$

Thus we obtain $f(t) \in \mathcal{T}_a$ and $f(t) \leftrightarrow F(p), \sigma > a$. Since

$$f(t) = (\frac{d}{dt} - b)^k f_1(t)$$

and via (1.8) we obtain representation (1.3), Theorem 5 is proved.   □

2. Sobolev spaces $\mathbf{H}^k(\mathbf{R}^+)$

2.1. Weighted Sobolev spaces with integer order.

DEFINITION 3. *We say that the function $f(x)$ belongs to the Lebesgue $\mathbf{L}^p(\mathbf{R}^+)$ with $1 < p < +\infty$ space if there exists the Lebesgue integral*

$$\int_0^{+\infty} |f(x)|^p \, dx < +\infty.$$

We denote the norm of the Lebesgue space $\|f\|_{\mathbf{L}^p} = \left(\int_0^{+\infty} |f(x)|^p \, dx\right)^{\frac{1}{p}}$ for $1 < p < +\infty$, $\|f\|_{\mathbf{L}^\infty} = \operatorname{ess\,sup}_{x \geq 0} |f(x)|$.

DEFINITION 4. *Let $k \geq 0$ be an integer. We say that the function $f(x)$ belongs to the Sobolev space $\mathbf{H}_p^k(\mathbf{R}^+)$ if $\frac{d^j}{dx^j} f(x) \in \mathbf{L}^p(\mathbf{R}^+)$, for $j \leq k$. We denote the norm of the Sobolev space $\|f\|_{\mathbf{H}_p^k} = \sum_{j=0}^{k} \left\| \frac{d^j}{dx^j} f(x) \right\|_{\mathbf{L}^p}$.*

DEFINITION 5. *Let $k \geq 0$ be an integer. We say that the function $f(x)$ belongs to the weighted Sobolev space $\mathbf{H}_p^{k,m}(\mathbf{R}^+)$ if $(1+x^m)\frac{d^j}{dx^j} f(x) \in \mathbf{L}^p(\mathbf{R}^+), j = 0, ..., k$. We denote $\|f\|_{\mathbf{H}_p^{m,k}} = \left\| (1 + \cdot)^k f(\cdot) \right\|_{\mathbf{H}_p^m} = \left\| \langle \cdot \rangle^k f(\cdot) \right\|_{\mathbf{H}_p^m}$.*

THEOREM 6. *If $k \geqslant 1$ then $\mathbf{H}_2^k(\mathbf{R}^+) \subset \mathbf{L}^\infty(\mathbf{R}^+)$.*

PROOF. Since

$$f^2(x) = -\int_x^{+\infty} df^2 = -2 \int_x^{+\infty} f f_x dx,$$

we easily obtain $\|f\|_{\mathbf{L}^\infty} \leq \sqrt{2} \|f\|_{\mathbf{L}^2}^{\frac{1}{2}} \|f_x\|_{\mathbf{L}^2}^{\frac{1}{2}} \leq \sqrt{2} \|f\|_{\mathbf{H}_2^1}$. Theorem 6 is proved.   □

THEOREM 7. *If $m > \frac{1}{2}$, then $\mathbf{H}_2^{0,m}(\mathbf{R}^+) \subset \mathbf{L}^1(\mathbf{R}^+)$.*

PROOF. We have

$$\|f\|_{\mathbf{L}^1} = \int_0^{+\infty} |f(x)| \, dx = \int_0^{+\infty} (1 + x)^m |f(x)| \frac{1}{(1 + x)^m} dx \leq C \|f\|_{\mathbf{H}_2^{0,m}}.$$

          □

THEOREM 8. *The Sobolev space $\mathbf{H}_p^{m,k}(\mathbf{R}^+)$ is complete.*

2.2. **Sobolev spaces of noninteger order.** We denote by $\mathcal{J}_+$ the space of functions $f(t) \in \mathbf{C}^\infty(\mathbf{R}^+)$ with semi norms

$$\sup_{x>0} (1+x)\,|D^\alpha f(x)| < +\infty, \alpha \leqslant k.$$

The space $\mathcal{J}_+$ is dense in $\mathbf{H}_p^k(\mathbf{R}^+)$.

Let $f(x) \in \mathcal{J}_+$. We define a mapping Ef by

$$Ef = \begin{cases} f(x), x > 0 \\ \sum_{j=1}^N a_j f(-jx), x < 0 \end{cases}$$

where N is an integer .

THEOREM 9. *For some choice of constants $a_j, j = 1, ..., N$ the mapping E has a unique linear continuous extension*

$$E : \mathbf{H}_p^k(\mathbf{R}^+) \to \mathbf{H}_p^k(\mathbf{R}), \ k \leqslant N - 1.$$

PROOF. To prove this result we need to obtain estimates for Ef in $\mathbf{H}_p^k(\mathbf{R})$, when $f \in \mathcal{J}_+$. It is possible if there exist derivatives Ef of order $\alpha \leqslant N-1$. Thus we get a linear system of N equations for a_j

$$\sum_{j=1}^N (-j)^l a_j = 1, \ l = 0, 1, ..., N - 1.$$

Theorem 9 is proved. □

DEFINITION 6. *We denote the intermediate space*

$$[E_1, E_2]_\theta = \{u(\theta) : u \in \mathcal{H}_{E_1,E_2}(\Omega)\},$$

where

$$\mathcal{H}_{E_1,E_2}(\Omega) = \{u : u(z) \text{ is holomorphic in } \Omega,$$
$$\text{bounded and continuous function into } E_1 + E_2,$$
$$\|u(iy)\|_{E_1} < \infty, \|u(1+iy)\|_{E_2} < \infty\},$$

and

$$\Omega = \{z \in \mathbb{C} \ 0 < \operatorname{Re} z < 1\}.$$

We define the Sobolev space $\mathbf{H}_p^s(\mathbf{R}^+)$ as interpolation

$$\mathbf{H}_p^s(\mathbf{R}^+) = [\mathbf{L}^p(\mathbf{R}^+), \mathbf{H}_p^k(\mathbf{R}^+)]_\theta, \ s < k, \ s = \theta k.$$

CHAPTER 3

General Theory

In this chapter we define pseudodifferential operators on a half-line and study their properties. We introduce a general class of symbols $\mathcal{K}^\alpha$. Then we study the linear initial-boundary value problem with pseudodifferential operator. Finally we give some sufficient conditions for the local and global existence of solutions of the initial- boundary value problem for nonlinear nonlocal equations.

1. Pseudodifferential Operator on a Half-Line

1.1. Laplace integral representation and a class of symbols. Let us denote

$$D^n u(x) = \frac{d^n}{dx^n} u.$$

We have for the Laplace inverse transform

$$(1.1) \qquad D^n u(x) = \frac{1}{2\pi i} \int_{a-i\infty}^{a+i\infty} e^{px} p^n \left(\widehat{u}(p) - \sum_{j=1}^{n} D^{j-1} u(0) p^{-j} \right) dp,$$

where $\widehat{u}(p)$ is the direct Laplace transform defined by

$$\widehat{u}(p) = \int_0^{+\infty} e^{-px} u(x) dx.$$

Using the integral representation (1.1) of inverse Laplace transformation, we introduce pseudodifferential operators on a half-line for a general class of symbols $\mathcal{K}^\alpha$, which we now define.

We denote by $[s]$ the largest integer less than s. In what follows we consistently make a cut along the negative real axis $(-\infty, 0)$ in the complex plane. By p^α we denote the main branch of the complex analytic function, so that $1^\alpha = 1$.

DEFINITION 7. *We define a class $\mathcal{K}^\alpha$ of symbols $K(p)$ as a set of complex functions $K(p)$ such that $K(p)$ is analytic for all $\operatorname{Re} p \geq a \geq 0$. Let $K'(p) \neq 0$ for all $\operatorname{Re} p \geq a$, and $\operatorname{Re} K(p) \geq 0$ for $\operatorname{Re} p = a$, $|p| > C > 0$ and the asymptotics representation is true*

$$K(p) = \sum_j C_j p^{\alpha_j} + O(p^\beta)$$

as $p \to \infty$, where $\alpha_j > 0$, $\beta < 1$ and $\alpha = \max \alpha_j$.

DEFINITION 8. *We define a pseudodifferential operator* $\mathbb{K}$ *on a half-line as follows*

$$\mathbb{K}u = \frac{1}{2\pi i} \sum_j C_j \int_{-i\infty+a}^{i\infty+a} e^{px} p^{\alpha_j} \left(\widehat{u}(p) - \sum_{l=1}^{[\alpha_j]} p^{-l} \partial_x^{l-1} u(0) \right) dp$$

$$+ \frac{1}{2\pi i} \int_{-i\infty+a}^{i\infty+a} e^{px} \left(K(p) - \sum_j C_j p^{\alpha_j} \right) \widehat{u}(p) dp$$

$$(1.2) \qquad = \frac{1}{2\pi i} \int_{-i\infty+a}^{i\infty+a} e^{px} \left(K(p)\widehat{u}(p) - \sum_j \sum_{l=1}^{[\alpha_j]} C_j p^{\alpha_j - l} \partial_x^{l-1} u(0) \right) dp.$$

Here the sum of the form $\sum_{l=1}^{0}$ is identically equal to zero. Function $K(p) \in \mathcal{K}^\alpha$ is called a symbol of operator $\mathbb{K}$, and $\alpha \geq 0$ is the order of the operator $\mathbb{K}$.

DEFINITION 9. *We say that a pseudodifferential operator* $\mathbb{K}$ *is dissipative and write* $K(p) \in \mathcal{K}_{diss}^\alpha$ *if* $K(p) \in \mathcal{K}^\alpha$ *and* $\mathrm{Re}\, K(p) > 0$ *for* $\mathrm{Re}\, p = a$. *Pseudodifferential operator* $\mathbb{K}$ *is called by dispersive* $K(p) \in \mathcal{K}_{disp}^\alpha$ *if* $K(p) \in \mathcal{K}^\alpha$ *and* $\mathrm{Re}\, K(p) = 0$ *for* $\mathrm{Re}\, p = a$.

Now we state some properties of pseudodifferential operators $\mathbb{K}$.

THEOREM 10. *If* $K(p) \in \mathcal{K}^\alpha$, *then operator* $\mathbb{K}$ *is a continuous operator*

$$\mathbb{K} : \mathbf{H}_1^{0,[\alpha]+m+2}(\mathbf{R}^+) \to \mathbf{C}^m(\mathbf{R}^+), \ m \geq 0.$$

Proof. Integrating by parts we have

$$\widehat{u}(p) = \sum_{j=1}^{n} \partial_x^{j-1} u(0) p^{-j} + \frac{1}{p^n} \int_0^{+\infty} e^{-px} D^n u(x) dx.$$

Thus if $K(p) \in \mathcal{K}^\alpha$ and $u(x) \in \mathbf{H}_1^{0,[\alpha]+m+2}(\mathbf{R}^+)$, then integrating $m+1$ times by parts with respect to p in the domain $|p| > 1$ we obtain

$$\sum_j C_j \int_{|p|>1, \mathrm{Re}\, p=a} e^{px} \frac{d^{m+1}}{dp^{m+1}} \left(p^{\alpha_j+m} \left(\widehat{u}(p) - \sum_{l=1}^{[\alpha_j]} \partial_x^{l-1} u(0) p^{-l} \right) \right) dp$$

$$= \sum_j C_j \int_{|p|>1, \mathrm{Re}\, p=a} e^{px} \frac{d^{m+1}}{dp^{m+1}} \left(p^{\alpha_j+m} \left(\sum_{l=[\alpha_j]+1}^{[\alpha_j]+m+2} \partial_x^{l-1} u(0) p^{-l} \right. \right.$$

$$\left. \left. + \frac{1}{p^{[\alpha_j]+m+2}} \int_0^{+\infty} e^{-px} D^{[\alpha_j]+m+2} u(x) dx \right) \right) dp.$$

Therefore the integral representing $D^m \mathbb{K} u$

$$D^m \mathbb{K} u = \frac{1}{2\pi i} \sum_j C_j \int_{-i\infty+a}^{i\infty+a} e^{px} p^{\alpha_j + m} \left(\widehat{u}(p) - \sum_{l=1}^{[\alpha_j]} \partial_x^{l-1} u(0) p^{-l} \right) dp$$

$$+ \frac{1}{2\pi i} \int_{-i\infty+a}^{i\infty+a} e^{px} p^m \left(K(p) - \sum_j C_j p^{\alpha_j} \right) \widehat{u}(p) dp.$$

converges absolutely, thus the result follows. Theorem 10 is proved.

1.2.   Asymptotics of inverse functions $K^{-1}(-\xi)$. Now we obtain some results about asymptotic behavior of inverse functions $\phi_j(\xi) = K^{-1}(-\xi)$ for symbol $K(p) \in \mathcal{K}^\alpha$, which we will use later to solve linear initial-boundary value problem. Let us denote $m = \left[\frac{\alpha+1}{2} \right]$.

THEOREM 11. *Let* $K(p) \in \mathcal{K}^\alpha$, α *is not equal to an odd integer. Then there exist* $m = \left[\frac{\alpha+1}{2} \right]$ *different inverse functions* $\phi_j(\xi) = K^{-1}(-\xi)$, *such that for* $\operatorname{Re} \xi > b$

$$\operatorname{Re} \phi_j(\xi) > 0$$

for $j = 1, 2, ..., m$. *Moreover the asymptotics*

$$\phi_l(\xi) = e^{i(\pi + 2\pi l) \frac{1}{\alpha}} \left(C_1^{-1} \xi \right)^{\frac{1}{\alpha}} + O\left(\xi^{-\frac{1+\gamma}{\alpha}} \right)$$

is true as $\xi \to \infty$.

Proof. In the case $\alpha \in (0,1)$ we have $\operatorname{Re} K(p) \geq 0$ for all $\operatorname{Re} p \geq a \geq 0$, so there are no inverse functions $\phi_j(\xi) = K^{-1}(-\xi)$ defined in domain $\operatorname{Re} \xi \geq b$, where $b \geq 0$. Consider now $\alpha \geq 1$. Since $K'(p) \neq 0$ the transformation $K(p) = -\xi$ is conformal for all $\operatorname{Re} p \geq a$. Also if $K(p) \in \mathcal{K}^\alpha$, we have the asymptotics

$$K(p) = C_1 p^\alpha \left(1 + O(p^{-\gamma}) \right)$$

as $p \to \infty$. Here C_1 is such that $\operatorname{Re} K(a + iy) \geq 0$ for all $y \in \mathbf{R}$, $|y|$ is sufficiently large. The function $\xi = -C_1 p^\alpha$ defined in $\operatorname{Re} p \geq a$ has $m = \left[\frac{\alpha+1}{2} \right]$ different inverse functions

$$\left| \frac{\xi}{C_1} \right|^{\frac{1}{\alpha}} \exp\left(\frac{i(\pi + 2\pi l)}{\alpha} + \frac{i}{\alpha} \arg \xi - \frac{i}{\alpha} \arg C_1 \right)$$

in the domain $\operatorname{Re} \xi \geq b$, where l is an integer, such that

(1.3) $$-\frac{\pi}{2} \leq \frac{1}{\alpha} \left(\pi + 2\pi l + \arg \xi - \arg C_1 \right) \leq \frac{\pi}{2}$$

for all $-\frac{\pi}{2} \leq \arg \xi \leq \frac{\pi}{2}$. From (1.3) we get the following estimate

$$-\frac{\alpha+2}{4} + \phi \leq l \leq \frac{\alpha+2}{4} - 1 + \phi$$

for $\phi = -\frac{\arg \xi}{2\pi} + \frac{\arg C_1}{2\pi}$. The value $\arg C_1$ is defined by the dissipation condition $\operatorname{Re} K(p) > 0$ for $\operatorname{Re} p = a$, when $|p|$ is large, i.e. $\cos\left(\arg C_1 \pm \frac{\pi}{2}\alpha\right) > 0$. This implies

(1.4)
$$\left\{ \begin{array}{c} \frac{\alpha+1}{4} - \left\{\frac{\alpha+1}{2}\right\} \le \frac{\arg C_1}{2\pi} \le \frac{\alpha+1}{4} - \left\{\frac{\alpha+1}{2}\right\} + \frac{1}{2} \\ m + \frac{\alpha+1}{4} - \frac{1}{2} \le \frac{\arg C_1}{2\pi} \le m + \frac{\alpha+1}{4}, \end{array} \right.$$

here by $\{x\} = x - [x]$ we denote the fractional part of x. We have $m = 0$, then for $\psi = \frac{\arg C_1}{2\pi} - \frac{\alpha+1}{4}$ we get

$$\max\left(-\frac{1}{2}, -\left\{\frac{\alpha+1}{2}\right\}\right) \le \psi \le \min\left(0, \frac{1}{2} - \left\{\frac{\alpha+1}{2}\right\}\right).$$

Thus integers l satisfy inequalities

$$-\frac{\arg \xi}{2\pi} - \frac{1}{4} + \psi \le l \le \frac{\alpha}{2} - \frac{1}{4} - \frac{\arg \xi}{2\pi} + \psi$$

for all $\frac{\arg \xi}{2\pi} \in \left[-\frac{1}{4}, \frac{1}{4}\right]$. Therefore

$$\psi \le l \le \frac{\alpha-1}{2} + \psi,$$

where

$$\max\left(-\frac{1}{2}, -\left\{\frac{\alpha-1}{2}\right\}\right) \le \psi \le \min\left(0, \frac{1}{2} - \left\{\frac{\alpha-1}{2}\right\}\right)$$

since $\left\{\frac{\alpha+1}{2}\right\} = \left\{\frac{\alpha-1}{2}\right\}$. We see that $l = 0, ..., \left[\frac{\alpha+1}{2}\right]$. Hence there exist $m = \left[\frac{\alpha+1}{2}\right]$ different inverse functions $\phi_l(\xi) = K^{-1}(-\xi)$, which are analytic in $\operatorname{Re}\xi \ge b$ and the asymptotics

$$\phi_l(\xi) = K^{-1}(-\xi) = e^{i(\pi + 2\pi l)\frac{1}{\alpha}} \left(C_1^{-1}\xi\right)^{\frac{1}{\alpha}} + O\left(\xi^{-\frac{1+\gamma}{\alpha}}\right)$$

is true for $\xi \to +\infty$. Theorem 11 is proved.

REMARK 1. *Note that if α is an odd integer the number of inverse functions $\phi_l(\xi)$ depends on the sign of $\arg C_1$. Solving inequality (1.4) we can choose $m = 0$, then $\arg C_1 = \pi\frac{\alpha+1}{2}$ and $m = 1$, then $\arg C_1 = \pi\frac{\alpha+3}{2}$. In the first case we have $\left[\frac{\alpha+1}{2}\right]$ inverse functions $\phi_l(\xi)$, and in the second case we have $\left[\frac{\alpha}{2}\right]$ inverse functions $\phi_l(\xi)$.*

2. Boundary Value Problem on a Half-Line

2.1. Definitions. Let us consider the following initial-boundary value problem on a half-line for nonlinear nonlocal equation:

(2.1)
$$\left\{ \begin{array}{c} u_t + \mathrm{N}(u) + \mathbb{K}u = f, \ x > 0, \ t > 0 \\ u(x, 0) = u_0(x), \ x > 0, \\ \partial_x^{j-1} u(0, t) = h_j(t), \ j = 1, 2, ..., N, \end{array} \right.$$

where the nonlinear term $\mathrm{N}(u)$ depends on the unknown function $u(x, t)$ and its derivatives; $\mathbb{K}$ is a pseudodifferential operator on a half-line (see Section 1 above). The number $N \ge 0$ of boundary data, which we need to include in problem (2.1) for its well posedness, depends essentially on the properties of the operator $\mathbb{K}$ and will be calculated below in subsection 2.2.

Let $\mathbf{B}$ be a Banach space; we then denote

$$\mathbf{C}^k\left([0,T],\mathbf{B}\right)$$
$$= \ \left\{ f(t) \in \mathbf{B} : \lim_{t_1 \to t, t_1 \in [0,T]} \left\|\partial_t^k f(t_1) - \partial_t^k f(t)\right\|_{\mathbf{B}} = 0, \forall t \in [0,T] \right\}.$$

Now we define the well posedness of the problem (2.1).

DEFINITION 10. *Problem (2.1) is called well posed in a semiclassical sense if the following two properties are fulfilled. Firstly, if there exists a unique solution $u(x,t)$ belonging to a metric space*

$$\mathbf{C}^0\left([0,T],\mathbf{M}_1\right) \cap \mathbf{C}^1\left((0,T],\mathbf{M}_2\right),$$

which satisfy equation $u_t + \mathrm{N}(u) + \mathbb{K}u = f$ in the generalized sense. Boundary and initial conditions are fulfilled in the classical sense

$$\lim_{t \to 0} u(x,t) \ = \ u_0\left(x\right) \ \textit{in } \mathbf{M}_1 \ \textit{and}$$
$$\lim_{x \to 0} \partial_x^{j-1} u(x,t) \ = \ h_j(t) \ \textit{in } \mathbf{C}^0([0,T]) \ \textit{for all } j = 1,2,...,N.$$

Secondly, if solution $u(x,t)$ is stable with respect to the initial data $u_0(x)$, boundary data $h_j(t)$ and a source $f(x,t)$. The function $u(x,t)$ we call a semiclassical solution. If $T = +\infty$ the function $u(x,t)$ is called a global semiclassical solution.

We say that solution $u(x,t)$ has asymptotics

$$u(x,t) = \Lambda(x,t) + O(\phi(x,t))$$

for large time $t \to +\infty$ uniformly with respect to $x > 0$ if there exist some $T_0 > 0$ and a constant $C > 0$ which do not depend on x,t and such that the estimate

$$|u(x,t) - \Lambda(x,t)| \le C\,|\phi(x,t)|$$

is true for all $t > T_0$.

2.2. Linear problem. In this section we solve the following lineal initial-boundary value problem

$$(2.2) \qquad \begin{cases} u_t + \mathbb{K}u = f, \ t > 0, x > 0, \\ u(x,0) = u_0(x), \ x > 0, \\ \partial_x^{j-1} u(0,t) = h_j\left(t\right), \ t > 0, \ j = 1,2,...,N. \end{cases}$$

As above in Section 1.2 we denote by $\phi_j(\xi) = K^{-1}(-\xi)$ the inverse functions for the operator $\mathbb{K}$ with a symbol $K(p) \in \mathcal{K}^\alpha$, $\alpha > 0$. Let $m \ge 0$ be the number of "positive" inverse functions, such that

$$\mathrm{Re}\,\phi_j(\xi) > 0$$

for all $\mathrm{Re}\,\xi > b \ge 0$ and $j = 1,2,...,m$. We will prove below that $N = [\alpha] - m$.

DEFINITION 11. *We call the function $\mathcal{J}(p,t)$*

$$\mathcal{J}(p,t) = \widehat{u}_0(p) + \int_0^t e^{K(p)\tau} \left(\widehat{f}(p,\tau) + \sum_j \sum_{l=1}^{\min([\alpha_j],N)} C_j h_l(\tau) p^{\alpha_j - l} \right) d\tau$$

as the input function for problem (2.2).

We denote $m \times m$ matrix $A(\xi) = (a_{lj}(\xi))_{1 \le l, j \le m}$ with elements

$$a_{lj}(\xi) = \sum_{[\alpha_k] \ge N+j} C_k \phi_l^{\alpha_k - N - j}(\xi).$$

We define the operator $\mathbb{A}$ as

(2.3)
$$\mathbb{A}\vec{\mathcal{J}} = \frac{1}{2\pi i} \int_{-i\infty+b}^{i\infty+b} e^{\xi t} A^{-1} \vec{\mathcal{J}} d\xi,$$

where the input vector

$$\vec{\mathcal{J}} = -(\mathcal{J}(\phi_1, +\infty), \mathcal{J}(\phi_2, +\infty), ..., \mathcal{J}(\phi_m, +\infty))^T.$$

We use a norm

$$\left\| \mathbb{A}\vec{\mathcal{J}} \right\| = \sum_{j=1}^m \left| \left(\mathbb{A}\vec{\mathcal{J}} \right)_j \right|.$$

We introduce the norm

$$\| \mathcal{J}(p,t) \|_{\mathbf{S}_M}$$

$$= \sum_{s=0}^{[\alpha]-1} \sup_{t \in [0,T]} t^{\delta_s} \lim_{p \to +\infty, \operatorname{Re} p = a} p^{s+1+\gamma} \left| e^{-K(p)t} \mathcal{J}(p,t) - \sum_{j=1}^N h_j(t) p^{-j} \right|$$

$$+ \sup_{t \in [0,T]} \sup_{\operatorname{Re} p = a} | \mathcal{J}(p,t) |,$$

where $\delta_s = \max(0, \frac{s-M}{\alpha})$, $M \ge 0$.

THEOREM 12. *Let $K(p) \in \mathcal{K}^\alpha$, $\alpha > 1$. Let input function $\mathcal{J}(p,t)$ be such that*

$$\| \mathcal{J}(p,t) \|_{\mathbf{S}_M} = \lambda.$$

Let

(2.4)
$$\sup_{t \in [0,T]} t^\upsilon \left\| \mathbb{A}\vec{\mathcal{J}} \right\| \le C\lambda,$$

where $0 \le \nu < 1$ and

(2.5)
$$\sup_{t,\tau \in [0,T]} t^{\upsilon+\kappa} |t - \tau|^{-\kappa} \sum_{j=1}^m \left| \left(\mathbb{A}\vec{\mathcal{J}} \right)_j (\tau) - \left(\mathbb{A}\vec{\mathcal{J}} \right)_j (t) \right| \le C\lambda,$$

where $0 < \kappa < 1$. Then the problem (2.2) is well posed in

$$\mathbf{C}^0([0,T], \mathbf{H}_2^M \cap \mathbf{C}^M) \cap \mathbf{C}^0((0,T], \mathbf{H}_2^{M_1} \cap \mathbf{C}^{M_1}),$$

where $M_1 = \min([\alpha] - 1, N + [\kappa\alpha])$ and

$$\sup_{t \in [0,T]} \sum_{j=0}^{M_1} t^{\delta_j} \left(\left\| \partial_x^j u(x,t) \right\|_{\mathbf{L}^\infty} + \left\| \partial_x^j u(x,t) \right\|_{\mathbf{L}^2} \right) \leq C\lambda,$$

where $\delta_j = \max\left(0, \nu + \frac{j-N}{\alpha}, \frac{j-M}{\alpha}\right)$. Moreover the solution has the following representation

$$
\begin{aligned}
(2.6) \qquad u(x,t) \;=\; & \frac{1}{2\pi i} \int_{a-i\infty}^{a+i\infty} e^{px - K(p)t} \left(\mathscr{J}(p,t) \right. \\
& \left. + \sum_k \sum_{j=1}^{[\alpha_k]-N} C_k p^{\alpha_k - N - j} \int_0^t e^{K(p)\tau} (\mathbb{A}\vec{\mathscr{J}})_j(\tau) d\tau \right) dp.
\end{aligned}
$$

PROOF. Taking the Laplace transformation of the problem (2.2) with respect to the space variable x we get

$$\hat{u}_t(p,t) + K(p)\,\hat{u}(p,t) = f_1(p,t),$$

where

$$f_1(p,t) = \sum_k \sum_{l=1}^{[\alpha_k]} C_k \partial_x^{l-1} u(0,t) p^{\alpha_k - l} + \hat{f}(p,t)$$

and

$$\hat{u}(p,t) = \int_0^{+\infty} e^{-px} u(x,t)\, dx.$$

Integrating with respect to time this equation we have the following representation for the Laplace transform of the solution

$$(2.7) \qquad \hat{u}(p,t) = e^{-K(p)t} \hat{u}_0(p) + \int_0^t e^{-K(p)(t-\tau)} f_1(p,\tau) d\tau.$$

In order to attain the integral formula for solutions of (2.2), we find $\partial_x^{j-1} u(0,t)$. The condition

$$(2.8) \qquad |\hat{u}(p,t)| \leq C(1+|p|)^\beta \text{ for all Re } p \geq a > 0,$$

with some M, $\beta > 0$ is necessary and sufficient for the existence of the inverse Laplace transformation (see Chapter 2, Section 1, Theorem 5). Clearly condition (2.8) is fulfilled in domains Re $K(p) \geq -b$ of the right half-complex plane Re $p \geq a \geq 0$. In domains of the right half-complex plane Re $p \geq a$, where Re $K(p) < -b$, we rewrite formula (2.7) as

$$
\begin{aligned}
\hat{u}(p,t) \;=\; & e^{-K(p)t} \left(\hat{u}_0(p) + \int_0^{+\infty} e^{K(p)\tau} f_1(p,\tau) d\tau \right) \\
& - \int_t^{+\infty} e^{K(p)(t-\tau)} f_1(p,\tau) d\tau.
\end{aligned}
$$

It is clear that the last integral

$$\int_t^{+\infty} e^{K(p)(t-\tau)} f_1(p,\tau)d\tau$$

satisfies the condition (2.8) for all Re $p \geq a$ such that Re $K(p) < -b$. However the first summand with the exponentially growing factor $e^{-K(p)t}$ does not satisfy condition (2.8); therefore, in order to satisfy (2.8), we must put

$$(2.9) \qquad \hat{u}_0(p) + \int_0^{+\infty} e^{K(p)\tau} f_1(p,\tau)d\tau = 0$$

for all Re $p \geq a$, where Re $K(p) < -b$. We use (2.9) to find the boundary functions $\partial_x^j u(0,t)$ involved in (2.7). Making the change of independent variable $K(p) = -\xi$ we transform domains Re $K(p) < -b$ of the right half-complex plane Re $p \geq a$ to the half-complex plane Re $\xi > b$.

Since $K(p) \in \mathcal{K}^\alpha$ the equation $K(p) = -\xi$ has m different "positive" roots $\phi_1(\xi)$, $\phi_2(\xi)$,..., $\phi_m(\xi)$, which are analytic functions for Re $\xi > b$ and transform the half-complex plane Re $\xi > b$ to domains, where

$$\text{Re } \phi_j(\xi) > 0, \ j = 1, 2, ..., m.$$

Condition (2.9) can be written as a system of m equations in the half-complex plane Re $\xi > b$

$$\widehat{u}_0(\phi_l) + \widehat{\widehat{f}}(\phi_l, \xi) + \sum_k \sum_{j=1}^{[\alpha_k]} C_k \phi_l^{\alpha_k - j} \int_0^{+\infty} e^{-\xi\tau} u_x^{(j-1)}(0,\tau)d\tau = 0,$$

for $l = 1, 2, ..., m$, where

$$\widehat{u}_0(\phi_l) \quad = \quad \int_0^{+\infty} e^{-\phi_l y} u_0(y)dy,$$

$$\widehat{\widehat{f}}(\phi_l, \xi) \quad = \quad \int_0^{+\infty} \int_0^{+\infty} e^{-(\phi_l y + \xi t)} f(y,t)dydt.$$

We have m equations with $[\alpha]$ unknowns $u_x^{(j-1)}(0,t)$, so we must include $[\alpha] - m = N$ boundary data into the problem. For example, if we use Dirichlet type boundary data $u_x^{(j-1)}(0,t) = h_j(t)$, $j = 1, 2, ..., N$, then we can find the rest of the Laplace transforms of the boundary values

$$\hat{v}_j(\xi) \equiv \int_0^{+\infty} e^{-\xi t} \partial_x^{(j+N-1)} u(0,t)dt, \ j = 1, ..., m$$

from the system of m equations

$$(2.10) \qquad\qquad A\vec{\hat{V}} = \vec{\mathcal{J}},$$

where A is $m \times m$ matrix $A = \|a_{ij}\|_m^m$ with elements

$$a_{ij} = \sum_{[\alpha_k] \geq N+j} C_k \phi_i^{\alpha_k - N - j},$$

the vector $\overrightarrow{V} = (\widehat{v}_1(\xi), ..., \widehat{v}_m(\xi))^T$ and $\overrightarrow{\mathcal{J}}$ is the input characteristic vector for problem (2.2) with components

$$\mathcal{J}(\phi_l, +\infty) = -(\widehat{u}_0(\phi_l) + \widehat{\overrightarrow{f}}(\phi_l, \xi)) - \sum_k \sum_{j=1}^{\min(N,[\alpha_k])} C_k \widehat{h}_j(\xi) \phi_l^{\alpha_k - j}.$$

Here

(2.11) $$\widehat{h}_j(\xi) = \int_0^{+\infty} e^{-\xi t} h_j(t) dt.$$

The determinant of system (2.10) is not equal to zero since all functions ϕ_l, $l = 1, ..., m$ are different for $\operatorname{Re} \xi > b$. Solving (2.10) and taking the inverse Laplace inverse transformation with respect to time we obtain

$$
\overrightarrow{V}(t) \;=\; \begin{pmatrix} v_1(t) \\ ... \\ v_m(t) \end{pmatrix} = \begin{pmatrix} \partial_x^{(N)} u(0,t) \\ ... \\ \partial_x^{([\alpha]-1)} u(0,t) \end{pmatrix}
$$

$$
= \; \frac{1}{2\pi i} \int_{-i\infty+b}^{i\infty+b} e^{\xi t} A^{-1} \overrightarrow{\mathcal{J}} \, d\xi.
$$

By (2.7) we get

$$
\begin{aligned}
\widehat{u}(p,t) \;=\;& e^{-K(p)t} \widehat{u}_0(p) + \int_0^t e^{-K(p)(t-\tau)} \widehat{f}(p,\tau) d\tau \\
& + \sum_k C_k p^{\alpha_k} \int_0^t e^{-K(p)(t-\tau)} \\
& \times \left(\sum_{j=1}^{\min([\alpha_k],N)} h_j(\tau) p^{-j} + \sum_{j=1}^{[\alpha_k]-N} v_j(\tau) p^{-N-j} \right) d\tau.
\end{aligned}
$$

Thus we get

(2.12) $$\widehat{u}(p,t) = e^{-K(p)t} \mathcal{J}(p,t) + \sum_k \sum_{j=1}^{[\alpha_k]-N} C_k p^{\alpha_k - N - j} \int_0^t e^{-K(p)(t-\tau)} v_j(\tau) d\tau,$$

where

$$v_j(t) = \frac{1}{2\pi i} \int_{-i\infty+b}^{i\infty+b} e^{\xi t} \left(A^{-1} \overrightarrow{\mathcal{J}} \right)_j (\xi) d\xi = \left(A \overrightarrow{\mathcal{J}} \right)_j.$$

Now we prove the following asymptotic representation

(2.13) $$\widehat{u}(p,t) = \sum_{j=1}^N h_j(t) p^{-j} + \sum_{j=1}^m v_j(t) p^{-N-j} + \lambda O \left(p^{-s-1-\gamma} t^{-\delta_s} \right)$$

for all $|p| > 1, \operatorname{Re} p = 0$ and $s = 0, 1, ..., M_1$, where $M_1 = \min([\alpha] - 1, N + [\kappa\alpha])$, $\delta_s = \max \left(0, \nu + \frac{s-N}{\alpha}, \frac{s-M}{\alpha} \right).$

Since $\mathcal{J}(p,t) \in \mathbf{S}_M$, using conditions (2.4) and (2.5), we rewrite the last integral in formula (2.12) as follows

$$\sum_k \sum_{j=1}^{[\alpha_k]-N} C_k p^{\alpha_k-N-j} \int_0^t e^{-K(p)(t-\tau)} v_j(\tau) d\tau = \sum_{j=1}^m v_j(t) p^{-N-j} + R(p,t)$$

where

$$R(p,t) = \sum_k C_k p^{\alpha_k} \int_0^t e^{-K(p)(t-\tau)} \left(\sum_{j=1}^{[\alpha_k]-N} \left(v_j(\tau) - v_j(t) \right) p^{-N-j} \right.$$

$$\left. - \sum_{j=[\alpha_k]-N+1}^m v_j(t) p^{-N-j} \right) d\tau$$

$$- \sum_{j=1}^m v_j(t) p^{-N-j} \left(1 - \sum_k C_k p^{\alpha_k} \frac{1-e^{-K(p)t}}{K(p)} \right).$$

Using (2.4) we get

$$\sup_{t\in[0,T]} t^v \sum_{j=1}^m |v_j(t)| = \sup_{t\in[0,T]} t^v \left\| \mathbb{A}\vec{\mathcal{J}} \right\| < C\lambda$$

for $0 \le v < 1$, and by (2.5) we find

$$\sup_{t,\tau\in[0,T]} t^{v+\kappa} |t-\tau|^{-\kappa} \sum_{j=1}^m |v_j(\tau) - v_j(t)| \le C\lambda$$

for $\kappa > 0$. Thus we obtain

$$R(p,t) \le C\lambda p^{\alpha-N-1} \int_0^t e^{-\operatorname{Re} K(p)(t-\tau)} t^{-v-\kappa}(t-\tau)^\kappa d\tau$$

$$+ C\lambda p^{\alpha_k-[\alpha_k]-1} \int_0^t e^{-\operatorname{Re} K(p)(t-\tau)} t^{-v} d\tau$$

$$\le C\lambda p^{-s-1} t^{-\overline{\delta}_s}$$

for $\overline{\delta}_s = \max\left(0, v + \frac{s-N}{\alpha}\right)$, $s < \min\left(N + \kappa\alpha, \alpha - 1\right)$. Since $\|\mathcal{J}(p,t)\|_{\mathbf{S}_M} = \lambda$ now we get (2.13).

Also from (2.12) using (2.4) and we easily obtain the following estimate (2.14)

$$\sup_{p\in[a-i,a+i]} |\hat{u}(p,t)| \le C \left(\sup_{p\in[a-i,a+i]} \left| e^{-K(p)t} \mathcal{J} \right| + \sum_{j=1}^m \int_0^t |v_j(\tau)| d\tau \right) \le C\lambda.$$

Now we prove that the solution $u(x,t)$ is given by the inverse Laplace transform of $\hat{u}(p,t)$

$$u(x,t) = \frac{1}{2\pi i} \int_{a-i\infty}^{a+i\infty} e^{px} \hat{u}(p,t) dp.$$

Using (2.13) and (2.14) we see that the integral is converging. Let us now prove that $u(x,t) = 0$ for $x < 0$. Since $\hat{u}(p,t)$ is an analytic function in the right-half complex plane $\operatorname{Re} p \geq a$ we get

$$\int_{a-i\infty}^{a+i\infty} e^{px}\hat{u}(p,t)dp = -\lim_{R\to+\infty}\int_{\Gamma_R} e^{px}\hat{u}(p,t)dp,$$

where Γ_R is a circumference $p = a + R\,e^{i\phi}$, $\phi \in \left(-\frac{\pi}{2}, \frac{\pi}{2}\right)$, $R > 0$. We denote $\Gamma_R = \Gamma_1 + \Gamma_2$, where $\Gamma_1 = \{p = a + Re^{i\phi}, \phi \in \left(-\frac{\pi}{2} + \varkappa, \frac{\pi}{2} - \varkappa\right)\}$, and $\Gamma_2 = \{p = a + Re^{i\phi}, \phi \in \left(-\frac{\pi}{2}, -\frac{\pi}{2} + \varkappa\right) \cup \left(\frac{\pi}{2} - \varkappa, \frac{\pi}{2}\right)\}$; here $\varkappa > 0$ is such that $\operatorname{Re} K(p) > 0$ for $p \in \Gamma_2$ (such small values $\varkappa$ exist since $\operatorname{Re} K(p) > 0$ on the imaginary axis $p \in (a - i\infty, a + i\infty)$). By (2.13) through (2.14) we have

$$|\hat{u}(p,t| \leq C(1+|p|)^C$$

for $\operatorname{Re} p > 0$, with some $C > 0$. Since

$$|e^{px}| \leq C\exp(ax - \sin(\varkappa R|x|)) \leq C(\varkappa R|x|)^{-1-C-\gamma}$$

for $x < 0$, $p \in \Gamma_1$, $\gamma > 0$ we have with $\varkappa = R^{-\frac{\gamma}{2(1+C+\gamma)}}$

$$\left|\lim_{R\to+\infty}\int_{\Gamma_1} e^{px}\hat{u}(p,t)dp\right|$$

$$\leq \ C\lim_{R\to+\infty}\int_{\Gamma_1} (1+|p|)^C \exp(-\sin(\varkappa R|x|))|dp|$$

$$\leq \ C\lim_{R\to+\infty} (\varkappa|x|)^{-1-C-\gamma}R^{-\gamma} = 0.$$

Since $\operatorname{Re} K(p) > 0$ for $p \in \Gamma_2$ from the asymptotic representation (2.13) we obtain

$$\hat{u}(p,t) = \frac{u(0,t)}{p} + O\left(\lambda|p|^{-1-\gamma}\right).$$

Therefore,

$$\left|\lim_{R\to+\infty}\int_{\Gamma_2} e^{px}\hat{u}(p,t)dp\right| \leq C\lim_{R\to+\infty} (\varkappa|u(0,t)| + \lambda\varkappa R^{-\gamma}) = 0.$$

Thus,

$$\int_{a-i\infty}^{a+i\infty} e^{px}\hat{u}(p,t)dp = 0$$

for all $x < 0$. Via (2.13), we have

$$\partial_x^j u = \frac{1}{2\pi i}\int_{a-i\infty}^{a+i\infty} e^{px}p^j\left(\hat{u}(p,t) - \sum_{k=1}^{j}\frac{\partial_x^{k-1}u(0,t)}{p^k}\right)dp$$

$$= \partial_x^j u(0,t)\frac{1}{2\pi i}\int_{|p|\geq 1, \operatorname{Re} p = a} e^{px}\frac{dp}{p}$$

$$+Ct^{-\delta_j}\int_{|p|\geq 1, \operatorname{Re} p = a} e^{px}O(\lambda|p|^{-1-\gamma})dp + C\lambda t^{-\delta_j}.$$

Therefore we see that the derivatives $\partial_x^j u(x,t)$, $j = 0, 1, ..., M_1$, $M_1 = \min([\alpha] - 1, N + [\kappa\alpha])$ are continuous with respect to x and the boundary data are fulfilled in the classical sense

$$\partial_x^{j-1} u(x,t) \to h_j(t)$$

as $x \to 0$ for all $j = 0, 1, ..., N$ and $t > 0$. Moreover by (2.13) and (2.14) we have the estimates

$$\sup_{t\in[0,T]} \sum_{s=0}^{M_1} t^{\delta_s} \|\partial_x^s u\|_{\mathbf{L}^2}$$

(2.15)
$$\leq C \sum_{s=0}^{M_1} \sup_{t\in[0,T]} \left(\lambda + \left(\int_{|p|>1, \operatorname{Re} p=a} O\left(\frac{\lambda}{|p|^2}\right) |dp| \right)^{\frac{1}{2}} \right) \leq C\lambda$$

and

$$\sup_{t\in[0,T]} \sum_{s=0}^{M_1} t^{\delta_s} \|\partial_x^s u\|_{\mathbf{L}^\infty}$$

$$\leq C \sum_{s=0}^{M_1} \sup_{t\in[0,T]} \left(\lambda + t^{\delta_s} \left| \int_{a-i\infty}^{a+i\infty} e^{-px} \partial_x^s u(0,t) \frac{dp}{p} \right| \right.$$

(2.16)
$$\left. + \int_{|p|>1, \operatorname{Re} p=a} O\left(\lambda |p|^{-1-\gamma}\right) dp \right)$$

$$\leq C\lambda.$$

We now prove the uniqueness of the solution. On the contrary we consider two different solutions u_1 and u_2. Then the difference $u_1 - u_2$ satisfies linear problem (2.2) with homogeneous data $f = 0$, $u_0 = 0$ and $h_j = 0$. Then by estimate (2.15) we get $\|u_1 - u_2\|_{\mathbf{L}^2} = 0$; hence $u_1 = u_2$. Theorem 12 is proved. □

2.3. Green operator. We denote by $F(x,y,t)$ the following function
(2.17)

$$F(x,y,t) = \frac{1}{2\pi i} \int_{a-i\infty}^{a+i\infty} e^{px} \left(e^{-py-K(p)t} - \sum_k C_k \sum_{j=1}^{[\alpha_k]-N} \Theta_j p^{\alpha_k - N - j} \right) dp,$$

where the functions Θ_j are components of the vector Θ given by

$$\Theta = \frac{1}{2\pi i} \int_{b-i\infty}^{b+i\infty} \frac{e^{\xi t}}{K(p)+\xi} A^{-1} \begin{pmatrix} e^{-\phi_1(\xi)y} \\ ... \\ e^{-\phi_m(\xi)y} \end{pmatrix} d\xi.$$

We define operator $\mathcal{H}\left[\vec{h}\right](x,t)$ as

$$\mathcal{H}\left[\vec{h}\right](x,t) = \frac{1}{2\pi i} \sum_k \sum_{j=1}^{[\alpha_k]} C_k \int_0^t H_j(\tau)d\tau \int_{a-i\infty}^{a+i\infty} e^{px-K(p)(t-\tau)} p^{\alpha_k - j} dp,$$

where the functions H_j are

$$H_j(t) = h_j(t)$$

for $j = 1, ..., N$ and

$$H_j(t) = -\sum_k \sum_{l=1}^{\min(N,[\alpha_k])} \frac{C_k}{2\pi i} \int_{-i\infty+b}^{i\infty+b} d\xi e^{\xi t} \widehat{h}_l(\xi) \left(A^{-1} \begin{pmatrix} \phi_1^{\alpha_k-l}(\xi) \\ .... \\ \phi_m^{\alpha_k-l}(\xi) \end{pmatrix} \right)_j$$

for $j = N+1, ..., [\alpha]$. Note that the integrals in the definition of operator $\mathcal{H}\left[\overrightarrow{h}\right](x,t)$ converge absolutely for all $x > 0$ if $K(p) \in \mathcal{K}^\alpha$. In fact, since

$$\frac{1}{|z|} \le \frac{1}{|\mathrm{Re}\, z|^\nu |\mathrm{Im}\, z|^{1-\nu}}$$

for any $z \in \mathbf{C}$, where $\nu \in [0,1]$, we have the estimate with $K(p) \in \mathcal{K}^\alpha$

$$\frac{1}{|K(p) + \xi|} \le |p|^{-2\nu} |\xi + Cp^\alpha|^{-1+\nu}$$

for all $\mathrm{Re}\, p = 0$, $\mathrm{Re}\, \xi = b$, where $\nu \in [0,1]$. Therefore, for example, integrating by parts with respect to p in the last integral of the right-hand side of (2.17) we get

$$\left| \frac{1}{4\pi^2} \sum_k \sum_{j=1}^{[\alpha_k]-N} \int_{-i\infty+b}^{i\infty+b} e^{\xi t} A^{-1} \begin{pmatrix} e^{-\phi_1(\xi)y} \\ ... \\ e^{-\phi_m(\xi)y} \end{pmatrix} \right.$$
$$\left. \times \int_{a-i\infty}^{a+i\infty} \frac{e^{px}}{K(p)+\xi} p^{\alpha_k-N-j} d\xi dp \right|$$
$$< \frac{C}{x} \int_{\mathrm{Re}\,\xi=b} \left\| A^{-1} \begin{pmatrix} e^{-\phi_1(\xi)y} \\ ... \\ e^{-\phi_m(\xi)y} \end{pmatrix} \right\| d|\xi|$$
$$\times \int_{\mathrm{Re}\,p=a} \frac{1}{|p|} \left(\frac{|p^{2\alpha-N-2}|}{|K(p)+\xi|^2} + \frac{|p^{\alpha-N-2}|}{|K(p)+\xi|} \right) d|p| < \frac{C}{x}.$$

THEOREM 13. *Let*

$$u_0 \in \mathbf{L}^1\left(\mathbf{R}^+\right), \; f \in \mathbf{L}^q\left(0,T;\mathbf{L}^1\left(\mathbf{R}^+\right)\right), \; h_j \in \mathbf{L}^q\left(0,t\right), \; j = 1,...,N$$

with $q > 2$. Then the solution of the problem (2.2) has the following form:

$$u(x,t) = \int_0^{+\infty} F(x,y,t)u_0(y)dy + \int_0^t \int_0^{+\infty} F(x,y,t-\tau)f(y,\tau)dyd\tau$$

$$(2.18) \qquad + \mathcal{H}\left[\overrightarrow{h}\right](x,t).$$

PROOF. By (2.6) we write a formula for the solution of problem (2.2)

$$(2.19) \quad u(x,t) = \frac{1}{2\pi i} \int_{a-i\infty}^{a+i\infty} e^{px-K(p)t} \left(\mathcal{J}(p,t) \right.$$

$$+ \sum_k \sum_{j=1}^{[\alpha_k]-N} C_k p^{\alpha_k - N - j} \int_0^t e^{K(p)\tau} (\mathbb{A}\vec{\mathcal{J}})_j(\tau) d\tau \left. \right) dp.$$

We consider the last term in the right-hand side of (2.19)

$$\sum_k \sum_{j=1}^{[\alpha_k]-N} \frac{C_k}{2\pi i} \int_{a-i\infty}^{a+i\infty} e^{px-K(p)t} p^{\alpha_k - N - j} \int_0^t e^{K(p)\tau} (\mathbb{A}\vec{\mathcal{J}})_j(\tau) d\tau dp$$

$$= \int_0^{+\infty} dy \left(u_0(y) + \int_0^t f(y,\tau) d\tau \right)$$

$$\times \frac{1}{4\pi^2} \int_{a-i\infty}^{a+i\infty} e^{px} e^{-K(p)t} \int_{b-i\infty}^{b+i\infty} \frac{e^{(\xi+K(p))t}-1}{K(p)+\xi} G(p,\xi,y) \, d\xi dp$$

$$- \frac{1}{2\pi i} \sum_k \sum_{j=1}^{[\alpha_k]-N} C_k \int_0^t \Psi_j(\tau) d\tau \int_{a-i\infty}^{a+i\infty} e^{px-K(p)(t-\tau)} p^{\alpha_k - N - j} dp,$$

where

$$G(p,\xi,y) = - \sum_k \sum_{j=1}^{[\alpha_k]-N} C_k p^{\alpha_k - N - j} \left(A^{-1} \begin{pmatrix} e^{-\phi_1(\xi)y} \\ \cdots \\ e^{-\phi_m(\xi)y} \end{pmatrix} \right)_j$$

and

$$\Psi_j(t) = \frac{1}{2\pi i} \sum_r \sum_{l=1}^{\min([\alpha_r],N)} C_r \int_{-i\infty+b}^{i\infty+b} e^{\xi t} \widehat{h_l}(\xi)$$

$$\times \left(A^{-1} \begin{pmatrix} \phi_1^{\alpha_r - l}(\xi) \\ \cdots \\ \phi_m^{\alpha_r - l}(\xi) \end{pmatrix} \right)_j d\xi .$$

The change of the order of integration in formula (2.19) is justified by the Fubini Theorem under the conditions

$$u_0 \in \mathbf{L}^1 (\mathbf{R}^+), \ f \in \mathbf{L}^q (0,t; \mathbf{L}^1 (\mathbf{R}^+)), \ h_j \in \mathbf{L}^q (0,t), \ \text{for } j = 1, ..., N$$

with $q > 2$. Note that $G(p,\xi,y)$ is analytic in $\operatorname{Re}\xi \geq b$ for any $\operatorname{Re}p = a$, $y > 0$ fixed, and by virtue of Theorem 11 the estimate

$$|G(p,\xi,y)| \leq C(p) e^{-Cy|\xi|^{\frac{1}{\alpha}}}$$

is true for all $\operatorname{Re}\xi \geq b$. Hence by the Cauchy Theorem we have

$$\int_{b-i\infty}^{b+i\infty} \frac{G(p,\xi,y)}{K(p)+\xi} d\xi = 0$$

for all $\operatorname{Re} p = a, y > 0$, since $\operatorname{Re} K(p) > 0$. Also for the first summand in (2.19) we have

$$\frac{1}{2\pi i} \int_{a-i\infty}^{a+i\infty} e^{px - K(p)t} \mathcal{J}(p,t) dp$$

$$= \frac{1}{2\pi i} \int_0^{+\infty} dy u_0(y) \int_{a-i\infty}^{a+i\infty} e^{p(x-y)} e^{-(K(p)t+py)} dp$$

$$+ \frac{1}{2\pi i} \int_0^t \int_0^{+\infty} dy \int_{a-i\infty}^{a+i\infty} e^{p(x-y)} e^{-(K(p)(t-\tau))} f(y,\tau) dp d\tau$$

$$+ \frac{1}{2\pi i} \sum_k \sum_{j=1}^{\min([\alpha_k],N)} \int_0^t \int_{a-i\infty}^{a+i\infty} e^{px} e^{-(K(p)(t-\tau))} C_k h_j(\tau) p^{\alpha_k - j} dp d\tau.$$

Hence we obtain the integral representation (2.18) for the solution $u(x,t)$ of the problem (2.2). Theorem 12 is proved. $\qquad\square$

Now we consider two examples.

EXAMPLE 1.

Let $K(p) = E_\alpha p^\alpha$, $E_\alpha = e^{i\pi\left[\frac{\alpha+1}{2}\right]}$, where $\alpha > 1$ is not equal to an odd integer, $K(p) \in \mathcal{K}_{diss}^\alpha$. Since $K(p)$ is analytic for $\operatorname{Re} p > 0$ and $K'(p) \neq 0$ due to definition (1.2) we write pseudodifferential operator $\mathbb{K}u$ on a half-line as follows:

$$\mathbb{K}u = \frac{1}{2\pi i} \int_{-i\infty}^{i\infty} e^{px} K(p) \left(\widehat{u}(p) - \sum_{l=1}^{[\alpha]} \partial_x^{l-1} u(0) p^{-l} \right) dp.$$

The equation $K(p) = -\xi$ has $m = \left[\frac{\alpha}{2}\right]$ roots $\phi_1(\xi), \phi_2(\xi),..., \phi_m(\xi)$, which are analytic functions for $\operatorname{Re} \xi > 0$ and

$$p = \phi_j(\xi) = \left(\frac{\xi}{E_\alpha} \exp\left(2i\pi j\right) \right)^{\frac{1}{\alpha}}$$

transforms the half-complex plane $\operatorname{Re} \xi > 0$ to domains, where

$$\operatorname{Re} \phi_j(\xi) > 0, j = 1, 2, ..., m.$$

Thus we have $N = [\alpha] - m = \left[\frac{\alpha+1}{2}\right]$.

In this case the characteristic matrix A of the operator $\mathbb{K}$ has the following form:

$$A = -\xi \left\| \phi_i^{-N-j} \right\|_m^m.$$

By Theorem 13 we have

$$u(x,t) = \int_0^{+\infty} F(x,y,t) u_0(y) dy + \int_0^t \int_0^{+\infty} F(x,y,t-\tau) f(y,\tau) dy d\tau$$

$$+ \mathcal{H}\left[\overrightarrow{h}\right](x,t),$$

where $F(x, y, t)$ is

$$(2.20) \qquad F(x, y, t) = \frac{1}{2\pi i} \int_{-i\infty}^{i\infty} e^{px} \left(e^{-K(p)t - py} - K(p) \sum_{j=1}^{m} \Theta_j p^{-N-j} \right) dp,$$

and the functions Θ_j are components of vector Θ

$$\Theta = \frac{1}{2\pi i} \int_{-i\infty}^{i\infty} \frac{e^{\xi t}}{K(p) + \xi} A^{-1} \left(\begin{array}{c} e^{-\phi_1(\xi)y} \\ \dots \\ e^{-\phi_m(\xi)y} \end{array} \right) d\xi$$

$$(2.21) \qquad = -\frac{1}{2\pi i} \int_{-i\infty}^{i\infty} \frac{e^{\xi t}}{(K(p) + \xi)\xi} \tilde{A}^{-1} \left(\begin{array}{c} e^{-\phi_1(\xi)y} \\ \dots \\ e^{-\phi_m(\xi)y} \end{array} \right) d\xi,$$

where $m \times m$ matrix $\tilde{A} = \left\| \phi_i^{-N-j} \right\|_m^m$. Putting the representation (2.21) into (2.20) we obtain for the Green function

$$F(x, y, t) = \frac{1}{2\pi i} \int_{-i\infty}^{i\infty} e^{-K(p)t + p(x-y)} dp$$

$$(2.22) \qquad + \frac{1}{4\pi^2} \sum_{j=1}^{m} \int_{-i\infty}^{i\infty} e^{\xi t} \frac{\theta_j(\xi, y)}{\xi} \int_{-i\infty}^{i\infty} \frac{e^{px} K(p)}{K(p) + \xi} p^{-N-j} d\xi dp,$$

where the functions $\theta_j(\xi)$ are equal to

$$\theta_j(\xi, y) = \left(\tilde{A}^{-1} \left(\begin{array}{c} e^{-\phi_1(\xi)y} \\ \dots \\ e^{-\phi_m(\xi)y} \end{array} \right) \right)_j,$$

for $j = N + 1, ..., [\alpha]$.

Operator $\mathcal{H} \left[\vec{h} \right]$ is

$$\mathcal{H} \left[\vec{h} \right] (x, t) = \frac{1}{2\pi i} \sum_{j=1}^{[\alpha]} \int_0^t d\tau \, H_j(\tau) \int_{-i\infty}^{i\infty} e^{px - K(p)(t-\tau)} K(p) p^{-j} dp,$$

where the functions H_j are

$$H_j(t) = h_j(t)$$

for $j = 1, ..., N$ and

$$H_j(t) = -\frac{1}{2\pi i} \sum_{l=1}^{N} \int_{-i\infty}^{i\infty} e^{\xi t} \widehat{h}_l(\xi) \tilde{A}^{-1} \left(\begin{array}{c} \phi_1^{-l}(\xi) \\ \dots \\ \phi_m^{-l}(\xi) \end{array} \right) d\xi.$$

EXAMPLE 2.

As further example let us consider $K(p) = Cp^n$, where n is natural. Thus $K(p)$ is holomorphic function in $\mathbf{C}$. In this particular case we consider all roots $p = \phi_j(\xi)$, $j = 1, 2, ..., n$ of the equation $K(p) = -\xi$. (Not only "positive" roots

$$\operatorname{Re} \phi_l(\xi) > 0 \text{ for } \operatorname{Re} \xi > 0, \ l = 1, 2, ..., m$$

as above). By the fact that

$$K(\phi_j(\xi)) = -\xi$$

and

$$\phi_j'(\xi) = -\frac{1}{K'(\phi_j)}$$

via the Cauchy Theorem we get for $x > 0$

(2.23) $$\int_{-i\infty}^{i\infty} \frac{e^{px} K(p)}{K(p) + \xi} p^{-N-j} dp = 2\pi i \sum_{l=1}^{N} e^{\phi_{m+l}(\xi)x} \xi \phi_{m+l}^{-N-j}(\xi) \phi_{m+l}'(\xi),$$

where $\phi_{m+l}(\xi)$, $l = 1, ..., N$, $N = n - m$ are "negative" inverse functions $K^{-1}(-\xi)$ such that

$$\operatorname{Re} \phi_{m+l}(\xi) < 0, \text{ for } \operatorname{Re} \xi > 0.$$

Also since

(2.24) $$\int_{-i\infty}^{i\infty} e^{-K(p)t+p(x-y)} dp = \sum_{l=1}^{N} \int_{-i\infty}^{i\infty} e^{\xi t + \phi_{m+l}(\xi)(x-y)} \phi_{m+l}'(\xi) d\xi$$

using (2.23), (2.24) in (2.22) we have for the Green function

$$F(x, y, t) \quad = \quad \frac{1}{2\pi i} \sum_{l=1}^{N} \int_{-i\infty}^{i\infty} e^{\xi t + \phi_{m+l}(\xi)x} \phi_{m+l}'(\xi)$$

(2.25) $$\times \left(e^{-\phi_{m+l}(\xi)y} - \sum_{j=1}^{m} \theta_j(\xi, y) \phi_{m+l}^{-N-j}(\xi) \right) d\xi,$$

where

$$\theta_j(\xi, y) = \left(\begin{pmatrix} \phi_1^{-N-1}(\xi) & \cdots & \phi_1^{-n}(\xi) \\ \cdots & \cdots & \cdots \\ \phi_m^{-N-1}(\xi) & \cdots & \phi_m^{-n}(\xi) \end{pmatrix}^{-1} \begin{pmatrix} e^{-\phi_1(\xi)y} \\ \cdots \\ e^{-\phi_m(\xi)y} \end{pmatrix} \right)_j.$$

In particular, if $K(p) = -p^2$ we have two inverse functions

$$\phi_j(\xi) = K^{-1}(\xi),$$

which are analytic in domain $\operatorname{Re} \xi > 0$

$$\phi_1(\xi) = \sqrt{\xi} \text{ and } \phi_2(\xi) = -\sqrt{\xi}.$$

Thus $N = 1$ and

$$\theta_1(\xi, y) = \xi e^{-\phi_1(\xi)y}.$$

We have for $q = \frac{z}{\sqrt{t-\tau}}$

$$\int_{-i\infty}^{i\infty} e^{z^2 t} e^{-zx} dz = i \int_{-\infty}^{+\infty} e^{-z^2 t} e^{-izx} dz$$

(2.26)

$$= \frac{i}{\sqrt{t-\tau}} e^{-\frac{x^2}{4(t-\tau)}} \int_{-\infty}^{+\infty} \exp\left(-\left(q + \frac{ix}{2\sqrt{t-\tau}}\right)^2\right) dz$$

$$= i\sqrt{\frac{\pi}{t-\tau}} \exp\left(-\frac{x^2}{4(t-\tau)}\right).$$

Taking $b = 0$, using $\phi_1 + \phi_2 = 0$ and changing $p = \phi_2(\xi)$ via (2.26) we obtain

$$F(x, y, t)$$

$$= \frac{1}{2\pi i} \int_{-i\infty}^{i\infty} e^{\xi t + \phi_2(\xi)x} \left(e^{-\phi_2(\xi)y} - e^{-\phi_1(\xi)y}\xi\phi_2^{-2}(\xi)\right) \phi_2'(\xi)d\xi$$

$$= \frac{1}{2\pi i} \left(\int_{-i\infty}^{i\infty} e^{\xi t + \phi_2(\xi)(x-y)}\phi_2'(\xi)d\xi - \int_{-i\infty}^{i\infty} e^{\xi t + \phi_2(\xi)(x+y)}\phi_2'(\xi)d\xi\right)$$

$$= \frac{1}{2\pi i} \left(\int_{-i\infty}^{i\infty} e^{p^2 t + p(x-y)}dp - \int_{-i\infty}^{i\infty} e^{p^2 t + p(x+y)}dp\right)$$

$$= \frac{1}{\sqrt{4\pi t}} \left(e^{-\frac{(x-y)^2}{4t}} - e^{-\frac{(x+y)^2}{4t}}\right).$$

The operator

(2.27)
$$\mathcal{H}\left[\overrightarrow{h}\right](x, t) = -\frac{1}{2\pi i} \sum_{j=1}^{2} \int_0^t d\tau H_j(\tau) \int_{-i\infty}^{i\infty} e^{px + p^2(t-\tau)} p^{2-j} dp,$$

where the functions H_j are

$$H_1(t) = h_1(t)$$

and

$$H_2(t) = -\frac{1}{2\pi i} \int_{-i\infty}^{i\infty} e^{\xi t}\widehat{h}_1(\xi)\phi_1(\xi)\, d\xi.$$

We consider the second summand of formula (2.27). Using the Cauchy's theory we attain by changing the variable $z = \phi_1(\xi)$

$$-\frac{1}{2\pi i} \int_0^t d\tau H_2(\tau) \int_{-i\infty}^{i\infty} e^{px + p^2(t-\tau)} dp$$

$$= \frac{1}{4\pi^2} \int_{-i\infty}^{i\infty} e^{\xi t}\widehat{h}_1(\xi)\phi_1(\xi)\, d\xi \int_{-i\infty}^{i\infty} e^{px} \frac{1}{(p-\phi_1)(p-\phi_2)}dp =$$

$$= \frac{1}{2\pi i} \int_0^t h_1(\tau)d\tau \int_{-i\infty}^{i\infty} e^{z^2(t-\tau)} e^{-zx} z dz$$

$$= -\frac{1}{2\pi i} \int_0^t h_1(\tau)d\tau \partial_x \left(\int_{-i\infty}^{i\infty} e^{z^2(t-\tau)} e^{-zx} dz\right).$$

With (2.27) we obtain the well-known formula

$$\mathcal{H}\left[\overrightarrow{h}\right](x,t) = -2\int_0^t d\tau h_1(\tau)\frac{1}{\sqrt{4\pi(t-\tau)}}\partial_x\left(e^{-\frac{x^2}{4(t-\tau)}}\right)$$

$$= \int_0^t h_1(\tau)\partial_y F(x,0,t-\tau)\,d\tau.$$

2.4. A general lemma. In this section we give some sufficient conditions for local and global existence of solutions to the initial-boundary value problem for the nonlinear nonlocal equation

(2.28)
$$\begin{cases} u_t + \mathbb{N}(u) + \mathbb{K}u = f, \ x > 0, \ t > 0 \\ u(x,0) = u_0(x), \ x > 0, \\ \partial_x^{j-1}u(0,t) = h_j(t), \ t > 0, \ j = 1, 2, ..., N. \end{cases}$$

Let us denote $\mathbf{X}_T$ and $\mathbf{Y}_T$ some complete metric spaces of functions defined in $\mathbf{R}\times(0,T)$; $\mathbf{Z}$ is a complete metric space of functions defined on $\mathbf{R}$, and $\mathbf{B}_T$ is a complete metric space of functions defined on $(0,T)$. We denote the Green operator

$$\mathcal{F}[v(\tau)](x,t) = \int_0^{+\infty} F(x,y,t)v(y,\tau)dy.$$

We have the following result.

LEMMA 1. *Suppose that*

$$\left|\left|\left|\mathcal{F}[u_0](x,t) + \mathcal{H}\left[\overrightarrow{h}\right](x,t)\right|\right|\right|_{\mathbf{X}_T} \le C\|u_0\|_{\mathbf{Z}} + C\left\|\overrightarrow{h}\right\|_{\mathbf{B}_T},$$

$$\left|\left|\left|\int_0^t \mathcal{F}[w(\tau)](x,t-\tau)\,d\tau\right|\right|\right|_{\mathbf{X}_T} \le C\min(1,T^\mu)\,|||w|||_{\mathbf{Y}_T}$$

$$\left|\left|\left|\mathcal{H}\left[\overrightarrow{h}\right](x,t)\right|\right|\right|_{\mathbf{X}_T} \le C\min(1,T^\mu)\,|||w|||_{\mathbf{Y}_T}$$

and

$$|||\mathbb{N}(v)|||_{\mathbf{Y}_T} < C\,|||v|||_{\mathbf{X}_T}^\delta,$$

where $\mu \in (0,1]$, $\delta > 1$. Then

1) if $T > 0$ is sufficiently small, then for any initial data $u_0 \in \mathbf{Z}$, boundary data $\overrightarrow{h} \in \mathbf{B}_T$ and a source $f \in \mathbf{Y}_T$ there exists a unique solution $u(x,t) \in \mathbf{X}_T$ of problem (2.28),

2) if initial data $u_0 \in \mathbf{Z}$, boundary data $\overrightarrow{h} \in \mathbf{B}_\infty$ and a source $f \in \mathbf{Y}_{+\infty}$ are sufficiently small, that is the norm $\|u_0\|_{\mathbf{Z}} + \left\|\overrightarrow{h}\right\|_{\mathbf{B}_\infty} + |||f|||_{\mathbf{Y}_\infty}$ is small, then there exists a unique global solution $u(x,t) \in \mathbf{X}_\infty$ of problem (2.28).

REMARK 2. *We take $0 < \mu \le 1$ since the estimates of $\mathcal{F}[w(\tau)](x,t-\tau)$ in some norms (including derivatives) could have an integrable singularity for $t \to 0$.*

PROOF. Using the Green operator $\mathcal{F}\left[v(\tau)\right](x,t)$ we rewrite the initial-boundary value problem (2.28) as an integral equation

$$(2.29) \qquad u(x,t) \;=\; \mathcal{F}\left[u_0\right](x,t) + \mathcal{H}\left[\overrightarrow{h}\right](x,t)$$

$$+ \int_0^t \mathcal{F}\left[\mathbb{N}(u\left(\tau\right)) + f\right](x,t-\tau)\,d\tau.$$

We define the mapping $\mathbb{M}$ by

$$u \;=\; \mathbb{M}v = \mathcal{F}\left[u_0\right](x,t) + \mathcal{H}\left[\overrightarrow{h}\right](x,t)$$

$$+ \int_0^t \mathcal{F}\left[\mathbb{N}(v\left(\tau\right)) + f\right](x,t-\tau)\,d\tau$$

for $v \in X_{T,\rho}$, where $\rho > 0$ and

$$\mathbf{X}_{T,\rho} = \left\{ v \in X_T; |||v|||_{\mathbf{X}_T} \leq \rho \right\}.$$

Therefore we have, taking $\rho = \frac{1}{2C}\left(\|u_0\|_{\mathbf{Z}} + \left\|\overrightarrow{h}\right\|_{\mathbf{B}_T} + |||f|||_{\mathbf{Y}_T} \right)$

$$|||u|||_{\mathbf{X}_T} \leq C \|u_0\|_{\mathbf{Z}} + C \left\|\overrightarrow{h}\right\|_{\mathbf{B}_T}$$

$$+ C \left|\left|\left| \int_0^t \mathcal{F}\left[\mathbb{N}(v\left(\tau\right)) + f\right](x,t-\tau)\,d\tau \right|\right|\right|_{\mathbf{X}_T}$$

$$\leq \; C \|u_0\|_{\mathbf{Z}} + C \left\|\overrightarrow{h}\right\|_{\mathbf{B}_T} + \min\left(1, T^\gamma\right) |||\mathbb{N}(v\left(\tau\right)) + f|||_{\mathbf{Y}_T}$$

$$(2.30) \qquad \leq \; C \|u_0\|_{\mathbf{Z}} + C \left\|\overrightarrow{h}\right\|_{\mathbf{B}_T}$$

$$+ C \min\left(1, T^\gamma\right) |||v|||_{\mathbf{X}_T}^\delta + C \min\left(1, T^\gamma\right) |||f|||_{\mathbf{Y}_T}$$

$$\leq \; C\frac{\rho}{2C} + C \min\left(1, T^\gamma\right) \rho^\delta < \rho,$$

if $T > 0$ is small or $\rho > 0$ is small. We introduce a distance in the space $\mathbf{X}_T$ such that

$$d\left(u_1, u_2\right) = |||u_1 - u_2|||_{\mathbf{X}_T}.$$

Then as in the proof of (2.30) we have

$$(2.31) \qquad d\left(u_1, u_2\right) = d\left(\mathbb{M}v_1, \mathbb{M}v_2\right) \leq \frac{1}{2} d\left(v_1, v_2\right),$$

where

$$\begin{cases} (u_j)_t + \mathbb{K}u = -\mathbb{N}(v_j) + f, & t > 0, x > 0, \\ \quad u_j(x,0) = u_0(x), & x > 0, \\ \partial_x^{j-1} u\left(0,t\right) = h_j(t), & t > 0, \; j = 1,...,N. \end{cases}$$

The estimates (2.30) and (2.31) show that $\mathbb{M}$ is a contraction mapping from $\mathbf{X}_T$ into itself. Therefore there exists a unique solution u of problem (2.28) satisfying the estimate (2.30). This completes the proof of Lemma 1.   □

CHAPTER 4

Nonlinear Schrödinger Type Equations

1. Setting of the problem

In this chapter we study the initial-boundary value problem for nonlinear nonlocal Schrödinger equations

(1.1)
$$\begin{cases} u_t + \mathbb{N}(u) + \mathbb{K}u = 0, (t, x) \in \mathbf{R}^+ \times \mathbf{R}^+, \\ u(x, 0) = u_0(x), x > 0, \\ \partial_x^{j-1} u(0, t) = h_j(t), t > 0, j = 1, ..., N \end{cases}$$

with the compatibility conditions $\partial_x^{j-1} u_0(0) = h_j(0)$, $j = 1, ..., N$, where $N = \left[\frac{\alpha}{2}\right]$. By $[s]$ we denote the largest integer less than s. We take the nonlinear term of the form

$$\mathbb{N}(u) = ia(t)|u|^\rho u, \rho > 1,$$

with the coefficient $a(t) \in \mathbf{C}^1$. We are interested in the case, when the symbol of the operator $\mathbb{K}$ has the following form

$$K(p) = Ep^\alpha,$$

where the constant E is such that the operator $\mathbb{K}$ is dissipative, i. e. $\operatorname{Re} K(p) > 0$ for all p on the imaginary axis $\operatorname{Re} p = 0$. The dissipation condition implies that α is not equal to an odd integer. Also we assume that $\alpha > 1$. (We take p^α as the main branch of the complex analytic function so that $1^\alpha = 1$. We make a cut along the negative real axis $(+\infty, 0)$ in the complex plane of variable p).

In this case the pseudodifferential operator $\mathbb{K}$ is defined as follows (see 3):

$$\mathbb{K}u = \frac{1}{2\pi i} \int_{-i\infty}^{i\infty} e^{px} K(p) \left(\hat{u}(p, t) - \sum_{j=1}^{[\alpha]} \frac{\partial_x^{j-1} u(0, t)}{p^j} \right) dp,$$

where $\hat{u}(p, t) = \int_0^{+\infty} e^{-px} u(x, t) dx$ is the Laplace transform with respect to x of the function $u(x, t)$.

The initial-boundary value problem (1.1) is of great interest from the physical point of view, since it describes many physical phenomena, such as the focusing of laser beams, waves on water and others [103]. Note that in the particular case $\alpha = 2, a = e^{-2t}$ and $\rho = 2$, problem (1.1) contains the initial-value problem for the well-known Landau-Ginzburg equation $u_t + i|u|^2 u + u - u_{xx} = 0$ (see [103]). In fact, changing $u = e^t v$ we get for new function $v(x, t)$ the following equation

$$v_t + ie^{-2t} |v|^2 v - v_{xx} = 0.$$

Existence, uniqueness and some qualitative properties of the solutions to the Cauchy problems for some classes of nonlinear nonlocal dissipative equations were studied in [67] through [103]. Large time asymptotic behavior of solutions to the Cauchy problem for dissipative and dispersive nonlinear nonlocal equations was studied in [76]-[103] .

To state our results in this paper we give the following notations.

Let us denote $\mathbf{X} = \mathbf{H}_\infty^{[\alpha]}(\mathbf{R}^+) \cap \mathbf{H}_1^{[\alpha]+1}(\mathbf{R}^+)$ and $\mathbf{Y} = \mathbf{H}_\infty^1(\mathbf{R}^+) \cap \mathbf{H}_1^2(\mathbf{R}^+)$, where $\mathbf{H}_p^k(\mathbf{R}^+)$ is the Sobolev space. We also introduce the following function space

$$\begin{aligned}
\mathbf{Z}_T \;=\; & \{\phi(x,t) \in \mathbf{C}^0\left([0,T], \mathbf{H}_2^N(\mathbf{R}^+) \cap \mathbf{C}^N(\mathbf{R}^+)\right) \\
& \cap \mathbf{C}^0\left((0,T]; \mathbf{H}_2^{[\alpha]-1}(\mathbf{R}^+) \cap \mathbf{C}^{[\alpha]-1}(\mathbf{R}^+)\right) : \\
& \|\phi\|_{\mathbf{Z}_T} < +\infty\}
\end{aligned}$$

with the norm

$$\|\phi\|_{\mathbf{Z}_T} = \|\phi\|_{\mathbf{L}^2} + \sup_{t \in [0,T]} \sum_{j=0}^{[\alpha]-1} t^{\delta_j} \left(\left\|\partial_x^j \phi\right\|_{\mathbf{L}^2} + \left\|\partial_x^j \phi\right\|_{\mathbf{L}^\infty}\right),$$

where $\delta_j = \max\left(0, \frac{j-N}{\alpha}\right)$ is small enough and $N = \left[\frac{\alpha}{2}\right]$. By the same letter C we denote different positive constants.

Now we state our results. First of all we formulate the local existence of solutions to the initial-boundary value problem (1.1). We consider the quasi classical solutions of the initial-boundary problem (1.1), that is we multiply equation (1.1) by any function $\varphi \in \mathbf{C}^2([0,T] \times (0, +\infty))$ such that $\varphi(x,T) = 0$ and $\partial_x^j \varphi(0,t) = 0$, $j = 0, 1, 2$ and integrate by parts in the domain $[0,T] \times (0, +\infty)$. Then the linear operator $\mathbb{K}$ make sense since we can represent it in the following form:

$$\begin{aligned}
\mathbb{K}u \;=\; & \partial_x^3 \frac{1}{2\pi i} \int_{\mathrm{Re}\,p=0, |p|>1} e^{px} \frac{K(p)}{p^3} \left(\hat{u}(p,t) - \sum_{j=1}^{[\alpha]} \frac{\partial_x^{j-1}u(0,t)}{p^j}\right) dp \\
& + \frac{1}{2\pi i} \int_{\mathrm{Re}\,p=0, |p|\leq 1} e^{px} K(p) \left(\hat{u}(p,t) - \sum_{j=1}^{[\alpha]} \frac{\partial_x^{j-1}u(0,t)}{p^j}\right) dp.
\end{aligned}$$

Thus we see that the integrals converge uniformly with respect to $x > 0$ since

$$\frac{K(p)}{p^{[\alpha]+2}} \in \mathbf{L}^2((-i\infty, -i] \cup [i, i\infty))$$

and

$$p^{[\alpha]-1} \left(\hat{u}(p,t) - \sum_{j=1}^{[\alpha]} \frac{\partial_x^{j-1}u(0,t)}{p^j}\right) \in \mathbf{C}\left(\mathbf{R}^+; \mathbf{L}^2(\mathbf{R}^+)\right) :$$

for the solution $u(x,t) \in \mathbf{C}\left(\mathbf{R}^+; \mathbf{H}_2^{[\alpha]-1}(\mathbf{R}^+)\right)$.

THEOREM 14. *Let $\alpha > 1$ and not be equal to an odd integer. Suppose that the initial data $u_0(x) \in \mathbf{X}$ and the boundary data $h_j(t) \in \mathbf{Y}$, $j = 1, ..., [\frac{\alpha+1}{2}]$ (in the case $\alpha \geq 2$). Then for some $T > 0$ there exists a unique solution $u(x, t) \in \mathbf{Z}_T$ of the initial-boundary value problem (1.1).*

Now we give some sufficient conditions for the global existence of the solutions. First we consider the case, when the asymptotics of solutions for large time is determined by decaying properties of the boundary data.

THEOREM 15. *Let $\alpha \geq 2$ and not be equal to an odd integer. Let the coefficient in the nonlinearity $a(t) \in \mathbf{C}^1(\mathbf{R}^+)$ satisfy the estimate $|a(t)| \leq C(1 + t)^{-\eta}$ for all $t > 0$, where $\eta \in \mathbf{R}$. Let the initial data $u_0 \in \mathbf{X}$ and the norm $\|u_0\|_{\mathbf{X}} \leq \epsilon$, and the boundary data $h_j(t)$ satisfy the following conditions $\sum_{j=1}^{[\frac{\alpha}{2}]} \|h_j\|_{\mathbf{Y}} \leq \epsilon$ and*

$$(1.2) \qquad h_j(t) = A_j t^{-\chi - \frac{j-1}{\alpha}} + \phi_j(t)$$

for $j = 1, ..., N$, where

$$|\phi_j(t)| \leq C\epsilon t^{-\chi - \frac{j-1}{\alpha} - \gamma}$$

and

$$|\phi_1'(t)| = O\left(\epsilon t^{-\chi - \gamma}\right),$$

$\max\left(0, \frac{1-\eta+1/\alpha}{\rho+1}\right) < \chi < \frac{1}{\alpha}$. We suppose that the constants $\epsilon > 0$ and $\gamma > 0$ are small enough. Then there exists a unique solution

$$u(x, t) \quad \in \quad \mathbf{C}\left([0, +\infty); \mathbf{L}^2\left(\mathbf{R}^+\right) \cap \mathbf{L}^\infty\left(\mathbf{R}^+\right)\right)$$
$$\cap \mathbf{C}\left(\mathbf{R}^+; \mathbf{H}_2^{[\alpha]-1}\left(\mathbf{R}^+\right) \cap \mathbf{C}^{[\alpha]-1}\left(\mathbf{R}^+\right)\right)$$

of the initial-boundary value problem (1.1). Moreover this solution has the following asymptotics for large time uniformly with respect to $x > 0$:

$$(1.3) \qquad u(x, t) = \frac{1}{2\pi i t^\chi} \sum_{j=1}^{N} A_j G_j\left(xt^{-1/\alpha}\right) + O\left(t^{-\chi - \gamma}\right),$$

where

$$G_j(q) = \int_{-i\infty}^{i\infty} dy\, y^{\alpha-j} \int_0^1 dz\, e^{yq - Ey^\alpha(1-z)} z^{-\chi - \frac{j-1}{\alpha}}, E = e^{i\pi[\frac{\alpha+1}{2}]}.$$

REMARK 3. *In the case $A_j = 0$ the asymptotic formula (1.3) gives only the estimate of the solution: $\|u\|_{\mathbf{L}^\infty} \leq Ct^{-\chi - \gamma}$.*

In the following theorem we consider the case, when the boundary data decay with time sufficiently rapidly and we show that the character of the large time asymptotics of the solutions is defined by the initial data.

THEOREM 16. *Let $\alpha > 1$ and not be equal to an odd integer. Let the coefficient in the nonlinearity $a(t) \in \mathbf{C}^1(\mathbf{R}^+)$ satisfy the estimate*

$$|a(t)| \leq C(1 + t)^{-\eta}$$

for all $t > 0$, where $\eta \in \mathbf{R}$. Suppose that $\eta > 1 - \rho/\alpha$. Let the initial data $u_0 \in \mathbf{X}$ be such that $x^\delta u_0 \in \mathbf{L}^1(\mathbf{R}^+)$, with $0 < \delta < \frac{1}{2}$ and the norm $\|u_0\|_{\mathbf{X}} \le \epsilon$, where $\epsilon > 0$ is sufficiently small. Let the boundary data $h_j \in \mathbf{Y}$ for $j = 1, ..., [\frac{\alpha}{2}]$, (in the case $\alpha \ge 2$) satisfy condition (1.2) with $\chi = \frac{1}{\alpha}$ and the following estimates:

$$\sum_{j=1}^{N} \|h_j\|_{\mathbf{Y}} \le \epsilon$$

and

$$|\widehat{h}'(\xi)| = O\left(\epsilon |\xi|^{-2}\right)$$

for all $|\xi| > 1$, $\mathrm{Re}\,\xi = 0$. When α is an integer we also suppose that $|\widehat{h}''_j(\xi)| = O\left(\epsilon |\xi|^{-2}\right)$ for all $|\xi| > 1$, $\mathrm{Re}\,\xi = 0$. Then there exists a unique solution

$$u(x,t) \in \mathbf{C}\left([0,+\infty); \mathbf{L}^2\left(\mathbf{R}^+\right)\right) \cap \mathbf{C}\left(\mathbf{R}^+; \mathbf{H}_2^{[\alpha]-1}\left(\mathbf{R}^+\right) \cap \mathbf{C}^{[\alpha]-1}\left(\mathbf{R}^+\right)\right)$$

of the initial-boundary value problem (1.1). (In the case $\alpha \ge 2$ we have

$$u(x,t) \quad \in \quad \mathbf{C}\left([0,+\infty); \mathbf{L}^2\left(\mathbf{R}^+\right) \cap \mathbf{L}^\infty\left(\mathbf{R}^+\right)\right)$$
$$\cap \mathbf{C}\left(\mathbf{R}^+; \mathbf{H}_2^{[\alpha]-1}\left(\mathbf{R}^+\right) \cap \mathbf{C}^{[\alpha]-1}\left(\mathbf{R}^+\right)\right).)$$

This solution has the following asymptotics for large time uniformly with respect to $x > 0$

(1.4) $$u(x,t) = t^{-\frac{1}{\alpha}} \sum_{j=0}^{N} B_j G_j\left(x/t^{1/\alpha}\right) + O\left(t^{-\frac{1}{\alpha}-\gamma}\right),$$

where $N = [\alpha]$ if α is not integer and $N = \alpha - 1$ if α is integer,

$$G_0(q) = \int_{-i\infty}^{i\infty} e^{qy - Ey^\alpha} dy$$

and

$$G_j(q) = \int_{-i\infty}^{i\infty} dy\, y^{\alpha-j} \int_0^1 dz\, e^{yq - Ey^\alpha(1-z)} z^{-\frac{j}{\alpha}}.$$

$E = e^{i\pi\left[\frac{\alpha+1}{2}\right]}$; the constants B_j we define below in Section 5.

As an example of an application of our theory we consider the initial-boundary value problem for the well-known Landau-Ginzburg equation $u_t + i|u|^2 u + u - u_{xx} = 0$. By changing $u = e^{-t}v$ we get for the new function $v(x,t)$ the following equation

$$v_t + ie^{-2t}|v|^2 v - v_{xx} = 0.$$

In this case we have $K(p) = p^2$, $b = 1$, $a = 1$ and $\rho = 2$.

Since $\alpha = 2$ we need only one boundary condition, so we obtain the following initial-boundary value problem

(1.5) $$\begin{cases} v_t + ie^{-2t}|v|^2 v - v_{xx} = 0, (t,x) \in \mathbf{R}^+ \times \mathbf{R}^+, \\ v(x,0) = u_0(x), x > 0, \ v(0,t) = e^t h(t), t > 0. \end{cases}$$

We suppose that the initial data $u_0(x) \in \mathbf{H}^2_\infty (\mathbf{R}^+) \cap \mathbf{H}^3_1 (\mathbf{R}^+)$ and the boundary data $h \in \mathbf{H}^1_\infty (\mathbf{R}^+) \cap \mathbf{H}^2_1 (\mathbf{R}^+)$ satisfy the compatibility condition $u_0(0) = h(0)$. Then for some time $T > 0$ the initial-boundary value problem (1.5) has a unique solution

$$u(x,t) \in \mathbf{C}\left((0,T]; \mathbf{H}^1_{\frac{1}{2}} (\mathbf{R}^+) \cap \mathbf{C}^1 (\mathbf{R}^+)\right) \cap \mathbf{C}\left([0,T]; \mathbf{L}^2 (\mathbf{R}^+) \cap \mathbf{L}^\infty (\mathbf{R}^+)\right).$$

If we assume in addition that the initial and boundary data are small enough and the boundary data have the following asymptotics for large time

$$h(t) = Ae^{-t}t^{-\chi} + O\left(e^{-t}t^{-\chi-\gamma}\right),$$

where $0 \le \chi < \frac{1}{2}$, then there exists a unique global in time solution

$$u(x,t) \in \mathbf{C}\left([0,+\infty); \mathbf{L}^2 (\mathbf{R}^+) \cap \mathbf{L}^\infty (\mathbf{R}^+)\right) \cap \mathbf{C}\left(\mathbf{R}^+; \mathbf{H}^1_{\frac{1}{2}} (\mathbf{R}^+) \cap \mathbf{C}^1 (\mathbf{R}^+)\right)$$

of the initial-boundary value problem (1.5). This solution has the following large time asymptotics uniformly with respect to $x > 0$

$$u(x,t) = \frac{Ae^{-t}}{2\pi i t^\chi} G\left(x/\sqrt{t}\right) + O\left(e^{-t}t^{-\chi-\gamma}\right),$$

where

$$G(q) = \int_{-i\infty}^{i\infty} y\, dy \int_0^1 e^{yq+y^2(1-z)} z^{-\chi} dz.$$

Finally, if we suppose that the boundary and initial data are small enough and the boundary data decay with time more rapidly

$$h(t) = Ae^{-t}t^{-\frac{1}{2}} + O\left(e^{-t}t^{-\frac{1}{2}-\gamma}\right),$$

then the character of the asymptotic behavior of solutions is defined by the initial data

$$u(x,t) = e^{-t}t^{-\frac{1}{2}} \sum_{j=0}^1 B_j G_j(x/\sqrt{t}) + O\left(e^{-t}t^{-\frac{1}{2}-\gamma}\right),$$

where

$$G_0(q) = \int_{-i\infty}^{i\infty} e^{qy+y^2}\, dy$$

and

$$G_1(q) = \int_{-i\infty}^{i\infty} dy\, y \int_0^1 e^{yq+y^2(1-z)} \frac{dz}{\sqrt{z}}.$$

The constants B_j are defined below in Section 5.

We organize this chapter as follows. In Section 2 we consider the linear initial-boundary value problem corresponding to nonlinear problem (1.1). We discuss an important question on the amount of the necessary boundary data to be posed for the correct resolution of the initial-value problem. In Theorem 17 we prove the local existence of solutions to the linear problem. Section 3 is devoted to the proof of Theorem 14. In Sections 4 and 5 we prove Theorem 15 and Theorem 16 respectively. The results of this chapter were published in paper [71].

2. Linear problem

In this section we consider the following linear initial-boundary value problem

(2.1)
$$
\begin{cases}
u_t + \mathbb{K}u = f(x,t), (t,x) > 0, \\
u(x,0) = u_0(x), x > 0, \\
\partial_x^{j-1} u(0,t) = h_j(t), t > 0, j = 1, 2, ..., N,
\end{cases}
$$

with the compatibility conditions $\partial_x^{j-1} u_0(0) = h_j(0), j = 1, ..., N$, where symbol of the pseudodifferential operator $\mathbb{K}$ defined by formula

$$
K(p) = E p^\alpha, \alpha > 1
$$

and via Definition 8 of the pseudodifferential operator we have

$$
\mathbb{K}u = \frac{1}{2\pi i} \int_{-i\infty}^{i\infty} e^{px} K(p) \left(\hat{u}(p,t) - \sum_{j=1}^{[\alpha]} \frac{\partial_x^{j-1} u(0,t)}{p^j} \right) dp.
$$

Due to Theorem 11 (see Chapter 3) there exist $m = \left[\frac{\alpha+1}{2} \right]$ different inverse functions $\phi_j(\xi) = K^{-1}(-\xi)$, such that for $\text{Re }\xi > 0$

$$
\text{Re } \phi_j(\xi) > 0, j = 1, 2, ..., m
$$

and $N = [\alpha] - m = \left[\frac{\alpha}{2} \right]$ different inverse functions such that

$$
\text{Re } \phi_j(\xi) < 0, j = m+1, ..., [\alpha].
$$

Also A is a characteristic matrix for operator $\mathbb{K}u$ (see Chapter 3)

(2.2)
$$
A = -\xi \begin{pmatrix}
\phi_1^{-N-1}(\xi) & \cdots & \phi_1^{-[\alpha]}(\xi) \\
\cdots & \cdots & \cdots \\
\phi_m^{-N-1}(\xi) & \cdots & \phi_m^{-[\alpha]}(\xi)
\end{pmatrix}.
$$

The input function $J(p,t)$ for problem (2.1) was defined in Chapter 3 (see Definition 11)

(2.3)
$$
J(p,t) = \hat{u}_0(p) + \int_0^t e^{K(p)\tau} \left(\hat{f}(p,\tau) + K(p) \sum_{j=1}^N h_j(\tau) p^{-j} \right) d\tau
$$

and input vector $\overrightarrow{J}$ is

(2.4)
$$
\overrightarrow{J} = -(J(\phi_1, +\infty), J(\phi_2, +\infty), ..., J(\phi_m, +\infty)).
$$

Also we define the operator $\mathbb{A}$ as

(2.5)
$$
\mathbb{A}\overrightarrow{J} = \frac{1}{2\pi i} \int_{-i\infty+b}^{i\infty+b} e^{\xi t} A^{-1} \overrightarrow{J} d\xi
$$

with norm

$$
\left\| \mathbb{A}\overrightarrow{J} \right\| = \sum_{j=1}^m \left| \mathbb{A}\overrightarrow{J} \right|_j
$$

The determinant of matrix A is equal to

$$(2.6) \qquad W(\xi) = \det(-\xi) \begin{pmatrix} \phi_1^{-N-1}(\xi) & \cdots & \phi_1^{-[\alpha]}(\xi) \\ \cdots & \cdots & \cdots \\ \phi_m^{-N-1}(\xi) & \cdots & \phi_m^{-[\alpha]}(\xi) \end{pmatrix}$$

$$= \frac{\det \tilde{W}}{\xi^{\frac{m(2N+m+1)}{2\alpha}}} \prod_{j=1}^{m} r_j^{-[\alpha]},$$

where the matrix

$$\tilde{W} = \begin{pmatrix} r_1^{m-1} & \cdots & 1 \\ \cdots & \cdots & \cdots \\ r_m^{m-1} & \cdots & 1 \end{pmatrix}.$$

It is not equal to zero since all the constants r_l are different.

Now we prove the following result.

THEOREM 17. *Let $\alpha > 1$ not be equal to an odd number. Then for some $T > 0$ there exists a unique solution $u(x,t) \in \mathbf{Z}_T$ of the initial-boundary value problem (2.1) such that*

$$\|u\|_{\mathbf{Z}_T} \le C\lambda,$$

where

$$\lambda = \|u_0\|_{\mathbf{X}} + T^\mu \sup_{t \in [0,T]} t^{\bar{\gamma}} \|f(t)\|_{\mathbf{L}^1}$$

for the case $\alpha \in (1,2)$ and

$$\lambda = \|u_0\|_{\mathbf{X}} + \sum_{j=1}^{N} \|h_j\|_{\mathbf{Y}} + T^\mu \sup_{t \in [0,T]} \left(t^{\bar{\gamma}} \|f(t)\|_{\mathbf{H}_1^1} + \|f(t)\|_{\mathbf{L}^\infty} \right)$$

for the case $\alpha \ge 2$; here $N = [\frac{\alpha}{2}]$, $\mu = 1 - \frac{1}{\alpha} - \bar{\gamma} > 0$, $\bar{\gamma} > 0$ is small enough. We assume that the initial data u_0, boundary data h_j, $j = 1, 2, ..., N$ and the force $f(x,t)$ in (2.1) are such that the value $\lambda < +\infty$.

Before proving Theorem 17 we consider the following function

$$F(x,t) = \int_{-i\infty}^{+i\infty} e^{\xi t - x\Theta\xi^\mu} \xi^{-\beta} d\xi$$

for $x \ge 0$, $t \in \mathbf{R} \backslash 0$, where Θ is a complex constant such that $\mathrm{Re}(\pm i)^\mu \Theta > 0$ and $0 < \beta < 1$, $\mu > 0$. First of all we prove an estimate and the Hölder conditions for the function $F(x,t)$ with respect to $x \ge 0$ and $t \in \mathbf{R} \backslash 0$.

LEMMA 2. *We have the following estimates*

$$|F(x,t)| \le C|t|^{\beta-1},$$

$$|F(x,t) - F(y,t)| \le C|x-y|^\omega |t|^{\beta-1-\omega\mu}$$

and

$$|F(x,t) - F(x,\tau)| \le C|t-\tau|^\nu (|\tau|^{\beta-1-\nu} + |t|^{\beta-1-\nu})$$

for all $x, y \ge 0, t, \tau \in \mathbf{R} \backslash 0$, where $0 \le \nu < \min(1, \frac{\beta}{\mu})$, $0 \le \omega < \min(1, \frac{\beta}{\mu})$.

PROOF. In the case of $x > 1$ we can differentiate with respect to x and t to obtain

$$F_t(x,t) = \int_{-i\infty}^{+i\infty} e^{\xi t - x\Theta\xi^\mu} \xi^{1-\beta} d\xi$$

and

$$F_x(x,t) = -\Theta \int_{-i\infty}^{+i\infty} e^{\xi t - x\Theta\xi^\mu} \xi^{\mu-\beta} d\xi,$$

whence using the estimate

$$\left| \xi^\sigma e^{-x\Theta\xi^\mu} \right| \le C|\xi|^{-2}$$

for $|\xi| > 1$ and any $\sigma \in \mathbf{R}$ we easily get

$$|F(x,t)| + |F_x(x,t)| + |F_t(x,t)| \le C$$

uniformly with respect to $x > 1$.   □

Now consider the case $0 \le x \le 1$. If $t, \tau > 0$ we change the variable of integration $\xi = q/t$ and denote $z = x/t^\mu$. Then we divide the domain of integration in three parts as follows

$$F(x,t) = t^{\beta-1}(F_1(z) + F_2(z) + F_3(z)),$$

where

$$F_1(z) = \int_{-i}^{+i} e^{q - z\Theta q^\mu} q^{-\beta} dq, \quad F_2(z) = \int_{i}^{+i\infty} e^{q - z\Theta q^\mu} q^{-\beta} dq$$

and $F_3(z) = \int_{-i\infty}^{-i} e^{q - z\Theta q^\mu} q^{-\beta} dq$. Clearly $|F_1(z)| < C$ and

$$|F_1'(z)| = C \left| \int_{-i}^{+i} e^{q - z\Theta q^\mu} q^{-\beta+\mu} dq \right| \le C \text{ for all } z \ge 0.$$

Therefore $F_1(z)$ satisfies the Hölder condition. Let $z \ge z' > 0$. Integration by parts with respect to q yields

$$|F_2(z)| \le \left| e^{q - z\Theta q^\mu} q^{-\beta} \Big|_i^{i\infty} \right| + C \left| \int_i^{i\infty} e^{q - \Theta q^\mu z} z \frac{dq}{q^{1+\beta-\mu}} \right| + C \left| \int_i^{i\infty} e^{q - \Theta q^\mu z} z \frac{dq}{q^{1+\beta}} \right|$$

and

$$
\begin{aligned}
|F_2(z) - F_2(z')| &\le \left| e^{q - z'\Theta q^\mu}(e^{-\Theta q^\mu(z-z')} - 1)q^{-\beta} \Big|_i^{i\infty} \right| \\
&\quad + C \left| \int_i^{i\infty} e^{q - \Theta q^\mu z'} z'(1 - e^{-\Theta q^\mu(z-z')})q^{\mu-1-\beta} dq \right| \\
&\quad + (z - z')C \left| \int_i^{i\infty} e^{q - \Theta q^\mu z} z \frac{dq}{q^{1+\beta-\mu}} \right| \\
&\quad + C \left| \int_i^{i\infty} e^{q - \Theta q^\mu z'}(1 - e^{-\Theta q^\mu(z-z')}) \frac{dq}{q^{1+\beta}} \right|.
\end{aligned}
$$

Now using estimates

$$\left|z'q^{\mu}e^{-\Theta q^{\mu}z'}\right| \leq C,$$

$$\left|1 - e^{-\Theta q^{\mu}(z-z')}\right| \leq C\min(1, |q|^{\mu}|z-z'|)$$

and

$$\left|(z-z')q^{\mu}e^{-\Theta q^{\mu}z}\right| \leq C$$

we get for F_2 the following estimates $|F_2(z)| \leq C$ and

$$|F_2(z) - F_2(z')| \leq C(z-z')^{\nu}\left(1 + \int_1^{+\infty} q^{\mu\nu-1-\beta}dq\right) \leq C|z-z'|^{\nu},$$

since $0 \leq \nu < \min(1, \beta/\mu)$. The integral $F_3(z)$ is considered analogously. Thus we get

$$|F(x,t)| \leq Ct^{\beta-1}, |F(x,t) - F(y,t)| \leq C|x-y|^{\omega}|t|^{\beta-1-\omega\mu}$$

and

$$|F(x,t) - F(x,\tau)| \leq C|t-\tau|^{\nu}(\tau^{\beta-1-\nu} + t^{\beta-1-\nu})$$

for all $0 \leq x \leq 1, t > 0, \tau > 0$, since

$$|t^{\sigma} - \tau^{\sigma}| \leq C|t-\tau|^{\nu}(t^{\sigma-\nu} + \tau^{\sigma-\nu})$$

for any $\sigma \in \mathbf{R}$.

The case $t, \tau < 0$ is considered similarly. In addition, in the cases $t > 0, \tau < 0$ or $t < 0, \tau > 0$ we have $|t - \tau| = |t| + |\tau|$ so it is sufficient to use the estimate

$$|F(x,t)| \leq C|t|^{\beta-1}$$

to obtain

$$|F(x,t) - F(x,\tau)| \leq C\left(|t|^{\beta-1-\nu} + |\tau|^{\beta-1-\nu}\right)|t-\tau|^{\nu}.$$

Lemma 2 is proved.

Proof of Theorem 17. From (2.2) through (2.6) we have for the Laplace transforms $\hat{v}_j(\xi) = \overrightarrow{(A\mathcal{J})_j}$ relationship

$$\hat{v}_j(\xi) = \frac{C_1}{\xi^{1-\frac{N+j}{\alpha}}} \sum_{l=1}^{m} r_l^{[\alpha]} M_{j,l}\left(\widehat{u}_0(\phi_l(\xi))\right)$$

(2.7)
$$-\xi \sum_{k=1}^{N} \frac{\widehat{h_k}(\xi)}{\phi_l^k(\xi)} - \hat{\hat{f}}(\phi_l(\xi), \xi)\Bigg),$$

where $C_1 = (\det\tilde{W})^{-1}$ and $M_{j,l}$ are the algebraic minors of the matrix $\tilde{W}$ (see (2.6)) and $\hat{\hat{f}}(x,t)$ is the Laplace transform with respect to the space and time of the force f

$$\hat{\hat{f}}(\phi_l(\xi), \xi) = \int_0^{+\infty} e^{-\phi_l(\xi)x}dx \int_0^{T} e^{-\xi\tau}f(x,\tau)d\tau.$$

We remind the reader that we consider the symbol $K(p) = Ep^{\alpha}$ with $E = e^{i\pi[\frac{\alpha+1}{2}]}$, so $m = [\frac{\alpha}{2}], \phi_l(\xi) = r_l\xi^{1/\alpha}$, and $r_l = e^{\frac{i}{\alpha}(2\pi l - \pi[\frac{\alpha-1}{2}])}$ are some constants, such that

$\operatorname{Re} \phi_l(\xi) > 0$ on the imaginary axis $\xi \in (-i\infty, +i\infty)$. (Note that the sum $\sum_1^0$, which appears in (3.3) and below in the case $\alpha \in (1, 2)$ is assumed to be identically zero.) We denote

$$\lambda = \|u_0\|_{\mathbf{X}} + T^\mu \sup_{t\in[0,T]} t^{\bar{\gamma}} \|f(t)\|_{\mathbf{L}^1}$$

for the case $\alpha \in (1, 2)$ and

$$\lambda = \|u_0\|_{\mathbf{X}} + \sum_{j=1}^N \|h_j\|_{\mathbf{Y}} + T^\mu \sup_{t\in[0,T]} \left(t^{\bar{\gamma}} \|f(t)\|_{\mathbf{H}_1^1} + \|f(t)\|_{\mathbf{L}^\infty} \right)$$

for the case $\alpha \geq 2$, where $\mu = 1 - \frac{1}{\alpha} - \bar{\gamma} > 0$, $\bar{\gamma} > 0$ is small enough.

First of all we estimate $v_j = (\mathbb{A}\vec{\mathcal{J}})_j(\tau)$ of the solution.

LEMMA 3. *The following estimate*

$$\sup_{t\in[0,T]} \sum_{j=1}^m t^{\frac{j-1}{\alpha}} |v_j(t)| \leq C\lambda$$

is valid.

PROOF. Integrating $N + m$ times by parts in the Laplace transform of the initial data and one time in the Laplace transform of the boundary data, taking into account the compatibility conditions $h_j(0) = \partial_x^{j-1} u_0(0)$ for $j = 1, ..., N$, we get

$$\widehat{u_0}(\phi_l) - \xi \sum_{k=1}^N \frac{\widehat{h_k}(\xi)}{\phi_l^k}$$

$$= \sum_{k=1}^m \frac{\partial_x^{N-1+k} u_0(0)}{\phi_l^{N+k}} + \frac{1}{\phi_l^{N+m}} \int_0^{+\infty} e^{-\phi_l(\xi)x} \partial_x^{N+m} u_0(x) dx$$

$$- \sum_{k=1}^N \phi_l^{-k} \int_0^{+\infty} e^{-\xi\tau} h_k'(\tau) d\tau$$

$$= \sum_{k=1}^m \frac{\partial_x^{N-1+k} u_0(0)}{\phi_l^{N+k}} + O\left(\lambda|\xi|^{-\frac{N+1+m}{\alpha}}\right)$$

for $|\xi| > 1$. Therefore via (2.7) we get

$$(2.8) \quad \hat{v}_j(\xi) = \sum_{k=1}^m \frac{C_{k,j} \partial_x^{N-1+k} u_0(0)}{\xi^{1+\frac{k-j}{\alpha}}} - C_1 \sum_{l=1}^m M_{j,l} r_l^{[\alpha]} I_{j,l} + O\left(\lambda|\xi|^{-1-\frac{m+1-j}{\alpha}}\right)$$

for $j = 1, ..., m$, where we denote

$$(2.9) \qquad I_{j,l} = \xi^{\frac{N+j}{\alpha}-1} \hat{\hat{f}}(\phi_l(\xi), \xi)$$

and

$$C_{k,j} = C_1 \sum_{l=1}^m r_l^{m-k} M_{j,l}$$

are some constants. From (2.8) we get

$$\hat{v}_j(\xi)$$

$$= \sum_{k=1}^{m} \frac{C_{k,j} \partial_x^{N-1+k} u_0(0)}{\xi^{1+\frac{k-j}{\alpha}}}$$

(2.10)
$$+ \frac{C_1}{\xi^{1-\frac{N-1+j}{\alpha}}} \sum_{l=1}^{m} M_{j,l} r_l^{[\alpha]} \int_0^{+\infty} dx e^{-\phi_l(\xi)x} \int_0^T e^{-\xi\tau} f(x,\tau) d\tau.$$

Changing the contour of integration with respect to ξ to $\Gamma_1 = \{\xi \in \mathbf{C} : \arg \xi = \pm \left(\frac{\pi}{2} + \varepsilon\right)\}$ and $\Gamma_2 = \{\xi \in \mathbf{C} : \arg \xi = \pm \left(\frac{\pi}{2} - \varepsilon\right)\}$ we have

$$\left| \int_{-i\infty}^{i\infty} \frac{1}{\xi^{1-\frac{N+j}{\alpha}}} d\xi \int_0^{+\infty} dx e^{-x\phi_l(\xi)} \int_0^T e^{\xi(t-\tau)} f(x,\tau) d\tau \right|$$

$$= \left| \int_{\Gamma_1} \frac{1}{\xi^{1-\frac{N+j}{\alpha}}} d\xi \int_0^{+\infty} dx e^{-x\phi_l(\xi)} \int_0^t e^{\xi(t-\tau)} f(x,\tau) d\tau \right|$$

$$+ \left| \int_{\Gamma_2} \frac{1}{\xi^{1-\frac{N+j}{\alpha}}} d\xi \int_0^{+\infty} dx e^{-x\phi_l(\xi)} \int_t^T e^{\xi(t-\tau)} f(x,\tau) d\tau \right|$$

$$\leq C \sup_{t \in [0,T]} \|f(t)\|_{\mathbf{L}^\infty} \left(\int_0^t d\tau \left| \int_{\Gamma_1} e^{\operatorname{Re}\xi(t-\tau)} |\xi|^{\frac{N+j-1}{\alpha}-1} d\xi \right| \right.$$

$$\left. + \int_t^T d\tau \left| \int_{\Gamma_2} e^{\operatorname{Re}\xi(t-\tau)} |\xi|^{\frac{N+j-1}{\alpha}-1} d\xi \right| \right)$$

$$\leq C \sup_{t \in [0,T]} \|f(t)\|_{\mathbf{L}^\infty} \int_0^t |t-\tau|^{-\frac{N+j-1}{\alpha}} d\tau$$

(2.11)
$$\leq CT^\mu \sup_{t \in [0,T]} \|f(t)\|_{\mathbf{L}^\infty},$$

where $\mu = 1 - \frac{1}{\alpha} - \bar{\gamma}$, $z = x|t-\tau|^{-\frac{1}{\alpha}}$, $j, l = 1, ..., m$. From (2.7) we clearly see that

$$|\hat{v}_j(t)| \leq C\lambda|\xi|^{\frac{N+j}{\alpha}-1}$$

for $|\xi| \leq 1$. Therefore the substitution of (2.11) into (2.10) yields

$$|v_j(t)| = C \left| \int_{-i\infty}^{+i\infty} e^{\xi t} \hat{v}_j(\xi) d\xi \right| \leq C\|u_0\|_\mathbf{x} \sum_{k=1}^{m} \left| \int_{|\xi|>1, \operatorname{Re}\xi=0} e^{\xi t} \xi^{-1-\frac{k-j}{\alpha}} d\xi \right|$$

$$+ CT^\mu \sup_{t \in [0,T]} t^{\bar{\gamma}} \|f_x(t)\|_{\mathbf{L}^1} + \lambda \int_{|\xi|>1, \operatorname{Re}\xi=0} O\left(|\xi|^{-1-\frac{m-j+1}{\alpha}} \right) d\xi$$

$$+ \left| \int_{-i}^{i} e^{\xi t} \hat{v}_j(\xi) d\xi \right|$$

$$\leq C\lambda t^{-\frac{j-1}{\alpha}}$$

for all $t \in (0, T]$, $j = 1, ..., m$. $\square$

We now consider the case $\alpha \in (1, 2)$. Writing $I_{1,1} = I$ in (2.9), we have

$$\left| \int_{-i\infty}^{i\infty} e^{\xi t} I(\xi) d\xi \right|$$

$$\leq C \int_0^T d\tau \int_0^{+\infty} dx |f(x, \tau)| \left| \int_{-i\infty}^{+i\infty} e^{\xi(t-\tau) - x\xi^{\frac{1}{\alpha}}} \xi^{1 - \frac{1}{\alpha}} d\xi \right|$$

$$\leq C \int_0^T d\tau \|f(\tau)\|_{\mathbf{L}^1} \sup_{x > 0} |F(x, t - \tau)|.$$

By virtue of Lemma 2 we have

$$|F(x, t)| \leq C|t|^{-\frac{1}{\alpha}}$$

for $t \in \mathbf{R} \backslash 0$. Therefore we get

$$\left| \int_{-i\infty}^{i\infty} e^{\xi t} I(\xi) d\xi \right| \leq C \sup_{t \in [0, T]} t^{\bar{\gamma}} \|f(t)\|_{\mathbf{L}^1} \int_0^T \tau^{-\bar{\gamma}} |t - \tau|^{-\frac{1}{\alpha}} d\tau \leq C\lambda.$$

Thus we obtain

$$|v_1(t)| = C \left| \int_{-i\infty}^{+i\infty} e^{\xi t} \hat{v}_1(\xi) d\xi \right| \leq C \|u_0\|_{\mathbf{x}} \left| \int_{|\xi| > 1, \mathrm{Re}\, \xi = 0} e^{\xi t} \xi^{-1} d\xi \right|$$

$$+ C\lambda + C\lambda \int_{|\xi| > 1, \mathrm{Re}\, \xi = 0} O\left(|\xi|^{-1 - \frac{1}{\alpha}} \right) d\xi + \left| \int_{-i}^{i} e^{\xi t} \hat{v}_1(\xi) d\xi \right|$$

$$\leq C\lambda.$$

Lemma 3 is proved.

In the following lemma we obtain the Hölder conditions for $v_j(t) = (\mathbb{A}\vec{\mathcal{J}})_j(t)$.

LEMMA 4. *We have*

$$|v_j(\tau) - v_j(t)| \leq C\lambda(t - \tau)^\nu t^{-\nu - \frac{j-1}{\alpha}}$$

for all $0 < t/2 \leq \tau \leq t \leq T$ and $j = 1, ..., m$, where $\nu \in (0, \frac{m-j+1}{\alpha})$.

PROOF. Taking the inverse Laplace transformation of (2.10) we have in the case $\alpha \geq 2$

$$
\begin{aligned}
|v_j(\tau) - v_j(t)| \quad &\leq \quad C\sum_{k=1}^{m} |\partial_x^{N-1+k} u_0(0)| \left| \tau^{\frac{k-j}{\alpha}} - t^{\frac{k-j}{\alpha}} \right| \\
&+ C(t-\tau)\|u_0\|_{\mathbf{X}} \sum_{k=1}^{m} \left| \int_{-i}^{i} \frac{d\xi}{\xi^{\frac{k-j}{\alpha}}} \right| \\
&+ C\sum_{l=1}^{m} \int_0^T d\tau' \int_0^{+\infty} |f_x(x,\tau')| dx \\
&\quad \times \left| \int_{|\xi| \geq 1, \mathrm{Re}\,\xi = 0} \frac{(e^{\xi(\tau - \tau')} - e^{\xi(t-\tau')})e^{-\phi_l(\xi)x} d\xi}{\xi^{1 - \frac{N-1+j}{\alpha}}} \right| \\
&+ C\lambda \left| \int_{|\xi| \geq 1} (e^{\xi\tau} - e^{\xi t}) |\xi|^{-1 - \frac{m-j+1}{\alpha}} d\xi \right| \\
&+ C \left| \int_{-i}^{i} |e^{\xi\tau} - e^{\xi t}| |\hat{v}_j(\xi)| d\xi \right| \\
&= \quad C\sum_{j=1}^{5} J_j.
\end{aligned}
$$

(2.12)

We estimate each summand in (2.12). We have

$$
|\tau^\eta - t^\eta| \leq Ct^{\eta - \nu}(t - \tau)^\nu
$$

for $0 < t/2 \leq \tau \leq t \leq T$, where $\eta \in \mathbf{R}$ and $0 \leq \nu \leq 1$. Therefore taking $\eta = \frac{k-j}{\alpha}$ we see that the first two summands in (2.12) are less than

$$
C\|u_0\|_{\mathbf{X}}(t-\tau)^\nu \sum_{k=1}^{m} t^{\frac{k-j}{\alpha} - \nu} \leq C\lambda(t-\tau)^\nu t^{-\nu - \frac{j-1}{\alpha}}.
$$

Using Lemma 2 with $\beta = 1 - \frac{N-1+j}{\alpha}$, $\mu = \frac{1}{\alpha}$, $\Theta = r_l$ and $0 \leq \nu < \frac{m+1-j}{\alpha}$ we get for the third summand in (2.12)

$$
\begin{aligned}
J_3 &= \sum_{l=1}^{m} \int_0^T d\tau' \int_0^{+\infty} |f_x(x,\tau')| dx \left| \int_{|\xi|\geq 1, \mathrm{Re}\,\xi=0} \frac{(e^{\xi(\tau-\tau')} - e^{\xi(t-\tau')})e^{-x\phi_l(\xi)} d\xi}{\xi^{1-\frac{N-1+j}{\alpha}}} \right| \\
&\leq C \sum_{l=1}^{m} \int_0^T d\tau' \int_0^{+\infty} dx\, |f_x(x,\tau')| |F(x,\tau-\tau') - F(x,t-\tau')| \\
&\quad + CT^{1-\bar\gamma}(t-\tau) \sup_{t\in[0,T]} t^{\bar\gamma} \|f_x(t)\|_{\mathbf{L}^1} \\
&\leq C(t-\tau)^\nu \sup_{t\in[0,T]} t^{\bar\gamma} \|f(t)\|_{\mathbf{H}_1^1} \\
&\quad \times \left(T^{1-\bar\gamma} + \int_0^T \left(|t-\tau'|^{-\frac{N-1+j}{\alpha}-\nu} + |\tau-\tau'|^{-\frac{N-1+j}{\alpha}-\nu} \right) d\tau' \right) \\
&\leq C\lambda(t-\tau)^\nu.
\end{aligned}
$$

For the last two summands in (2.12) we obtain

$$
\begin{aligned}
|J_4 + J_5| &\leq \lambda \left| \int_{|\xi|\geq 1, \mathrm{Re}\,\xi=0} \frac{|e^{\xi\tau} - e^{\xi t}| d\xi}{|\xi|^{1+\frac{m-j+1}{\alpha}}} \right| \\
&\quad + \left| \int_{-i}^{i} |e^{\xi\tau} - e^{\xi t}| |\hat v_j(\xi)| d\xi \right| \\
&\leq C\lambda(t-\tau)^\nu
\end{aligned}
$$

since $0 \leq \nu < \frac{m-j+1}{\alpha}$. In the case $\alpha \in (1,2)$ we have by virtue of (2.8)

$$
\begin{aligned}
|v_1(\tau) &- v_1(t)| \\
&\leq C \int_0^T d\tau' \int_0^{+\infty} |f(x,\tau')| |F(x,\tau-\tau') - F(x,t-\tau')| dx \\
&\quad + C\lambda \left| \int_{|\xi|\geq 1, \mathrm{Re}\,\xi=0} (e^{\xi\tau} - e^{\xi t}) |\xi|^{-1-\frac{1}{\alpha}} d\xi \right| + C \left| \int_{-i}^{i} |e^{\xi\tau} - e^{\xi t}| |\hat v_1(\xi)| d\xi \right|,
\end{aligned}
$$

where

$$
F(x,t) = \int_{-i\infty}^{i\infty} e^{\xi t - \xi^{\frac{1}{\alpha}} x} \xi^{\frac{1}{\alpha}-1} d\xi.
$$

Using Lemmas 2 and 3 we obtain for $\tau \in [t/2, t]$

$$
\begin{aligned}
|v_1(\tau) &- v_1(t)| \\
&\leq C\lambda(t-\tau)^\nu t^{-\nu} \\
&\quad + C \sup_{t\in[0,T]} t^{\bar\gamma} \|f(t)\|_{\mathbf{L}^1}(t-\tau)^\nu \int_0^T \tau^{\bar\gamma}(|\tau-\tau'|^{-\frac{1}{\alpha}-\nu} + |t-\tau'|^{-\frac{1}{\alpha}-\nu}) d\tau' \\
&\leq C\lambda(t-\tau)^\nu t^{-\nu},
\end{aligned}
$$

where $\nu \in (0, \frac{1}{\alpha})$. Thus we get the desired estimate of Lemma 4. $\qquad\square$

We now prove the following estimate for $\delta_s = \max(0, \frac{s-N}{\alpha})$

(2.13)
$$\sup_{t \in [0,T]} \lim_{p \to \infty, \text{Re } p=0} \sum_{s=0}^{[\alpha]-1} p^{s+1+\gamma} t^{\delta_s} \left| e^{-K(p)t} \mathcal{J}(p,t) - \sum_{j=1}^{N} h_j(t) p^{-j} \right| = \lambda < +\infty,$$

where

$$e^{-K(p)t} \mathcal{J}(p,t) = e^{-K(p)t} \widehat{u}_0(p) + K(p) \sum_{j=1}^{N} p^{-j} \int_0^t e^{-K(p)(t-\tau)} h_j(\tau) d\tau$$

(2.14)
$$+ \int_0^t e^{-K(p)(t-\tau)} \widehat{f}(p,\tau) d\tau.$$

Integrating by parts in the second summand of the right-hand side of (2.14) we get

(2.15)
$$e^{-K(p)t} \mathcal{J}(p,t) = \sum_{j=1}^{N} \frac{h_j(t)}{p^j} + R(p,t),$$

where

$$R(p,t) = e^{-K(p)t} \left(\widehat{u}_0(p) - \sum_{j=1}^{N} \frac{h_j(0)}{p^j} \right)$$

$$- \sum_{j=1}^{N} p^{-j} \int_0^t e^{-K(p)(t-\tau)} h_j'(\tau) d\tau + \int_0^t e^{-K(p)(t-\tau)} \widehat{f}(p,\tau) d\tau$$

(2.16)
$$= \sum_{j=1}^{3} I_j.$$

We have the following inequality

$$e^{-\text{Re } K(p)t} \leq \frac{C}{(|p|^\alpha t)^\nu}$$

for all $p \in (-i\infty, +i\infty)$ and any $\nu \geq 0$. Therefore choosing $\nu = \max(0, \frac{s-N}{\alpha})$ and using the compatibility conditions, we get for the first summand in (2.16)

$$|I_1| \leq \frac{e^{-\text{Re } K(p)t}}{|p|^{N+1}} \left(|\partial_x^N u_0(0)| + \left| \int_0^{+\infty} e^{-px} \partial_x^{N+1} u_0(x) dx \right| \right)$$

(2.17)
$$\leq \frac{\lambda}{|p|^{N+1+\nu\alpha} t^\nu} \leq O(\lambda |p|^{-s-1-\gamma} t^{-\delta_s}).$$

By virtue of Lemma 3 we get for the second and third summands in (2.16)

(2.18)
$$|I_2| = O\left(\lambda |p|^{-s-1-\gamma} t^{-\delta_s} \right).$$

In the case $\alpha \geq 2$ integration by parts in the Laplace transform of the right-hand side of (2.1) yields

$$|\hat{f}(p,t)| = \left|p^{-1}\left(f(0,t) + \int_0^{+\infty} e^{-px} f_x(x,t)dx\right)\right| \leq C\lambda|p|^{-1}t^{-\bar{\gamma}}$$

for all $|p| \geq 1, p \in (-i\infty, i\infty)$ and also

$$\left\|\hat{f}(t)\right\|_{\mathbf{L}^\infty} \leq C\lambda t^{-\bar{\gamma}}.$$

Thus we obtain

$$(2.19) \qquad |I_3| \leq C\lambda|p|^{-1} \int_0^t e^{-\operatorname{Re}(K(p)(t-\tau))} t^{-\bar{\gamma}}d\tau = O\left(\lambda|p|^{-s-1-\gamma}\right).$$

In the case $\alpha \in (1,2)$ we have

$$|I_3| \leq \sup_{t\in[0,T]} t^{\bar{\gamma}}\|f(t)\|_{\mathbf{L}^1} \int_0^t \tau^{-\bar{\gamma}} e^{-\operatorname{Re} K(p)(t-\tau)}d\tau \leq C\lambda|p|^{-\alpha}.$$

Putting estimates (2.17) through (2.19) into representation (2.16) gives us the estimate

$$(2.20) \qquad R(p,t) = O\left(\lambda|p|^{-s-1-\gamma}t^{-\delta_s}\right).$$

We have

$$\sup_{p\in[-i,i]} |e^{-K(p)t} \mathcal{J}(p,t)| \leq C\left(\|u_0\|_{\mathbf{L}^1} + \sum_{j=1}^N \int_0^t |h_j(\tau)|d\tau\right.$$

$$(2.21) \qquad \left. + \int_0^t \|f(\tau)\|_{\mathbf{L}^1}d\tau\right) \leq C\lambda.$$

From estimates (2.17) through (2.21) we have (2.13). Also we easily get

$$e^{-K(p)t} \mathcal{J}(p,t) \in \mathbf{C}^0([0,T], \mathbf{L}^\infty(\mathbf{R}^+)) \cap \mathbf{C}^0([0,T], \mathbf{H}_1^{0,N}(\mathbf{R}^+))$$
$$\cap \mathbf{C}^0((0,T], \mathbf{H}_1^{0,[\alpha]-1}(\mathbf{R}^+)).$$

From Lemma 3 through 4 we have for $v = \frac{m-1}{\alpha} < 1$

$$\sup_{t\in[0,T]} t^v \left\|\mathbb{A}\vec{\mathcal{J}}\right\| < C\lambda$$

and for $0 < \kappa = \frac{m-1}{\alpha} < 1$

$$\sup_{t,\tau\in[0,T]} t^{v+\kappa}(t-\tau)^{-\kappa} \sum_{j=1}^m |\left(\mathbb{A}\vec{\mathcal{J}}\right)_j(\tau) - \left(\mathbb{A}\vec{\mathcal{J}}\right)_j(t)| < C\lambda.$$

Via Theorem 12 we then obtain a unique solution to problem (2.1)

$$u(x,t) \in \mathbf{C}^0([0,T], \mathbf{H}_2^N(\mathbf{R}^+) \cap \mathbf{C}^N(\mathbf{R}^+))$$
$$\cap \mathbf{C}^0((0,T], \mathbf{H}_2^{[\alpha]-1}(\mathbf{R}^+) \cap \mathbf{C}^{[\alpha]-1}(\mathbf{R}^+)),$$

and for $\delta_j = \max(0, \frac{j-N}{\alpha})$

$$\sup_{t\in[0,T]} \sum_{j=0}^{[\alpha]-1} t^{\delta_j} \left(\left\| \partial_x^j u(t) \right\|_{\mathbf{L}^\infty} + \left\| \partial_x^j u(t) \right\|_{\mathbf{L}^2} \right) \leq C\lambda.$$

Moreover the solution has the following form

$$u(x,t) = \frac{1}{2\pi i} \int_{-i\infty}^{i\infty} e^{px - K(p)t} \left(\mathcal{J}(p,t) \right.$$
$$\left. + K(p) \sum_{j=1}^{m} p^{-N-j} \int_0^t e^{K(p)\tau} (\mathbb{A}\vec{\mathcal{J}})_j(\tau) d\tau \right) dp.$$

Theorem 17 is proved.

3. Local existence for nonlinear problem

Proof of Theorem 14. We prove local existence of solutions by the contraction mapping principle. We define u as a solution of the following linear problem

(3.1)
$$\begin{cases} u_t + \mathbb{N}(w) + \mathbb{K}u = 0, t > 0, x > 0, \\ u(x,t)|_{t=0} = u_0(x), x > 0 \\ \partial_x^{j-1} u(x,t)\big|_{x=0} = h_j(t), t > 0, j = 1, ..., N, \end{cases}$$

with the compatibility conditions $h_j(0) = \partial_x^{j-1} u_0(0)$ for $j = 1, ..., N$. The term $\mathbb{N}(w) = ia(t)|w|^p w$ is known since $w(x,t)$ is fixed from the space $\mathbf{Z}_T$ and satisfies the initial and boundary conditions of the problem (3.1), $N = [\frac{\alpha+1}{2}]$ (in the case $\alpha \in (1,2)$ the boundary data are absent.) Note that the initial-boundary value problem (3.1) defines a mapping $u = \mathbb{M}(w)$, and we will show that $\mathbb{M}$ is a contraction mapping from $\mathbf{Z}_T$ to $\mathbf{Z}_T$.

As we know from Theorem 17, problem (3.1) has a unique solution which can be represented by the Laplace transformation in the following manner

$$\hat{u}(p,t) = e^{-K(p)t} \widehat{u_0}(p)$$

(3.2)
$$+ \int_0^t e^{-K(p)(t-\tau)} \left(K(p) \left(\sum_{j=1}^{N} \frac{h_j(\tau)}{p^j} + \sum_{j=1}^{m} \frac{v_j(\tau)}{p^{N+j}} \right) \right.$$
$$\left. + \widehat{\mathbb{N}(w)}(p,\tau) \right) d\tau,$$

where the Laplace transforms $\hat{v}_j(\xi)$ of the boundary values $v_j(t) = \partial_x^{N+j-1} u(0,t)$ of the solution are defined from the system (2.7) taking $\widehat{\mathbb{N}(w)}(\phi_l(\xi),\xi)$ instead of

$$\hat{f}(\phi_l(\xi), \xi)$$

(3.3)
$$\hat{v}_j(\xi) = \frac{C_1}{\xi^{1-\frac{N+j}{\alpha}}} \sum_{l=1}^{m} r_l^{[\alpha]} M_{j,l} \left(\widehat{u}_0(\phi_l(\xi))\right)$$
$$-\xi \sum_{k=1}^{N} \frac{\widehat{h}_k(\xi)}{\phi_l^k(\xi)} - \widehat{\mathbb{N}(w)}(\phi_l(\xi), \xi) \right),$$

where $h = (\det \tilde{W})^{-1}$ and $M_{j,l}$ are the algebraic minors of the matrix $\tilde{W}$ (see (2.6));
$\widehat{\mathbb{N}(w)}$ is also the Laplace transform of the nonlinearity with respect to the space and time

$$\widehat{\mathbb{N}(w)}(\phi_l(\xi), \xi) = \int_0^{+\infty} e^{-\phi_l(\xi)x} dx \int_0^T e^{-\xi\tau} \mathbb{N}(w)(x, \tau) d\tau.$$

Here $m = [\frac{\alpha}{2}]$, $\phi_l(\xi) = r_l \xi^{1/\alpha}$, $r_l = e^{\frac{i}{\alpha}(2\pi l - \pi [\frac{\alpha-1}{2}])}$ are some constants such that $\mathrm{Re}\,\phi_l(\xi) > 0$ on the imaginary axis $\xi \in (-i\infty, +i\infty)$. (Note that the sum $\sum_1^0$ which appears in (3.3) and below in the case $\alpha \in (1, 2)$ is assumed to be identically zero.) Since $w \in \mathbf{Z}_T$ we have for the nonlinear term in the case $\alpha \geq 2$

(3.4)
$$\sup_{t \in [0,T]} t^{\bar{\gamma}} \int_0^{+\infty} |\partial_x \mathbb{N}(w)(x, t)| dx$$
$$\leq C \sup_{t \in [0,T]} |a(t)| t^{\bar{\gamma}} \|\partial_x w(t)\|_{\mathbf{L}^2} \|w(t)\|_{\mathbf{L}^\infty}^{\rho-1} \|w(t)\|_{\mathbf{L}^2} \leq C \|w\|_{\mathbf{Z}_T}^{\rho+1},$$

$$\sup_{t \in [0,T]} \|\mathbb{N}(w(t))\|_{\mathbf{L}^\infty} \leq C \sup_{t \in [0,T]} |a(t)| \|w(t)\|_{\mathbf{L}^\infty}^{\rho+1} \leq C \|w\|_{\mathbf{Z}_T}^{\rho+1}$$

and

(3.5)
$$\sup_{t \in [0,T]} t^{\bar{\gamma}} \int_0^{+\infty} |\mathbb{N}(w)(x, t)| dx$$
$$\leq C \sup_{t \in [0,T]} |a(t)| t^{\bar{\gamma}} \|w(t)\|_{\mathbf{L}^\infty}^{\rho-1} \|w(t)\|_{\mathbf{L}^2}^2 \leq C \|w\|_{\mathbf{Z}_T}^{\rho+1}.$$

Via Theorem 17 we have the following estimate for the solution
(3.6)
$$\|u\|_{\mathbf{Z}_T} = \sup_{t \in [0,T]} \left(\|u(t)\|_{\mathbf{L}^2} + \sum_{j=0}^{[\alpha]-1} t^{\delta_j} \left(\|\partial_x^j u(t)\|_{\mathbf{L}^2} + \|\partial_x^j u(t)\|_{\mathbf{L}^\infty} \right) \right) \leq C\lambda,$$

where

$$\lambda = \|u_0\|_{\mathbf{X}} + \sum_{j=1}^{N} \|h_j\|_{\mathbf{Y}} + T^\mu \sup_{t \in [0,T]} t^{\bar{\gamma}} \|\mathbb{N}(w)\|_{\mathbf{H}_1^1}$$

for $\alpha \geq 2$ and

$$\lambda = \|u_0\|_{\mathbf{X}} + T^\mu \sup_{t \in [0,T]} t^{\bar{\gamma}} \|\mathbb{N}(w)(t)\|_{\mathbf{L}^1}$$

for $\alpha \in (1, 2)$. Here $\delta_j = \max(0, \frac{j-N}{\alpha})$, $\mu = 1 - \frac{1}{\alpha} - \bar\gamma > 0$, $\bar\gamma = \frac{(\rho-1)\gamma}{\alpha} > 0$, $\gamma > 0$ is small enough. Choosing sufficiently small $T > 0$ we obtain

$$
\begin{aligned}
\|u\|_{\mathbf{Z}_T} &= \sup_{t \in [0,T]} \left(\|u(t)\|_{\mathbf{L}^2} + \sum_{j=0}^{[\alpha]-1} t^{\delta_j} \left(\|\partial_x^j u(t)\|_{\mathbf{L}^\infty} + \|\partial_x^j u(t)\|_{\mathbf{L}^2} \right) \right) \\
&\leq C \left(\|u_0\|_{\mathbf{X}} + \sum_{j=1}^{N} \|h_j\|_{\mathbf{Y}} \right).
\end{aligned}
$$

Thus the mapping $\mathbb{M}$ transforms the closed ball in the space $\mathbf{Z}_T$ with a center at the origin and a radius $C(\|u_0\|_{\mathbf{X}} + \sum_{j=1}^{N} \|h_j\|_{\mathbf{Y}})$ to itself. Analogously we can prove the estimate

$$\|u - \tilde u\|_{\mathbf{Z}_T} \leq \frac{1}{2} \|w - \tilde w\|_{\mathbf{Z}_T}.$$

In fact, the initial and boundary data for the function $u - \tilde u$ are equal to zero. By choosing small $T > 0$ such that $(\|w\|_{\mathbf{Z}_T} + \|\tilde w\|_{\mathbf{Z}_T})^\rho T^\mu \leq 1/2$ we find

$$\|u - \tilde u\|_{\mathbf{Z}_T} \leq \frac{1}{2} \|w - \tilde w\|_{\mathbf{Z}_T}$$

from (3.4) and (3.5). Therefore $\mathbb{M}$ is a contraction mapping and there exists the unique solution $u(x, t)$ of the initial-boundary value problem (1.1). Theorem 14 is proved.

REMARK 4. *If we can obtain the following a priori estimate of the solution*

$$\|u(t)\|_{\mathbf{L}^2} + \|u(t)\|_{\mathbf{L}^\infty} < \infty$$

for all $t \in (0, +\infty)$, *then via estimate (3.4) or (3.5) by the standard continuation argument we can prove a global existence of a unique solution* $u \in \mathbf{C}((0, +\infty), \mathbf{Z}_\infty)$.

REMARK 5. *From (3.5) and (3.6) we see that if the norm of the initial data* $\|u_0\|_{\mathbf{X}} < \epsilon$ *and the norm of the boundary data* $\sum_{j=1}^{N} \|h_j\|_{\mathbf{Y}} < \epsilon$, *then there exists a time* $T > 1$ *such that the solution is also sufficiently small*

$$\sup_{t \in [0,T]} (t^\delta \|u(t)\|_{\mathbf{L}^\infty} + \|u(t)\|_{\mathbf{L}^2}) < C\epsilon,$$

where $\delta = \max(0, \frac{\gamma - [\alpha/2]}{\alpha})$, $\gamma > 0$ *is small enough.*

4. Asymptotics determined by the boundary data

This section is devoted to the proof of Theorem 15. Here we consider the initial-boundary value problem (1.1) with $\alpha \geq 2$, which is not equal to an odd integer and possesses small initial and boundary data $\|u_0\|_{\mathbf{X}} < \epsilon$, and $\sum_{j=1}^{N} \|h_j\|_{\mathbf{Y}} \leq \epsilon$. Also we suppose that the boundary data have the following asymptotics

$$h_j(t) = A_j t^{-\chi - \frac{j-1}{\alpha}} + O\left(\epsilon t^{-\chi - \frac{j-1}{\alpha} - \gamma}\right)$$

as $t \to +\infty$ for all $j = 1, ..., N$, where the coefficients A_j do not equal zero simultaneously and are small: $|A_j| \leq \epsilon$, the value $\epsilon > 0$ is sufficiently small. Here $\frac{\max(0, 1 - \eta + \frac{1}{\alpha})}{\rho + 1} < \chi < \frac{1}{\alpha}, \eta \geq 0$.

Before proving Theorem 15 we offer some preliminary estimates in Lemmas 5 through 11. First we consider the following function

$$G(x) = \int_{-i\infty}^{i\infty} e^{xy} y^\delta dy \int_0^1 e^{-Ey^\alpha(1-z)} z^\beta dz,$$

where $-1 < \beta < 0, -1 < \delta \leq \alpha - 1$ and E is a constant such that $\mathrm{Re}(E(\pm i)^\alpha) > 0$. In the next lemma we prove that the function $G(x)$ is bounded for all $x \geq 0$.

LEMMA 5. *There exists a constant $C > 0$ such that $|G(x)| \leq C$ for all $x > 0$, and there exists a limit $\lim_{x \to +0} G(x)$.*

PROOF. We write the following representation

$$
\begin{aligned}
G(x) &= \int_0^{1/2} dz z^\beta \int_{-i\infty}^{i\infty} e^{yx - Ey^\alpha(1-z)} y^\delta dy \\
&+ \int_{-i}^{i} e^{yx} y^\delta dy \int_{1/2}^1 e^{-Ey^\alpha(1-z)} z^\beta dz \\
&+ \int_{|y| \geq 1, \mathrm{Re}\, y = 0} e^{yx} y^\delta dy \int_{1/2}^1 e^{-Ey^\alpha(1-z)} z^\beta dz \\
&= J_1 + J_2 + J_3.
\end{aligned}
$$

Taking $\theta = \min(\mathrm{Re}\, E(-i)^\alpha, \mathrm{Re}\, Ei^\alpha) > 0$, and changing $y = iq$ we get

$$|J_1| \leq 2 \int_0^{1/2} dz z^\beta \int_0^{+\infty} q^\delta e^{-\theta q^\alpha / 2} dq \leq C$$

and

$$|J_2| \leq 2 \int_0^1 q^\delta dq \int_{1/2}^1 z^\beta dz \leq C$$

since $-1 < \beta$ and $-1 < \delta$. Also note that J_1 and J_2 are continuous with respect to $x \geq 0$. Integrating by parts with respect to z, we get

$$
\begin{aligned}
|J_3| &\leq C \left| \int_{|y| > 1, \mathrm{Re}\, y = 0} e^{xy} y^{\delta - \alpha} dy \right| \\
&+ C \left| \int_{|y| > 1, \mathrm{Re}\, y = 0} e^{xy - \frac{Ey^\alpha}{2}} y^{\delta - \alpha} dy \right| \\
&+ C \int_{|y| > 1, \mathrm{Re}\, y = 0} |y|^{\delta - \alpha - \nu\alpha} dy \int_{1/2}^1 z^{\beta - 1}(1 - z)^{-\nu} dz \\
&\leq C,
\end{aligned}
$$

where $\frac{\delta+1-\alpha}{\alpha} < \nu < 1$. Also we see that each term in the representation of J_3 has a limit for $x \to +0$. In the case $\delta - \alpha = -1$ by virtue of the identity

$$\text{VP} \int_{|y|>1, \text{Re}\, y=0} e^{xy}\frac{dy}{y} = 2\pi i - 2i \lim_{\epsilon \to +0} \int_{\epsilon}^{1} \frac{\sin xy}{y}dy$$

we integrate by parts to get

$$\left| \text{VP} \int_{|y|>1, \text{Re}\, y=0} e^{xy}\frac{dy}{y} - 2\pi i \right| \le 2|x| \left(1 - \lim_{\epsilon \to +0} \epsilon \log \epsilon \right) = 2|x|,$$

whence we have $\lim_{x \to +0} \text{VP} \int_{|y|>1, \text{Re}\, y=0} e^{xy}\frac{dy}{y} = 2\pi i$. Lemma 5 is proved.   □

Now we consider the function

$$F(x,t) = \int_{-i\infty}^{+i\infty} e^{\xi t - \Theta \xi^{\mu} x} d\xi$$

for all $x > 0$ and $t > 0$, where $\mu \in (0,1)$. Let Θ be a constant such that $\text{Re}\, \Theta(\pm i)^{\mu} > 0$.

LEMMA 6. *We have the following estimate* $|F(x,t)| \le Ct^{-\delta\mu-1}x^{\delta}$ *for all* $x > 0, t > 0$ *and for any* $\delta \in \left[-\frac{1}{\mu}, 1\right]$.

PROOF. By changing the variable $\xi t = q$ and taking $z = xt^{-\mu}$ we get

$$F(x,t) = \frac{1}{t} \int_{-i\infty}^{i\infty} e^{q - \Theta q^{\mu} z} dq.$$

Integration by parts in the case $0 < z < 1$ yields

$$|F| \le Czt^{-1} \left| \int_{-i\infty}^{i\infty} e^{q - \Theta q^{\mu} z} q^{\mu-1} dq \right| \le Czt^{-1} \le Ct^{-1}z^{\delta},$$

where $\delta \in \left[-\frac{1}{\mu}, 1\right]$. In the case $z > 1$ by changing the variable $y = qz^{1/\mu}$ we obtain

$$|F| \le Ct^{-1}z^{-1/\mu} \left| \int_{-i\infty}^{i\infty} e^{yz^{-1/\mu} - \Theta y^{\mu}} dy \right| \le Ct^{-1}z^{-1/\mu} \le Ct^{-1}z^{\delta},$$

with any $\delta \in \left[-\frac{1}{\mu}, 1\right]$. Lemma 6 is proved.   □

LEMMA 7. *Suppose that* $\|u_0\|_{\mathbf{L}^1} \le \epsilon$ *and* $\sum_{j=1}^{N} \|h_j\|_{\mathbf{H}_1^1} \le \epsilon$ *and* $\hat{h}'_j(\xi) = O\left(\frac{\epsilon}{|\xi|^2}\right)$ *for* $|\xi| > 1, j = 1, ..., N$. *We assume that the following estimates for the solution of (1.1) are valid*

(4.1)
$$\sup_{t>0}(1+t)^{\chi}\left((1+t)^{-\frac{1}{2\alpha}}\|u\|_{\mathbf{L}^2} + \|u\|_{\mathbf{L}^{\infty}}\right) \le \epsilon_1.$$

Then for the solutions v_j *of system (2.7) we have*

(4.2)
$$\sup_{t>1} t^{\frac{N+j}{\alpha}}|v_j(t)| \le C(\epsilon + \epsilon_1^{\rho+1})$$

for $j = 1, ..., m$.

PROOF. According to (3.3) we write the solutions $v_j(t)$ of system (2.7) in the form

$$(4.3) \qquad v_j(t) = \frac{C_1}{2\pi i} \sum_{l=1}^{m} r_l^{[\alpha]} M_{j,l}(I_1 + I_2 + I_3),$$

where

$$I_1 = \int_{-i\infty}^{i\infty} d\xi e^{\xi t} \xi^{-1 + \frac{N+j}{\alpha}} \int_0^{+\infty} e^{-\phi_l(\xi)x} u_0(x) dx,$$

$$I_2 = \sum_{k=1}^{N} \int_{-i\infty}^{i\infty} e^{\xi t} \xi^{\frac{N+j}{\alpha}} \phi_l^{-k}(\xi) \hat{u}_k(\xi) d\xi$$

and

$$I_3 = \int_{-i\infty}^{i\infty} d\xi e^{\xi t} \xi^{\frac{N+j}{\alpha} - 1} \int_0^{+\infty} d\tau \int_0^{+\infty} dx \mathbb{N}(u)(x, \tau) e^{-\xi \tau - \phi_l(\xi)x}.$$

Now we estimate each summand of (4.3). Since $u_0 \in \mathbf{L}^1(\mathbf{R}^+)$ and $\operatorname{Re} r_l(\pm i)^{\frac{1}{\alpha}} > 0$ by changing the variable of integration $y = \xi$ in the first summand I_1 we have by virtue of Lemma 6 with $\delta = 0$

$$(4.4) \qquad I_1 = \int_0^{+\infty} u_0(x) dx \int_{-i\infty}^{+i\infty} e^{yt - r_l y^{\frac{1}{\alpha}} x} y^{\frac{N+j}{\alpha} - 1} dy = O\left(\epsilon t^{-\frac{N+j}{\alpha}}\right),$$

for all $j, l = 1, ..., m$. Integrating by parts in the second summand I_2 in (4.3) we get

$$I_2 = \sum_{k=1}^{N} \int_{-i\infty}^{i\infty} d\xi e^{\xi t} \widehat{h}_k(\xi) \xi^{\frac{N+j}{\alpha}} \phi_l^{-k}(\xi) d\xi$$

$$= t^{-1} \sum_{k=1}^{N} \left(e^{\xi t} \frac{\widehat{h}_k(\xi) \xi^{\frac{N+j}{\alpha}}}{\phi_l^k} \bigg|_{-i\infty}^{i\infty} - \int_{-i\infty}^{i\infty} e^{\xi t} \left(\frac{C\widehat{h}_k(\xi)}{\xi^{1 - \frac{N+j}{\alpha}} \phi_l^k(\xi)} \right. \right.$$

$$\left. \left. + \frac{C\widehat{h}_k(\xi)}{\xi^{1 - \frac{N+j-k}{\alpha}}} + \frac{C\widehat{h}'_k(\xi) \xi^{\frac{N+j}{\alpha}}}{\phi_l^k(\xi)} \right) d\xi \right)$$

$$(4.5) \qquad = O\left(\epsilon t^{-1}\right),$$

since $\widehat{h}_k(\xi) = O(\epsilon |\xi|^{-1})$ and $\widehat{h}'_k(\xi) = O(\epsilon |\xi|^{-2})$ for all $|\xi| > 1, \operatorname{Re} \xi = 0, k = 1, ..., N$. Via (4.1) we have

$$\int_0^{+\infty} (1 + x^{-\gamma}) |\mathbb{N}(u)(x, t)| dx \leq |a(t)| \|u(t)\|_{\mathbf{L}^\infty}^{\rho - 1} \left(\|u(t)\|_{\mathbf{L}^2}^2 + \|u(t)\|_{\mathbf{L}^\infty}^2 \right)$$

$$(4.6) \qquad\qquad\qquad\qquad\qquad \leq \epsilon_1^{\rho+1} (1 + t)^{-\chi(\rho+1) + 1/\alpha - \eta},$$

where $\gamma > 0$ is small enough. Therefore interchanging the order of integration, changing the variable of integration $y = \xi(t - \tau)$, and using the condition $\chi(\rho +$

1) $-\frac{1}{\alpha} + \eta > 1$ by virtue of Lemma 6 we have for the third summand I_3 in (4.3)

$$|I_3| \leq C \int_0^{+\infty} |t - \tau|^{-\frac{N+j}{\alpha}} d\tau \int_0^{+\infty} |\mathbb{N}(u)(x,\tau)| dx$$

$$\times \left| \int_{-i\infty}^{i\infty} e^{y - r_l y^{\frac{1}{\alpha}} x(t-\tau)^{-1/\alpha}} y^{\frac{N+j}{\alpha} - 1} dy \right|$$

$$\leq C \epsilon_1^{\rho+1} \int_0^{+\infty} \frac{d\tau}{|t - \tau|^{\frac{N+j}{\alpha}} (1 + \tau)^{\chi(\rho+1) - 1/\alpha + \eta}} d\tau$$

$$(4.7) \qquad = O(\epsilon_1^{\rho+1} t^{-\frac{N+j}{\alpha}})$$

for $j = 1, ..., m$, if $\alpha \neq [\alpha]$ and for $j = 1, ..., m - 1$, if $\alpha = [\alpha]$. In the case $\alpha = [\alpha]$ for $j = m$ using (4.6) and Lemma 6 with $\Theta = r_l, l = 1, ..., m$ and $\delta = \gamma$ in the domain $|t - \tau| \leq 1$ and $\delta = 0$ in the domain $|t - \tau| > 1$ we have for I_3

$$|I_3| = C \left| \int_0^{+\infty} d\tau \int_0^{+\infty} F(x, t - \tau) \mathbb{N}(u)(x,\tau) dx \right|$$

$$\leq C \left(\int_{|t-\tau| \leq 1} \frac{d\tau}{|t - \tau|^{1 - \frac{\gamma}{\alpha}}} \int_0^{+\infty} \frac{|\mathbb{N}(u)(x,\tau)|}{x^\gamma} dx \right.$$

$$\left. + \int_{|t-\tau| > 1} \frac{d\tau}{|t - \tau|} \int_0^{+\infty} |\mathbb{N}(u)(x,\tau)| dx \right)$$

$$= O\left(\epsilon_1^{\rho+1} t^{-1} \right).$$

Putting (4.4), (4.5) and (4.7) into (4.3) yields estimate (4.2). Lemma 7 is proved.
□

LEMMA 8. *Let the initial data* $u_0 \in \mathbf{L}^1(\mathbf{R}^+)$ *be small* $\|u_0\|_{\mathbf{L}^1} \leq \epsilon$. *Then*

$$\left\| \int_{-i\infty}^{i\infty} e^{p \cdot - K(p)t} \widehat{u}_0(p) dp \right\|_{\mathbf{L}^s} = O\left(\epsilon(1 + t)^{-\frac{1}{\alpha} + \frac{1}{s\alpha}} \right)$$

for all $t > 1$, *where* $s = 2, +\infty$.

PROOF. By changing the variable $iy = pt^{\frac{1}{\alpha}}$ we have

$$\left\| \int_{-i\infty}^{i\infty} e^{p \cdot - K(p)t} \widehat{u}_0(p) dp \right\|_{\mathbf{L}^\infty}$$

$$\leq t^{-\frac{1}{\alpha}} \|\widehat{u}_0\|_{\mathbf{L}^\infty} \int_0^{+\infty} e^{-\Theta y^\alpha} dy = O\left(\epsilon(1 + t)^{-\frac{1}{\alpha}} \right),$$

and

$$\left\| \int_{-i\infty}^{i\infty} e^{p\cdot - K(p)t} \widehat{u}_0(p) dp \right\|_{\mathbf{L}^2}$$

$$\leq \left(\int_{-i\infty}^{i\infty} e^{-2Ep^\alpha t} |\widehat{u}_0|^2 dp \right)^{\frac{1}{2}}$$

$$\leq \frac{C\epsilon}{(1+t)^{\frac{1}{2\alpha}}} \left(\int_0^{+\infty} e^{-2\Theta y^\alpha} dy \right)^{\frac{1}{2}} < \frac{C\epsilon}{(1+t)^{\frac{1}{2\alpha}}}$$

since $\Theta = \min(\operatorname{Re} E(\pm i)^\alpha) > 0$. Lemma 8 is proved. $\qquad\square$

LEMMA 9. *Let the boundary data have the large time representation*

(4.8) $$h_j(t) = A_j t^{-\chi - \frac{j-1}{\alpha}} + \phi_j(t)$$

for $j = 1, ..., N$ and $t > 0$, where

$$|\phi_j(t)| \leq C\epsilon t^{-\chi - \frac{j-1}{\alpha} - \gamma}$$

and

$$|\phi_1'(t)| \leq C\epsilon t^{-\chi - \frac{j-1}{\alpha} - \gamma}.$$

A_j *are some constants which do not equal zero simultaneously, and $|A_j| \leq \epsilon$, here $\gamma, \epsilon > 0$ are small enough. χ is defined in the beginning of this section. Then we have the following estimate*

$$\left\| \int_{-i\infty}^{i\infty} dp e^{p\cdot} K(p) p^{-j} \int_0^t e^{-K(p)(t-\tau)} h_j(\tau) d\tau \right\|_{\mathbf{L}^s} = O\left(\epsilon t^{-\chi - \frac{1}{s\alpha}} \right)$$

with $s = 2, +\infty$ and the asymptotics

$$I \equiv \frac{1}{2\pi i} \int_{-i\infty}^{i\infty} dp e^{px} K(p) p^{-j} \int_0^t e^{-K(p)(t-\tau)} h_j(\tau) d\tau$$

(4.9) $$= \frac{E}{2\pi i} t^{-\chi} A_j G_j \left(\frac{x}{t^{1/\alpha}} \right) + O(\epsilon t^{-\chi - \gamma}),$$

for all $t > 1$ uniformly with respect to x, where

$$G_j(q) = \int_{-i\infty}^{i\infty} e^{yq} y^{\alpha-j} dy \int_0^1 e^{-Ey^\alpha(1-z)} z^{-\chi - \frac{j-1}{\alpha}} dz,$$

and E is the constant from the definition of the symbol $K(p)$ of the pseudodifferential operator (such that $\operatorname{Re} E(\pm i)^\alpha > 0$).

PROOF. Putting (4.8) into I yields

(4.10) $$I \equiv \frac{EA_j}{2\pi i} \int_{-i\infty}^{i\infty} dp e^{px} p^{\alpha-j} \int_0^t e^{-Ep^\alpha(t-\tau)} \tau^{-\chi - \frac{j-1}{\alpha}} d\tau + R(x,t),$$

where

$$R(x,t) = \epsilon \int_{-i\infty}^{i\infty} e^{px} p^{\alpha-j} dp \int_0^t e^{-Ep^\alpha(t-\tau)} \phi_j(\tau) d\tau.$$

Since $|\phi_j(t)| \le C\epsilon t^{-\chi - \frac{j-1}{\alpha} - \gamma}$ and $\chi < \frac{1}{\alpha}$ by changing the variable of integration $p(t-\tau)^{1/\alpha} = y$, we get for $j = 2, ..., N$

$$|R(x,t)| \le C\epsilon \int_0^t \frac{d\tau}{\tau^{\chi + \frac{j-1}{\alpha} + \gamma}(t-\tau)^{1-\frac{j-1}{\alpha}}} \int_0^{+\infty} e^{-\Theta y^\alpha} y^{\alpha - j} dy$$

$$= O\left(\epsilon t^{-\chi - \gamma}\right),$$

where $\Theta = \min(\operatorname{Re} E(i)^\alpha, \operatorname{Re} E(-i)^\alpha) > 0$. In the case $j = 1$ by changing the variables $\tau = tz$ and $pt^{1/\alpha} = y$ we have

$$R(x,t) = \int_0^{\frac{1}{2}} \phi_1(tz)dz \int_{-i\infty}^{i\infty} e^{yq} y^{\alpha - 1} e^{-Ey^\alpha(1-z)} dy$$

$$+ \int_{-i}^{i} e^{yq} y^{\alpha - 1} dy \int_{\frac{1}{2}}^{1} e^{-Ey^\alpha(1-z)} \phi_1(tz)dz$$

$$+ \int_{|y| \ge 1} e^{yq} y^{\alpha - 1} dy \int_{\frac{1}{2}}^{1} e^{-Ey^\alpha(1-z)} \phi_1(tz)dz,$$

where $q = xt^{-\frac{1}{\alpha}}$.

By analogy with the proof of Lemma 5 we get for small $\nu > 0$

$$|R(x,t)| \le C\left(\int_0^1 |\phi_1(tz)|d\tau + \int_0^{\frac{1}{2}} |\phi_1(tz)|(1-z)^{-\nu} dz \right.$$

$$\left. + |\phi_1(t)| + |\phi_1(t/2)| + \int_{1/2}^1 |\phi_1'(tz)|(1-z)^{-\nu} dz \right).$$

Then using the conditions of the lemma for $\phi_1(t)$ and $\phi_1'(t)$ we obtain in the case $j = 1$

$$R(x,t) = O\left(\epsilon t^{-\chi - \gamma}\right).$$

Therefore by changing the variables $\tau = tz$ and $pt^{1/\alpha} = y$ in the first summand of (4.10) we get (4.9). Since $\chi < \frac{1}{\alpha}$ by changing the variable of integration $iy = p(t-\tau)^{\frac{1}{\alpha}}$ and using (4.8) we obtain

$$\left\| \int_{-i\infty}^{i\infty} dp e^{p \cdot} K(p) p^{-j} \int_0^t e^{-K(p)(t-\tau)} h_j(\tau)d\tau \right\|_{L^2}$$

$$\le C \int_0^t |h_j(\tau)|d\tau \left(\int_{-i\infty}^{+i\infty} e^{-2\operatorname{Re} Ep^\alpha(t-\tau)} |p|^{2\alpha - 2j} |dp| \right)^{\frac{1}{2}}$$

$$\le C\epsilon \int_0^t \frac{d\tau}{\tau^{\chi + \frac{j-1}{\alpha}}(t-\tau)^{1 - \frac{j}{\alpha} + 1/2\alpha}} \left(\int_0^{+\infty} e^{-\Theta y^\alpha} y^{2\alpha - 2j} dy \right)^{\frac{1}{2}}$$

$$< \frac{C\epsilon}{(1+t)^{\chi - 1/2\alpha}}.$$

From Lemmas 5, 6 and asymptotics (4.9) we easily derive

$$\left\| \int_{-i\infty}^{i\infty} dp e^{p \cdot} K(p) p^{-j} \int_0^t e^{-K(p)(t-\tau)} h_j(\tau) d\tau \right\|_{\mathbf{L}^\infty} < C\epsilon (1+t)^{-\chi}.$$

Therefore Lemma 9 is proved. $\qquad\qquad\qquad\qquad\qquad\qquad\qquad\qquad\qquad\qquad\qquad$ □

LEMMA 10. *Let estimate (4.2) be valid. Then the following estimates*

$$\left\| \int_{-i\infty}^{i\infty} e^{p \cdot} K(p) p^{-N-j} dp \int_0^t e^{-K(p)(t-\tau)} v_j(\tau) d\tau \right\|_{\mathbf{L}^s}$$
$$= O\left((\epsilon + \epsilon_1^{p+1}) t^{-\chi-\gamma+\frac{1}{s\alpha}} \right)$$

are true for all $j = 1, ..., m$, *where* $s = 2$ *or* $+\infty$, $\gamma > 0$ *is small.*

PROOF. Via the estimate of Lemma 6 we have $t^{\frac{j-1}{\alpha}} |v_j(t)| \leq \epsilon + \epsilon_1^{p+1}$ for $t \in [0, 1]$, $j = 1, ..., m$. Therefore by virtue of (4.2) by changing the variables $tz = \tau$ and $y = p t^{\frac{1}{\alpha}}$ we obtain

$$\left\| \int_{-i\infty}^{i\infty} e^{p \cdot} K(p) p^{-N-j} dp \int_0^t e^{-K(p)(t-\tau)} v_j(\tau) d\tau \right\|_{\mathbf{L}^\infty}$$
$$\leq Ct^{\frac{N+j-1}{\alpha}} \left| \int_{-i\infty}^{i\infty} e^{xy/t^{1/\alpha}} y^{\alpha-N-j} dy \int_0^1 e^{-Ey^\alpha(1-z)} v_j(tz) dz \right|$$
$$\leq C(\epsilon + \epsilon_1^{p+1}) t^{-\frac{1}{\alpha}} \int_0^{+\infty} y^{\alpha-N-j} dy \int_{1/t}^1 e^{-\Theta y^\alpha(1-z)} z^{-\frac{N+j}{\alpha}} dz$$
$$+ C(\epsilon + \epsilon_1^{p+1}) t^{\frac{N}{\alpha}} \left| \int_0^{1/t} \frac{dz}{z^{\frac{j-1}{\alpha}}} \right|$$
$$= O\left((\epsilon + \epsilon_1^{p+1}) t^{-\frac{1}{\alpha}} \log(2+t) \right)$$
$$= O\left((\epsilon + \epsilon_1^{p+1}) t^{-\chi-\gamma} \right)$$

with $\Theta = \min(\operatorname{Re}(i)^\alpha E, \operatorname{Re}(-i)^\alpha E) > 0$, where $\gamma > 0$ is small enough. Similarly by changing the variables $p(t-\tau)^{\frac{1}{\alpha}} = y$ we get

$$\left\| \int_{-i\infty}^{i\infty} e^{p\cdot} K(p) p^{-N-j} dp \int_0^t e^{-K(p)(t-\tau)} v_j(\tau) d\tau \right\|_{L^2}$$

$$= C \left\| K(p) p^{-N-j} dp \int_0^t e^{-K(p)(t-\tau)} v_j(\tau) d\tau \right\|_{L^2}$$

$$\leq C \left(\int_0^{+\infty} dy \, e^{-\Theta y^\alpha} y^{2\alpha - 2N - 2j} \right)^{\frac{1}{2}} \int_0^t \frac{|v_j(\tau)| d\tau}{|t-\tau|^{1+\frac{1}{2\alpha} - \frac{N+j}{\alpha}}}$$

$$\leq C(\epsilon + \epsilon_1^{\rho+1}) \int_0^1 \frac{d\tau}{\tau^{\frac{j-1}{\alpha}} (t-\tau)^{1+1/2\alpha - \frac{N+j}{\alpha}}}$$

$$+ C(\epsilon + \epsilon_1^{\rho+1}) \int_1^t \frac{d\tau}{\tau^{\frac{N+j}{\alpha}} (t-\tau)^{1+\frac{1}{2\alpha} - \frac{N+j}{\alpha}}}$$

$$\leq \frac{C(\epsilon + \epsilon_1^{\rho+1})}{(1+t)^{\frac{1}{2\alpha}}} \leq \frac{C(\epsilon + \epsilon_1^{\rho+1})}{(1+t)^{\chi + \gamma - \frac{1}{2\alpha}}}$$

for all $j = 1, ..., m$. Lemma 10 is proved. $\qquad\square$

LEMMA 11. *Let estimates (4.1) be valid. Then we have*

$$\left\| \int_{-i\infty}^{i\infty} e^{p\cdot} dp \int_0^t e^{-K(p)(t-\tau)} \widehat{N(u)}(p,\tau) d\tau \right\|_{L^s} = O\left(\epsilon_1^{\rho+1} t^{\frac{1}{s\alpha} - \chi - \gamma} \right)$$

for all $t > 1$, where $s = 2$ or $+\infty$, $\gamma > 0$ is small.

PROOF. Via (4.1) we have

$$\left\| \widehat{N(u)}(t) \right\|_{L^\infty(\operatorname{Re} p=0)} = \sup_{p:\operatorname{Re} p=0} \left| \int_0^{+\infty} e^{-px} a(t) |u|^\rho u \, dx \right|$$

$$\leq |a(t)| \int_0^{+\infty} |u|^{\rho+1} dx$$

$$\leq |a(t)| \|u(t)\|_{L^\infty}^{\rho-1} \|u(t)\|_{L^2}^2 \leq \epsilon_1^{\rho+1} (1+t)^{-\beta},$$

where $\beta = \chi(\rho+1) - 1/\alpha + \eta > 1$ by virtue of the condition of Theorem 15. Therefore by changing the variables $p(t-\tau)^{\frac{1}{\alpha}} = iy$ we obtain

$$\left\| \int_{-i\infty}^{i\infty} e^{p\cdot} dp \int_0^t e^{-K(p)(t-\tau)} \widehat{N}(u)(p,\tau) d\tau \right\|_{L^\infty}$$

$$\leq C\epsilon_1^{\rho+1} \int_0^t \frac{d\tau}{(1+\tau)^\beta (t-\tau)^{1/\alpha}} \int_0^{+\infty} e^{-\Theta y^\alpha} dy = O(\epsilon_1^{\rho+1} t^{-\frac{1}{\alpha}})$$

for $t > 1$, where $\Theta = \min(\operatorname{Re} E(+i)^{\alpha}, \operatorname{Re} E(-i)^{\alpha}) > 0$. Similarly

$$\left\| \int_{-i\infty}^{i\infty} e^{p \cdot} dp \int_0^t e^{-K(p)(t-\tau)} \widehat{N(u)}(p, \tau) d\tau \right\|_{\mathbf{L}^2}$$

$$\leq C \int_0^t \frac{\|\widehat{N(u)}(t)\|_{\mathbf{L}^\infty(\operatorname{Re} p=0)}}{(t-\tau)^{\frac{1}{2\alpha}}} \left(\int_0^{+\infty} e^{-2\Theta y^\alpha} dy \right)^{\frac{1}{2}}$$

$$\leq C \epsilon_1^{\rho+1} \int_0^t \frac{d\tau}{(1+\tau)^\beta (t-\tau)^{\frac{1}{2\alpha}}}$$

$$\leq C \epsilon_1^{\rho+1} (1+t)^{-1/2\alpha} < C \epsilon_1^{\rho+1} (1+t)^{-\chi + \frac{1}{2\alpha} - \gamma}.$$

Lemma 11 is proved. $\qquad\qquad\square$

Proof of Theorem 15. Let us prove the following estimate

$$(4.11) \qquad (1+t)^\chi \left(\|u(t)\|_{\mathbf{L}^\infty} + (1+t)^{-\frac{1}{2\alpha}} \|u(t)\|_{\mathbf{L}^2} \right) < \epsilon_1$$

for all $t > 0$. On the contrary we suppose that the estimate (4.11) is broken for some time. By Theorem 14 the norms $\|u(t)\|_{\mathbf{L}^2}$ and $\|u(t)\|_{\mathbf{L}^\infty}$ are continuous. Therefore there exists a maximal time $T > 0$ such that the nonstrict estimate (4.11) is valid on $[0, T]$. We have by formula (3.2)

$$
\begin{aligned}
u(x,t) \;=\; & \frac{1}{2\pi i} \int_{-i\infty}^{i\infty} dp e^{px} \left(\widehat{u}_0(p) e^{-K(p)t} \right. \\
& + \sum_{j=1}^{N} \frac{K(p)}{p^j} \int_0^t e^{-K(p)(t-\tau)} h_j(\tau) d\tau \\
& + \sum_{j=1}^{m} K(p) p^{-N-j} \int_0^t e^{-K(p)(t-\tau)} v_j(\tau) d\tau \\
(4.12) \quad & \left. + \int_0^t e^{-K(p)(t-\tau)} \widehat{N(u)}(p, \tau) d\tau \right).
\end{aligned}
$$

Thus by Lemmas 7 through 11 we get

$$(1+t)^\chi \left(\|u(t)\|_{\mathbf{L}^\infty} + (1+t)^{-\frac{1}{2\alpha}} \|u(t)\|_{\mathbf{L}^2} \right) < \epsilon_1$$

for all $t \in [0, T]$ if $\epsilon > 0$ is sufficiently small. The contradiction obtained proves estimate (4.11). Then from (4.12), by virtue of Remark 4, we see that the solution

$$
\begin{aligned}
u(x,t) \;\in\; & \mathbf{C}\left([0, +\infty); \mathbf{L}^2\left(\mathbf{R}^+\right) \cap \mathbf{L}^\infty\left(\mathbf{R}^+\right)\right) \\
& \cap \mathbf{C}\left(\mathbf{R}^+; \mathbf{H}_2^{[\alpha]-1}\left(\mathbf{R}^+\right) \cap \mathbf{C}^{[\alpha]-1}\left(\mathbf{R}^+\right)\right).
\end{aligned}
$$

By virtue of Lemmas 5 through 11 the solution has asymptotics (1.3) for $t > 1$ uniformly with respect to $x \geq 0$. Theorem 15 is proved.

5. Asymptotics determined by nonlinearity

This section is devoted to the proof of Theorem 16. We consider the case, when $\alpha > 1$ is not equal to an odd integer number. Before proving Theorem 16 we prepare some estimates in Lemmas 12 through 16. We suppose that we already have the following estimates for the solution

$$(5.1) \qquad \sup_{t \in [0,T]} (1+t)^{\frac{1}{\alpha}} \left((1+t)^{-1/2\alpha} \|u(t)\|_{\mathbf{L}^2} + t^{\gamma}(1+t)^{-\gamma}\|u(t)\|_{\mathbf{L}^\infty} \right) < \epsilon_1,$$

where $\epsilon_1 > 0$ is some small constant and $T > 0$. Throughout the book we suppose the sum of the form $\sum_{j=1}^{0}$ to be identically zero (It appears for the case $\alpha \in (1,2)$ so $N = 0$ and the boundary data are absent). Also any condition in which j varies from 1 to $N = 0$ we assume to be absent.

LEMMA 12. *Let the initial data* $u_0 \in \mathbf{L}^\infty(\mathbf{R}^+)$ *and* $x^\delta u_0 \in \mathbf{L}^1(\mathbf{R}^+)$, $\delta \in [0, \frac{1}{2}]$ *and the boundary data* $h_j \in \mathbf{Y}$, $j = 1, ..., N$. *The latter satisfy condition (1.2) with* $\chi = \frac{1}{\alpha}$ *as follows*

$$h_j(t) = A_j t^{-\frac{j}{\alpha}} + \phi_j(t),$$

where

$$|\phi_j(t)| \le C\epsilon t^{-\frac{j}{\alpha}-\gamma}$$

and

$$\phi_1'(t) = O\left(\epsilon t^{-\frac{j}{\alpha}-\gamma}\right).$$

Suppose that the estimate (5.1) is valid and $\frac{\rho}{\alpha} + \eta > 1$. *Then we have the following estimate* $\|x^\delta u\|_{\mathbf{L}^2} \le C$ *for all* $t > 1$.

PROOF. By formula (3.2) for the solution we get

$$
\begin{aligned}
\|(\cdot)^\delta u(\cdot,t)\|_{\mathbf{L}^2} &= C\left\|\partial_p^\delta \hat{u}(t)\right\|_{\mathbf{L}^2(\mathrm{Re}\,p=0)} \le C\left(\left\|\partial_p^\delta \left(\hat{u}_0(p)e^{-K(p)t}\right)\right\|_{\mathbf{L}^2(\mathrm{Re}\,p=0)} \right. \\
&+ \sum_{j=1}^{N} \int_0^t |h_j(\tau)| \left\|\partial_p^\delta \left(e^{-K(p)(t-\tau)}p^{\alpha-j}\right)\right\|_{\mathbf{L}^2(\mathrm{Re}\,p=0)} d\tau \\
&+ \sum_{j=1}^{m} \int_0^t |v_j(\tau)| \left\|\partial_p^\delta \left(e^{-K(p)(t-\tau)}p^{\alpha-N-j}\right)\right\|_{\mathbf{L}^2(\mathrm{Re}\,p=0)} d\tau \\
&+ \left. \int_0^t \left\|\partial_p^\delta \left(e^{-K(p)(t-\tau)}\widehat{N(u)}(p,\tau)\right)\right\|_{\mathbf{L}^2(\mathrm{Re}\,p=0)} d\tau \right)
\end{aligned}
$$

$$(5.2) \qquad\qquad\equiv C\sum_{j=1}^{4} J_j,$$

where $\partial_p^\delta \phi = \mathcal{L}^{-1}(p^\delta \mathcal{L}\phi)$. With $\mathcal{L}$ we denote the Laplace transformation. We have the Sobolev embedding inequality (see, for example, [38])

$$(5.3) \qquad \|\partial_p^\delta \phi\|_{\mathbf{L}^2} \le \|\phi\|_{\mathbf{L}^2}^{1-2\delta} \|\partial_p \phi\|_{\mathbf{L}^1}^{2\delta}.$$

Consider the function $G(x,t) = \int_{-i\infty}^{i\infty} e^{px - Ep^\alpha t} dp$. Note that the Laplace transform $\widehat{G}(p,t) = 2\pi i e^{-Ep^\alpha t}$. Then by changing the variable of integration $y = pt^{1/\alpha}$ we easily obtain

(5.4) $$\|G(t)\|_{\mathbf{L}^2(R^+)} = \|\widehat{G}(t)\|_{\mathbf{L}^2(\mathrm{Re}\,p=0)} \leq Ct^{-\frac{1}{2\alpha}}$$

and

$$
\begin{aligned}
(5.5) \quad \|(\cdot)^\delta G(\cdot,t)\|_{\mathbf{L}^2} &= C\|\partial_p^\delta \widehat{G}(t)\|_{\mathbf{L}^2(\mathrm{Re}\,p=0)} \\
&\leq C\|\widehat{G}(t)\|_{\mathbf{L}^2(\mathrm{Re}\,p=0)}^{1-2\delta} \|\partial_p \widehat{G}(t)\|_{\mathbf{L}^1(\mathrm{Re}\,p=0)}^{2\delta} \\
&\leq Ct^{\frac{2\delta-1}{2\alpha}}
\end{aligned}
$$

for $t > 0$. Changing the order of integration in the first summand in (5.2) we get

$$
\begin{aligned}
J_1 &= \left\| (\cdot)^\delta \int_{-i\infty}^{i\infty} dp\, e^{p\cdot - Ep^\alpha t} \int_0^{+\infty} e^{-py} u_0(y) dy \right\|_{\mathbf{L}^2} \\
&= \left\| (\cdot)^\delta \int_0^{+\infty} u_0(y) G(\cdot - y, t) dy \right\|_{\mathbf{L}^2}.
\end{aligned}
$$

Since

$$\left\| \int_0^{+\infty} \phi(\cdot - y)\psi(y) dy \right\|_{\mathbf{L}^2} \leq \|\phi\|_{\mathbf{L}^2} \|\psi\|_{\mathbf{L}^1},$$

by virtue of (5.4) and (5.5) we have

$$
\begin{aligned}
J_1 &\leq \left\| \int_0^{+\infty} y^\delta u_0(y) G(\cdot - y, t) dy \right\|_{\mathbf{L}^2} + \left\| \int_0^{+\infty} |\cdot - y|^\delta u_0(y) G(\cdot - y, t) dy \right\|_{\mathbf{L}^2} \\
(5.6) \quad &\leq \|(\cdot)^\delta u_0(\cdot)\|_{\mathbf{L}^1} \|G(t)\|_{\mathbf{L}^2} + \|u_0\|_{\mathbf{L}^1} \|(\cdot)^\delta G(\cdot,t)\|_{\mathbf{L}^2} \leq C
\end{aligned}
$$

for all $t > 1$. Similarly we can estimate J_4. In fact, from (5.1) we have

$$
\begin{aligned}
\|(\cdot)^\delta N(u)(\cdot,t)\|_{\mathbf{L}^1} &\leq C|a(t)| \|(\cdot)^\delta u(\cdot,t)\|_{\mathbf{L}^2} \|u(t)\|_{\mathbf{L}^2} \|u(t)\|_{\mathbf{L}^\infty}^{\rho-1} \\
&\leq Ct^{-\gamma}(1+t)^{-\frac{2\rho-1}{2\alpha} - \eta + \gamma} \|(\cdot)^\delta u(\cdot,t)\|_{\mathbf{L}^2}
\end{aligned}
$$

and

$$\|N(u)(t)\|_{\mathbf{L}^1} \leq C|a(t)| \|u(t)\|_{\mathbf{L}^2}^2 \|u(t)\|_{\mathbf{L}^\infty}^\rho \leq Ct^{-\gamma}(1+t)^{-\frac{\rho}{\alpha} - \eta + \gamma}.$$

Therefore using (5.4) and (5.5) we get

$$
\begin{aligned}
J_4 &\leq \int_0^t \left(\|(\cdot)^\delta N(u)(\cdot,t)\|_{\mathbf{L}^1} \|G(t-\tau)\|_{\mathbf{L}^2} + \|N(u)(t)\|_{\mathbf{L}^1} \|(\cdot)^\delta G(\cdot, t-\tau)\|_{\mathbf{L}^2} \right) d\tau \\
(5.7) \quad &\leq C + \int_0^t \|(\cdot)^\delta u(\cdot,t)\|_{\mathbf{L}^2} (t-\tau)^{-\frac{1}{2\alpha}} \tau^{-\gamma}(1+\tau)^{-\frac{2\rho-1}{2\alpha} - \eta + \gamma} d\tau
\end{aligned}
$$

since $\frac{\ell}{\alpha} + \eta > 1$. Applying the condition $t^{\frac{j}{\alpha}}|h_j(t)| \leq C$ of the lemma, by changing the variable of integration $y = p(t-\tau)^{1/\alpha}$ via (5.3) we obtain

$$J_2 \ \leq \ C\sum_{j=1}^{N}\int_0^t |h_j(\tau)|\||e^{-Ep^\alpha(t-\tau)}p^{\alpha-j}\|_{L^2(\mathrm{Re}\,p=0)}^{1-2\delta}$$

$$\times \left(\||e^{-Ep^\alpha(t-\tau)}p^{\alpha-j-1}\|_{L^1(\mathrm{Re}\,p=0)}^{2\delta}\right.$$

$$\left. +(t-\tau)^{2\delta}\||e^{-Ep^\alpha(t-\tau)}p^{2\alpha-j-1}\|_{L^1(\mathrm{Re}\,p=0)}^{2\delta}\right)$$

$$(5.8) \qquad\qquad \leq \ C\sum_{j=1}^{N}\int_0^t \tau^{-\frac{j}{\alpha}}(t-\tau)^{-1+\frac{j}{\alpha}+\frac{2\delta-1}{2\alpha}}\,d\tau \leq C$$

for all $t \geq 1$. By Lemma 7 we have $t^{\frac{j-1}{\alpha}}(1+t)^{\frac{N+1}{\alpha}}|v_j(t)| \leq C$. Then in the same way we obtain the following estimate for J_3

$$(5.9) \qquad J_3 \leq C\sum_{j=1}^{m}\int_0^t \tau^{\frac{1-j}{\alpha}}(1+\tau)^{-\frac{N+1}{\alpha}}(t-\tau)^{-1+\frac{j+N}{\alpha}-\frac{1-2\delta}{2\alpha}}\,d\tau \leq C$$

for all $t \geq 1$. Putting (5.6) through (5.9) into (5.2) yields

$$\|\partial_p^\delta \hat{u}(t)\|_{L^2(\mathrm{Re}\,p=0)} \leq C + \int_0^t \tau^{-\gamma}(1+\tau)^{-\frac{2\rho-1}{2\alpha}-\eta+\gamma}(t-\tau)^{-\frac{1}{2\alpha}}\|\partial_p^\delta \hat{u}(\tau)\|_{L^2(\mathrm{Re}\,p=0)}\,d\tau.$$

Therefore via the Gronwall inequality we have

$$\|\partial_p^\delta \hat{u}(t)\|_{L^2(\mathrm{Re}\,p=0)} = \|(\cdot)^\delta u(\cdot,t)\|_{L^2} \leq C$$

since $\frac{\ell}{\alpha} + \eta > 1$. Lemma 12 is proved. $\square$

LEMMA 13. *Let the initial data $u_0 \in \mathbf{X}$, $xu_0 \in \mathbf{L}^1(\mathbf{R}^+)$, and boundary data $h_j \in \mathbf{Y}$, $j = 1, ..., N$ be small*

$$\|u_0\|_{\mathbf{X}} + \|(\cdot)u_0(\cdot)\|_{\mathbf{L}^1} + \sum_{j=1}^{N}\|h_j\|_{\mathbf{Y}} \leq \epsilon.$$

Moreover let the following estimate for the Laplace transform of the boundary data $\hat{h}'_j = O(\frac{\epsilon}{|\xi|^2})$ be valid for $|\xi| > 1, j = 1, ..., N$. When $\alpha > 1$ is an integer we also suppose that $\hat{h}''_j = O\left(\frac{\epsilon}{|\xi|^2}\right)$ for $|\xi| > 1$, $j = 1, ..., N$. Let estimate (5.1) be valid and $\frac{\ell}{\alpha} + \eta > 1$. Then the following asymptotics for the solutions $v_j(t)$ of system (2.7) (see (3.3))

$$(5.10) \qquad v_j(t) = B_j t^{-\frac{N+j}{\alpha}} + O\left((\epsilon + \epsilon_1^{\rho+1})t^{-\frac{N+j}{\alpha}-\gamma}\right)$$

is valid, where

$$B_j = C_1\Gamma\left(1 - \frac{N+j}{\alpha}\right)\sum_{l=1}^{m}M_{j,l}r_l^{[\alpha]}\left(\hat{u}_0(0) + \int_0^{+\infty}\int_0^{+\infty}\mathcal{N}(u)(x,\tau)\,dx\,d\tau\right),$$

for $j = 1, ..., m$, $C_1 = (\det \tilde{W})^{-1}$. $M_{j,l}$ are the algebraic minors of matrix $\tilde{W}$ (see (2.6)) and $\gamma > 0$ is some small constant. When $\alpha > 1$ is an integer we denote $B_m = 0$. Γ is the Euler Gamma-function.

PROOF. According to (3.3) we write the solutions $v_j(t)$ of the system (2.7) in the form

$$(5.11) \qquad v_j(t) = C_1 \sum_{l=1}^{m} r_l^{[\alpha]} M_{j,l}(I_1 + I_2 + I_3),$$

where

$$I_1 = \frac{1}{2\pi i} \int_{-i\infty}^{i\infty} d\xi e^{\xi t} \xi^{-1+\frac{N+j}{\alpha}} \int_0^{+\infty} e^{-r_l \xi^{\frac{1}{\alpha}} x} u_0(x) dx,$$

$$I_2 = \frac{1}{2\pi i} \sum_{k=1}^{N} \int_{-i\infty}^{i\infty} e^{\xi t} \xi^{\frac{N+j-k}{\alpha}} r_l^{-k} \widehat{h}_k(\xi) d\xi$$

and

$$I_3 = \frac{1}{2\pi i} \int_{-i\infty}^{i\infty} d\xi e^{\xi t} \xi^{-1+\frac{N+j}{\alpha}} \int_0^{+\infty} d\tau \int_0^{+\infty} dx \mathbb{N}(u)(x, \tau) e^{-\xi\tau - r_l \xi^{\frac{1}{\alpha}} x}.$$

Now we estimate each summand in representation (5.11). We rewrite the first integral in (5.11) as follows

$$(5.12) \quad I_1 = \frac{\widehat{u}_0(0)}{2\pi i} \int_{-i\infty}^{i\infty} e^{\xi t} \xi^{\frac{N+j}{\alpha} - 1} d\xi + R = \widehat{u}_0(0) t^{-\frac{N+j}{\alpha}} \Gamma\left(1 - \frac{N+j}{\alpha}\right) + R$$

for $j = 1, ..., m$, if $\alpha \neq [\alpha]$ and $j = 1, ..., m-1$, if $\alpha = [\alpha]$, where

$$\begin{aligned}
R = \ & C\left(\left|\int_{|\xi|\leq 1/t, \text{Re } \xi=0} e^{\xi t}(\widehat{u}_0(r_l \xi^{\frac{1}{\alpha}}) - \widehat{u}_0(0))\xi^{\frac{N+j}{\alpha} - 1} d\xi\right| \right. \\
& + \left|\int_{1\geq|\xi|\geq 1/t, \text{Re } \xi=0} e^{\xi t}(\widehat{u}_0(r_l \xi^{\frac{1}{\alpha}}) - \widehat{u}_0(0))\xi^{\frac{N+j}{\alpha} - 1} d\xi\right| \\
& + \left.\left|\int_{|\xi|>1, \text{Re } \xi=0} e^{\xi t}(\widehat{u}_0(r_l \xi^{\frac{1}{\alpha}}) + \widehat{u}_0(0))\xi^{\frac{N+j}{\alpha} - 1} d\xi\right|\right).
\end{aligned}$$

(5.13)

Since $xu_0 \in \mathbf{L}^1(\mathbf{R}^+)$, we have $|\widehat{u}_0(r_l \xi^{\frac{1}{\alpha}}) - \widehat{u}_0(0)| \leq C\epsilon|\xi|^{1/\alpha}$. Then for the first summand in (5.13) we get

$$\left|\int_{|\xi|\leq 1/t, \text{Re } \xi=0} e^{\xi t}(\widehat{u}_0(r_l \xi^{\frac{1}{\alpha}}) - \widehat{u}_0(0))\xi^{\frac{N+j}{\alpha} - 1} d\xi\right|$$

$$(5.14) \qquad \leq C\epsilon \int_0^{1/t} \xi^{-1+\frac{N+j+1}{\alpha}} d\xi = O\left(\epsilon t^{-\frac{N+j}{\alpha} - \gamma}\right),$$

and since $|\widehat{u}_0'| \leq \epsilon$ integrating by parts we have for the second summand in (5.13)

$$\left| \int_{i/t}^{i} e^{\xi t}(\widehat{u}_0(r_l\xi^{\frac{1}{\alpha}}) - \widehat{u}_0(0))\xi^{\frac{N+j}{\alpha}-1}d\xi \right|$$

$$\leq \frac{1}{t}\left(\left| (\widehat{u}_0(r_l\xi^{\frac{1}{\alpha}}) - \widehat{u}_0(0)) \xi^{\frac{N+j}{\alpha}-1} \Big|_{i/t}^{i} \right| \right.$$

$$+ \left| \int_{i/t}^{i} e^{\xi t}(\widehat{u}_0(r_l\xi^{\frac{1}{\alpha}}) - \widehat{u}_0(0))\xi^{\frac{N+j}{\alpha}-2}d\xi \right|$$

$$\left. + \left| \int_{i/t}^{i} e^{\xi t}\widehat{u}_0'(r_l\xi^{\frac{1}{\alpha}})\xi^{\frac{N+1+j}{\alpha}-2}d\xi \right| \right)$$

(5.15) $$= O\left(\epsilon t^{-\frac{N+j}{\alpha}-\gamma} \right),$$

where $\gamma = \frac{\alpha - [\alpha]}{\alpha}$. Since $\|u_0\|_{\mathbf{L}^1} \leq \epsilon$ and $e^{-\operatorname{Re}\phi_l(\xi)x} \leq C(\phi_l(\xi)x)^{-1}$ we obtain the estimate

$$\left| \frac{d}{d\xi}\widehat{u}_0(\phi_l(\xi)) \right| \leq \left| \int_{-i\infty}^{i\infty} e^{-\operatorname{Re}\phi_l(\xi)x}\phi_l'(\xi)xu_0(x)dx \right| \leq C\epsilon|\xi|^{-1}.$$

Therefore integrating by parts we get for the third summand in (5.13)

$$\left| \int_{|\xi|>1} e^{\xi t}(\widehat{u}_0(r_l\xi^{\frac{1}{\alpha}}) + \widehat{u}_0(0))\xi^{\frac{N+j}{\alpha}-1}d\xi \right|$$

$$\leq \frac{1}{t}\left(\left| e^{\xi t}(\widehat{u}_0(r_l\xi^{1/\alpha}) + \widehat{u}_0(0)) \xi^{\frac{N+j}{\alpha}-1}\Big|_{i}^{i\infty} \right| \right.$$

$$+ \left| \int_{i}^{i\infty} e^{\xi t}(\widehat{u}_0(r_l\xi^{\frac{1}{\alpha}}) + \widehat{u}_0(0))\xi^{\frac{N+j}{\alpha}-2}d\xi \right|$$

$$\left. + \left| \int_{i}^{i\infty} e^{\xi t}\xi^{\frac{N+j}{\alpha}-1}\frac{d}{d\xi}\widehat{u}_0(r_l\xi^{\frac{1}{\alpha}})d\xi \right| \right)$$

(5.16) $$= O\left(\epsilon t^{-\frac{N+j}{\alpha}-\gamma} \right).$$

Putting (5.14) through (5.16) into (5.13) we obtain for the first summand in (5.11)

(5.17) $$I_1 = \widehat{u}_0(0)\Gamma\left(1 - \frac{N+j}{\alpha} \right)t^{-\frac{N+j}{\alpha}} + O(\epsilon t^{-\frac{N+j}{\alpha}-\gamma})$$

for all $j = 1, ..., m$, if $\alpha \neq [\alpha]$ and $j = 1, ..., m-1$, if $\alpha = [\alpha]$. In the case $\alpha = [\alpha]$ for $j = m$ using Lemma 6 with $\delta = \gamma\alpha \leq 1$ we get

(5.18) $$I_1 = Ct^{-1-\gamma}\int_0^{+\infty} x^{\alpha\gamma}|u_0(x)|dx = O(\epsilon t^{-1-\gamma}).$$

By estimate (4.5) from Lemma 7 we have

(5.19) $$I_2 = O(\epsilon t^{-1}).$$

In the case $\alpha = [\alpha]$, using the condition $\hat{u}''(\xi) = O(\epsilon|\xi|^{-2})$ for $|\xi| > 1$ we integrate in (4.5) one more time by parts with respect to ξ to obtain $I_2 = O(\epsilon t^{-2})$. Now let us estimate the third integral I_3 in the representation (5.11). Changing the order of integration we obtain

$$(5.20) \qquad I_3 = -\Gamma\left(1 - \frac{N+j}{\alpha}\right) t^{-\frac{N+j}{\alpha}} \int_0^{+\infty} d\tau \int_0^{+\infty} \mathbb{N}(u)(x, \tau)dx + R$$

for $j = 1, ..., m$, if α is non integer and $j = 1, ..., m - 1$, if α is integer, where

$$R = \int_0^t d\tau \int_0^{+\infty} dx \mathbb{N}(x, \tau) \left((F_l(0, t - \tau) - F_l(0, t)) \right.$$
$$\left. + (F_l(x, t - \tau) - F_l(0, t - \tau)) \right),$$

here we write

$$F_l(x, t) = \int_{-i\infty}^{i\infty} e^{\xi t - r_l \xi^{\frac{1}{\alpha}} x} \xi^{\frac{N+j}{\alpha} - 1} d\xi.$$

From (5.1) and Theorem 14 we have

$$(5.21) \qquad \int_0^{+\infty} (1 + x^{-\delta})|\mathbb{N}(u)(x, \tau)|dx \leq \epsilon_1^{p+1} t^{-\gamma}(1 + t)^{-\frac{p}{\alpha} - \eta + \gamma},$$

where $0 < \delta < 1$, and by the Hölder inequality with $p = \frac{1}{\gamma\alpha}$, $q = \frac{1}{1-\gamma\alpha}$ we get

$$(5.22) \qquad \int_0^{+\infty} |x^{\gamma\alpha}\mathbb{N}(u)(x, t)|dx \leq \epsilon_1^{p+1} t^{-\gamma}(1 + t)^{-\frac{p}{\alpha} - \eta + 2\gamma}.$$

Then using Lemma 6 with $\beta = 1 - \frac{N+j}{\alpha}$, $\mu = \frac{1}{\alpha}$, $\omega = \frac{1}{2} - \gamma\alpha$ and $\nu = \gamma > 0$ is small (we can choose $2\gamma < 1 - \eta - \frac{p}{\alpha}$ since $\frac{p}{\alpha} + \eta > 1$) we have

$$R \leq C\int_0^{+\infty} d\tau \tau^\gamma \left(|\tau|^{-\frac{N+j}{\alpha} - \gamma} + |t - \tau|^{-\frac{N+j}{\alpha} - \gamma}\right) \int_0^{+\infty} |\mathbb{N}(u)(x, \tau)|dx$$
$$+ C\int_0^{+\infty} d\tau |t - \tau|^{-\frac{N+j}{\alpha} - \gamma} \int_0^{+\infty} x^{\gamma\alpha}|\mathbb{N}(u)(x, \tau)|dx$$
$$= O(\epsilon_1^{p+1} t^{-\frac{N+j}{\alpha} - \gamma}).$$

In the case $\alpha = [\alpha]$ we interchange the order of integration and use Lemma 6 with $\delta = -\gamma\alpha$ in the domain $|t - \tau| \leq 1$ and $\delta = \gamma\alpha$ in the domain $|t - \tau| > 1$. Then by virtue of (5.21) and (5.22) we have

$$I_3 = C\int_0^{+\infty} d\tau \int_0^{+\infty} \mathbb{N}(u)(x, \tau)dx \int_{-i\infty}^{i\infty} e^{\xi(t-\tau) - r_l \xi^{\frac{1}{\alpha}} x} d\xi$$
$$\leq C\int_{|t-\tau|\leq 1} d\tau |t - \tau|^{\gamma - 1} \int_0^{+\infty} |x^{-\gamma\alpha}\mathbb{N}(u)(x, \tau)|dx$$
$$+ C\int_{|t-\tau|\geq 1} d\tau |t - \tau|^{-1-\gamma} \int_0^{+\infty} |x^{\gamma\alpha}\mathbb{N}(u)(x, \tau)|dx$$
$$(5.23) \qquad = O\left(\epsilon_1^{p+1} t^{-1-\gamma}\right).$$

From (5.11), (5.17) through (5.20) and (5.23) we get (5.10). Lemma 13 is proved.
□

LEMMA 14. *Let* $\|\langle\cdot\rangle^\delta u_0(\cdot)\|_{\mathbf{L}^1} \le \epsilon$, *where* $0 < \delta < \frac{1}{2}$. *Then the estimate*

$$\left\|\int_{-i\infty}^{i\infty} e^{px-K(p)t}\widehat{u}_0(p)dp\right\|_{\mathbf{L}^2} = O\left(\epsilon t^{-\frac{1}{2\alpha}}\right)$$

is valid for $t > 1$, *and the following asymptotics as* $t \to +\infty$ *uniformly with respect to* $x > 0$

$$\int_{-i\infty}^{i\infty} e^{px-K(p)t}\widehat{u}_0(p)dp = G_0(xt^{-\frac{1}{\alpha}})\widehat{u}_0(0)t^{-\frac{1}{\alpha}} + O\left(\epsilon t^{-\frac{1}{\alpha}-\gamma}\right),$$

is true, where $G_0(q) = \int_{-i\infty}^{+i\infty} e^{yq-Ey^\alpha}dy$.

PROOF. By changing the variable of integration $y = ipt^{1/\alpha}$ and using the conditions of the lemma we get

$$\left\|\int_{-i\infty}^{i\infty} e^{p\cdot-K(p)t}\widehat{u}_0(p)dp\right\|_{\mathbf{L}^2}$$

$$\le \left(\int_{-i\infty}^{i\infty} e^{-2\operatorname{Re} Ep^\alpha t}|\widehat{u}_0|^2|dp|\right)^{\frac{1}{2}}$$

$$\le C\epsilon t^{-\frac{1}{2\alpha}}\left(\int_0^{+\infty} e^{-2\Theta y^\alpha}dy\right)^{\frac{1}{2}} = O\left(\epsilon t^{-\frac{1}{2\alpha}}\right)$$

for $t > 1$, where $\Theta = \min(\operatorname{Re} E(+i)^\alpha, \operatorname{Re} E(-i)^\alpha) > 0$. We write the representation

$$\int_{-i\infty}^{i\infty} e^{px-K(p)t}\widehat{u}_0(p)dp$$

$$= \widehat{u}_0(0)\int_{-i}^{i} e^{px-K(p)t}dp + \int_{-i}^{i} e^{px-K(p)t}(\widehat{u}_0(p) - \widehat{u}_0(0))dp$$

$$+ \int_{|p|\ge 1, \operatorname{Re} p=0} e^{px-K(p)t}\widehat{u}_0(p)dp$$

$$= J_1 + J_2 + J_3.$$

Changing the variable of integration $y = pt^{1/\alpha}$ in the first integral J_1 we get

$$J_1 = \widehat{u}_0(0)G_0(xt^{-\frac{1}{\alpha}})t^{-\frac{1}{\alpha}} + O\left(\epsilon t^{-\frac{1}{\alpha}-\gamma}\right).$$

Since

$$|\widehat{u}_0(p) - \widehat{u}_0(0)| = \left|\int_0^{+\infty}(e^{px} - 1)u_0(x)dx\right|$$

$$\le |p|^{\gamma\alpha}\int_0^{+\infty} x^{\gamma\alpha}|u_0|dx \le \epsilon|p|^{\gamma\alpha},$$

by changing the variable of integration $y = pt^{1/\alpha}$ we easily obtain for the second integral

$$|J_2| \leq C \int_{-i}^{i} e^{-\Theta|p|^\alpha t} |p|^{\gamma \alpha} |dp| = O\left(\epsilon t^{-\frac{1}{\alpha}-\gamma}\right).$$

Finally since $\|\hat{u}_0\|_{\mathbf{L}^\infty} \leq \epsilon$ we have

$$|J_3| \leq C e^{-\frac{\Theta t}{2}} \int_{|p| \geq 1, \operatorname{Re} p = 0} e^{-\frac{\Theta|p|^\alpha t}{2}} |\hat{u}_0(p)| |dp| = O\left(\epsilon t^{-\frac{1}{\alpha}-\gamma}\right).$$

Lemma 14 is proved.   □

LEMMA 15. *Let the functions $v_j(t)$, $j = 1, ..., m$ have asymptotics (5.14) for $t \to +\infty$. Then the estimate*

$$\left\| \int_{-i\infty}^{i\infty} e^{p \cdot} dp \int_0^t e^{-K(p)(t-\tau)} K(p) \frac{v_j(\tau)}{p^{N+j}} d\tau \right\|_{\mathbf{L}^2} = O\left(\left(\epsilon + \epsilon_1^{p+1}\right) t^{-\frac{1}{2\alpha}}\right)$$

is valid for $t > 1$, and the following asymptotics

$$\int_{-i\infty}^{i\infty} e^{px} dp \int_0^t e^{-K(p)(t-\tau)} K(p) \frac{v_j(\tau)}{p^{N+j}} d\tau$$

$$= E B_j t^{-\frac{1}{\alpha}} G_j(x/t^{\frac{1}{\alpha}}) + O\left(\left(\epsilon + \epsilon_1^{p+1}\right) t^{-\frac{1}{\alpha}-\gamma}\right)$$

is true for $t \to +\infty$ uniformly with respect to $x > 0$, where

$$G_j(q) = \int_{-i\infty}^{+i\infty} dy e^{yq} y^{\alpha - N - j} dy \int_0^1 e^{-Ey^\alpha(1-z)} z^{-\frac{N+j}{\alpha}} dz$$

for $j = 1, ..., m$ in the case α is non integer and $j = 1, ..., m - 1$ if α is integer. In the case of integer α we denote

$$G_m(q) = G_0(q) \int_0^{+\infty} v_m(\tau) d\tau, \, G_0(q) = \int_{-i\infty}^{i\infty} e^{qy - Ey^\alpha} dy,$$

and v_m is the boundary value of the last derivative of the solution defined by (5.11).

PROOF. By virtue of (5.10) and Lemma 6 we have

$$v_j = O\left(\left(\epsilon + \epsilon_1^{p+1}\right) t^{-\frac{j-1}{\alpha}}(1+t)^{-\frac{N+1}{\alpha}}\right)$$

for $j = 1, ..., m$. Then by changing the variable of integration $y = p(t - \tau)^{1/\alpha}$ we get

$$\left\| \int_{-i\infty}^{i\infty} e^{p \cdot} dp \int_0^t e^{-K(p)(t-\tau)} K(p) \frac{v_j(\tau)}{p^{N+j}} d\tau \right\|_{\mathbf{L}^2}$$

$$\leq C\left(\epsilon + \epsilon_1^{p+1}\right) \left(\int_0^1 (t - \tau)^{-\frac{1}{2\alpha}-1+\frac{N+j}{\alpha}} \tau^{-\frac{j-1}{\alpha}} d\tau \right.$$

$$\left. + \int_1^t (t - \tau)^{-\frac{1}{2\alpha}-1+\frac{N+j}{\alpha}} \tau^{-\frac{N+j}{\alpha}} d\tau \right)$$

$$= O\left(\left(\epsilon + \epsilon_1^{p+1}\right) t^{-\frac{1}{2\alpha}}\right).$$

By changing the variables of integration $\tau = tz$ and $y = pt^{1/\alpha}$ we get

$$I = E \int_{-i\infty}^{+i\infty} e^{px} p^{\alpha-N-j} dp \int_0^t e^{-K(p)(t-\tau)} v_j(\tau) d\tau$$

$$(5.24) \qquad = E t^{\frac{N-1+j}{\alpha}} \int_{-i\infty}^{+i\infty} e^{yxt^{-\frac{1}{\alpha}}} y^{\alpha-N-j} dy \int_0^1 e^{-Ey^\alpha(1-z)} v_j(tz) dz$$

for $j = 1, 2..., m$, if $\alpha \neq [\alpha]$ and for $j = 1, ..., m-1$, if $\alpha = [\alpha]$. Putting (5.10) into (5.24) yields

$$(5.25) \qquad I = EB_j t^{-\frac{1}{\alpha}} \int_{-i\infty}^{+i\infty} e^{yxt^{-\frac{1}{\alpha}}} y^{\alpha-N-j} dy \int_0^1 e^{-Ey^\alpha(1-z)} z^{-\frac{N+j}{\alpha}} dz + R,$$

where

$$R = C t^{\frac{N-1+j}{\alpha}} \int_{-i\infty}^{+i\infty} e^{yxt^{-\frac{1}{\alpha}}} y^{\alpha-N-j} dy \int_0^1 e^{-Ey^\alpha(1-z)} O\left(\frac{\epsilon + \epsilon_1^{p+1}}{(tz)^{\frac{N+j}{\alpha}+\gamma}}\right) dz.$$

Similarly to the proof of Lemma 5 we see that

$$R = O\left(\frac{\epsilon + \epsilon_1^{p+1}}{t^{\frac{1}{\alpha}+\gamma}}\right).$$

In the case $\alpha = [\alpha]$ we have

$$I = E \int_{-i\infty}^{i\infty} dp\, e^{px} \int_0^t e^{-Ep^\alpha(t-\tau)} v_m(\tau) d\tau.$$

By virtue of Theorem 14 and (5.12) the following estimate

$$|v_m(t)| \leq C(\epsilon + \epsilon_1^{p+1}) t^{-\frac{m-1}{\alpha}} (1+t)^{-1+\frac{m-1}{\alpha}-\gamma}$$

is true. Then by interchanging the order of integration and by changing the variable of integration $y = pt^{1/\alpha}$ we get

$$I = E t^{-\frac{1}{\alpha}} \left(F(q,0) \int_0^{+\infty} v_m(\tau) d\tau + R\right),$$

where

$$R = \int_0^{t/2} v_m(\tau)(F(q,z) - F(q,0)) d\tau - F(q,0) \int_{t/2}^{+\infty} v_m(\tau) d\tau$$

$$+ \int_{t/2}^t v_m(\tau) F(q,z) d\tau$$

and

$$F(q,z) = \int_{-i\infty}^{i\infty} e^{yq-Ey^\alpha(1-z)} dy, q = xt^{-\frac{1}{\alpha}}, z = \tau t^{-1}.$$

We have

$$|F(q,z)| \leq \frac{C}{(1-z)^{\frac{1}{\alpha}}}.$$

and

$$|F_z(q,z)| = |E\int_{-i\infty}^{i\infty} e^{yq-Ey^\alpha(1-z)}y^\alpha dy| \le \frac{C}{(1-z)^{1+\frac{1}{\alpha}}}.$$

Therefore we obtain $R = O\left((\epsilon + \epsilon_1^\rho)t^{-\gamma}\right)$. Lemma 15 is proved. $\qquad\square$

LEMMA 16. *Let estimate (5.1) be true. Then the estimate*

$$\left\|\int_{-i\infty}^{i\infty} e^{p\cdot} dp \int_0^t e^{-K(p)(t-\tau)}\widehat{N(u)}(p,\tau)d\tau\right\|_{\mathbf{L}^2} = O\left(\epsilon_1^{\rho+1}t^{-\frac{1}{2\alpha}}\right).$$

is valid for $t > 1$ and the following asymptotics

$$\int_{-i\infty}^{+i\infty} e^{px} dp \int_0^t e^{-K(p)(t-\tau)}\widehat{N(u)}(p,\tau)d\tau$$

$$(5.26) \qquad = t^{-\frac{1}{\alpha}}G_0(xt^{-\frac{1}{\alpha}})\int_0^{+\infty}\widehat{N(u)}(0,\tau)d\tau + O\left(\epsilon_1^\rho t^{-\frac{1}{\alpha}-\gamma}\right)$$

is true as $t \to +\infty$ uniformly with respect to $x > 0$, where

$$\widehat{N(u)}(0,t) = \int_0^{+\infty} N(u)dx,$$

and

$$G_0(q) = \int_{-i\infty}^{+i\infty} e^{yq-Ey^\alpha} dy.$$

PROOF. By virtue of (5.1) we have

$$\|\widehat{N(u)}(t)\|_{\mathbf{L}^\infty} \le C\epsilon_1^{\rho+1}t^{-\gamma}(1+t)^{-\frac{\rho}{\alpha}-\eta+\gamma}.$$

Therefore since $\frac{\rho}{\alpha} + \eta > 1$ we get

$$\left\|\int_{-i\infty}^{+i\infty} e^{p\cdot} dp \int_0^t e^{-K(p)(t-\tau)}\widehat{N(u)}(p,\tau)d\tau\right\|_{\mathbf{L}^2}$$

$$\le \int_0^t \|\widehat{N(u)}(t)\|_{\mathbf{L}^\infty(\text{Re }p=0)}\left(\int_{-i\infty}^{+i\infty} e^{-2\Theta|p|^\alpha(t-\tau)}|dp|\right)^{\frac{1}{2}} d\tau$$

$$\le C\epsilon_1^{\rho+1}\int_0^t \frac{\tau^{-\gamma}d\tau}{(1+\tau)^{\frac{\rho}{\alpha}+\eta-\gamma}(t-\tau)^{\frac{1}{2\alpha}}} = O\left(\epsilon_1^{\rho+1}t^{-\frac{1}{2\alpha}}\right).$$

We write the representation

$$\int_{-i\infty}^{i\infty} e^{px} dp \int_0^t e^{-K(p)(t-\tau)}\widehat{N(u)}(p,\tau)d\tau$$

$$(5.27) \qquad = \int_{-i}^{i} e^{px-K(p)t}dp \int_0^{\frac{t}{2}}\widehat{N(u)}(0,\tau)d\tau + R,$$

where

$$
R = \int_{-i}^{i} e^{px} dp \int_{0}^{\frac{t}{2}} (e^{-K(p)(t-\tau)} - e^{-K(p)t}) \widehat{N(u)}(0,\tau) d\tau
$$

$$
+ \int_{-i}^{i} e^{px} dp \int_{0}^{\frac{t}{2}} e^{-K(p)(t-\tau)} (\widehat{N(u)}(p,\tau) - \widehat{N(u)}(0,\tau)) d\tau
$$

$$
+ \int_{-i}^{i} e^{px} dp \int_{\frac{t}{2}}^{t} e^{-K(p)(t-\tau)} \widehat{N(u)}(p,\tau) d\tau
$$

$$
+ \int_{|p|>1,\mathrm{Re}\,p=0} e^{px} dp \int_{0}^{t} e^{-K(p)(t-\tau)} \widehat{N(u)}(p,\tau) d\tau
$$

$$
= \sum_{j=1}^{4} J_j.
$$

By changing the variable of integration $y = pt^{1/\alpha}$ in the first summand of representation (5.27) we get

$$
\int_{-i}^{i} e^{px - K(p)t} dp \int_{0}^{t} \widehat{N(u)}(0,\tau) d\tau
$$

$$
(5.28) \qquad = t^{-\frac{1}{\alpha}} G_0(x/t^{\frac{1}{\alpha}}) \int_{0}^{+\infty} \widehat{N(u)}(0,\tau) d\tau + O\left(\epsilon_1^{\rho+1} t^{-\frac{1}{\alpha}-\gamma}\right).
$$

We have

$$
\left| e^{-Ep^{\alpha}(t-\tau)} - e^{-Ep^{\alpha}t} \right| \leq C e^{-\Theta|p|^{\alpha}(t-\tau)} |p|^{\alpha\gamma} \tau^{\gamma}
$$

for all $p \in [-i, i]$, since $\Theta = \min(\mathrm{Re}\,E(+i)^{\alpha}, \mathrm{Re}\,E(-i)^{\alpha}) > 0$. Therefore by changing the variable of integration $y = p(t-\tau)^{1/\alpha}$ we get via (5.1)

$$
J_1 \leq C \int_{0}^{\frac{t}{2}} |\widehat{N(u)}(0,\tau)| d\tau \int_{-i}^{i} e^{-\Theta|p|^{\alpha}(t-\tau)} |p|^{\alpha\gamma} \tau^{\gamma} |dp|
$$

$$
(5.29) \qquad \leq C \epsilon_1^{\rho+1} \int_{0}^{\frac{t}{2}} \frac{d\tau}{(1+\tau)^{\frac{\rho}{\alpha}+\eta-\gamma}(t-\tau)^{\frac{1}{\alpha}+\gamma}} = O\left(\epsilon_1^{\rho+1} t^{-\frac{1}{\alpha}-\gamma}\right).
$$

By the estimate of Lemma 12

$$
\|(\cdot)^{\delta} u(t)\|_{\mathbf{L}^2} \leq C
$$

for $0 < \delta \leq \frac{1}{2}$ we get

$$|\widehat{N(u)}(p,\tau) - \widehat{N(u)}(0,\tau)|$$

$$= \left| \int_0^{+\infty} (e^{-px} - 1)\mathbb{N}(u)(x,\tau)dx \right|$$

$$\leq \sqrt{|p|} \int_0^{+\infty} \sqrt{x}|\mathbb{N}(u)(x,\tau)|dx$$

$$\leq Ct^{-\eta}\sqrt{|p|}\|u(t)\|_{\mathbf{L}^\infty}^{\rho-1}\|u(t)\|_{\mathbf{L}^2}\|\sqrt{\cdot}u(\cdot,t)\|_{\mathbf{L}^2}$$

$$\leq C\epsilon_1^{\rho+1}\sqrt{|p|}(1+t)^{-\frac{2\rho-1}{2\alpha}-\eta}$$

for all $p \in [-i,i]$. By changing the variable of integration $y = p(t-\tau)^{1/\alpha}$ we obtain

$$(5.30) \quad J_2 \leq C\epsilon_1^{\rho+1} \int_0^{\frac{t}{2}} \frac{\tau^{-\gamma}d\tau}{(1+\tau)^{\frac{2\rho-1}{2\alpha}+\eta-\gamma}(t-\tau)^{\frac{1}{\alpha}+\frac{1}{2\alpha}}} \int_0^{+\infty} e^{-\Theta y^\alpha}\sqrt{y}dy$$

$$= O\left(\frac{\epsilon_1^{\rho+1}}{t^{\frac{1}{\alpha}+\gamma}}\right).$$

Via (5.1) we get

$$\|\widehat{N(u)}(t)\|_{\mathbf{L}^\infty(\mathrm{Re}\,p=0)} \leq \epsilon_1^{\rho+1}|a(t)|\|u(t)\|_{\mathbf{L}^\infty}^{\rho-1}\|u(t)\|_{\mathbf{L}^2}^2 \leq C\epsilon_1^{\rho+1}(1+t)^{-\frac{\rho}{\alpha}-\eta},$$

whence by changing the variable of integration $y = p(t-\tau)^{1/\alpha}$ we have

$$|J_3| \leq \int_{\frac{t}{2}}^t \|\widehat{N(u)}(\tau)\|_{\mathbf{L}^\infty(\mathrm{Re}\,p=0)}d\tau \int_{-i}^i e^{-\Theta|p|^\alpha(t-\tau)}|dp|$$

$$(5.31) \qquad \leq C\epsilon_1^{\rho+1} \int_{\frac{t}{2}}^t \frac{d\tau}{(1+\tau)^{\frac{\rho}{\alpha}+\eta}(t-\tau)^{\frac{1}{\alpha}}} \int_0^{+\infty} e^{-\Theta y^\alpha}dy$$

$$= O\left(\epsilon_1^{\rho+1}t^{-\frac{1}{\alpha}-\gamma}\right).$$

For the last integral J_4 we easily obtain

$$|J_4| \leq \int_0^t e^{-\frac{\Theta(t-\tau)}{2}}\|\widehat{N(u)}(\tau)\|_{\mathbf{L}^\infty(\mathrm{Re}\,p=0)}d\tau \int_{|p|\geq 1,\mathrm{Re}\,p=0} e^{-\Theta|p|^\alpha(t-\tau)/2}|dp|$$

$$(5.32) \quad \leq \int_0^t \frac{e^{-\frac{\Theta(t-\tau)}{2}}}{(1+\tau)^{\frac{\rho}{\alpha}+\eta}(t-\tau)^{\frac{1}{\alpha}}}d\tau \int_0^{+\infty} e^{-\Theta y^\alpha/2}dy$$

$$= O\left(\epsilon_1^{\rho+1}t^{-\frac{1}{\alpha}-\gamma}\right).$$

From estimates (5.28) through (5.32) the result of Lemma 16 follows. $\qquad \square$

Proof of Theorem 16. Let us prove the following estimate

$$(5.33) \qquad (1+t)^{\frac{1}{\alpha}}\left((1+t)^{-\frac{1}{2\alpha}}\|u(t)\|_{\mathbf{L}^2} + t^\gamma(1+t)^{-\gamma}\|u(t)\|_{\mathbf{L}^\infty}\right) < \epsilon_1$$

for all $t > 0$. On the contrary we suppose that estimate (5.33) is broken for some time. By Theorem 14 the left-hand side of (5.33) is continuous. Therefore there exists a maximal time $T > 1$ such that the nonstrict estimate (5.33) is valid for all $t \in [0, T]$. Thus the supposition (5.1) is valid for the time interval $[0, T]$, and we can apply Lemmas 14 through 16 to the representation (3.2) of the solution. Hence we get estimate (5.33) for all $[0, T]$. The contradiction obtained proves estimate (5.33) for all $t > 0$. Moreover by virtue of Remark 4, Lemmas 14 through 16 and Lemma 9 with $\chi = \frac{1}{\alpha}$ we see that the solution $u(x, t) \in \mathbf{C}([0, +\infty); \mathbf{L}^2(\mathbf{R}^+))$ $\cap \mathbf{C}\left(\mathbf{R}^+; \mathbf{H}_2^{[\alpha]-1}(\mathbf{R}^+) \cap \mathbf{C}^{[\alpha]-1}(\mathbf{R}^+)\right)$ has asymptotics (1.4) with the coefficient

$$B_0 = \frac{1}{2\pi i}\left(\widehat{u}_0(0) + \int_0^{+\infty} d\tau \int_0^{+\infty} \mathbb{N}(u)(x, \tau)dx + R\right),$$

where $R = \int_0^{+\infty} v_m(\tau)d\tau$ in the case of integer α and $R = 0$ otherwise α. (Note that v_m is the boundary value of the derivative of the solution of order $N + m$, it is defined by (5.11)). The coefficients $B_j = \frac{A_j}{2\pi i}$, for $j = 1, ..., N$ and

$$B_j = \Gamma\left(1 - \frac{N + j}{\alpha}\right) \sum_{l=1}^{m} \frac{C_1 r_l^{[\alpha]} M_{j,l}}{2\pi i}$$

$$\times \left(\widehat{u}_0(0) - \int_0^{+\infty} d\tau \int_0^{+\infty} \mathbb{N}(u)(x, \tau)dx\right)$$

for $j = N + 1, ..., N + m$. Here $C_1 = (\det \tilde{W})^{-1}$. $M_{j,l}$ are the algebraic minors of matrix $\tilde{W}$ (see (2.6)). Integrals B_j converge in view of estimate (5.33) (see estimate (5.21) of Lemma 13 for details). Theorem 16 is proved.

CHAPTER 5

Whitham Equation

1. Introduction

We study the following initial-boundary value problem for the nonlinear non-local Whitham equation

(1.1)
$$\begin{cases} u_t + u u_x + \mathbb{K}u = 0, & (t,x) \in \mathbf{R}^+ \times \mathbf{R}^+, \\ u(x,0) = u_0(x), & x \in \mathbf{R}^+, \\ \partial_x^{j-1} u(0,t) = h_j(t), & \text{for} \quad j = 1, 2, ..., N, \end{cases}$$

where the number $N = \left[\frac{\alpha}{2}\right]$ of the necessary boundary data depends essentially on the order α of the pseudodifferential operator $\mathbb{K}$ on a half-line. For Mathematics and Physics, the initial-boundary value problem (1.1) holds a great interest since it describes many physical phenomena, for example, the focusing of the laser beams, waves on water and others. Equation (1.1) involves local and nonlocal nonlinear partial differential equations, such as, Korteweg-de Vries-Burgers equation and Ott-Sudan-Ostrovskiy equation (see [**103**]). In Chapter 4 the nonlinear nonlocal Schrödinger equation on a half-line was studied. Here we are interested in the case of the nonlinearity of shallow water type: $u u_x$. This nonlinearity contains the derivative of the unknown function and so represents the so-called derivative loss. Therefore the methods of Chapter 4 do not work directly. We must use the smoothing properties of the strongly dissipative operator $\mathbb{K}$. Also we adopt here a more direct approach based on the estimates of the Green function.

In this chapter we consider the dissipative case in which for simplicity the symbol of the operator $\mathbb{K}$ has the following form $K(p) = -p^\alpha$ and $\alpha \in (1,2)$. (We denote by p^α the main branch of the complex analytic function so that $1^\alpha = 1$. We cut along the negative real axis $(+\infty, 0)$ in the complex plane of variable p).

As we showed in the case $\alpha \in (1,2)$ we do not need to take any boundary data into consideration for problem (1.1); therefore we study the following problem

(1.2)
$$\begin{cases} u_t + u u_x + \mathbb{K}u = 0, & (x,t) \in \mathbf{R}^+ \times \mathbf{R}^+, \\ u(x,0) = u_0(x), & x \in \mathbf{R}^+, \end{cases}$$

where

$$\mathbb{K}u = \frac{1}{2\pi i} \int_{-i\infty}^{+i\infty} e^{px} K(p) \left(\hat{u}(p,t) - \frac{u(0,t)}{p} \right) dp.$$

The boundary data of (1.2) are determined by the initial data and the solution of (1.2) (see Chapter 3).

The Cauchy problem for the nonlocal nonlinear Schrödinger equation was studied in [**70**] and the existence, uniqueness and property of solutions in the case of strongly dissipative pseudodifferential operator $\mathbb{K}$. Large time asymptotics of solutions to the Cauchy problem for the nonlinear nonlocal Schrödinger equation with a source was found in paper [**76**]. As far as we know the initial-boundary value problem for the nonlocal Whitham equation (1.2) on a half-line was not considered previously. In the case $\mathbb{K}u = -\Delta u$, which corresponds to the dissipation of the heat equation, the initial-boundary value problem was studied extensively (see, for example, [**3**], [**4**]).

To state the results of the present chapter precisely we give some notations. Let us denote

$$\mathbf{X} = \{\varphi \in \mathbf{L}^1\left(\mathbf{R}^+\right), \varphi_x \in \mathbf{L}^1\left(\mathbf{R}^+\right); \|\varphi\|_{\mathbf{X}} = \|\varphi\|_{\mathbf{L}^1} + \|\varphi_x\|_{\mathbf{L}^1} < +\infty\}$$

and

$$\mathbf{H}_2^s\left(\mathbf{R}^+\right) = \{f \in \mathbf{L}^2\left(\mathbf{R}^+\right); \|f\|_{\mathbf{H}_2^s} = \|(1 - \partial_x^2)^{s/2} f\|_{\mathbf{L}^2} < +\infty\}$$

is the Sobolev space. We also introduce the following function space

$$\mathbf{Z}_{T,\gamma} = \{\phi(x,t) \in \mathbf{C}([0,T]; \mathbf{L}^2\left(\mathbf{R}^+\right)) \cap \mathbf{C}((0,T]; \mathbf{H}_2^1\left(\mathbf{R}^+\right)); \|\phi\|_{\mathbf{Z}_{T,\gamma}} < +\infty\}$$

with the norm

$$\|\phi\|_{\mathbf{Z}_{T,\gamma}} = \sup_{t\in[0,T]} \left(\|\phi\left(t\right)\|_{\mathbf{L}^2} + t^\gamma (\|\phi\left(t\right)\|_{\mathbf{L}^\infty} + t^{\frac{1}{2\alpha}} \|\phi_x\left(t\right)\|_{\mathbf{L}^2})\right).$$

Since the symbol of the operator $\mathbb{K}$ has the form $K(p) = -p^\alpha$, where $\alpha \in (1,2)$ we easily see that there exists a positive constant C_α such that $\operatorname{Re} K(p) > C_\alpha |p|^\alpha$ for all p on the imaginary axis $\operatorname{Re} p = 0$ since $\operatorname{Re} K(p) = -|p|^\alpha \cos \frac{\alpha\pi}{2}$. The inequality $\operatorname{Re} K(p) > C_\alpha |p|^\alpha$ means that the equation satisfies the dissipation condition. By the same letter C we denote different positive constants.

We consider the generalized solutions of the initial-boundary value (1.2), that is we multiply equation (1.2) by any function $\phi \in \mathbf{C}^2([0,T] \times (0,+\infty))$ such that $\phi(x,T) = 0$ and $\phi(0,t) = 0$ and integrate by parts in the domain $[0,T] \times (0,+\infty)$. Then the linear operator $\mathbb{K}$ makes sense since we can represent it in the following form

$$\begin{aligned}
\mathbb{K}u &= \partial_x^3 \frac{1}{2\pi i} \int_{\operatorname{Re} p=0, |p|>1} e^{px} \frac{K(p)}{p^3} \left(\hat{u}(p,t) - \frac{u(0,t)}{p}\right) dp \\
&+ \frac{1}{2\pi i} \int_{\operatorname{Re} p=0, |p|\leq 1} e^{px} K(p) \left(\hat{u}(p,t) - \frac{u(0,t)}{p}\right) dp.
\end{aligned}$$

Hence we see that the integrals are convergent uniformly with respect to $x > 0$ since $\frac{K(p)}{p^3} \in \mathbf{L}^2((-i\infty, -i] \cup [i, i\infty))$ and $\left(\hat{u}(p,t) - \frac{u(0,t)}{p}\right) \in \mathbf{C}(\mathbf{R}^+; \mathbf{L}^2\left(\mathbf{R}^+\right))$ for the solution $u(x,t) \in \mathbf{C}(\mathbf{R}^+, \mathbf{L}^2\left(\mathbf{R}^+\right))$.

Now we state the results of this chapter (see paper [**53**]). Firstly we formulate the local existence of the solutions of the initial-boundary value problem (1.2).

THEOREM 18. *Let $\alpha \in (1,2)$. Suppose that the initial data $u_0(x) \in \mathbf{X}$. Then there exists a unique solution $u(x,t) \in \mathbf{Z}_{T,\gamma}$ for some $T > 0$ and $\gamma > 0$ is small enough.*

REMARK 6. *If the initial data u_0 are small, i.e. the norm $\|u_0\|_{\mathbf{X}} \le \epsilon$, where $\epsilon > 0$ is sufficiently small, then there exists $T > 1$, such that the solution is also sufficiently small: $\sup_{t \le T}(\|u(t)\|_{\mathbf{L}^2} + t^\gamma\|u(t)\|_{\mathbf{L}^\infty} + t^{\frac{1}{2\alpha}+\gamma}\|u_x(t)\|_{\mathbf{L}^2}) < C\epsilon$.*

Now we give some sufficient conditions for global existence of solutions.

THEOREM 19. *Let $\alpha \in (1,2)$ and $\gamma \in (0, \frac{2-\alpha}{2\alpha}]$. Suppose that the initial data satisfy $\|u_0\|_{\mathbf{X}} \le \epsilon_1$ and $\|(\cdot)^{\gamma\alpha}u_0(\cdot)\|_{\mathbf{L}^1} \le \epsilon_1$, where $\epsilon_1 > 0$ is small enough. Then there exists a unique solution u of (1.2) such that $u \in \mathbf{C}([0,+\infty); \mathbf{L}^2(\mathbf{R}^+)) \cap \mathbf{C}(\mathbf{R}^+; \mathbf{H}_2^1(\mathbf{R}^+))$ and*

$$\sup_{t \ge 0}(1+t)^{\frac{1}{2\alpha}-\gamma}\left(\|u(t)\|_{\mathbf{L}^2} + t^\gamma(1+t)^{\frac{1}{2\alpha}}\|u(t)\|_{\mathbf{L}^\infty}\right.$$
$$\left. + t^{\frac{1}{2\alpha}+\gamma}(1+t)^{\frac{1}{2\alpha}-\gamma}\|u_x(t)\|_{\mathbf{L}^2}\right)$$
$$< +\infty.$$

Furthermore the solution u has the following asymptotics for large time

$$u(x,t) = at^{-\frac{1}{\alpha}}\Lambda(x/t^{\frac{1}{\alpha}}) + O\left(t^{-\frac{1}{\alpha}-\gamma}\right),$$

where

$$\Lambda(s) = \frac{1}{2\pi i}\int_{-i\infty}^{+i\infty} e^{zs+z^\alpha}dz$$
$$+ \frac{1}{4\pi^2}\int_{-i\infty}^{+i\infty}\frac{dz}{z^{1-\alpha}}e^{zs}\int_{-i\infty}^{+i\infty}e^q q^{\frac{1}{\alpha}-1}\frac{dq}{z^\alpha - q},$$
$$a = \int_0^{+\infty} u_0(x)dx + \int_0^{+\infty}d\tau\int_0^{+\infty} uu_x dx.$$

We organize this chapter as follows. In Section 2 we consider the linear initial-boundary value problem (1.1) with pseudodifferential operator $\mathbb{K}$. In Section 3 we prove some preliminary estimates in Lemmas 17 through 19. Section 4 is devoted to the proof of Theorem 18. We prove Theorem 19 in Section 5.

2. Linear problem

Consider the following linear initial-boundary value problem

(2.1)
$$\begin{cases} u_t + \mathbb{K}u = f(x,t), & (t,x) > 0, \\ u(x,t)|_{t=0} = u_0(x), & x \in \mathbf{R}^+. \end{cases}$$

Here $\mathbb{K}u$ is the pseudodifferential operator defined as follows

$$\mathbb{K}u = \frac{1}{2\pi i}\int_{-i\infty}^{+i\infty} e^{px}K(p)\left(\hat{u}(p,t) - \frac{u(0,t)}{p}\right)dp,$$

where $K(p) = -p^\alpha$, $\alpha \in (1,2)$.

We have the following result from the result of Chapter 4 (see Theorem 17).

THEOREM 20. *Let $\alpha \in (1,2)$. Then for some $T > 0$ there exists a unique solution $u \in \mathbf{Z}_{T,\gamma}$ of the initial-boundary value problem (2.1) such that*

$$\|u\|_{\mathbf{Z}_{T,\gamma}} \leq C\lambda$$

provided that $\lambda < +\infty$, where $\lambda = \|u_0\|_{\mathbf{X}} + T^\mu \sup_{t \in [0,t]} t^\gamma \|f(t)\|_{\mathbf{L}^1}$.

Moreover from Theorem 13 if $u_0 \in \mathbf{L}^1(\mathbf{R}^+)$, $f \in \mathbf{L}^2(0,T;\mathbf{L}^1(\mathbf{R}^+))$, the solution of the problem (2.1) has the following form

$$(2.2) \qquad u(x,t) = \mathcal{F}[u_0](x,t) + \int_0^t \mathcal{F}[f(\tau)](x,t-\tau)\, d\tau,$$

where

$$\mathcal{F}[\phi](x,t) = \int_0^{+\infty} (F(x,y,t) + G(x-y,t))\,\phi(y)\,dy,$$

$$G(x,t) = \frac{1}{2\pi i}\int_{-i\infty}^{+i\infty} e^{-K(p)t+px}\,dp,$$

$$F(x,y,t) = \frac{1}{4\pi^2}\int_{-i\infty}^{+i\infty} dp\, e^{px} p^{\alpha-1}\int_{-i\infty}^{+i\infty} e^{\xi t - \xi^{\frac{1}{\alpha}}y}\xi^{\frac{1}{\alpha}-1}(p^\alpha - \xi)^{-1}d\xi.$$

Indeed in the general case when $K(p) = Ep^\alpha$, $\alpha > 1$ is not an odd integer (see example 1 of Chapter 3) we have

$$u(x,t) = \mathcal{F}[u_0](x,t) + \int_0^t \mathcal{F}[f(\tau)](x,t-\tau)\, d\tau,$$

where $F(x,y,t)$ is defined by

$$F(x,y,t) = \frac{1}{2\pi i}\int_{-i\infty}^{+i\infty} e^{-K(p)t+p(x-y)}\,dp$$

$$+ \frac{1}{4\pi^2}\sum_{j=1}^m \int_{-i\infty}^{+i\infty} e^{\xi t}\frac{\theta_j(\xi)}{\xi}\int_{-i\infty}^{+i\infty}\frac{e^{px}K(p)}{K(p)+\xi}p^{-N-j}d\xi dp,$$

where the functions $\theta_j(\xi)$ are equal to

$$\theta_j(\xi) = \left(\tilde{A}^{-1}\begin{pmatrix} e^{-\phi_1(\xi)y} \\ \cdots \\ e^{-\phi_m(\xi)y}\end{pmatrix}\right)_j.$$

Here $\tilde{A} = \left\|\phi_i^{-N-j}\right\|_m^m$, $N = \left[\frac{\alpha}{2}\right]$, $m = \left[\frac{\alpha+1}{2}\right]$, $\phi_i(\xi)$ are m different positive inverse functions defined as $\phi_i(\xi) = K^{-1}(-\xi)$, such that

$$\operatorname{Re}\phi_j(\xi) > 0, \quad j = 1,2,...,m$$

for $\operatorname{Re}\xi > 0$ (see Chapter 2).

Clearly in our case $N = 0, m = 1$ we have only one positive inverse function $\phi_1(\xi) = \xi^{\frac{1}{\alpha}}$ and so we get

$$\theta_1(\xi) = \phi_1(\xi)e^{-\phi_1(\xi)y} = \xi^{-\frac{1}{\alpha}}e^{-\xi^{\frac{1}{\alpha}}y}.$$

3. Preliminaries

Consider the following functions for $\alpha \in (1, 2)$, $x, y, t \geq 0$

(3.1) $$G(x, t) = \frac{1}{2\pi i}\int_{-i\infty}^{+i\infty} e^{px+p^\alpha t}dp$$

and

$$F(x, y, t) = \frac{1}{4\pi^2}\int_{-i\infty}^{+i\infty} dp e^{px} p^{\alpha-1}\int_{-i\infty}^{+i\infty} e^{\xi t - \xi^{\frac{1}{\alpha}}y}\xi^{\frac{1}{\alpha}-1}(p^\alpha - \xi)^{-1}d\xi.$$

We prove some estimates for the function $F(x, y, t)$ and $G(x, t)$.

LEMMA 17. *We have for* $\mu_1 \in [0, \frac{1}{\alpha})$, $\delta \in [0, \frac{1}{2})$ *and* $\gamma > 0$

$$\sup_{t,y\geq 0} y^{\mu_1 \alpha} t^{\frac{1}{2\alpha}+\gamma-\mu_1}(1+t)^{-2\gamma}\left(\|F(\cdot, y, t)\|_{\mathbf{L}^2} + t^{\frac{1}{\alpha}}\|F_x(\cdot, y, t)\|_{\mathbf{L}^2}\right) \leq C,$$

$$\sup_{t\geq 0} t^{\frac{1-2\delta}{2\alpha}+\gamma}(1+t)^{-2\gamma}\|(\cdot)^\delta F(\cdot, y, t)\|_{\mathbf{L}^2} \leq C$$

and

$$\sup_{t\geq 0} t^{\frac{1}{\alpha}-\mu_1}y^{\mu_1\alpha}\|F(\cdot, y, t)\|_{\mathbf{L}^\infty} \leq C.$$

PROOF. We have

$$\frac{z^{\alpha-1}}{z^\alpha - q} = \frac{1}{z+1} + \frac{q}{(z+1)(z^\alpha - q)} + \frac{z^{\alpha-1}}{(z+1)(z^\alpha - q)}$$

and $\int_{-i\infty}^{i\infty} e^{zx}(z+1)^{-1}dz = 2\pi i e^{-x}$. Then changing the variables $p^\alpha t = z^\alpha$ and $\xi t = q$ we get

$$4\pi^2 F(x, y, t)$$

$$= t^{-\frac{1}{\alpha}}\left(2\pi i e^{-\tilde{x}}\int_{-i\infty}^{+i\infty} e^{q-q^{\frac{1}{\alpha}}\tilde{y}}q^{\frac{1}{\alpha}-1}dq\right.$$

$$+ \int_{-i\infty}^{+i\infty} dz e^{z\tilde{x}}(z+1)^{-1}\int_{-i\infty}^{+i\infty} e^{q-q^{\frac{1}{\alpha}}\tilde{y}}q^{\frac{1}{\alpha}}(z^\alpha - q)^{-1}dq$$

(3.2) $$+ \left.\int_{-i\infty}^{+i\infty} dz e^{z\tilde{x}}z^{\alpha-1}(z+1)^{-1}\int_{-i\infty}^{+i\infty} e^{q-q^{\frac{1}{\alpha}}\tilde{y}}q^{\frac{1}{\alpha}-1}(z^\alpha - q)^{-1}dq\right),$$

where $\tilde{x} = xt^{-\frac{1}{\alpha}}$, $\tilde{y} = yt^{-\frac{1}{\alpha}}$. Obviously we can change the contour of integration into $\mathcal{C}_1 = \{z = \rho e^{\pm i\phi_1}, \rho \geq 0, \phi_1 = \frac{\pi}{2} + \epsilon_1\}$ and $\mathcal{C}_2 = \{q = \rho e^{\pm i\phi_2}, \rho \geq 0, \phi_2 = $

$\frac{\pi}{2} + \epsilon_2\}$, where $\epsilon_1 \in (0, \frac{\pi(3-\alpha)}{2})$ and $\epsilon_2 \in (0, \frac{\pi(\alpha-1)}{2})$ are the fixed constants. Then since $\operatorname{Re} q, \operatorname{Re} z < 0$ and $\operatorname{Re} q^{\frac{1}{\alpha}} \tilde{y} > 0$ for all $z \in \mathcal{C}_1$, $q \in \mathcal{C}_2$ we have

$$(3.3) \qquad e^{\operatorname{Re} z\tilde{x}} \leq C|z|^{-\mu_2\alpha}\tilde{x}^{-\mu_2\alpha}$$

$$(3.4) \qquad e^{\operatorname{Re} q} \leq C|q|^{-\gamma}$$

and

$$(3.5) \qquad e^{-\operatorname{Re} q^{\frac{1}{\alpha}}\tilde{y}} \leq C|q|^{-\mu_1}\tilde{y}^{-\mu_1\alpha},$$

where $\mu_1, \mu_2, \gamma \geq 0$. Also it is easy to see that for all $z \in \mathcal{C}_1$, $q \in \mathcal{C}_2$ and $\nu \in [0, 1]$

$$(3.6) \qquad |z^\alpha - q|^{-1} \leq C|z|^{-\nu\alpha}|q|^{\nu-1}.$$

Therefore for $\mu_1 \in [0, \frac{1}{\alpha})$ and $\mu_2 \in [0, \frac{1}{\alpha})$, choosing $\nu \in [0, 1]$ such that $1 - \frac{1}{\alpha} < \mu_2 + \nu < \frac{1}{\alpha}$ and $\gamma > 0$ such that $\mu_1 + \gamma < \frac{1}{\alpha}$ for $|q| \leq 1$ and $\mu_1 + \gamma > \frac{1}{\alpha} + \nu + 1$ for $|q| > 1$ we have

$$|F(x, y, t)| \leq Ct^{-\frac{1}{\alpha}+\mu_1+\mu_2}y^{-\mu_1\alpha}x^{-\mu_2\alpha}\left(\int_{\mathcal{C}_2}|q|^{\frac{1}{\alpha}-1-\mu_1-\gamma}|dq|\right.$$
$$+ \int_{\mathcal{C}_1}|dz||z+1|^{-1}|z|^{-\nu\alpha-\mu_2\alpha}\int_{\mathcal{C}_2}|q|^{\frac{1}{\alpha}+\nu-1-\mu_1-\gamma}|dq|$$
$$\left. + \int_{\mathcal{C}_1}|dz||z|^{\alpha-1-\nu\alpha-\mu_2\alpha}|z+1|^{-1}\int_{\mathcal{C}_2}|q|^{\frac{1}{\alpha}-2+\nu-\mu_1-\gamma}|dq|\right)$$
$$(3.7) \qquad \leq Ct^{-\frac{1}{\alpha}+\mu_1+\mu_2}y^{-\mu_1\alpha}x^{-\mu_2\alpha}.$$

Differentiating (3.2) with respect to x and using (3.3)-(3.6) we get

$$|F_x(x, y, t)|$$
$$\leq Ct^{-\frac{2}{\alpha}+\mu_1+\mu_2}y^{-\mu_1\alpha}x^{-\mu_2\alpha}\left(\int_{\mathcal{C}_2}|q|^{\frac{1}{\alpha}-1-\mu_1-\gamma}|dq|\right.$$
$$+ \int_{\mathcal{C}_1}|dz||z+1|^{-1}|z|^{1-\nu\alpha-\mu_2\alpha}\int_{\mathcal{C}_2}|q|^{\frac{1}{\alpha}+\nu-1-\mu_1-\gamma}|dq|$$
$$(3.8) \qquad \left. + \int_{\mathcal{C}_1}|dz||z|^{\alpha-\nu\alpha-\mu_2\alpha}|z+1|^{-1}\int_{\mathcal{C}_2}|q|^{\frac{1}{\alpha}-2+\nu-\mu_1-\gamma}|dq|\right).$$

For any $\mu_1 \in [0, \frac{1}{\alpha})$ and $\mu_2 \in (0, \frac{1}{\alpha})$ we can choose $\nu \in (1 - \frac{1}{\alpha}, 1)$ and $\gamma > 0$ in such a way that the inequalities $1 < \mu_2 + \nu < 1 + \frac{1}{\alpha}$ and $\mu_1 + \gamma < \frac{1}{\alpha}$ in the domain $|q| \leq 1$ or $\mu_1 + \gamma > \frac{1}{\alpha} + \nu + 1$ in the domains $|q| > 1$ are valid. Then we obtain

$$(3.9) \qquad |F_x(x, y, t)| \leq Ct^{-\frac{2}{\alpha}+\mu_1+\mu_2}y^{-\mu_1\alpha}x^{-\mu_2\alpha}.$$

Choosing $\mu_2\alpha = 1/2 - \gamma$ in domains $|x| \leq 1$ and $\mu_2\alpha = 1/2 + \gamma$ in domains $|x| > 1$ in formulas (3.7) and (3.9) we obtain for $\gamma > 0$

$$\sup_{t, y \geq 0} y^{\mu_1\alpha}t^{\frac{1}{2\alpha}+\gamma-\mu_1}(1+t)^{-2\gamma}\left(\|F(\cdot, y, t)\|_{\mathbf{L}^2} + t^{\frac{1}{\alpha}}\|F_x(\cdot, y, t)\|_{\mathbf{L}^2}\right) \leq C.$$

By virtue of (3.7) with $\mu_1 = 0$, $\mu_2\alpha + \delta = 1/2 - \gamma$ in the domains $|x| \leq 1$ and $\mu_2\alpha + \delta = 1/2 + \gamma$ in the domains $|x| > 1$ we obtain for $\delta \in [0, \frac{1}{2})$

$$\sup_{t \geq 0} t^{\frac{1-2\delta}{2\alpha}+\gamma}(1+t)^{-2\gamma}\|(\cdot)^\delta F(\cdot, y, t)\|_{\mathbf{L}^2} \leq C.$$

Choosing in (3.7) $\mu_2 = 0$ we get

$$\sup_{t \geq 0} t^{\frac{1}{\alpha}-\mu_1} y^{\mu_1\alpha} \|F(\cdot, y, t)\|_{\mathbf{L}^\infty} \leq C.$$

Lemma 17 is now proved. □

LEMMA 18. *There exists some constant $C > 0$ such that*

$$\sup_{t \geq 0} \left(\|G(t)\|_{\mathbf{L}^1} + t^{\frac{1}{\alpha} - \frac{1}{s\alpha}} \|G(t)\|_{\mathbf{L}^s} \right) \leq C,$$

$$\sup_{t \geq 0} t^{\frac{1}{\alpha}+\gamma}(1+t)^{-2\gamma} \left(t^{\frac{1}{2\alpha}} \|G_x(t)\|_{\mathbf{L}^2} + \|G_x(t)\|_{\mathbf{L}^1} \right) \leq C$$

and

$$\sup_{t \geq 0} t^{\frac{1-2\delta}{2\alpha}+\gamma}(1+t)^{-2\gamma}\|(\cdot)^\delta G(\cdot, t)\|_{\mathbf{L}^2} \leq C$$

for $s = 2, +\infty$, $\gamma > 0$ and $\delta \in (0, \frac{1}{2})$.

PROOF. We have $\operatorname{Re} y^\alpha < 0$ for $y \in (-i\infty, i\infty)$ and $\alpha \in (1, 2)$. Therefore by changing variable $y^\alpha = p^\alpha t$ we obtain

$$\|G(t)\|_{\mathbf{L}^\infty} \leq Ct^{-\frac{1}{\alpha}} \int_0^{+\infty} e^{\operatorname{Re} y^\alpha}|dy| \leq Ct^{-\frac{1}{\alpha}}.$$

By the identity

$$e^{p^\alpha t + px} = \partial_p \left(\left(\frac{1}{x} - \frac{\alpha t}{x^2} p^{\alpha-1} \right) e^{p^\alpha t + px} \right)$$

$$+ \left(\frac{\alpha(\alpha-1)}{x^2} p^{\alpha-2} t + \frac{\alpha^2}{x^2} p^{2\alpha-2} t^2 \right) e^{p^\alpha t + px}$$

and a change of variables $y^\alpha = p^\alpha t$ we have for $x > 0$

$$|G(x, t)| = \left| x^{-2} \int_{-i\infty}^{i\infty} e^{px + p^\alpha t} \left(\alpha(\alpha-1)p^{\alpha-2}t + \alpha^2 p^{2\alpha-2}t^2 \right) dp \right|$$

$$\leq Ct^{\frac{1}{\alpha}} x^{-2} \int_0^{+\infty} e^{\operatorname{Re} y^\alpha} \left(y^{\alpha-2} + y^{2\alpha-2} \right) dy = O(t^{\frac{1}{\alpha}} x^{-2}).$$

Therefore we get

$$\|G(t)\|_{\mathbf{L}^1} \leq C \int_0^{t^{\frac{1}{\alpha}}} dx \int_0^{+\infty} e^{\operatorname{Re} y^\alpha t} dy + Ct^{\frac{1}{\alpha}} \int_{t^{\frac{1}{\alpha}}}^{+\infty} x^{-2} dx$$

$$\leq C \int_0^{+\infty} e^{\operatorname{Re} y^\alpha} dy + Ct^{\frac{1}{\alpha}} \int_{t^{\frac{1}{\alpha}}}^{+\infty} x^{-2} dx \leq C.$$

□

The Laplace transform of the function $G(x,t)$ is equal to $\widehat{G}(p,t) = e^{p^\alpha t}$. By the same change of the variables we get

$$\|G(t)\|_{\mathbf{L}^2} = \|\widehat{G}(t)\|_{\mathbf{L}^2(\operatorname{Re} p=0)} \leq Ct^{-\frac{1}{2\alpha}} \left(\int_0^{+\infty} e^{\operatorname{Re} y^\alpha} |dy| \right)^{\frac{1}{2}} \leq Ct^{-\frac{1}{2\alpha}}.$$

These two estimates imply the first estimate of the lemma. We have

$$G_x(x,t) = \int_{-i\infty}^{i\infty} e^{px + p^\alpha t} p\, dp.$$

Then by changing the variable $z = pt^{-\frac{1}{\alpha}}$ and changing the contour of integration into $\mathcal{C}_1$ such that $\operatorname{Re} z^\alpha < 0$ and $\operatorname{Re} z < 0$ and so for $\mu, \mu_1 \geq 0$ we have

$$e^{\operatorname{Re}(z^\alpha + zxt^{-\frac{1}{\alpha}})} \leq C|z|^{-\mu - \mu_1\alpha} x^{-\mu} t^{\frac{\mu}{\alpha}}.$$

Therefore we obtain

$$|G_x(x,t)| \leq t^{\frac{-2+\mu}{\alpha}} x^{-\mu} \int_{\mathcal{C}_1} |z|^{1-\mu-\mu_1\alpha} |dz| \leq Ct^{\frac{-2+\mu}{\alpha}} x^{-\mu},$$

where $\mu \in [0,2)$ and we choose μ_1 such that $\mu + \mu_1\alpha < 2$ for $|z| \leq 1$ and $\mu + \mu_1\alpha > 2$ for $|z| \geq 1$.

Choosing $\mu = 1/2 - \alpha\gamma$ in domains $|x| \leq 1$ and $\mu = 1/2 + \alpha\gamma$ in domains $|x| > 1$ we obtain

$$\sup_{t \geq 0} t^{\frac{3}{2\alpha}+\gamma}(1+t)^{-2\gamma}\|G_x(t)\|_{\mathbf{L}^2} \leq C$$

and choosing $\mu = 1 - \alpha\gamma$ in domains $|x| \leq 1$ and $\mu = 1 + \alpha\gamma$ in domains $|x| > 1$ we obtain

$$\sup_{t \geq 0} t^{\frac{1}{\alpha}+\gamma}(1+t)^{-2\gamma}\|G_x(t)\|_{\mathbf{L}^1} \leq C.$$

Therefore the second estimate of the lemma is proved. The last estimate of the lemma is proved in the same way. Lemma 18 is proved.

LEMMA 19. *We have for $\mu_1 \in (0,1]$, $\mu_2 > 0$*

$$\begin{aligned} G(x-y, t-\tau) &= t^{-\frac{1}{\alpha}}\Lambda_1(x/t^{\frac{1}{\alpha}}) + y^{\mu_1}O(t^{-\frac{1+\mu_1}{\alpha}}) \\ &\quad + \tau^{\frac{\mu_2}{\alpha}}O\left((t-\tau)^{-\frac{1+\mu_2}{\alpha}}\right) \end{aligned}$$

and

$$\begin{aligned} F(x,y,t-\tau) &= t^{-\frac{1}{\alpha}}\Lambda_2(x/t^{\frac{1}{\alpha}}) + y^{\mu_1}O(t^{-\frac{1+\mu_1}{\alpha}}) \\ &\quad + \tau^{\frac{2\mu_2}{\alpha}}O\left((t-\tau)^{-\frac{1+\mu_2}{\alpha}}\right), \end{aligned}$$

where

$$\Lambda_1(s) = \frac{1}{2\pi i}\int_{-i\infty}^{+i\infty} e^{zs + z^\alpha}\, dz$$

and

$$\Lambda_2(s) = \frac{1}{4\pi^2}\int_{-i\infty}^{i\infty} \frac{dz}{z^{1-\alpha}} e^{zs} \int_{-i\infty}^{+i\infty} e^q q^{\frac{1}{\alpha}-1}(z^\alpha - q)^{-1}\, dq.$$

PROOF. We write the representation for function $G(x, t)$ as follows

$$G(x - y, t - \tau) = G(x, t) + (G(x - y, t) - G(x, t))$$
$$+ (G(x - y, t - \tau) - G(x - y, t)).$$

By changing the variables $p^\alpha t = z^\alpha$ we easily find that

$$G(x, t) = t^{-\frac{1}{\alpha}} \Lambda_1(x/t^{\frac{1}{\alpha}}).$$

Using the estimate $|e^{-py} - 1| \le C|py|^{\mu_1}$ for $y > 0$ and $p \in (-i\infty, i\infty)$ and making the same change of the variables we get for $\mu \in (0, 1)$

$$|G(x - y, t) - G(x, t)| \le \left| \int_{-i\infty}^{+i\infty} e^{p^\alpha t + px} (1 - e^{-py}) dp \right|$$

$$\le C y^{\mu_1} t^{-\frac{1+\mu_1}{\alpha}} \int_{-i\infty}^{+i\infty} e^{\text{Re} z^\alpha} |z|^{\mu_1} |dz| \le C y^{\mu_1} t^{-\frac{1+\mu_1}{\alpha}}.$$

Since $\text{Re} \, p^\alpha < 0$ for $p \in (-i\infty, i\infty)$ we have for $\mu_2 > 0$ $|e^{p^\alpha \tau} - 1| \le |p^\alpha \tau|^{\frac{\mu_2}{\alpha}}$. Therefore by changing the variables $p^\alpha(t - \tau) = z^\alpha$ we get

$$|G(x - y, t - \tau) - G(x - y, t)|$$
$$\le \left| \int_{-i\infty}^{+i\infty} e^{p^\alpha(t-\tau) + p(x-y)} (1 - e^{p^\alpha \tau}) dp \right|$$
$$\le C(t - \tau)^{-\frac{1+\mu_2}{\alpha}} \tau^{\frac{\mu_2}{\alpha}} \left| \int_{-i\infty}^{+i\infty} e^{\text{Re} z^\alpha} |z|^{\mu_2} dz \right| \le C(t - \tau)^{-\frac{1+\mu_2}{\alpha}} \tau^{\frac{\mu_2}{\alpha}}.$$

We write the representation of the function $F(x, y, t)$

$$F(x, y, t - \tau) = F(x, 0, t) + (F(x, y, t) - F(x, 0, t))$$
$$+ (F(x, y, t - \tau) - F(x, y, t)).$$

Changing the variables $\xi t = q$ and $p^\alpha t = z^\alpha$ we get

$$F(x, 0, t) = t^{-\frac{1}{\alpha}} \Lambda_2(\frac{x}{t^{\frac{1}{\alpha}}}).$$

Differentiating the representation (3.2) of the function $F(x, y, t)$ with respect to y we get

$$F_y(x, y, t)$$
$$= -t^{-\frac{2}{\alpha}} \left(2\pi i e^{-\tilde{x}} \int_{-i\infty}^{+i\infty} e^{q - q^{\frac{1}{\alpha}} \tilde{y}} q^{\frac{2}{\alpha} - 1} dq \right.$$
$$+ \int_{-i\infty}^{+i\infty} e^{z\tilde{x}} (z + 1)^{-1} dz \int_{-i\infty}^{+i\infty} e^{q - q^{\frac{1}{\alpha}} \tilde{y}} q^{\frac{2}{\alpha}} (z^\alpha - q)^{-1} dq$$
$$+ \left. \int_{-i\infty}^{+i\infty} e^{z\tilde{x}} z^{\alpha - 1} (z + 1)^{-1} dz \int_{-i\infty}^{+i\infty} e^{q - q^{\frac{1}{\alpha}} \tilde{y}} q^{\frac{2}{\alpha} - 1} (z^\alpha - q)^{-1} dq \right).$$

Changing the contour of the integration with respect to q into $C_2 = \{q = \rho e^{\pm i\phi_2}, \rho \geq 0, \phi_2 = \frac{\pi}{2} + \epsilon_2\}$, where $\epsilon_2 \in (0, \frac{\pi(\alpha-1)}{2})$ is the fixed constant and using the inequalities (3.4) through (3.6) with $\mu_1, \mu_2 = 0$, $1 - \frac{1}{\alpha} < \nu < \frac{1}{\alpha}$ and $0 < \gamma < \frac{2}{\alpha} + \nu - 1$ for $|q| \leq 1$ and $\gamma > \frac{2}{\alpha} + \nu$ for $|q| > 1$ we get

$$\|F_y(\cdot, \cdot, t)\|_{\mathbf{L}^\infty}$$

$$\leq Ct^{-\frac{2}{\alpha}} \left(\int_{C_2} |q|^{\frac{2}{\alpha}-1-\gamma} |dq| \right.$$

$$+ \int_{-i\infty}^{+i\infty} |z|^{-\alpha\nu} |z+1|^{-1} |dz| \int_{C_2} |q|^{\frac{2}{\alpha}-1+\nu-\gamma} |dq|$$

$$\left. + \int_{-i\infty}^{+i\infty} |z|^{\alpha-1-\alpha\nu} |z+1|^{-1} |dz| \int_{C_2} |q|^{\frac{2}{\alpha}-2+\nu-\gamma} |dq| \right)$$

$$\leq Ct^{-\frac{2}{\alpha}}.$$

Via Lemma 17 with $\mu_1 = 0$ we have $\|F(\cdot, \cdot, t)\|_{\mathbf{L}^\infty} \leq Ct^{-\frac{1}{\alpha}}$, so we get

$$|F(x, y, t) - F(x, 0, t)|$$

$$\leq C\|F_y(\cdot, \cdot, t)\|_{\mathbf{L}^\infty}^{\mu_1} \|F(\cdot, \cdot, t)\|_{\mathbf{L}^\infty}^{1-\mu_1} y^{\mu_1} \leq Ct^{-\frac{1+\mu_1}{\alpha}} y^{\mu_1}$$

for $\mu_1 \in [0, 1]$. Since

$$\frac{p^{\alpha-1}}{p^\alpha - \xi} = \frac{1}{p+1} + \frac{\xi}{(p^\alpha - \xi)(p+1)} + \frac{p^{\alpha-1}}{(p^\alpha - \xi)(p+1)}$$

we have

$$F(x, y, t) - F(x, y, t - \tau)$$

$$= 2\pi i e^{-x} \int_{-i\infty}^{+i\infty} e^{\xi(t-\tau) - \xi^{\frac{1}{\alpha}} y} (1 - e^{\xi\tau}) \xi^{\frac{1}{\alpha}-1} d\xi$$

$$+ \int_{-i\infty}^{+i\infty} e^{px}(p+1)^{-1} dp \int_{-i\infty}^{+i\infty} \frac{e^{\xi(t-\tau) - \xi^{\frac{1}{\alpha}} y} (1 - e^{\xi\tau}) \xi^{\frac{1}{\alpha}}}{p^\alpha - \xi} d\xi$$

$$+ \int_{-i\infty}^{+i\infty} e^{px} p^{\alpha-1}(p+1)^{-1} dp \int_{-i\infty}^{+i\infty} \frac{e^{\xi(t-\tau) - \xi^{\frac{1}{\alpha}} y} (1 - e^{\xi\tau}) \xi^{\frac{1}{\alpha}-1}}{p^\alpha - \xi} d\xi.$$

Changing the contour of integration into C_2 we have for $\gamma, \mu \geq 0$ $|e^{\xi\tau} - 1| \leq C|\xi|^\gamma \tau^\gamma$ and $e^{\operatorname{Re} \xi(t-\tau)} \leq C|\xi(t-\tau)|^{-\mu}$, where $\xi \in C_2$. Therefore using (3.6) we obtain

$$|F(x, y, t - \tau) - F(x, y, t)|$$

$$\leq C(t-\tau)^{-\mu} \tau^\gamma \left(\int_{C_2} |\xi|^{\frac{1}{\alpha}-1-\mu+\gamma} |d\xi| \right.$$

$$+ \int_{-i\infty}^{+i\infty} |p+1|^{-1} |p|^{-\nu\alpha} |dp| \int_{C_2} |\xi|^{\frac{1}{\alpha}-1+\nu-\mu+\gamma} |d\xi|$$

$$\left. + \int_{-i\infty}^{+i\infty} |p+1|^{-1} |p|^{\alpha-1-\nu\alpha} |dp| \int_{C_2} |\xi|^{\frac{1}{\alpha}-2+\nu-\mu+\gamma} |d\xi| \right).$$

Then choosing $\mu = \frac{1+\mu_2}{\alpha}$ and $\gamma, \nu > 0$ such that for $|\xi| \geq 1$ $\nu + \gamma < \frac{1}{\alpha}$ and for $|\xi| \leq 1$ $\nu + \gamma > \frac{\mu_2}{\alpha} + 1$ we get for $\mu_2 > 0$ and $\tau > 1$

$$|F(x, y, t - \tau) - F(x, y, t)| \leq C(t - \tau)^{-\frac{1+\mu_2}{\alpha}} \tau^{\frac{2\mu_2}{\alpha}},$$

where $\mu_2 \in [0, \frac{1}{\alpha})$. Lemma 19 is proved.   □

4. Local existence

In this section we prove Theorem 18 by the contraction mapping principle. We let u as a solution of the following linear problem

(4.1)
$$\begin{cases} u_t + \mathbb{N}(w) + \mathbb{K}u = 0, & t > 0, x > 0, \\ u(x, 0) = u_0(x), & x > 0, \end{cases}$$

where $\mathbb{N}(w) = iww_x$ is well-defined since $w \in \mathbf{Z}_{T,\gamma,\rho}$, where

$$\mathbf{Z}_{T,\gamma,\rho} = \{w \in \mathbf{Z}_{T,\gamma}; \|w\|_{\mathbf{Z}_{T,\gamma}} \leq \rho\}.$$

Note that initial value problem (4.1) defines a mapping $\mathbb{M}$ by $u = \mathbb{M}(w)$ and we will show that $\mathbb{M}$ is the contraction mapping from $\mathbf{Z}_{T,\gamma,\rho}$ into itself for a sufficiently small $T > 0$.

From Section 2 we get

$$u(x, t)$$
$$= \int_0^{+\infty} dy u_0(y) \left(G(x - y, t) + F(x, y, t)\right)$$

(4.2)
$$+ \int_0^t d\tau \int_0^{+\infty} dy \mathbb{N}(w)(y, \tau) \left(G(x - y, t - \tau) + F(x, y, t - \tau)\right),$$

where the functions $F(x, y, t)$ and $G(x, t)$ are defined in (2.2) and (3.1). Here we have used the facts that for $\alpha \in (1, 2)$

$$\int_{-i\infty}^{+i\infty} dp e^{px} p^{\alpha-1} \int_{-i\infty}^{+i\infty} d\xi e^{-\xi \frac{1}{\alpha} y} \xi^{\frac{1}{\alpha}-1} (p^\alpha - \xi)^{-1} = 0$$

and for $t - \tau' < 0$

$$\int_{-i\infty}^{+i\infty} dp e^{px} p^{\alpha-1} \int_{-i\infty}^{+i\infty} d\xi e^{\xi(t-\tau')-\xi \frac{1}{\alpha} y} \xi^{\frac{1}{\alpha}-1} (p^\alpha - \xi)^{-1} = 0.$$

We will prove the following estimate

(4.3) $\|u\|_{\mathbf{Z}_{T,\gamma}} = \sup\limits_{t \in [0,T]} \left(t^\gamma \left(\|u(t)\|_{\mathbf{L}^\infty} + t^{\frac{1}{2\alpha}} \|u_x(t)\|_{\mathbf{L}^2}\right) + \|u(t)\|_{\mathbf{L}^2}\right) \leq C\lambda,$

where

$$\lambda = \|u_0\|_{\mathbf{X}} + T^\mu \|w\|_{\mathbf{Z}_{T,\gamma}}^2,$$

$0 < \mu < 1 - \frac{1}{\alpha}$ and $\gamma > 0$.
By the inequality

$$\left\| \int_0^{+\infty} f(x - y)\phi(y) dy \right\|_{\mathbf{L}^2} \leq C \|f\|_{\mathbf{L}^2} \|\phi\|_{\mathbf{L}^1}$$

and (4.2) we get

$$\|u(t)\|_{\mathbf{L}^2} \leq C\|u_0\|_{\mathbf{L}^2}\|G(t)\|_{\mathbf{L}^1} + C\int_0^{+\infty} |u_0(y)|\, \|F(\cdot,y,t)\|_{\mathbf{L}^2}dy$$

$$+C\int_0^t \|\mathbb{N}(w)(\tau)\|_{\mathbf{L}^2}\|G(t-\tau)\|_{\mathbf{L}^1}d\tau$$

$$(4.4) \qquad +C\int_0^t d\tau \int_0^{+\infty} |\mathbb{N}(w)(y,\tau)|\,\|F(\cdot,y,t-\tau)\|_{\mathbf{L}^2}dy.$$

From Lemma 17 for $\mu_1 \in [0,\frac{1}{\alpha})$ and $t \leq 1$ we have

$$(4.5) \qquad \|F(\cdot,y,t)\|_{\mathbf{L}^2} \leq Ct^{-\frac{1}{2\alpha}+\mu_1-\gamma}y^{-\mu_1\alpha}.$$

Therefore using (4.5) with $\mu_1\alpha = \frac{1}{2} - \gamma\alpha$, where $0 < \gamma < \frac{1}{2\alpha}$ we obtain

$$\int_0^{+\infty} |u_0(y)|\,\|F(\cdot,y,t)\|_{\mathbf{L}^2}dy$$

$$\leq C\int_0^{+\infty} |u_0(y)||y|^{-\frac{1}{2}+\gamma\alpha}dy$$

$$(4.6) \qquad \leq C\left(\|u_0\|_{\mathbf{L}^\infty}\int_0^1 y^{-\frac{1}{2}+\gamma\alpha}dy + \|u_0\|_{\mathbf{L}^1}\right) \leq C\|u_0\|_X.$$

Since $w \in \mathbf{Z}_{T,\gamma}$ we have for $t \leq 1$

$$(4.7) \qquad \|\mathbb{N}(w)(t)\|_{\mathbf{L}^1} \leq \int_0^{+\infty} |w(x,t)w_x(x,t)|dx$$

$$\leq C\|w(t)\|_{\mathbf{L}^2}\|w_x(t)\|_{\mathbf{L}^2} \leq C\|w\|_{\mathbf{Z}_{T,\gamma}}^2 t^{-\frac{1}{2\alpha}-\gamma}$$

and

$$(4.8) \qquad \|\mathbb{N}(w)(t)\|_{\mathbf{L}^2} \leq \left(\int_0^{+\infty} |w(x,t)w_x(x,t)|^2dx\right)^{\frac{1}{2}}$$

$$\leq C\|w(t)\|_{\mathbf{L}^\infty}\|w_x(t)\|_{\mathbf{L}^2} \leq C\|w\|_{\mathbf{Z}_{T,\gamma}}^2 t^{-\frac{1}{2\alpha}-2\gamma}.$$

Therefore using (4.5) with $\mu_1 = 0$, we obtain

$$\int_0^t d\tau \int_0^{+\infty} |\mathbb{N}(w)(y,\tau)|\,\|F(\cdot,y,t-\tau)\|_{\mathbf{L}^2}dy$$

$$\leq C\int_0^t \left((t-\tau)^{-\gamma-1/2\alpha}\|\mathbb{N}(w)(\tau)\|_{\mathbf{L}^1}\right)d\tau$$

$$(4.9) \qquad \leq C\|w\|_{\mathbf{Z}_{T,\gamma}}^2 \int_0^t \tau^{-\frac{1}{2\alpha}-\gamma}(t-\tau)^{-\frac{1}{2\alpha}-\gamma}d\tau \leq C\|w\|_{\mathbf{Z}_{T,\gamma}}^2 T^\mu,$$

where $0 < \mu < 1 - \frac{1}{\alpha}$. From Lemma 18 we have $\|G(t)\|_{\mathbf{L}^1} \leq C$. Therefore, using (4.8) we get

$$\int_0^t \|\mathbb{N}(w)(\tau)\|_{\mathbf{L}^2} \|G(t - \tau)\|_{\mathbf{L}^1} d\tau$$

$$(4.10) \qquad \leq C\|w\|_{\mathbf{Z}_{T,\gamma}}^2 \int_0^t \tau^{-\frac{1}{2\alpha} - 2\gamma} d\tau \leq CT^\mu \|w\|_{\mathbf{Z}_{T,\gamma}}^2$$

and

$$(4.11) \qquad \|u_0\|_{\mathbf{L}^2} \|G(t)\|_{\mathbf{L}^1} \leq C\|u_0\|_X.$$

Putting (4.6) and (4.9) through (4.11) into (4.4) we obtain

$$(4.12) \qquad \|u(t)\|_{\mathbf{L}^2} \leq C\left(\|u_0\|_X + \|w\|_{\mathbf{Z}_{T,\gamma}}^2 T^\mu\right) \leq C\lambda,$$

where $0 < \mu < 1 - \frac{1}{\alpha}$. From (4.2) it follows that

$$\|u(t)\|_{\mathbf{L}^\infty} \leq C\left(\|u_0\|_{\mathbf{L}^\infty} \|G(t)\|_{\mathbf{L}^1} + \int_0^{+\infty} |u_0(y)| \|F(\cdot, y, t)\|_{\mathbf{L}^\infty} dy\right.$$

$$+ \int_0^t \|\mathbb{N}(w)(\tau)\|_{\mathbf{L}^2} \|G(t - \tau)\|_{\mathbf{L}^2} d\tau$$

$$(4.13) \qquad \left. + \int_0^t d\tau \int_0^{+\infty} |\mathbb{N}(w)(y, \tau)| \|F(\cdot, y, t - \tau)\|_{\mathbf{L}^\infty} dy\right).$$

Via Lemma 18 we have

$$\sup_{t \in [0,T]} \left(\|G(t)\|_{\mathbf{L}^1} + t^{\frac{1}{2\alpha}} \|G(t)\|_{\mathbf{L}^2}\right) \leq C.$$

Therefore by virtue of (4.8) we obtain

$$(4.14) \qquad \sup_{t \in [0,T]} t^\gamma \left(\|u_0\|_{\mathbf{L}^\infty} \|G(t)\|_{\mathbf{L}^1} + \int_0^t \|\mathbb{N}(\tau)\|_{\mathbf{L}^2} \|G(t - \tau)\|_{\mathbf{L}^2} d\tau\right) \leq C\lambda.$$

Lemma 17 shows

$$\|F(\cdot, y, t)\|_{\mathbf{L}^\infty} \leq Ct^{-\frac{1}{\alpha} + \mu_1} y^{-\mu_1 \alpha}.$$

Then choosing $\mu_1 \alpha = 1 - \gamma\alpha$, $\gamma \in (0, \frac{1}{\alpha})$ we get

$$\sup_{t \in [0,T]} t^\gamma \int_0^{+\infty} |\bar{u}(y)| \|F(\cdot, y, t)\|_{\mathbf{L}^\infty} dy$$

$$(4.15) \qquad \leq C\left(\|u_0\|_{\mathbf{L}^\infty} \int_0^1 y^{-1 + \gamma\alpha} dy + \|u_0\|_{\mathbf{L}^1}\right) \leq C\|u_0\|_X.$$

Choosing $\mu_1\alpha = \frac{1}{2} - \gamma\alpha$, and $\gamma \in (0, \frac{1}{2\alpha})$ we obtain

$$\sup_{t\in[0,T]} t^\gamma \int_0^t d\tau \int_0^{+\infty} |\mathbb{N}(w)(y,\tau)| \, \|F(\cdot, y, t-\tau)\|_{\mathbf{L}^\infty} dy$$

$$\leq C \int_0^t d\tau (t-\tau)^{-\frac{1}{2\alpha}-\gamma} \int_0^{+\infty} |\mathbb{N}(w)(y,\tau)| \, y^{-\frac{1}{2}+\gamma\alpha} dy$$

$$(4.16) \qquad \leq C\|w\|_{\mathbf{Z}_{T,\gamma}}^2 \int_0^t \tau^{-\frac{1}{2\alpha}-\gamma}(t-\tau)^{-\frac{1}{2\alpha}-\gamma} d\tau \leq C\lambda.$$

From (4.13)-(4.16) we have

$$(4.17) \qquad \sup_{t\in[0,T]} t^\gamma \|u(t)\|_{\mathbf{L}^\infty} \leq C\lambda.$$

Differentiating (4.2) with respect to x we have

$$\|u_x\|_{\mathbf{L}^2} \leq C \left\| \int_0^{+\infty} u_0(y) G_x(x-y,t) dy \right\|_{\mathbf{L}^2}$$

$$+ C \int_0^{+\infty} |u_0(y)| \, \|F_x(\cdot, y, t)\|_{\mathbf{L}^2} dy$$

$$+ C \int_0^t \|\mathbb{N}(w)(\tau)\|_{\mathbf{L}^2} \, \|G_x(t-\tau)\|_{\mathbf{L}^1} d\tau$$

$$(4.18) \qquad + C \int_0^t d\tau \int_0^{+\infty} |\mathbb{N}(w)(y,\tau)| \, \|F_x(\cdot, y, t-\tau)\|_{\mathbf{L}^2} dy.$$

Via Lemma 18 we have

$$(4.19) \qquad \sup_{t\in[0,T]} t^{\frac{1}{\alpha}+\gamma} \left(\|G_x(t)\|_{\mathbf{L}^1} + t^{\frac{1}{2\alpha}} \|G_x(t)\|_{\mathbf{L}^2} \right) \leq C.$$

Integration by parts yields

$$\sup_{t\in[0,T]} Ct^{\frac{1}{2\alpha}+\gamma} \left\| \int_0^{+\infty} u_0(y) G_x(\cdot - y, t) dy \right\|_{\mathbf{L}^2}$$

$$(4.20) \qquad \leq \sup_{t\in[0,T]} Ct^{\frac{1}{2\alpha}+\gamma} \|G(t)\|_{\mathbf{L}^2} \left(\|u_0\|_{\mathbf{L}^\infty} + \|u_{0x}\|_{\mathbf{L}^1} \right) \leq C\lambda.$$

Using (4.8) and (4.19) we obtain

$$\sup_{t\in[0,T]} t^{\frac{1}{2\alpha}+\gamma} \int_0^t \|\mathbb{N}(w)(\tau)\|_{\mathbf{L}^2} \|G_x(t-\tau)\|_{\mathbf{L}^1} d\tau$$

$$\leq C \sup_{t\in[0,T]} t^{\frac{1}{2\alpha}+\gamma} \|w\|_{\mathbf{Z}_{T,\gamma}}^2 \int_0^t \tau^{-\frac{1}{2\alpha}-2\gamma}(t-\tau)^{-\frac{1}{\alpha}-\gamma} d\tau$$

$$(4.21) \qquad \leq C\|w\|_{\mathbf{Z}_{T,\gamma}}^2 T^\mu \leq C\lambda.$$

Via Lemma 17 we have

$$(4.22) \qquad \|F_x(\cdot, y, t)\|_{\mathbf{L}^2} \leq Ct^{-\frac{3}{2\alpha}+\mu_1} y^{-\mu_1\alpha}.$$

and so by choosing $\mu_1\alpha = 1 - \gamma\alpha$ we get

$$(4.23) \qquad \sup_{t\in[0,T]} t^{\frac{1}{2\alpha}+\gamma} \int_0^{+\infty} |u_0(y)| \|F_x(\cdot,y,t)\|_{\mathbf{L}^2} dy$$

$$\leq C \int_0^{+\infty} |u_0(y)| |y|^{-1+\gamma\alpha} dy \leq C\lambda.$$

Using (4.22) with $\mu_1\alpha = \frac{1}{2} - \gamma\alpha$ we get

$$\sup_{t\in[0,T]} t^{\frac{1}{2\alpha}+\gamma} \int_0^t d\tau \int_0^{+\infty} |\mathbb{N}(w)(y,\tau)| \, \|F_x(\cdot,y,t-\tau)\|_{\mathbf{L}^2} dy$$

$$\leq C \sup_{t\in[0,T]} t^{\frac{1}{2\alpha}+\gamma} \int_0^t d\tau (t-\tau)^{-\frac{1}{\alpha}-\gamma} \int_0^{+\infty} |\mathbb{N}(w)| \, y^{-\frac{1}{2}+\gamma\alpha} dy$$

$$(4.24) \qquad \leq C \sup_{t\in[0,T]} t^{\frac{1}{2\alpha}+\gamma} \|w\|_{\mathbf{Z}_{T,\gamma}}^2 \int_0^t d\tau (t-\tau)^{-\frac{1}{\alpha}-\gamma} \tau^{-\frac{1}{2\alpha}-\gamma}$$

$$\leq C\|w\|_{\mathbf{Z}_{T,\gamma}}^2 T^\mu \leq C\lambda.$$

From (4.18), (4.20), (4.21), (4.23) and (4.24) we obtain

$$(4.25) \qquad \sup_{t\in[0,T]} t^{\frac{1}{2\alpha}+\gamma} \|u_x(t)\|_{\mathbf{L}^2} \leq C\lambda.$$

From (4.12), (4.17) and (4.25) we get (4.3):

$$\|u\|_{\mathbf{Z}_{T,\gamma}} = \sup_{t\in[0,T]} t^\gamma \left(\|u(t)\|_{\mathbf{L}^\infty} + t^{\frac{1}{2\alpha}} \|u_x(t)\|_{\mathbf{L}^2} \right) + \|u(t)\|_{\mathbf{L}^2}$$

$$\leq C \left(\|u_0\|_{\mathbf{X}} + T^\mu \|w\|_{\mathbf{Z}_{T,\gamma}}^2 \right).$$

We choose $T^\mu \leq \frac{1}{2C\rho}$ and $\|u_0\|_{\mathbf{X}} \leq \frac{\rho}{2C}$. Then we have $\|u\|_{\mathbf{Z}_{T,\gamma}} \leq \rho$. Thus the mapping $\mathbb{M}$ transforms the closed ball in $\mathbf{Z}_{T,\gamma}$ with a center at the origin and a radius ρ into itself. Analogously we can prove the estimate $\|u - \tilde{u}\|_{\mathbf{Z}_{T,\gamma}} < \|w - \tilde{w}\|_{\mathbf{Z}_{T,\gamma}}$ for small T. Therefore the mapping $\mathbb{M}$ is a contraction mapping in $\mathbf{Z}_{T,\gamma}$, and there exists the unique solution $u(x,t)$ of the initial-value problem (1.1). Theorem 18 is proved.

REMARK 7. *By virtue of estimates (4.3) we can see that if the norm of the initial data* $\|u_0\|_{\mathbf{X}} < \epsilon$, *then there exists a time* $T > 1$ *such that the solution is also sufficiently small*

$$\sup_{t\in[0,T]} \left(\|u(t)\|_{\mathbf{L}^2} + t^\gamma \|u(t)\|_{\mathbf{L}^\infty} + t^{\frac{1}{2\alpha}+\gamma} \|u_x(t)\|_{\mathbf{L}^2} \right) < C\epsilon.$$

5. Large time asymptotics

Consider the initial-boundary value problem (1.2) with small initial data

$$(5.1) \qquad \|u_0\|_{\mathbf{X}} < \epsilon_1,$$

where $\epsilon_1 > 0$ is sufficiently small. In Lemmas 20 - 21 below we suppose that we already have the following estimates for the solution of (1.2)

$$\sup_{t \geq 0}(1+t)^{\frac{1}{2\alpha}-\gamma}\left(\|u(t)\|_{\mathbf{L}^2} + t^\gamma(1+t)^{\frac{1}{2\alpha}}\|u(t)\|_{\mathbf{L}^\infty}\right.$$

$$\left. +t^{\frac{1}{2\alpha}+\gamma}(1+t)^{\frac{1}{2\alpha}-\gamma}\|u_x(t)\|_{\mathbf{L}^2}\right)$$

$$(5.2) \qquad\qquad \leq \epsilon,$$

where $\gamma \in (0, \frac{2-\alpha}{\alpha})$, and $\epsilon > 0$ is sufficiently small.

LEMMA 20.   Let $\|(\cdot)^\delta u_0\|_{\mathbf{L}^1} \leq \epsilon_1$, $\delta \in [0, \frac{1}{2})$ and (5.1) through (5.2) be valid. Then the following estimate is true

$$\sup_{t \geq 0} t^{\frac{1-2\delta}{2\alpha}+\gamma}(1+t)^{-2\gamma}\|(\cdot)^\delta u\|_{\mathbf{L}^2} \leq C\epsilon.$$

PROOF. We have for the solution of (1.2)

$$u(x,t)$$
$$= \int_0^{+\infty} dy u_0(y)\left(G(x-y,t)+F(x,y,t)\right)$$
$$(5.3) \qquad + \int_0^t d\tau \int_0^{+\infty} dy \mathbb{N}(u)(y,\tau)\left(G(x-y,t-\tau)+F(x,y,t-\tau)\right),$$

where

$$G(x,t) = \frac{1}{2\pi i}\int_{-i\infty}^{+i\infty} e^{px+p^\alpha t}dp$$

and

$$F(x,y,t) = \frac{1}{4\pi^2}\int_{-i\infty}^{+i\infty} e^{px}p^{\alpha-1}dp \int_{-i\infty}^{+i\infty} e^{\xi t - \xi^{\frac{1}{\alpha}}y}\xi^{\frac{1}{\alpha}-1}(p^\alpha - \xi)^{-1}d\xi.$$

Hence we get

$$\|(\cdot)^\delta u(\cdot,t)\|_{\mathbf{L}^2}$$
$$\leq C\left(\left\|(\cdot)^\delta \int_0^{+\infty} u_0(y)G(\cdot - y,t)dy\right\|_{\mathbf{L}^2}\right.$$
$$+ \int_0^t d\tau \left\|(\cdot)^\delta \int_0^{+\infty} \mathbb{N}(u)(y,\tau)G(\cdot - y,t-\tau)dy\right\|_{\mathbf{L}^2}$$
$$+ \left\|(\cdot)^\delta \int_0^{+\infty} u_0(y)F(\cdot,y,t)dy\right\|_{\mathbf{L}^2}$$
$$(5.4) \qquad + \int_0^t d\tau \left\|(\cdot)^\delta \int_0^{+\infty} \mathbb{N}(u)(y,\tau)F(\cdot,y,t-\tau)dy\right\|_{\mathbf{L}^2}.$$

By virtue of Lemma 18 we have

$$\sup_{t \geq 0} t^{\frac{1-2\delta}{2\alpha}+\gamma}(1+t)^{-2\gamma}\|(\cdot)^\delta G(\cdot,t)\|_{\mathbf{L}^2} \leq C$$

and
$$t^{\frac{1}{2\alpha}}\|G(t)\|_{\mathbf{L}^2} \le C.$$

Therefore we get

$$\sup_{t\ge0} t^{\frac{1-2\delta}{2\alpha}+\gamma}(1+t)^{-2\gamma}\left\|(\cdot)^\delta\int_0^{+\infty}u_0(y)G(\cdot-y,t)dy\right\|_{\mathbf{L}^2}$$

$$(5.5) \le \ C\sup_{t\ge0} t^{\frac{1-2\delta}{2\alpha}+\gamma}(1+t)^{-2\gamma}\left(\|(\cdot)^\delta u_0\|_{\mathbf{L}^1}\|G(t)\|_{\mathbf{L}^2}+\|u_0\|_{\mathbf{L}^1}\|(\cdot)^\delta G(\cdot,t)\|_{\mathbf{L}^2}\right)$$

$$\le \ C\epsilon_1,$$

where we have used the estimates

$$x^\delta \le C\left((x-y)^\delta+y^\delta\right)$$

and

$$\left\|\int_0^{+\infty}f(\cdot-y)g(y)dy\right\|_{\mathbf{L}^2} \le \|f\|_{\mathbf{L}^2}\|g\|_{\mathbf{L}^1}.$$

We have by virtue of (5.2)

$$(5.6) \qquad \|\mathbb{N}(u)(t)\|_{\mathbf{L}^1} \le C\|u(t)\|_{\mathbf{L}^2}\|u_x(t)\|_{\mathbf{L}^2} \le C\epsilon^2 t^{-\frac{1}{2\alpha}-\gamma}(1+t)^{-\frac{3}{2\alpha}+3\gamma}$$

and

$$\|(\cdot)^\delta\mathbb{N}(u)(\cdot,t)\|_{\mathbf{L}^1} \ \le \ C\epsilon\|u_x(t)\|_{\mathbf{L}^2}\|(\cdot)^\delta u(\cdot,t)\|_{\mathbf{L}^2}$$

$$\le \ C\epsilon t^{-\frac{1}{2\alpha}-\gamma}(1+t)^{-\frac{1}{\alpha}+2\gamma}\|(\cdot)^\delta u(\cdot,t)\|_{\mathbf{L}^2}.$$

Therefore in a similar way as in the proof of (5.5) we obtain

$$\sup_{t\ge0} t^{\frac{1-2\delta}{2\alpha}+\gamma}(1+t)^{-2\gamma}\int_0^t d\tau\left\|(\cdot)^\delta\int_0^{+\infty}\mathbb{N}(u)(y,\tau)G(\cdot-y,t-\tau)dy\right\|_{\mathbf{L}^2}$$

$$\le \ C\sup_{t\ge0} t^{\frac{1-2\delta}{2\alpha}+\gamma}(1+t)^{-2\gamma}\left(\int_0^t d\tau\|\mathbb{N}(u)(\tau)\|_{\mathbf{L}^1}\|(\cdot)^\delta G(\cdot,t-\tau)\|_{\mathbf{L}^2}\right.$$

$$\left.+\int_0^t d\tau\|(\cdot)^\delta\mathbb{N}(u)(\cdot,\tau)\|_{\mathbf{L}^1}\|G(t-\tau)\|_{\mathbf{L}^2}\right)$$

$$\le \ C\sup_{t\ge0} t^{\frac{1-2\delta}{2\alpha}+\gamma}(1+t)^{-2\gamma}\left(\epsilon^2\int_0^t(t-\tau)^{-\frac{1-2\delta}{\alpha}+\gamma}\tau^{-\frac{1}{2\alpha}-\gamma}(1+\tau)^{-\frac{3}{2\alpha}+3\gamma}d\tau\right.$$

$$\left.+\epsilon\int_0^t\|(\cdot)^\delta u(\cdot,\tau)\|_{\mathbf{L}^2}(t-\tau)^{-\frac{1}{2\alpha}}\tau^{-\frac{1}{\alpha}}(1+\tau)^{-\frac{1}{2\alpha}+\gamma}d\tau\right)$$

$$(5.7)\le \ C\epsilon^2+C\epsilon\sup_{t\ge0} t^{\frac{1-2\delta}{2\alpha}+\gamma}(1+t)^{-2\gamma}$$

$$\times\int_0^t\|(\cdot)^\delta u(\cdot,\tau)\|_{\mathbf{L}^2}(t-\tau)^{-\frac{1}{2\alpha}}\tau^{-\frac{1}{2\alpha}-\gamma}(1+\tau)^{-\frac{1}{\alpha}+2\gamma}d\tau.$$

From Lemma 17 we have

$$\sup_{t\ge0} t^{\frac{1-2\delta}{2\alpha}-\gamma}(1+t)^{-2\gamma}\|x^\delta F(\cdot,y,t)\|_{\mathbf{L}^2} \le C$$

and so we get

$$\sup_{t\geq 0} t^{\frac{1-2\delta}{2\alpha}+\gamma}(1+t)^{-2\gamma}\left\|(\cdot)^\delta \int_0^{+\infty} u_0(y)F(\cdot,y,t)dy\right\|_{\mathbf{L}^2}$$

(5.8)
$$\leq \sup_{t\geq 0} t^{\frac{1-2\delta}{2\alpha}+\gamma}(1+t)^{-2\gamma}\int_0^{+\infty}|u_0(y)|\|(\cdot)^\delta F(\cdot,y,t)\|_{\mathbf{L}^2}dy$$

$$\leq C\|u_0\|_{\mathbf{L}^1} \leq C\epsilon_1.$$

Using (5.6) in the same manner we get

$$\sup_{t\geq 0} t^{\frac{1-2\delta}{2\alpha}+\gamma}(1+t)^{-2\gamma}\int_0^t d\tau \left\|(\cdot)^\delta \int_0^{+\infty} \mathbb{N}(u)(y,\tau)F(\cdot,y,t-\tau)dy\right\|_{\mathbf{L}^2}$$

$$\leq C\sup_{t\geq 0} t^{\frac{1-2\delta}{2\alpha}+\gamma}(1+t)^{-2\gamma}\int_0^t \|\mathbb{N}(u)(\tau)\|_{\mathbf{L}^1}\|(\cdot)^\delta F(\cdot,y,t-\tau)\|_{\mathbf{L}^2}d\tau$$

(5.9) $$\leq C\epsilon^2\int_0^t (t-\tau)^{-\frac{1-2\delta}{2\alpha}+\gamma}\tau^{-\frac{1}{2\alpha}-\gamma}(1+\tau)^{-\frac{3}{2\alpha}+3\gamma}d\tau \leq C\epsilon^2.$$

Therefore from (5.4), (5.5), (5.7) through (5.9) we obtain

$$\|(\cdot)^\delta u(\cdot,t)\|_{\mathbf{L}^2}$$

$$\leq C\epsilon t^{-\frac{1-2\delta}{2\alpha}-\gamma}(1+t)^{2\gamma}$$

$$+C\epsilon\int_0^t \|(\cdot)^\delta u(\cdot,\tau)\|_{\mathbf{L}^2}(t-\tau)^{-\frac{1}{2\alpha}}\tau^{-\frac{1}{2\alpha}-\gamma}(1+\tau)^{-\frac{1}{\alpha}+2\gamma}d\tau$$

and via the Gronwall inequality we get

$$\|(\cdot)^\delta u(\cdot,t)\|_{\mathbf{L}^2} \leq C\epsilon t^{-\frac{1-2\delta}{2\alpha}-\gamma}(1+t)^{2\gamma}.$$

Lemma 20 is proved.   □

LEMMA 21. *Let the assumptions given in Lemma 20 be valid. Then the following asymptotics is valid*

$$u(x,t) = t^{-\frac{1}{\alpha}}a\Lambda\left(\frac{x}{t^{\frac{1}{\alpha}}}\right) + O(\epsilon^2 t^{-\frac{1}{\alpha}-\gamma}),$$

uniformly with respect to $x > 0$ as $t \to +\infty$, where

$$a = \int_0^{+\infty} u_0(y)dy + \int_0^{+\infty}\int_0^{+\infty} \mathbb{N}(u)(y,t)dydt$$

and $\Lambda(s) = \Lambda_1(s) + \Lambda_2(s)$ (see Lemma 19 for definitions of Λ_1 and Λ_2), $\gamma \in (0, \frac{2-\alpha}{2\alpha})$.

PROOF. From Lemma 19 we have

$$G(x-y,t-\tau) = t^{-\frac{1}{\alpha}}\Lambda_1(x/t^{\frac{1}{\alpha}}) + y^{\mu_1}O(t^{-\frac{1+\mu_1}{\alpha}})$$

$$+\tau^{\frac{\mu_2}{\alpha}}O\left((t-\tau)^{-\frac{1+\mu_2}{\alpha}}\right)$$

and

$$F(x, y, t - \tau) = t^{-\frac{1}{\alpha}} \Lambda_2(x/t^{\frac{1}{\alpha}}) + y^{\mu_1} O(t^{-\frac{1+\mu_1}{\alpha}})$$
$$+ \tau^{\frac{2\mu_2}{\alpha}} O\left((t - \tau)^{-\frac{1+\mu_2}{\alpha}}\right),$$

where $\mu_1 \in [0,1]$, $\mu_2 > 0$. Therefore by choosing $\mu_1 = \mu_2 = \gamma\alpha > 0$ in these identities we obtain from (5.3)

$$u(x, t) = t^{-\frac{1}{\alpha}} a\Lambda\left(\frac{x}{t^{\frac{1}{\alpha}}}\right) + R(x, t),$$

where

$$|R(x, t)| \leq Ct^{-\frac{1}{\alpha} - \gamma} \left(\|(\cdot)^{\gamma\alpha} u_0\|_{\mathbf{L}^1} + \int_0^t d\tau \|(\cdot)^{\gamma\alpha} \mathbb{N}(\cdot, \tau)\|_{\mathbf{L}^1} \right.$$
$$\left. + \int_0^t d\tau \tau^{2\gamma} (t - \tau)^{-\frac{1}{\alpha} - \gamma} \|\mathbb{N}(\tau)\|_{\mathbf{L}^1} \right).$$

Now we estimate $R(x, t)$. By virtue of Lemma 5.1 and (5.2) we have

$$\sup_{t \geq 0} t^{\frac{1}{\alpha} + \gamma} (1 + t)^{\frac{1}{\alpha} - 4\gamma} \|(\cdot)^{\gamma\alpha} \mathbb{N}(u)(\cdot, t)\|_{\mathbf{L}^1} \leq C\epsilon^2.$$

Therefore using estimate (5.6) and condition $\|x^{\gamma\alpha} u_0\|_{\mathbf{L}^1} \leq \epsilon_1$ we get

$$|R(x, t)| \leq Ct^{-\frac{1}{\alpha} - \gamma} \left(\epsilon_1 + \epsilon^2 \int_0^t \tau^{-\frac{1}{\alpha} - \gamma} (1 + \tau)^{-\frac{1}{\alpha} + 4\gamma} d\tau \right.$$
$$\left. + \epsilon^2 \int_0^t (t - \tau)^{-\frac{1}{\alpha} - \gamma} \tau^{-\frac{1}{2\alpha} + \gamma} (1 + \tau)^{-\frac{3}{2\alpha} + 3\gamma} d\tau \right)$$
$$\leq C\epsilon^2 t^{-\frac{1}{\alpha} - \gamma},$$

where $\gamma \in \left(0, \frac{2-\alpha}{2\alpha}\right)$. Lemma 21 is proved. □

Proof of Theorem 19. Let us prove the following estimate for all $t > 0$

(5.10)
$$(1 + t)^{\frac{1}{2\alpha} - \gamma} \left(\|u(t)\|_{\mathbf{L}^2} + t^\gamma (1 + t)^{\frac{1}{2\alpha}} \|u(t)\|_{\mathbf{L}^\infty} \right.$$
$$\left. + t^{\frac{1}{2\alpha} + \gamma} (1 + t)^{\frac{1}{2\alpha} - \gamma} \|u_x(t)\|_{\mathbf{L}^2} \right)$$
$$< \epsilon.$$

By Theorem 1.1 the norms $\|u(t)\|_{\mathbf{L}^2}$, $\|u(t)\|_{\mathbf{L}^\infty}$ and $\|u_x(t)\|_{\mathbf{L}^2}$ are continuous. Therefore via Remark 7 there exists a maximal time $T > 1$ such that the non strict estimate (5.10) is valid on $[0, T]$. We have by the formula (5.3)

$$
\begin{aligned}
\|u(t)\|_{\mathbf{L}^2} \leq\ & C\,(\|u_0\|_{\mathbf{L}^1}\|G(t)\|_{\mathbf{L}^2} \\
& + \int_0^{+\infty} dy\,|u_0(y)|\,\|F(\cdot, y, t)\|_{\mathbf{L}^2} \\
& + \int_0^t \|G(t-\tau)\|_{\mathbf{L}^2}\,\|\mathbb{N}(u)(\tau)\|_{\mathbf{L}^1}\,d\tau \\
& + \int_0^t d\tau \int_0^{+\infty} |\mathbb{N}(u)(y, \tau)|\,\|F(\cdot, y, t-\tau)\|_{\mathbf{L}^2}\,dy\Big).
\end{aligned}
$$

Since for all $t > 1$ $t^{\frac{1}{2\alpha}}(\|G(t)\|_{\mathbf{L}^2} + t^{-\gamma} \sup_{y\geq 0}\|F(\cdot, y, t)\|_{\mathbf{L}^2}) \leq C$, then by virtue of (5.6) and (5.1) we get

$$
\begin{aligned}
(5.11) \qquad \|u(t)\|_{\mathbf{L}^2} \leq\ & C\epsilon_1 t^{-\frac{1}{2\alpha}+\gamma} \\
& + C\epsilon^2 \int_0^t (t-\tau)^{-\frac{1}{2\alpha}-\gamma}\tau^{-\frac{1}{2\alpha}+\gamma}(1+\tau)^{-\frac{3}{2\alpha}+2\gamma}d\tau \\
< \ & C(\epsilon_1 + \epsilon^2)t^{-\frac{1}{2\alpha}+\gamma}.
\end{aligned}
$$

Now we estimate the norm $\|u_x(t)\|_{\mathbf{L}^2}$. We have by the formula (5.3)

$$
\begin{aligned}
(5.12) \qquad \|u_x\|_{\mathbf{L}^2} \leq\ & C\,(\|u_0\|_{\mathbf{L}^1}\|G_x(t)\|_{\mathbf{L}^2} \\
& + \int_0^{+\infty} d\tau\,|u_0|\,\|F_x(\cdot, y, t)\|_{\mathbf{L}^2} \\
& + \int_0^t d\tau \left\| \int_0^{+\infty} G_x(\cdot - y, t-\tau)\mathbb{N}(u)(y, \tau)dy \right\|_{\mathbf{L}^2} \\
& + \int_0^t d\tau \int_0^{+\infty} |\mathbb{N}(u)(y, \tau)|\,\|F_x(\cdot, y, t-\tau)\|_{\mathbf{L}^2}\,dy\,.
\end{aligned}
$$

Via Lemma 17 we have

$$
(5.13) \qquad \sup_{t\geq 0} t^{\frac{3}{2\alpha}+\gamma-\mu_1}(1+t)^{-2\gamma}\|F_x(\cdot, y, t)\|_{\mathbf{L}^2} \leq Cy^{\mu_1},
$$

and by Lemma 18

$$
(5.14) \qquad \sup_{t\geq 0} t^{\frac{1}{\alpha}+\gamma}(1+t)^{-2\gamma}\left(t^{\frac{1}{2\alpha}}\|G_x(t)\|_{\mathbf{L}^2} + t^{\frac{1}{\alpha}+\gamma}(1+t)^{-2\gamma}\|G_x(t)\|_{\mathbf{L}^1}\right) \leq C.
$$

Therefore using (5.13) with $\mu_1 = 0$ we get for $t > 1$

$$
(5.15) \qquad \|u_0\|_{\mathbf{L}^1}\|G_x\|_{\mathbf{L}^2} + \int_0^{+\infty} d\tau\,|u_0|\,\|F_x(\cdot, y, t)\|_{\mathbf{L}^2} \leq C\epsilon_1 t^{-\frac{3}{2\alpha}+\gamma}.
$$

First we consider the case $\alpha > 3/2$. Using (5.14) and (5.6) we get for $t > 1$

$$\int_0^t d\tau \left\| \int_0^{+\infty} G_x(\cdot - y, t - \tau) \mathbb{N}(u)(y, \tau) dy \right\|_{L^2}$$

$$\leq C \int_0^t d\tau \| G(t - \tau) \|_{L^2} \| \mathbb{N}(u)(\tau) \|_{L^1}$$

$$(5.16) \qquad \leq C\epsilon^2 \int_0^t d\tau (t - \tau)^{-\frac{3}{2\alpha} + \gamma} \tau^{-\frac{1}{2\alpha} - \gamma} (1 + \tau)^{-\frac{3}{2\alpha} + 3\gamma} \leq C\epsilon^2 t^{-\frac{3}{2\alpha} + \gamma},$$

where $\gamma \in (0, \frac{2-\alpha}{\alpha})$. Using (5.13) with $\mu_1 = 0$ we obtain for $t > 1$

$$\int_0^t d\tau \int_0^{+\infty} |\mathbb{N}(u)(y, \tau)| \, \| F_x(\cdot, y, t - \tau) \|_{L^2} \, dy$$

$$(5.17) \qquad \leq C\epsilon^2 \int_0^t (t - \tau)^{-\frac{3}{2\alpha} + \gamma} \, \tau^{-\frac{1}{2\alpha} - \gamma} (1 + \tau)^{-\frac{3}{2\alpha} + 3\gamma} d\tau \leq C\epsilon^2 t^{-\frac{3}{2\alpha} + \gamma}.$$

Now we consider the case $\alpha \leq 3/2$. Via (5.10) we have

$$\| \mathbb{N}(u)(t) \|_{L^2} \leq C \| u(t) \|_{L^\infty} \| u_x(t) \|_{L^2} \leq \epsilon^2 t^{-\frac{1}{2\alpha} - 2\gamma} (1 + t)^{-\frac{3}{2\alpha} + 4\gamma}.$$

Thus by virtue of (5.14) we get for $\gamma \in (0, \frac{2-\alpha}{\alpha})$

$$\int_0^t d\tau \left\| \int_0^{+\infty} G_x(\cdot - y, t - \tau) \mathbb{N}(u)(y, \tau) dy \right\|_{L^2}$$

$$\leq C \int_0^{t/2} d\tau \| G_x(t - \tau) \|_{L^2} \| \mathbb{N}(u)(\tau) \|_{L^1}$$

$$+ \int_{t/2}^t d\tau \| G_x(t - \tau) \|_{L^1} \| \mathbb{N}(u)(\tau) \|_{L^2}$$

$$\leq C\epsilon^2 \int_0^{t/2} d\tau (t - \tau)^{-\frac{3}{2\alpha} + \gamma} \tau^{-\frac{1}{2\alpha} - \gamma} (1 + \tau)^{-\frac{3}{2\alpha} + 3\gamma}$$

$$+ C\epsilon^2 \int_{t/2}^t d\tau (t - \tau)^{-\frac{1}{\alpha} + \gamma} \tau^{-\frac{1}{2\alpha} - 2\gamma} (1 + \tau)^{-\frac{3}{2\alpha} + 4\gamma}$$

$$(5.18) \qquad \leq C\epsilon^2 t^{-\frac{3}{2\alpha} + \gamma}.$$

Using (5.13) with $\mu_1 = 0$ in the interval $t \in [0, t/2)$ and with $\mu_1 = 1/2 - \gamma$ in the interval $t \in [t/2, t]$ we obtain

$$\int_0^t d\tau \int_0^{+\infty} |\mathbb{N}(u)(y, \tau)| \, \|F_x(\cdot, y, t - \tau)\|_{L^2} \, dy$$

$$\leq C \int_0^{t/2} d\tau (t - \tau)^{-\frac{3}{2\alpha} + \gamma} \|\mathbb{N}(u)(\tau)\|_{\mathbf{L^1}}$$

$$+ C \int_{t/2}^t d\tau (t - \tau)^{-\frac{1}{\alpha} + \gamma} \left(\|\mathbb{N}(u)(\tau)\|_{\mathbf{L^2}} + \|\mathbb{N}(u)(\tau)\|_{\mathbf{L^1}} \right)$$

$$\leq C\epsilon^2 \int_0^{t/2} d\tau (t - \tau)^{-\frac{3}{2\alpha} + \gamma} \tau^{-\frac{1}{2\alpha} - \gamma} (1 + \tau)^{-\frac{3}{2\alpha} + 3\gamma}$$

$$+ C\epsilon^2 \int_{t/2}^t d\tau (t - \tau)^{-\frac{1}{\alpha} + \gamma} \tau^{-\frac{1}{2\alpha} - 2\gamma} (1 + \tau)^{-\frac{3}{2\alpha} + 4\gamma}$$

(5.19) $$\leq C\epsilon^2 t^{-\frac{3}{2\alpha} + \gamma}.$$

We use (5.15) through (5.19) in the right-hand side of (5.12) to obtain

(5.20) $$\sup_{t>1} t^{\frac{3}{2\alpha} - \gamma} \|u_x(t)\|_{L^2} \leq C(\epsilon_1 + \epsilon^2).$$

From (5.2) and Lemma 21 we find that

(5.21) $$\sup_{t>1} t^{\frac{1}{\alpha}} \|u(t)\|_{\mathbf{L^\infty}} \leq C(\epsilon_1 + \epsilon^2).$$

From (5.11), (5.20) and (5.21) we get estimate (5.10) for all $t \in [0, T]$. The contradiction obtained proves the estimate (5.10) for all $t > 0$. Moreover by Lemma 21 the solution has asymptotics

$$u(x, t) = t^{-\frac{1}{\alpha}} a\Lambda \left(\frac{x}{t^{\frac{1}{\alpha}}} \right) + O(t^{-\frac{1}{\alpha} + \gamma})$$

uniformly with respect to $x > 0$ as $t \to +\infty$, where

$$a = \int_0^{+\infty} u_0(y) dy + \int_0^{+\infty} \int_0^{+\infty} \mathbb{N}(u)(y, t) dy dt$$

and $\Lambda(s) = \Lambda_1(s) + \Lambda_2(s)$, $\gamma \in (0, \frac{2-\alpha}{2\alpha})$. Theorem 19 is proved.

CHAPTER 6

Korteweg-de Vries-Burgers Equation

1. Introduction

We consider the initial-boundary value problem on a half-line for the Korteweg-de Vries-Burgers equation. The amount of the boundary data for this problem depends on the direction of a half-line: $x \geq 0$ or $x \leq 0$, so we are interested in the following two initial-boundary value problems

$$\begin{cases} u_t + uu_x - au_{xx} + bu_{xxx} = 0, & t > 0, x > 0, \\ u(x,0) = u_0(x), \ x > 0; \ u(0,t) = 0, \ t > 0, \end{cases}$$

and

$$\begin{cases} u_t + uu_x - au_{xx} + bu_{xxx} = 0, & t > 0, x < 0, \\ u(x,0) = u_0(x), \ x < 0; \ u(0,t) = u_x(0,t) = 0, \ t > 0, \end{cases}$$

where $a, b > 0$. By changing variables and introducing a parameter $\alpha = 0, 1$ we can combine these two problems as one problem on a half-line $x > 0$

(1.1) $$\begin{cases} u_t + (-1)^\alpha uu_x - u_{xx} + (-1)^\alpha u_{xxx} = 0, \ t > 0, x > 0, \\ u(x,0) = u_0(x), \ x > 0, \ \partial_x^j u(0,t) = 0, \ t > 0, \ j = 0, \alpha. \end{cases}$$

The Korteweg-de Vries-Burgers equation (1.1) is a simple universal model equation which appears as the first approximation in the description of the dispersive dissipative nonlinear waves.

The main goal of this chapter is to find the large time asymptotics of solutions to the problem (1.1). Some results on the decay estimates of the solutions in different norms to the Cauchy problems for some nonlinear evolution equations with strongly dissipative operator were obtained in papers [5], [7], [9].

For the nonlinear nonlocal Schrödinger equation on a half-line the initial-boundary value problem was studied in Chapter 4, where it was considered the pseudodifferential operator $\mathbb{K}$ with homogeneous symbol $K(p) = Cp^\beta$ except the case, where β is an odd integer. In the case of equation (1.1) we have the operator $\mathbb{K}u = -u_{xx} + (-1)^\alpha u_{xxx}$, i.e. the symbol $K_\alpha(p) = -p^2 + (-1)^\alpha p^3$ is not homogeneous and also the order of p in the second term is an odd integer. Thus we consider the critical case, where the number of the boundary data also depends on the sign of the highest derivative in the equation. Note that in the case of the Cauchy problem for the Korteweg-de Vries-Burgers equation the nonlinearity uu_x of the shallow water type is critical from the point of view of the large time asymptotic behavior of solutions since the nonlinearity in the equation has the same decay rate as the linear terms. Because of the homogeneous boundary data in the case of the initial-boundary value problem, the solution obtains an additional decay. As a result the nonlinear term in the boundary-value problem (1.1) appears to be super critical contrary to the corresponding Cauchy problem. The main difficulty in the boundary value problem (1.1) is in the evaluation of the contribution of the boundary data to the large time asymptotic formulas of the solutions. Our approach here is based on the $\mathbf{L}^p$ estimates of the Green function.

The symbol of the operator $\mathbb{K}$ has the form $K_\alpha(p) = -p^2 + (-1)^\alpha p^3$, so that it satisfies the dissipation condition $\operatorname{Re} K_\alpha(p) > 0$ for all p on the imaginary axis $\operatorname{Re} p = 0, p \neq 0$. By the same letter C we denote different positive constants. First of all we formulate the local existence of the solutions of the initial-boundary value problem (1.1).

Now we state the result of this chapter (see [60]).

THEOREM 21. *Suppose that the initial data* $u_0 \in \mathbf{H}_2^{1,\omega}(\mathbf{R}^+)$, *where* $\omega \in \left(1, \frac{3}{2}\right)$ *and the norm* $\|u_0\|_{\mathbf{H}_2^{1,\omega}} \leq \epsilon$, *where* $\epsilon > 0$ *is small enough. Then there exists a unique solution*

$$u \in \mathbf{C}([0, +\infty), \mathbf{H}_2^{0,\varkappa}(\mathbf{R}^+)) \cap \mathbf{C}\left((0, +\infty), \mathbf{H}_2^{1,\omega}(\mathbf{R}^+)\right) \cap \mathbf{C}\left((0, +\infty), \mathbf{H}_2^{2,0}(\mathbf{R}^+)\right)$$

of the initial-boundary value problem (1.1), where $\varkappa \in \left(0, \frac{1}{2}\right)$. *Moreover if the initial data are such that*

$$x^{1+\mu} u_0(x) \in \mathbf{L}^1(\mathbf{R}^+), \mu = \omega - \frac{1}{2},$$

then there exists a constant A *such that the solution has the following asymptotics*

$$u(x, t) = \frac{A}{t} \Phi_\alpha\left(\frac{x}{2\sqrt{t}}, t\right) + O\left(\min\left(\frac{x}{\sqrt{t}}, 1\right) t^{-1-\frac{\mu}{2}}\right)$$

for $t \to +\infty$ *uniformly with respect to* $x > 0$, *where* $\alpha = 0, 1$,

$$\Phi_0(q, t) = \frac{q}{\sqrt{\pi}} e^{-q^2}, \Phi_1(q, t) = \frac{1}{2\sqrt{\pi}\sqrt{t}}\left(e^{-q^2}\left(2q\sqrt{t} - 1\right) + e^{-2q\sqrt{t}}\right).$$

REMARK 8. *Note that the asymptotics of solutions to the problem (1.1) obtained in Theorem 21 has the decay rate* t^{-1} *which is more rapid in comparison with the case of the Cauchy problem* $t^{-\frac{1}{2}}$. *Similar effect takes place for the heat equation on the line and on the half-line with the homogeneous boundary data.*

We organize this chapter as follows. In Section 2 we consider the linear initial-boundary value problem corresponding to (1.1) and discuss the question of the number of the boundary data which is necessary to prove the existence and uniqueness of solutions. We construct the Green function of the solution of the linear problem and formulate Theorem 22 on the existence and uniqueness of the solution. In Section 3 we prove some preliminary estimates in Lemma 22. Section 4 is devoted to the proof of Theorem 22 for the linear problem. In Section 5 we prove Theorem 23 on the local existence of solutions to the nonlinear problem (1.1). Theorem 21 is proved in Section 6.

2. Linear problem

We consider the linear initial-boundary value problem corresponding to (1.1)

$$(2.1) \quad \begin{cases} u_t - u_{xx} + (-1)^\alpha u_{xxx} = f(x,t), & t > 0, x > 0, \\ u(x,0) = u_0(x), & x > 0, \\ u_x^j(0,t) = 0, \ t > 0, \ \text{for} \ \ j = 0, \alpha, \end{cases}$$

where $\alpha = 0, 1$.

In this section we obtain the explicit formula for the solution of the linear problem (2.1) under the condition

$$u_0 \in \mathbf{L}^1\left(\mathbf{R}^+\right), f \in \mathbf{L}^q\left(0, T; \mathbf{L}^1\left(\mathbf{R}^+\right)\right)$$

with $q > 2$.

We denote $K_\alpha(p) = p^2 + (-1)^\alpha p^3$. We prove the following integral representation of the solution $u(x,t)$ to the problem (2.1)

$$(2.2) \quad u(x,t) = \int_0^{+\infty} u_0(y) F_\alpha(x,y,t) dy + \int_0^t d\tau \int_0^{+\infty} f(y,\tau) F_\alpha(x,y,t-\tau) dy,$$

where

$$F_0(x,y,t) = -\frac{1}{4\pi^2} \int_{-i\infty}^{i\infty} dp e^{-py} \int_{-i\infty}^{i\infty} e^{\xi t} \frac{e^{\phi_3(\xi)x} + e^{px}}{K_0(p) + \xi} d\xi.$$

Here $\phi_3(\xi)$ is the root of the equation $K_0(p) = -\xi$, such that $\operatorname{Re}\phi_3(\xi) < 0$ for $\operatorname{Re}\xi > 0$. In the case $\alpha = 1$ we have

$$F_1(x,y,t) = \frac{1}{4\pi^2} \int_{-i\infty}^{i\infty} dp e^{px} \int_{-i\infty}^{i\infty} e^{\xi t} \frac{e^{-\tilde{\phi}_3(\xi)y} - e^{-py}}{K_1(p) + \xi} d\xi,$$

where $\tilde{\phi}_3(\xi)$ is the root of the equation $K_1(p) = -\xi$, such that $\operatorname{Re}\tilde{\phi}_3(\xi) > 0$ for $\operatorname{Re}\xi > 0$. Note that the integrals in the definition of functions F_0, F_1 are convergent absolutely for all $t > 0$. In fact, since

$$\frac{1}{|a + ib|} \leq \frac{1}{|a|^\nu |b|^{1-\nu}}$$

for any $a, b \in \mathbf{R}$, where $\nu \in [0, 1]$, we have

$$\frac{1}{|K_\alpha(p) + \xi|} \leq |p|^{-2\nu} \left|\xi + (-1)^\alpha p^3\right|^{-1+\nu}$$

for all $\operatorname{Re} p = 0, \operatorname{Re}\xi = 0, p, \xi \neq 0$, where $\nu \in [0,1]$. Therefore integrating by parts with respect to ξ in the domain $|\xi| > 1$we get

$$
\begin{aligned}
|F_\alpha| \;\leq\; & C \int_{-i\infty}^{i\infty} d|p| \int_{-i}^{i} \frac{|d\xi|}{|K_\alpha(p)+\xi|} \\
& + \frac{C}{t} \int_{-i\infty}^{+i\infty} |dp| \int_{\operatorname{Re}\xi=0,|\xi|>1} \frac{|d\xi|}{|K_\alpha(p)+\xi|^2} \\
& + \frac{C}{t} \int_{-i\infty}^{+i\infty} \frac{|dp|}{|K_\alpha(p)\pm i|} \\
\leq\; & C \max(t^{-1}, 1).
\end{aligned}
$$

In Chapter 3 we defined the Green operator as follows

$$
(2.3) \qquad \mathcal{F}\left[v(x,\tau)\right](x,t) = \int_0^{+\infty} F(x,y,t) v(y,\tau) dy.
$$

Using the result of Theorem 13 we have (since $\mathbf{h} = \mathbf{0}$)

$$
u(x,t) = \mathcal{F}[u_0](x,t) + \int_0^t \mathcal{F}[f(\tau)](x, t-\tau)\, d\tau,
$$

where $F(x,y,t)$ is the following function

$$
(2.4)\quad F(x,y,t) = \frac{1}{2\pi i} \int_{-i\infty}^{+i\infty} e^{px} \left(e^{-(K(p)t+py)} - \sum_k C_k \sum_{j=1}^{[\alpha_k]-N} \Theta_j p^{\alpha_k - N - j} \right) dp.
$$

The functions Θ_j are components of vector Θ

$$
\Theta = \frac{1}{2\pi i} \int_{-i\infty+b}^{+i\infty+b} \frac{e^{\xi t}}{K(p)+\xi} A^{-1} \begin{pmatrix} e^{-\phi_1(\xi)y} \\ \cdots \\ e^{-\phi_m(\xi)y} \end{pmatrix} d\xi
$$

where $\phi_i(\xi), i = 1, ..., m$ are inverse functions $\phi_i(\xi) = K^{-1}(-\xi)$ such that $\operatorname{Re}\phi_i(\xi) > 0$ for positive complex plane $\operatorname{Re}\xi > 0$ and

$$
A = \|a_{ij}\|_m^m, \quad a_{ij} = \sum_{[\alpha_k] \geq N+j} C_k \phi_i^{\alpha_k - N - j}
$$

for $N = [\alpha] - m$ is characteristic matrix of operator $\mathbb{K}$ (see Chapter 3).

Since $K_\alpha(p) = p^2 + (-1)^\alpha p^3$ in the case $\alpha = 0$ the equation

$$
-K_0(p) = p^2 - p^3 = \xi
$$

has two roots $\phi_1(\xi)$ and $\phi_2(\xi)$ such that $\phi_l(\xi)$ are analytic and $\operatorname{Re}\phi_l(\xi) > 0$, $l = 1, 2$, for all

$$
\xi \in D_0 \equiv \left\{ \xi \in \mathbf{C} : \operatorname{Re}\xi > 0, \xi \notin \left[0, \frac{4}{27}\right] \right\}.
$$

In the case $\alpha = 1$ the equation

$$
-K_1(p) = p^2 + p^3 = \xi
$$

has only one root $\widetilde{\phi}_3(\xi)$ such that $\operatorname{Re}\widetilde{\phi}_3(\xi) > 0$ for all

$$\xi \in D_1 \equiv \{\xi \in \mathbf{C} : \operatorname{Re}\xi > 0\}.$$

We prove formula (2.2) for a more difficult case $\alpha = 0$ only; for the case $\alpha = 1$ the proof is analogous.

Characteristic 2×2 matrix A_0 has the form

$$A_0 = \begin{vmatrix} -1 + \phi_1 & 1 \\ -1 + \phi_2 & 1 \end{vmatrix}.$$

By a direct calculation we get

$$\begin{aligned}
\Theta &= \frac{1}{2\pi i} \int_{\Gamma_0} \frac{e^{\xi t}}{K(p) + \xi} A_0^{-1} \begin{pmatrix} e^{-\phi_1(\xi)y} \\ e^{-\phi_2(\xi)y} \end{pmatrix} d\xi \\
&= \frac{1}{2\pi i} \int_{\Gamma_0} \frac{e^{\xi t}}{(K(p) + \xi)(\phi_1 - \phi_2)} \\
&\quad \times \begin{pmatrix} e^{-\phi_1 y} - e^{-\phi_2 y} \\ e^{-\phi_1 y} - e^{-\phi_2 y} + e^{-\phi_2 y}\phi_1 - e^{-\phi_1 y}\phi_2 \end{pmatrix} d\xi,
\end{aligned}$$

where

$$\Gamma_0 = (-i\infty, -i0) \cup \left[-i0, \frac{4}{27} - i0\right] \cup \left[\frac{4}{27} + i0, i0\right] \cup (i0, i\infty).$$

Therefore from the formula (2.4) we get in the case $\alpha = 0$

$$\begin{aligned}
F_0(x, y, t) &= \frac{1}{2\pi i} \int_{-i\infty}^{+i\infty} e^{px} \left(e^{-(K(p)t + py)} - \sum_k C_k \sum_{j=1}^{[\alpha_k] - N} \Theta_j p^{\alpha_k - N - j} \right) dp \\
&= \frac{1}{2\pi i} \int_{-i\infty}^{+i\infty} e^{px} \left(e^{-(K(p)t + py)} - (-\Theta_1 + \Theta_1 p + \Theta_2) \right) dp \\
&= \frac{1}{2\pi i} \int_{-i\infty}^{i\infty} e^{px} e^{-(K(p)t + py)} dp \\
&\quad + \frac{1}{4\pi^2} \int_{\Gamma_0} \frac{e^{\xi t}}{(K(p) + \xi)(\phi_1 - \phi_2)} \left(e^{-\phi_1 y}(p - \phi_2) \right. \\
&\quad \left. + e^{-\phi_2 y}(\phi_1 - p) \right) dp.
\end{aligned}$$

Let us denote

$$\widetilde{H}(p, \phi_1, \phi_2, y) = \frac{(p - \phi_2)e^{-\phi_1 y} + (\phi_1 - p)e^{-\phi_2 y}}{\phi_1 - \phi_2}.$$

Using the relations

$$\overline{\phi_1\left(\xi\right)} = \phi_2\left(\overline{\xi}\right)$$

for all

$$\xi \in D_0 \equiv \left\{\xi \in \mathbf{C} : \operatorname{Re}\xi > 0, \xi \notin \left[0, \frac{4}{27}\right]\right\}$$

and $\widetilde{H}(p, \phi_1, \phi_2, y) = \widetilde{H}(p, \phi_2, \phi_1, y)$, we can change the contour of integration Γ_0 to the imaginary axis $(-i\infty, i\infty)$ to get

$$\int_{\Gamma_0} \widetilde{H}(p, \phi_1, \phi_2, y) \frac{e^{\xi t}}{K_0(p) + \xi} d\xi$$

$$= \int_{-i\infty}^{i\infty} \widetilde{H}(p, \phi_1, \phi_2, y) \frac{e^{\xi t}}{K_0(p) + \xi} d\xi.$$

Applying the identities

$$K_0(p) + \xi = (p - \phi_1)(p - \phi_2)(p - \phi_3),$$

$$\frac{1}{(\phi_1 - \phi_2)(\phi_1 - \phi_3)} = \phi_1'(\xi),$$

$$\frac{1}{(\phi_2 - \phi_1)(\phi_2 - \phi_3)} = \phi_2'(\xi),$$

and using the theory of residues, we obtain

$$\int_{-i\infty}^{+i\infty} \frac{\widetilde{H}(p, \phi_1, \phi_2, y)}{K_0(p) + \xi} e^{px} dp$$

$$= 2\pi i \frac{\widetilde{H}(\phi_3, \phi_1, \phi_2, y)}{(\phi_3 - \phi_1)(\phi_3 - \phi_2)} e^{\phi_3(\xi)x}$$

$$= 2\pi i e^{\phi_3(\xi)x} \left(e^{-\phi_1(\xi)y} \phi_1'(\xi) + e^{-\phi_2(\xi)y} \phi_2'(\xi) \right)$$

$$= -e^{\phi_3(\xi)x} \int_{-i\infty}^{i\infty} \frac{1}{K_0(p) + \xi} e^{-py} dp.$$

Also since

$$e^{-K_0(p)t} = \frac{1}{2\pi i} \int_{-i\infty}^{i\infty} e^{\xi t} \frac{1}{K_0(p) + \xi} d\xi$$

we can rewrite

$$\frac{1}{2\pi i} \int_{-i\infty}^{i\infty} e^{px} e^{-(K_0(p)t + py)} dp$$

$$= -\frac{1}{4\pi^2} \int_{-i\infty}^{i\infty} e^{p(x-y)} dp \int_{-i\infty}^{i\infty} e^{\xi t} \frac{1}{K_0(p) + \xi} d\xi.$$

Therefore we have for function $F_0(x, y, t)$

$$F_0(x, y, t) = -\frac{1}{4\pi^2} \int_{-i\infty}^{i\infty} e^{-py} dp \int_{-i\infty}^{i\infty} e^{\xi t} \frac{e^{\phi_3(\xi)x} + e^{px}}{K_0(p) + \xi} d\xi.$$

Thus the formula (2.2) is proved in the case $\alpha = 0$.

Now we formulate the following result, which will be proved below in Section 4.

THEOREM 22. *Let*

$$u_0 \in \mathbf{H}_2^{1,\omega}\left(\mathbf{R}^+\right), t^\nu f \in \mathbf{L}^\infty\left(0, T; \mathbf{L}^1\left(\mathbf{R}^+\right)\right)$$

with $\omega \in \left(\frac{1}{2}, \frac{3}{2}\right)$, $\nu \in \left(0, \frac{3}{4}\right)$. *Then for some* $T > 0$ *there exists a unique solution*

$$u \in \mathbf{C}\left((0, T]; \mathbf{H}_2^{1,\omega}\left(\mathbf{R}^+\right)\right) \cap \mathbf{C}\left([0, T]; \mathbf{H}_2^{0,\varkappa}\left(\mathbf{R}^+\right)\right) \cap \mathbf{C}\left((0, T; \mathbf{H}_2^{2,0}\left(\mathbf{R}^+\right)\right)$$

of the initial-boundary value problem (2.1) such that

$$\sup_{t \in [0,T]}\left(\|u(t)\|_{\mathbf{H}_2^{0,\varkappa}} + \sum_{n=1}^2 t^{\frac{n}{4}}\|\partial_x^n u(t)\|_{\mathbf{L}^2} + \|(\cdot)^\omega u_x(\cdot, t)\|_{\mathbf{L}^2}\right) \leq C\lambda,$$

where

$$\lambda = \|u_0\|_{\mathbf{H}_2^{1,\omega}} + T^{\frac{3}{4}-\nu} \sup_{t \in [0,T]} t^\nu \left(\|f(t)\|_{\mathbf{H}_1^{0,\delta}} + \|\partial_x f(t)\|_{\mathbf{L}^1}\right)$$

and $\varkappa \in \left(0, \frac{1}{2}\right)$, $\delta \in \left[0, \frac{1}{2}\right)$.

To prove this result we prove some preliminary estimates in the next section.

3. Preliminaries

As in Section 2 we denote by $\phi_l(\xi)$, $l = 1, 2, 3$ the roots of the equation

$$-K_0\left(\xi\right) = p^2 - p^3 = \xi,$$

such that

$$\operatorname{Re}\phi_l(\xi) > 0, l = 1, 2, \operatorname{Re}\phi_3(\xi) < 0,$$

for all

$$\xi \in D_0 = \left\{\xi \in \mathbf{C} : \operatorname{Re}\xi \geq 0, \xi \notin \left[0, \frac{4}{27}\right]\right\}.$$

Note that the functions $\phi_l(\xi)$ are analytic in the domain

$$\left\{\xi \in \mathbf{C} : \xi \notin \left(+\infty, \frac{4}{27}\right]\right\}.$$

Note that if $p = |p|\, e^{i\pi/4}$, $|p| \to 0$, then

$$\xi = p^2 - p^3 = i\,|p|^2 + O\left(|p|^3\right) \to 0$$

and $\operatorname{Im}\xi > 0$. On the other hand if $p = 1 + i\rho$, $\rho \to 0$, then we have

$$\xi = p^2 - p^3 = -i\rho + O\left(\rho^2\right) \to 0$$

and $\operatorname{Im}\xi < 0$. Therefore we represent

$$p^2 = \frac{\xi}{1-p} \text{ or } p = 1 - \frac{\xi}{p^2} \text{for } |p| \leq 1 \text{ and}$$

$$p^3 = \frac{-\xi}{1 - \frac{1}{p}} \text{ for } |p| > 1;$$

hence we get the asymptotic representations

$$(3.1) \qquad \phi_1(\xi) = \begin{cases} \sqrt{\xi} + O(|\xi|), \xi \to 0, \operatorname{Im}\xi > 0, \\ 1 + O(|\xi|), \xi \to 0, \operatorname{Im}\xi < 0, \\ e^{i\frac{\pi}{3}}\sqrt[3]{\xi} + O(1), |\xi| \to +\infty, \end{cases}$$

$$(3.2) \qquad \phi_2(\xi) = \begin{cases} 1 + O(|\xi|), \xi \to 0, \operatorname{Im}\xi > 0, \\ \sqrt{\xi} + O(|\xi|), \xi \to 0, \operatorname{Im}\xi < 0, \\ e^{-i\frac{\pi}{3}}\sqrt[3]{\xi} + O(1), |\xi| \to +\infty, \end{cases}$$

and

$$(3.3) \qquad \phi_3(\xi) = \begin{cases} -\sqrt{\xi} + O(|\xi|), |\xi| \to 0, \\ -\sqrt[3]{\xi} + O(1), |\xi| \to +\infty, \end{cases}$$

for all

$$\xi \in \left\{ \mathbf{C} : \xi \notin \left(+\infty, \frac{4}{27}\right] \right\}$$

(by $\sqrt{\xi}$ and $\sqrt[3]{\xi}$ we denote the main value of the analytic function, i.e. $\sqrt{1} = \sqrt[3]{1} = 1$.) In the case $\alpha = 1$ the equation

$$-K_1(p) = p^2 + p^3 = \xi$$

has the roots

$$\widetilde{\phi}_l(\xi) = -\phi_l(\xi), l = 1, 2, 3,$$

such that

$$\operatorname{Re}\widetilde{\phi}_1(\xi) < 0, \operatorname{Re}\widetilde{\phi}_2(\xi) < 0, \operatorname{Re}\widetilde{\phi}_3(\xi) > 0$$

for all

$$\xi \in D_1 = \{\xi \in \mathbf{C} : \operatorname{Re}\xi \geq 0, \xi \neq 0\}.$$

Note that the functions $\widetilde{\phi}_l(\xi)$ are analytic in the domain

$$\{\xi \in \mathbf{C} : \xi \notin (+\infty, 0]\}$$

and have the following asymptotic representations

$$(3.4) \qquad \widetilde{\phi}_1(\xi) = \begin{cases} -\sqrt{\xi} + O(|\xi|), \xi \to 0, \operatorname{Im}\xi > 0, \\ -1 + O(|\xi|), \xi \to 0, \operatorname{Im}\xi < 0, \\ -e^{i\frac{\pi}{3}}\sqrt[3]{\xi} + O(1), |\xi| \to +\infty, \end{cases}$$

$$(3.5) \qquad \widetilde{\phi}_2(\xi) = \begin{cases} -1 + O(|\xi|), \xi \to 0, \operatorname{Im}\xi > 0, \\ -\sqrt{\xi} + O(|\xi|), \xi \to 0, \operatorname{Im}\xi < 0, \\ -e^{-i\frac{\pi}{3}}\sqrt[3]{\xi} + O(1), |\xi| \to +\infty, \end{cases}$$

and

$$(3.6) \qquad \widetilde{\phi}_3(\xi) = \begin{cases} \sqrt{\xi} + O(|\xi|), |\xi| \to 0, \\ \sqrt[3]{\xi} + O(1), |\xi| \to +\infty \end{cases}$$

for all

$$\xi \in \{\mathbf{C} : \xi \notin (+\infty, 0]\}.$$

As in the previous section we denote

$$F_0(x,y,t) = -\frac{1}{4\pi^2}\int_{-i\infty}^{i\infty}dp\,e^{-py}\int_{-i\infty}^{i\infty}d\xi\,e^{\xi t}\frac{e^{\phi_3(\xi)x}+e^{px}}{K_0(p)+\xi},$$

$$(3.7) \qquad F_1(x,y,t) = \frac{1}{4\pi^2}\int_{-i\infty}^{i\infty}dp\,e^{px}\int_{-i\infty}^{i\infty}d\xi\,e^{\xi t}\frac{e^{-\tilde\phi_3(\xi)y}-e^{-py}}{K_1(p)+\xi},$$

for all $x,y,t \geq 0$, where $K_0(p) = -p^2 + p^3$ and $K_1(p) = -p^2 - p^3$.

In the next lemma we prove the asymptotics of the functions F_α for large t.

LEMMA 22. *Let* $n = 0,1,2,3,\ \gamma > 0,\ \delta \geq 0$. *We have for* $t > 1$,

$$(3.8) \qquad \left\|(\cdot)^\delta\,\partial_x^n F_\alpha(\cdot,y,t)\right\|_{L^2} \leq Cyt^{-\frac{3}{4}+\frac{n+\delta}{2}+\gamma},$$

and

$$(3.9) \qquad F_\alpha(x,y,t) = t^{-1}\Lambda_\alpha(y)\Phi_\alpha\left(\frac{x}{2\sqrt{t}},t\right)$$
$$+y^{1+\mu}O\left(t^{-1-\frac{\mu}{2}}\min\left(\frac{x}{\sqrt{t}},1\right)\right)$$

for any $x,y,t > 0$, *where*

$$\mu \in (0,1),\ \alpha = 0,1,\ \Lambda_0(y) = e^{-y}-1+y,\ \Lambda_1(y) = y,$$

$$\Phi_0(q,t) = \frac{q}{\sqrt\pi}e^{-q^2},\ \Phi_1(q,t) = \frac{1}{2\sqrt{\pi t}}\left(e^{-q^2}\left(2q\sqrt{t}-1\right)+e^{-2q\sqrt{t}}\right).$$

First we show the estimate (3.9) for the function $F_1(x,y,t)$. Since $K_1'(p) = 0$ for $p = 0$ and $p = -\frac{2}{3}$, we have $\tilde\phi_1(\xi) \neq \tilde\phi_2(\xi)$ for all $\operatorname{Re}\xi = 0$. Also we effortlessly get

$$\sum_{j=1}^{3}\tilde\phi_j(\xi) = -1$$

and $\tilde\phi_1\tilde\phi_2\tilde\phi_3 = \xi$. Since

$$K_1'(\tilde\phi_j(\xi)) = -\frac{1}{\tilde\phi_j'(\xi)},$$

by the theory of residues we get

$$(3.10) \qquad \int_{-i\infty}^{+i\infty}\frac{e^{\xi t}}{K_1(p)+\xi}d\xi = 2\pi i e^{-K_1(p)t}$$

and

$$(3.11) \qquad \int_{-i\infty}^{+i\infty}\frac{e^{px}}{\xi+K_1(p)}dp = 2\pi i\sum_{j=1}^{2}\operatorname{res}_{p=\tilde\phi_j(\xi)}\frac{e^{px}}{\xi+K_1(p)}$$
$$= -2\pi i\sum_{j=1}^{2}e^{\tilde\phi_j(\xi)x}\tilde\phi_j'(\xi).$$

Also changing the order of integration and using the theory of residues we get for the function F_1

$$F_1(x,y,t) = \frac{1}{2\pi i} \left(\int_{-i\infty}^{+i\infty} d\xi e^{\xi t - \widetilde{\phi}_3(\xi)y} \sum_{j=1}^{2} e^{\widetilde{\phi}_j(\xi)x} \widetilde{\phi}_j'(\xi) \right.$$

(3.12)
$$\left. + \int_{-i\infty}^{+i\infty} dp e^{p(x-y)-K_1(p)t} \right) = M(x,y,t) + R(x,y,t),$$

where

$$M(x,y,t) = \frac{1}{2\pi i} \left(\int_{-i\infty}^{+i\infty} d\xi e^{\xi t} \left(1 - \widetilde{\phi}_3(\xi)y \right) \sum_{j=1}^{2} e^{\widetilde{\phi}_j(\xi)x} \widetilde{\phi}_j'(\xi) \right.$$

$$\left. + \int_{-i\infty}^{+i\infty} dp \, (1-py) \, e^{px - K_1(p)t} \right)$$

and

$$R(x,y,t)$$
$$= \frac{1}{2\pi i} \left(\int_{-i\infty}^{+i\infty} d\xi e^{\xi t} \left(e^{-\widetilde{\phi}_3(\xi)y} - 1 + \widetilde{\phi}_3(\xi)y \right) \sum_{j=1}^{2} e^{\widetilde{\phi}_j(\xi)x} \widetilde{\phi}_j'(\xi) \right.$$

$$\left. + \int_{-i\infty}^{+i\infty} dp \, \left(e^{-py} - 1 + py \right) e^{px - K_1(p)t} \right).$$

Using the formulas (3.10) and (3.11) and changing the order of integration we have

$$\int_{-i\infty}^{+i\infty} e^{px - K_1(p)t} (1 - py) dp$$

$$= \int_{-i\infty}^{+i\infty} e^{\xi t} d\xi \int_{-i\infty}^{+i\infty} \frac{e^{px}(1-py)}{K_1(p) + \xi} dp$$

$$= - \int_{-i\infty}^{+i\infty} e^{\xi t} \sum_{j=1}^{2} e^{\widetilde{\phi}_j(\xi)x}(1 - \widetilde{\phi}_j(\xi)y)\widetilde{\phi}_j'(\xi) d\xi.$$

Then via the identity

$$(\widetilde{\phi}_3(\xi) - \widetilde{\phi}_j(\xi))\widetilde{\phi}_j'(\xi) = \frac{(-1)^j}{\widetilde{\phi}_1 - \widetilde{\phi}_2},$$

$j = 1,2$, we write the function $M(x,y,t)$ as

$$M(x,y,t) = \frac{y}{2\pi i} \int_{-i\infty}^{i\infty} e^{\xi t} \frac{e^{\widetilde{\phi}_1(\xi)x} - e^{\widetilde{\phi}_2(\xi)x}}{\widetilde{\phi}_1 - \widetilde{\phi}_2} d\xi.$$

Similarly for the function $R(x, y, t)$ we obtain

$$\int_{-i\infty}^{i\infty} dp e^{-K_1(p)t} (e^{-py} - 1 + py)$$

$$= \frac{1}{2\pi i} \int_{-i\infty}^{i\infty} e^{\xi t} d\xi \int_{-i\infty}^{i\infty} \frac{e^{-py} - 1 + py}{K_1(p) + \xi} dp$$

$$= -\int_{-i\infty}^{i\infty} e^{\xi t} (e^{-\phi_3(\xi)y} - 1 + \phi_3(\xi)y) \sum_{j=1}^{2} \phi_j'(\xi) d\xi$$

since

$$\sum_{j=1}^{3} \widetilde{\phi}_j'(\xi) = 0, \operatorname{Re} \widetilde{\phi}_3 > 0$$

and $K_1(p) = O(|p|^3)$ as $|p| \to +\infty$. Therefore we can rewrite the function $R(x, y, t)$ as

$$
\begin{aligned}
R(x, y, t) = \ & \frac{1}{2\pi i} \left(\int_{-i\infty}^{+i\infty} d\xi e^{\xi t} (e^{-\widetilde{\phi}_3(\xi)y} - 1 + \widetilde{\phi}_3(\xi)y) \sum_{j=1}^{2} \left(e^{\widetilde{\phi}_j(\xi)x} - 1 \right) \widetilde{\phi}_j'(\xi) \right. \\
(3.13) \qquad & + \left. \int_{-i\infty}^{+i\infty} dp e^{-K_1(p)t} \left(e^{px} - 1 \right) \left(e^{-py} - 1 + py \right) \right).
\end{aligned}
$$

First we estimate the function R. We have $\operatorname{Re} \widetilde{\phi}_l(\xi) < 0$, $l = 1, 2$, $\operatorname{Re} \widetilde{\phi}_3(\xi) > 0$ on the imaginary axis $\operatorname{Re} \xi = 0$, $\xi \neq 0$. There is a contour in the complex left-half plane $\operatorname{Re} \xi < 0$ such that $\operatorname{Re} \widetilde{\phi}_l(\xi) = 0$; the contour is defined by equation $K_1(iy) = -\xi$, i.e. $-iy^3 - y^2 = \xi$ with $y = \operatorname{Im} \widetilde{\phi}_l(\xi) \in \mathbf{R}$. Therefore there exists a contour

$$\mathcal{C} = \left\{ \xi \in \mathbf{C}, \ \operatorname{Re} \xi < 0, \operatorname{Re} \xi = O\left(|\xi|\right), |\xi| \leq 1, \operatorname{Re} \xi = O\left(|\xi|^{\frac{2}{3}}\right), |\xi| > 1 \right\}$$

such that

$$\operatorname{Re} \widetilde{\phi}_l(\xi) \leq 0, l = 1, 2, \operatorname{Re} \widetilde{\phi}_3(\xi) \geq 0 \text{ for } \xi \in \mathcal{C}.$$

Note that asymptotic formulas (3.4)-(3.6) are valid on the contour $\mathcal{C}$. Then using (3.6) we have for $\xi \in \mathcal{C}$, $|\xi| < 1$

$$e^{-\widetilde{\phi}_3(\xi)y} - 1 + \widetilde{\phi}_3(\xi)y = y^{1+\mu} O\left(|\xi|^{\frac{1+\mu}{2}}\right),$$

where $\mu \in [0, 1]$. Since by (3.4), (3.5) we have

$$\widetilde{\phi}_j'(\xi) = O(|\xi|^{-\frac{1}{2}}) \text{ for } \xi \to -(-1)^j i0$$

and

$$\widetilde{\phi}_j'(\xi) = O(1) \text{ for } \xi \to (-1)^j i0, \xi \in \mathcal{C},$$

where $j = 1, 2$, we get in the case $|\xi| \leq 1$, $\xi \in \mathcal{C}$

$$\left| \sum_{j=1}^{2} (e^{\widetilde{\phi}_j(\xi)x} - 1)\widetilde{\phi}'_j(\xi) \right|$$

$$\leq \frac{C}{\sqrt{|\xi|}}(e^{-C\sqrt{|\xi|}x} - 1) + C(e^{-Cx} - 1)$$

$$\leq C \min \left(x, |\xi|^{-\frac{1}{2}} \right).$$

In the case $|\xi| > 1$, $\xi \in \mathcal{C}$ we have

$$\left| \sum_{j=1}^{2} (e^{\widetilde{\phi}_j(\xi)x} - 1)\widetilde{\phi}'_j(\xi) \right|$$

$$\leq C|\xi|^{-\frac{2}{3}}(e^{-C\sqrt[3]{|\xi|}x} - 1) \leq C \min \left(x|\xi|^{-\frac{1}{3}}, |\xi|^{-\frac{2}{3}} \right)$$

$$\leq C \min \left(x, |\xi|^{-\frac{1}{2}} \right).$$

Thus

$$\sum_{j=1}^{2} (e^{\widetilde{\phi}_j(\xi)x} - 1)\widetilde{\phi}'_j(\xi) = O \left(\min \left(x\sqrt{|\xi|}, 1 \right) |\xi|^{-\frac{1}{2}} \right)$$

for all $\xi \in \mathcal{C}$. Therefore we have

$$\left(e^{-\widetilde{\phi}_3(\xi)y} - 1 + \widetilde{\phi}_3(\xi)y \right) \sum_{j=1}^{2} (e^{\widetilde{\phi}_j(\xi)x} - 1)\widetilde{\phi}'_j(\xi)$$

$$= y^{1+\mu} O \left(\min \left(x\sqrt{|\xi|}, 1 \right) |\xi|^{\frac{\mu}{2}} \right)$$

for all $x, y > 0$, $\xi \in \mathcal{C}$. In the same manner we obtain

$$(e^{px} - 1)(e^{-py} - 1 + py) = y^{1+\mu} O \left(\min \left(x|p|, 1 \right) |p|^{1+\mu} \right)$$

for all $x, y > 0$, $\operatorname{Re} p = 0$, where $\mu \in [0, 1]$. Therefore making the changes of the variables of integration $p^2 t = z^2$ and $\xi t = q$ in the formula (3.13) we get

$$(3.14) \qquad R(x, y, t) = y^{1+\mu} O \left(\min \left(\frac{x}{\sqrt{t}}, 1 \right) t^{-1-\frac{\mu}{2}} \right).$$

Now we estimate the main term of asymptotics $M(x, y, t)$. Via (3.4) and (3.5) we can write in the case

$$x\sqrt{|\xi|} \leq 1, |\xi| < 1, \xi \in \mathcal{C}$$

$$\frac{e^{\tilde{\phi}_1(\xi)x} - e^{\tilde{\phi}_2(\xi)x}}{\tilde{\phi}_1(\xi) - \tilde{\phi}_2(\xi)}$$

$$= \frac{e^{-\sqrt{\xi}x + O(|\xi|x)} - e^{-x + O(|\xi|x)}}{1 - \sqrt{\xi} + O(|\xi|)}$$

$$= \left(e^{-\sqrt{\xi}x}\left(1 + O\left(|\xi|x\right)\right) - e^{-x}\left(1 + O\left(|\xi|x\right)\right)\right)\left(1 + \sqrt{\xi} + O\left(|\xi|\right)\right)$$

$$= \left(e^{-\sqrt{\xi}x} - e^{-x}\right)\left(1 + \sqrt{\xi}\right) + O\left(|\xi|x\right),$$

and in the case

$$x\sqrt{|\xi|} > 1, |\xi| < 1, \xi \in \mathcal{C}, \operatorname{Im}\xi > 0$$

(with the case $\operatorname{Im}\xi < 0$ considered in the same manner), we have

$$\left|e^{\tilde{\phi}_1(\xi)x} - e^{-\sqrt{\xi}x}\right| \le e^{-C\sqrt{|\xi|}x}O\left(|\xi|x\right) = O\left(\sqrt{|\xi|}\right);$$

and

$$\left|e^{\tilde{\phi}_2(\xi)x}\right| \le \frac{C}{x} \le C\sqrt{|\xi|}, \quad e^{-x} \le \frac{C}{x} \le C\sqrt{|\xi|},$$

therefore we obtain

(3.15)
$$\frac{e^{\tilde{\phi}_1(\xi)x} - e^{\tilde{\phi}_2(\xi)x}}{\tilde{\phi}_1(\xi) - \tilde{\phi}_2(\xi)} = \left(e^{-\sqrt{\xi}x} - e^{-x}\right)\left(1 + \sqrt{\xi}\right)$$

$$+ O\left(\min\left(\sqrt{|\xi|}x, 1\right)\sqrt{|\xi|}\right)$$

for all $|\xi| < 1, \xi \in \mathcal{C}$. Again by virtue of asymptotics (3.4) and (3.5) in the case $x\sqrt[3]{|\xi|} \le 1$ we get

$$\left|e^{\tilde{\phi}_1(\xi)x} - e^{\tilde{\phi}_2(\xi)x}\right| \le C\left|\tilde{\phi}_1(\xi) - \tilde{\phi}_2(\xi)\right|x \le Cx\sqrt[3]{|\xi|},$$

and if $x\sqrt[3]{|\xi|} \ge 1$ we have

$$\left|e^{\tilde{\phi}_1(\xi)x} - e^{\tilde{\phi}_2(\xi)x}\right| \le C;$$

hence

$$\left|\frac{e^{\tilde{\phi}_1(\xi)x} - e^{\tilde{\phi}_2(\xi)x}}{\tilde{\phi}_1(\xi) - \tilde{\phi}_2(\xi)}\right| \le \left|\frac{\tilde{\phi}_1(\xi) - \tilde{\phi}_2(\xi)}{O\left(\sqrt[3]{|\xi|}\right)}\right|x$$

$$\le C\min\left(x\sqrt[3]{|\xi|}, 1\right)\frac{1}{\sqrt[3]{|\xi|}}$$

$$\le C\sqrt{|\xi|}\min\left(\sqrt{|\xi|}x, 1\right)$$

for $|\xi| \ge 1, \xi \in \mathcal{C}$. We then see that representation (3.15) is valid for all $\xi \in \mathcal{C}$. By virtue of the estimate

$$\int_{\mathcal{C}} e^{\xi t} O\left(\min\left(\sqrt{|\xi|}x, 1\right)\sqrt{|\xi|}\right) d\xi = O\left(\min\left(\frac{x}{\sqrt{t}}, 1\right)t^{-\frac{3}{2}}\right)$$

we write the representation for the function $M(x, y, t)$

$$M(x, y, t) = \frac{y}{2\pi i} \int_C e^{\xi t} \left(e^{-\sqrt{\xi}x} - e^{-x} \right) \left(1 + \sqrt{\xi} \right) d\xi$$
$$+ yO \left(\min \left(\frac{x}{\sqrt{t}}, 1 \right) t^{-\frac{3}{2}} \right).$$

Since

$$\int_{-\infty}^{+\infty} e^{-q^2} dq = \sqrt{\pi}$$

and

$$\int_{-\infty}^{+\infty} e^{-q^2} q^2 dq = \frac{\sqrt{\pi}}{2},$$

changing the variables of integration $\xi t - \sqrt{\xi}x = -\eta^2$ and $\xi t = -q^2$ we obtain

$$\int_C e^{\xi t} \left(e^{-\sqrt{\xi}x} - e^{-x} \right) \left(1 + \sqrt{\xi} \right) d\xi$$
$$= \sqrt{\pi} i t^{-\frac{3}{2}} \left(e^{-\frac{x^2}{4t}} (x - 1) + e^{-x} \right).$$

Therefore we get

(3.16) $$M(x, y, t) = \frac{y}{2\sqrt{\pi}} t^{-\frac{3}{2}} \left(e^{-\frac{x^2}{4t}} (x - 1) + e^{-x} \right)$$
$$+ yO \left(\min \left(\frac{x}{\sqrt{t}}, 1 \right) t^{-\frac{3}{2}} \right).$$

From (3.12)-(3.16) we have the asymptotics (3.9) with $\alpha = 1$. Now we show the asymptotics (3.9) for the case $\alpha = 0$. Similarly to the case $\alpha = 1$ we rewrite the function $F_0(x, y, t)$ as

$$F_0(x, y, t) = \frac{1}{2\pi i} \left(\int_{-i\infty}^{+i\infty} d\xi e^{\xi t} e^{\phi_3(\xi)x} \sum_{j=1}^{2} e^{-\phi_j(\xi)y} \phi_j'(\xi) \right.$$

(3.17) $$\left. + \int_{-i\infty}^{+i\infty} dp e^{p(x-y) - K_0(p)t} \right).$$

Since

$$\sum_{j=1}^{3} \phi_j'(\xi) = 0, \sum_{j=1}^{3} \phi_j \phi_j'(\xi) = 0$$

and $K_0(p) = O(|p|^3)$ for $|p| \geq 1$ we have

$$\int_{-i\infty}^{+i\infty} d\xi e^{\xi t} e^{\phi_3(\xi)x} \sum_{j=1}^{2} (1 - \phi_j(\xi)y) \phi_j'(\xi)$$
$$= - \int_{-i\infty}^{+i\infty} dp e^{px - K_0(p)t} (1 - py)$$

and

$$\int_{-i\infty}^{+i\infty} d\xi e^{\xi t} \sum_{j=1}^{2} (e^{-\phi_j(\xi)y} - 1 + \phi_j(\xi)y)\phi_j'(\xi)$$

$$= -\int_{-i\infty}^{+i\infty} dp e^{-K_0(p)t}(e^{-py} - 1 + py).$$

Also we change the contour of integration with respect to ξ into $\mathcal{C}$. We denote

$$\mathcal{C}^+ = \{\xi \in \mathcal{C}, \operatorname{Im}\xi > 0\}$$

and

$$\mathcal{C}^- = \{\xi \in \mathcal{C}, \operatorname{Im}\xi < 0\}.$$

Therefore we can write the function $F_0(x, y, t) = M(x, y, t) + R(x, y, t)$, where

$$M(x, y, t)$$
$$= \frac{1}{2\pi i} \int_{\mathcal{C}^-} d\xi e^{\xi t} (e^{\phi_3(\xi)x} - 1)(e^{-\phi_1(\xi)y} - 1 + \phi_1(\xi)y)\phi_1'(\xi)$$
$$+ \frac{1}{2\pi i} \int_{\mathcal{C}^+} d\xi e^{\xi t} (e^{\phi_3(\xi)x} - 1)(e^{-\phi_2(\xi)y} - 1 + \phi_2(\xi)y)\phi_2'(\xi),$$

and the remainder

$$R(x, y, t)$$
$$= \frac{1}{2\pi i} \left(\int_{\mathcal{C}^-} d\xi e^{\xi t} (e^{\phi_3(\xi)x} - 1)(e^{-\phi_2(\xi)y} - 1 + \phi_2(\xi)y)\phi_2'(\xi) \right.$$
$$+ \int_{\mathcal{C}^+} d\xi e^{\xi t} (e^{\phi_3(\xi)x} - 1)(e^{-\phi_1(\xi)y} - 1 + \phi_1(\xi)y)\phi_1'(\xi)$$
$$\left. + \int_{-i\infty}^{i\infty} dp e^{-K_0(p)t}(e^{px} - 1)(e^{-py} - 1 + py) \right).$$

By virtue of (3.1) through (3.3) we have

$$\left(e^{\phi_3(\xi)x} - 1\right)\left(e^{-\phi_j(\xi)y} - 1 + \phi_j(\xi)y\right)\phi_j'(\xi)$$
$$= y^{1+\mu}O\left(\min\left(\sqrt{|\xi|}x, 1\right), |\xi|^{\frac{\mu}{2}}\right)$$

for $\xi \in \mathcal{C}^+$ if $j = 1$ and for $\xi \in \mathcal{C}^-$ if $j = 2$, where $\mu \in [0, 1]$. Also we have

$$(e^{px} - 1)\left(e^{-py} - 1 + py\right) = y^{1+\mu}O\left(\min\left(|p|\,x, 1\right), |p|^{1+\mu}\right)$$

for $\operatorname{Re} p = 0$. Therefore by an analogy to (3.14) we obtain

(3.18) $$R(x, y, t) = y^{1+\mu}O\left(\min\left(\frac{x}{\sqrt{t}}, 1\right)t^{-1-\frac{\mu}{2}}\right).$$

We use (3.1)-(3.3) to find that

$$\left(e^{\phi_3(\xi)x} - 1\right)\left(e^{-\phi_j(\xi)y} - 1 + \phi_j(\xi)y\right)\phi_j'(\xi)$$
$$= \left(e^{-\sqrt{\xi}x} - 1\right)\left(e^{-y} - 1 + y\right)$$
$$+ yO\left(\min\left(1, x\sqrt{|\xi|}\right), \sqrt{|\xi|}\right),$$

for $\xi \in C^+$ if $j = 1$ and for $\xi \in C^-$ if $j = 2$. Therefore similarly to (3.16) we get

$$
\begin{aligned}
M(x, y, t) &= \frac{e^{-y} - 1 + y}{2\pi i}\int_C d\xi e^{\xi t}\left(e^{-\sqrt{\xi}x} - 1\right) \\
&\quad + yO\left(\min\left(\frac{x}{\sqrt{t}}, 1\right)t^{-\frac{3}{2}}\right) \\
&= \frac{e^{-y} - 1 + y}{t\sqrt{\pi}}e^{-\frac{x^2}{4t}}\frac{x}{2\sqrt{t}} \\
&\quad + yO\left(\min\left(\frac{x}{\sqrt{t}}, 1\right)t^{-\frac{3}{2}}\right).
\end{aligned}
$$

(3.19)

By virtue of formulas (3.18) and (3.19) we obtain (3.9) with $\alpha = 0$.

Now we prove estimate (3.8) in the case $\alpha = 0$, $n = 0$ (the case $\alpha = 0, n = 1$ and cases $\alpha = 1, n = 0, 1$ are considered by analogy). Since

$$\int_{-i\infty}^{+i\infty} d\xi e^{\xi t}e^{\phi_3(\xi)x}\sum_{j=1}^{2}\left(1 - \phi_j(\xi)y\right)\phi_j'(\xi)$$
$$= -\int_{-i\infty}^{+i\infty} dp e^{px - K_0(p)t}(1 - py)$$

via (3.17) we have

$$F_0(x, y, t)$$
$$= \frac{1}{2\pi i}\left(\int_{-i\infty}^{+i\infty} d\xi e^{\xi t}e^{\phi_3(\xi)x}\sum_{j=1}^{2}\left(e^{-\phi_j(\xi)y} - 1 + \phi_j(\xi)y\right)\phi_j'(\xi)\right.$$
$$\left. + \int_{-i\infty}^{+i\infty} dp e^{px - K_0(p)t}\left(e^{-py} - 1 + py\right)\right).$$

By virtue of inequalities

$$\left|e^{\phi_3(\xi)x}\sum_{j=1}^{2}\left(e^{-\phi_j(\xi)y} - 1 + \phi_j(\xi)y\right)\phi_j'(\xi)\right|$$
$$\leq \begin{cases} C|\xi|^{-\frac{\mu}{2}}yx^{-\mu}, & \text{if } |\xi| \leq 1, \\ C|\xi|^{-\frac{\mu+1}{3}}yx^{-\mu}, & \text{if } |\xi| > 1 \end{cases}$$

and

$$\left\| (\cdot)^\delta \int_{-i\infty}^{+i\infty} dp e^{p\cdot - K_0(p)t} \left(e^{-py} - 1 + py \right) \right\|_{\mathbf{L}^2}$$

$$\leq \quad Cy \int_{-i\infty}^{+i\infty} e^{-\mathrm{Re}\, K_0(p)t} |p|^{1-\delta} |dp|$$

$$\leq \quad Cy t^{-\frac{3}{4}+\frac{\delta}{2}},$$

we get

$$\left\| (\cdot)^\delta F\left(\cdot, y, t\right) \right\|_{\mathbf{L}^2}$$

$$\leq \quad Cy \int_{\xi \in \mathcal{C}} |\xi|^{-\frac{1}{4}-\frac{\mu}{2}-\gamma} e^{\mathrm{Re}\,\xi t} d\xi + Cy t^{-\frac{3}{4}+\frac{\delta}{2}}$$

$$\leq \quad Cy t^{-\frac{3}{4}+\frac{\delta}{2}+\gamma}.$$

Lemma 22 is proved.

We consider the following functions for $x, y, t \geq 0$

$$(3.20) \qquad G_\alpha(x, t) = \frac{1}{2\pi i} \int_{-i\infty}^{i\infty} e^{px - K_\alpha(p)t} dp$$

and

$$(3.21) \qquad \mathcal{F}_0(x, y, t) \quad = \quad \frac{1}{2\pi i} \int_{-i\infty}^{+i\infty} d\xi e^{\xi t} e^{\phi_3(\xi)x} \sum_{j=1}^{2} e^{-\phi_j(\xi)y} \phi'_j(\xi),$$

$$\mathcal{F}_1(x, y, t) \quad = \quad \frac{1}{2\pi i} \int_{-i\infty}^{+i\infty} d\xi e^{\xi t - \tilde{\phi}_3(\xi)y} \sum_{j=1}^{2} e^{\tilde{\phi}_j(\xi)x} \tilde{\phi}'_j(\xi)$$

which are related to the initial boundary value problem (2.1). From (3.17),(3.17) we have

$$(3.22) \qquad F_\alpha(x, y, t) = G_\alpha(x - y, t) + \mathcal{F}_\alpha(x, y, t).$$

We prove some estimates.

LEMMA 23. *There exists some constant $C > 0$ such that for all $x, y, t > 0$*

$$\left\| (\cdot)^\delta \mathcal{F}_x^{(n)}(\cdot, y, t) \right\|_{\mathbf{L}^2} \leq C \left(1 + y^{-\theta} t^{\frac{2\theta + 2\delta - 2n - 2\gamma - 1}{6}} \right)$$

for $0 < t < 1$, where $\gamma > 0$, $0 \leq \delta < n + \frac{1}{2}$, $0 \leq \theta < n + \frac{1+2\delta}{2}$, $n = 0, 1, 2$.

PROOF. We prove Lemma 23 in the case $\alpha = 1$. Using (3.4)-(3.6) we obtain

$$e^{-\tilde{\phi}_3(\xi)y} \sum_{1}^{2} e^{\tilde{\phi}_j(\xi)x} \tilde{\phi}_j^n(\xi) \tilde{\phi}'_j(\xi)$$

$$= \quad O\left(x^{-\mu} y^{-\theta_1} |\xi|^{-\frac{1+\mu+\theta_1-n}{2}} + x^{-\mu} y^{-\theta_2} |\xi|^{-\frac{1+\theta_2}{2}} \right)$$

for $|\xi| < 1$ and

$$e^{-\widetilde{\phi}_3(\xi)y} \sum_{1}^{2} e^{\widetilde{\phi}_j(\xi)x} \widetilde{\phi}_j^n(\xi) \widetilde{\phi}_j'(\xi)$$

$$= O\left(x^{-\mu} y^{-\theta} |\xi|^{-\frac{2+\mu+\theta-n}{3}}\right)$$

for $|\xi| \geq 1$, where $n = 0, 1, 2$; μ, θ, θ_1, $\theta_2 \geq 0$. Therefore changing the contour of integration to the contour

$$\mathcal{C} = \left\{ \xi \in \mathbf{C}, \ \mathrm{Re}\,\xi < 0, \mathrm{Re}\,\xi = O\left(|\xi|\right), |\xi| \leq 1, \mathrm{Re}\,\xi = O\left(|\xi|^{\frac{2}{3}}\right), |\xi| > 1 \right\}$$

such that $\mathrm{Re}\,\widetilde{\phi}_l(\xi) \leq 0$, $l = 1, 2$ for $\xi \in \mathcal{C}$ we get

$$\left| \mathcal{F}_x^{(n)}(x, y, t) \right|$$

$$\leq \ Cx^{-\mu} \int_{\xi \in \mathcal{C}, |\xi| < 1} |\xi|^{-\frac{1+\mu-n}{2}} |d\xi|$$

$$+ Cx^{-\mu} \int_{\xi \in \mathcal{C}, |\xi| < 1} \frac{|d\xi|}{\sqrt{|\xi|}}$$

$$+ Cx^{-\mu} y^{-\theta} \int_{|\xi| > 1, \xi \in \mathcal{C}} e^{\mathrm{Re}\,\xi t} |\xi|^{-\frac{2+\theta+\mu-n}{3}} |d\xi|$$

$$= \ x^{-\mu} \left(O(1) + O\left(y^{-\theta} t^{-\frac{n-\mu-\theta+1}{3}}\right) \right),$$

where $\mu \in [0, 1 + n)$, $\theta \in [0, 1 + n - \mu)$. Therefore choosing

$$\mu = \frac{1}{2} + \delta \pm \gamma, \delta \in \left[0, \frac{1}{2} + n\right),$$

we obtain

(3.23) $$\left\| (\cdot)^\delta \, \mathcal{F}_x^{(n)}(\cdot, y, t) \right\|_{\mathbf{L}^2} \leq C \left(1 + y^{-\theta} t^{\frac{2\theta + 2\delta - 2n - 2\gamma - 1}{6}} \right)$$

for all $y > 0$, $t \in [0, 1]$. By virtue of (3.23) we get the results of Lemma 23. Lemma 23 is proved.   □

LEMMA 24. *There exists some constant $C > 0$ such that for all $t, x > 0$*

$$t^{\frac{2n+1}{4}} \| \partial_x^n G_\alpha(t) \|_{\mathbf{L}^2} + \| G_\alpha(t) \|_{\mathbf{L}^1} \leq C$$

and

$$\left| \partial_x^n G_\alpha(x, t) \right| \leq C t^{-\frac{1+n-\mu}{2}} x^{-\mu},$$

where $\mu > 0$.

PROOF. We have

$$\partial_p \left(\left(\frac{1}{x} + \frac{K_\alpha'(p)t}{x^2} \right) e^{-K_\alpha(p)t + px} \right) + \frac{e^{-K_\alpha(p)t + px}}{x^2} \left(K_\alpha''(p)t - (K_\alpha'(p))^2 t^2 \right)$$

$$= \ e^{-K(p)t + px}.$$

Therefore changing the contour of integration to $\mathcal{C}_1$ such that $\operatorname{Re} p < 0$, $\operatorname{Re} p^2 < 0$ and $\operatorname{Re} p^3 > 0$ we obtain

$$|G_\alpha(x,t)| \leq \frac{C}{x^2} \left(\int_{|p|\leq 1, p\in\mathcal{C}_1} e^{\operatorname{Re} p^2 t}(t + t^2|p|^2)|dp| \right.$$

$$\left. + \int_{|p|\leq 1, p\in\mathcal{C}_1} e^{-\operatorname{Re} p^3 t}(t|p| + t^2|p|^4)|dp| \right).$$

Then by changing the variables of integration $z^2 = p^2 t$ in the domain $|p| \leq 1$, and $z^3 = p^3 t$ in the domain $|p| \geq 1$, we get

$$|G_\alpha(x,t)| \leq Ct^{\frac{1}{3}}(1 + t)^{\frac{1}{6}}.$$

Therefore we obtain

$$\|G_\alpha(t)\|_{\mathbf{L}^1} \leq C.$$

We have

$$\|G_\alpha(t)\|_{\mathbf{L}^1} \leq C \int_0^{t^{\frac{1}{2}}} dx \int_0^{+\infty} e^{\operatorname{Re} y^2 t} dy + Ct^{\frac{1}{2}} \int_{t^{\frac{1}{2}}}^{+\infty} x^{-2} dx$$

$$\leq C \int_0^{+\infty} e^{\operatorname{Re} y^2} dy + Ct^{\frac{1}{2}} \int_{t^{\frac{1}{2}}}^{+\infty} x^{-2} dx \leq C$$

for $t > 1$. Similarly we obtain this estimate for $t \leq 1$. The Laplace transform of the function $\partial_x^n G_\alpha(x,t)$ is equal to $e^{-K_\alpha(p)t} p^n$, therefore by changing the variables $z^2 = p^2 t$ we get

$$\|\partial_x^n G_\alpha(t)\|_{\mathbf{L}^2} = \|e^{-K_\alpha(p)t} p^n\|_{\mathbf{L}^2}$$

$$\leq Ct^{-\frac{1}{4}-\frac{n}{2}} \left(\int_{-i\infty}^{+i\infty} e^{\operatorname{Re} z^2}|z|^{2n}|dz| \right)^{\frac{1}{2}} \leq Ct^{-\frac{1}{4}-\frac{n}{2}}.$$

These two estimates imply the first estimate of the lemma. Changing the contour of integration into $\mathcal{C}_1$ for $\mu \geq 0$ we have

$$|\partial_x^n G_\alpha(x,t)| \leq Cx^{-\mu} \int_{p\in\mathcal{C}_1} e^{-\operatorname{Re} p^2 t}|p|^{n-\mu}|dp| \leq Cx^{-\mu}t^{-\frac{1+n-\mu}{2}}.$$

Thus the second estimate of the lemma is proved, and Lemma 24 is proved. $\quad\square$

4. Local existence for linear case

In this section we prove Theorem 22. From Section 2 we see that the solution of problem (2.1) can be represented in the following manner

$$(4.1) \quad u(x,t) = \int_0^{+\infty} u_0(y) F_\alpha(x,y,t) dy + \int_0^t d\tau \int_0^{+\infty} dy f(y,\tau) F_\alpha(x,y,t-\tau),$$

where the functions $F_\alpha(x,y,t)$ are defined in (3.22). Let us prove the following estimate

$$(4.2) \qquad\qquad \|u\|_{\mathbf{X}_T^{*,\omega}} \leq \lambda,$$

where

$$\|u\|_{\mathbf{X}_T^{\varkappa,\omega}} = \sup_{t \in (0,T]} \left(\|\langle \cdot \rangle^\varkappa u(\cdot,t)\|_{\mathbf{L}^2} + \|(\cdot)^\omega\, u_x(\cdot,t)\|_{\mathbf{L}^2} + \sum_{n=0}^{2} t^{\frac{n}{4}} \|\partial_x^n u(t)\|_{\mathbf{L}^2} \right),$$

$$\lambda = \|u_0\|_{\mathbf{H}_2^{1,\omega}} + T^{\frac{3}{4}-\nu} \sup_{t \in [0,T]} t^\nu \left(\|\langle \cdot \rangle^\delta f(\cdot,t)\|_{\mathbf{L}^1} + \|\partial_x f(\cdot,t)\|_{\mathbf{L}^1} \right),$$

$$T > 0, \nu \in \left(0, \frac{3}{4}\right), \varkappa \in \left(0, \frac{1}{2}\right), \omega \in \left(\frac{1}{2}, \frac{3}{2}\right), \delta \in \left[0, \frac{3}{2}\right).$$

Here the norm is taken with respect to the space variable x which is denoted by the dot. From (4.1) we have

$$\| (\cdot)^\delta\, \partial_x^n u(\cdot,t) \|_{\mathbf{L}^2}$$

$$\leq\ C \left\| \int_0^{+\infty} u_0(y)\, (\cdot)^\delta\, \partial_x^n F_\alpha(\cdot,y,t) dy \right\|_{\mathbf{L}^2}$$

(4.3)
$$+C \int_0^t d\tau \int_0^{+\infty} |f(y,\tau)| \left\| (\cdot)^\delta\, \partial_x^n F_\alpha(\cdot,y,t-\tau) \right\|_{\mathbf{L}^2} dy,$$

for $\alpha = 0,1,\ n = 0,1,2,\ \delta \geq 0.$ We have

$$\partial_x^n F_\alpha(x,y,t) = \mathcal{F}_{\alpha x}^{(n)}(x,y,t) + G_{\alpha x}^{(n)}(x-y,t),$$

where by virtue of Lemmas (23)- (24)

(4.4)
$$\left\| (\cdot)^\delta\, \mathcal{F}_{\alpha x}^{(n)}(\cdot,y,t) \right\|_{\mathbf{L}^2} \leq C \left(1 + y^{\gamma-1} t^{\frac{1+2\delta-2n-2\gamma}{6}} \right)$$

(4.5)
$$\|G_\alpha(t)\|_{\mathbf{L}^1} + \left\| (\cdot)^\delta\, G_\alpha(\cdot,t) \right\|_{\mathbf{L}^1} \leq C$$

and

(4.6)
$$\left\| (\cdot)^\delta\, G_{\alpha x}^{(n)}(\cdot,t) \right\|_{\mathbf{L}^2} \leq C t^{-\frac{2n+1}{4}}$$

for all $t \in (0,T]$, where $\delta \in \left[0, \frac{1}{2} + n\right).$ Applying the Young inequality

$$\left\| \int_0^{+\infty} g(x-y) v(x) dx \right\|_{\mathbf{L}^2} \leq C \|g\|_{\mathbf{L}^2} \|v\|_{\mathbf{L}^1}$$

we have for $\nu < \frac{3}{4}, \delta \in \left[0, \frac{1}{2}\right)$

$$\| (\cdot)^\delta u(\cdot, t)\|_{\mathbf{L}^2} \leq C \int_0^{+\infty} |u_0(y)| \|| (\cdot)^\delta \mathcal{F}_\alpha(\cdot, y, t)\|_{\mathbf{L}^2} dy$$

$$+C \left\| \int_0^{+\infty} (\cdot)^\delta u_0(y) G_\alpha(\cdot - y, t) dy \right\|_{\mathbf{L}^2}$$

$$+C \sup_{t \in [0,T]} t^\nu \| \langle \cdot \rangle^\delta f(\cdot, t)\|_{\mathbf{L}^1} \int_0^t \frac{d\tau}{\tau^\nu (t - \tau)^{\frac{1}{4}}}$$

$$\leq C \left(t^{\frac{1 + 2\delta - 2\gamma}{6}} \int_0^{+\infty} |u_0(y)| \left(1 + y^{\gamma - 1}\right) dy + \|u_0\|_{\mathbf{L}^2} \left\| (\cdot)^\delta G_x^{(n)}(\cdot, t) \right\|_{L^1} \right.$$

$$\left. + \left\| (\cdot)^\delta u_0 \right\|_{\mathbf{L}^2} \left\| G_x^{(n)}(t) \right\|_{L^1} + T^{\frac{3}{4} - \nu} \sup_{t \in [0,T]} t^\nu \| \langle \cdot \rangle^\delta f(\cdot, t)\|_{\mathbf{L}^1} \right)$$

$$\leq C \left(\|u_0\|_{\mathbf{H}_2^{1, \omega}} + T^{\frac{3}{4} - \nu} \sup_{t \in [0,T]} t^\nu \| \langle \cdot \rangle^\delta f(\cdot, t)\|_{\mathbf{L}^1} \right) \leq C\lambda,$$

and for $\nu < \frac{3}{4}, \delta \in \left[0, \frac{3}{2}\right)$

$$\sup_{t \in [0,T]} t^{\frac{1}{6} - \gamma} \| (\cdot)^\delta u_x(\cdot, t)\|_{\mathbf{L}^2}$$

$$\leq C \sup_{t \in [0,T]} t^{\frac{1}{6} - \gamma} \int_0^{+\infty} |u_0(y)| \|| (\cdot)^\delta H_x(\cdot, y, t)\|_{\mathbf{L}^2} dy$$

$$+C \sup_{t \in [0,T]} t^{\frac{1}{6} - \gamma} \left\| \int_0^{+\infty} (\cdot)^\delta u_0(y) G_x(\cdot - y, t) dy \right\|_{\mathbf{L}^2}$$

$$+C \sup_{t \in [0,T]} t^\nu \| \langle \cdot \rangle^\delta f(\cdot, t)\|_{\mathbf{L}^1} \int_0^t \frac{d\tau}{\tau^\nu (t - \tau)^{\frac{3}{4}}}$$

$$\leq C \left(\int_0^{+\infty} |u_0(y)| \left(1 + y^{\gamma - 1}\right) dy + \|u_0\|_{\mathbf{L}^2} \left\| (\cdot)^\delta G_x^{(n)}(\cdot, t) \right\|_{L^1} \right.$$

$$\left. + \left\| (\cdot)^\delta u_0 \right\|_{\mathbf{L}^2} \left\| G_x^{(n)}(t) \right\|_{L^1} \right.$$

$$\left. + T^{\frac{3}{4} - 1} \sup_{t \in [0,T]} t^\nu \| \langle \cdot \rangle^\delta f(\cdot, t)\|_{\mathbf{L}^1} \right)$$

$$\leq C \left(\|u_0\|_{\mathbf{H}_2^{1, \omega}} + T^{\frac{3}{4} - \nu} \sup_{t \in [0,T]} t^\nu \| \langle \cdot \rangle^\delta f(\cdot, t)\|_{\mathbf{L}^1} \right) \leq C\lambda.$$

Integrating by parts we get for $\nu < \frac{3}{4}$

$$\sup_{t\in[0,T]} t^{\frac{1}{2}-\gamma} \|u_{xx}(t)\|_{\mathbf{L}^2} \leq C \sup_{t\in[0,T]} t^{\frac{1}{2}-\gamma} \int_0^{+\infty} |u_0(y)| \||H_{xx}(\cdot, y, t)\|_{\mathbf{L}^2} dy$$

$$+ C \sup_{t\in[0,T]} t^{\frac{1}{2}-\gamma} \left\| \int_0^{+\infty} u_0(y) G_{xx}(\cdot - y, t) dy \right\|_{\mathbf{L}^2}$$

$$+ C \sup_{t\in[0,T]} t^{\nu} \|f(t)\|_{\mathbf{L}^1} \int_0^{t/2} \frac{d\tau}{\tau^{\nu}(t-\tau)^{\frac{5}{4}}}$$

$$+ C \sup_{t\in[0,T]} t^{\nu} \|\partial_x f(t)\|_{\mathbf{L}^1} \int_{t/2}^{t} \frac{d\tau}{\tau^{\nu}(t-\tau)^{\frac{3}{4}}}$$

$$\leq C \int_0^{+\infty} |u_0(y)| \left(1 + y^{\gamma-1}\right) dy + C \|u_0\|_{\mathbf{L}^2} \|G_{xx}(t)\|_{L^1}$$

$$+ \left\| (\cdot)^{\delta} u_0 \right\|_{\mathbf{L}^2} \|G_{xx}(t)\|_{L^1} + T^{\frac{3}{4}-\nu} \sup_{t\in[0,T]} t^{\nu} \|\partial_x f(t)\|_{\mathbf{L}^1}$$

$$\leq C \|u_0\|_{\mathbf{H}_2^{1,\omega}} + C T^{\frac{3}{4}-\nu} \sup_{t\in[0,T]} t^{\nu} \left(\| \langle\cdot\rangle^{\delta} f(\cdot, t)\|_{\mathbf{L}^1} + \|\partial_x f(t)\|_{\mathbf{L}^1} \right) \leq C\lambda.$$

Thus we get (4.2). We then prove the uniqueness of the solution. Multiplying equation (2.1) with $f = 0$ by $u(x, t)$, integrating the result with respect to $x > 0$ and using identities

$$\int_0^{+\infty} uu_{xx} dx = - \int_0^{+\infty} u_x^2 dx$$

and

$$\int_0^{+\infty} uu_{xxx} dx = \frac{1}{2} u_x^2(0, t)$$

we have

$$\frac{1}{2} \frac{d}{dt} \|u(t)\|_{\mathbf{L}^2}^2 + \int_0^{+\infty} u_x^2 dx + \frac{(-1)^{\alpha}}{2} u_x^2(0, t) = 0.$$

Integrating the above equation with respect to t and using the homogeneous initial and boundary conditions we obtain $\|u(t)\|_{\mathbf{L}^2} = 0$. Thus the solution of the linear problem (2.1) is unique. Theorem 22 is proved.

5. Local existence for nonlinear problem

THEOREM 23. *Suppose that the initial data* $u_0(x) \in \mathbf{H}_2^{1,\omega}(\mathbf{R}^+)$, *where* $\omega \in \left(\frac{1}{2}, \frac{3}{2}\right)$. *Then for some* $T > 0$ *there exists a unique solution*

$$u(x, t) \in \mathbf{C}([0,T], \mathbf{H}_2^{0,\varkappa}(\mathbf{R}^+)) \cap \mathbf{C}((0,T], \mathbf{H}_2^{1,\omega}(\mathbf{R}^+)) \cap \mathbf{C}((0,T], \mathbf{H}_2^{2,0}(\mathbf{R}^+))$$

of the initial-boundary value problem (1.1), where $\varkappa \in \left(0, \frac{1}{2}\right)$.

PROOF. We prove Theorem 23 by the contraction mapping principle in the space

$$\mathbf{X}_{T,\rho}^{\varkappa,\omega} = \left\{ \phi \in \mathbf{L}^2 : \|\phi\|_{\mathbf{X}_T^{\varkappa,\omega}} < \rho \right\},$$

where

$$\|u\|_{\mathbf{X}_T^{\varkappa,\omega}} = \sup_{t\in(0,T]} \left(\|\langle\cdot\rangle^{\varkappa} u(\cdot,t)\|_{\mathbf{L}^2} + \|(\cdot)^{\omega} u_x(\cdot,t)\|_{\mathbf{L}^2} + \sum_{n=0}^{2} t^{\frac{n}{4}} \|\partial_x^n u(\cdot,t)\|_{\mathbf{L}^2} \right),$$

and

$$T > 0, \varkappa \in \left(0, \frac{1}{2} \right), \omega \in \left(\frac{1}{2}, \frac{3}{2} \right).$$

Let $u(x,t)$ be a solution of the following linear problem

(5.1)
$$\begin{cases} u_t + \mathbb{N}(w) - u_{xx} + (-1)^{\alpha} u_{xxx} = 0, & t > 0, x > 0, \\ u(x,0) = u_0(x), \ x > 0, \\ \partial_x^j u(0,t) = 0, j = 0, \alpha, \ t > 0, \end{cases}$$

where $\mathbb{N}(w) = w w_x$ is well-defined since $w \in \mathbf{X}_{T,\rho}^{\varkappa,\omega}$. Note that the initial-boundary value problem (5.1) defines a mapping $\mathbb{M}$ by $u = \mathbb{M}(w)$ and we will show that $\mathbb{M}$ is the contraction mapping from $\mathbf{X}_{T,\rho}^{\varkappa,\omega}$ into itself for a sufficiently small $T > 0$. Since $w \in \mathbf{X}_{T,\rho}^{\varkappa,\omega}$ we have

$$\sup_{t\in[0,T]} t^{\frac{1}{2}} \|\partial_x \mathbb{N}(w)(t)\|_{\mathbf{L}^1} \ \leq \ C\rho^2,$$

$$\sup_{t\in[0,T]} \|\langle\cdot\rangle^{\delta} \mathbb{N}(w)(\cdot,t)\|_{\mathbf{L}^1} \ \leq \ C\rho^2.$$

Via Theorem 22 problem (5.1) has a unique solution $u(x,t) \in \mathbf{X}_{T,\rho}^{\varkappa,\omega}$ with the norm

$$\|u\|_{\mathbf{X}_T^{\varkappa,\omega}} \leq C\lambda,$$

where

$$\lambda = \|u_0\|_{\mathbf{H}_2^{0,\omega}} + T^{\frac{1}{4}} \sup_{t\in[0,T]} t^{\frac{1}{2}} \left(\|\partial_x \mathbb{N}(w)(t)\|_{\mathbf{L}^1} + \|\langle\cdot\rangle^{\delta} \mathbb{N}(w)(\cdot,t)\|_{\mathbf{L}^1} \right)$$

and $\mu = \frac{1}{4} - \nu$. Therefore we obtain

(5.2)
$$\|u\|_{\mathbf{X}_T^{\varkappa,\omega}} \leq C\|u_0\|_{\mathbf{H}_2^{1,\omega}} + CT^{\frac{1}{4}}\rho^2,$$

whence we get $\|u\|_{\mathbf{X}_T^{\varkappa,\omega}} \leq \rho$ if $T > 0$ and $\|u_0\|_{\mathbf{H}_2^{1,\omega}}$ are sufficiently small. Thus the mapping $\mathbb{M}$ transforms the closed ball $\mathbf{X}_{T,\rho}^{\varkappa,\omega}$ with a center at the origin and a radius ρ into itself. Analogously we can prove the estimate $\sup_{t\in[0,T]} \|u-\tilde{u}\|_{\mathbf{X}_{T,\rho}^{\varkappa,\omega}} <$ $\sup_{t\in[0,T]} \|w - \tilde{w}\|_{\mathbf{X}_{T,\rho}^{\varkappa,\omega}}$ for small T. Therefore the mapping $\mathbb{M}$ is a contraction mapping in $\mathbf{X}_{T,\rho}^{\varkappa,\omega}$, and there exists a unique solution $u(x,t) \in \mathbf{X}_{T,\rho}^{\varkappa,\omega}$ of the initial-value problem (1.1). Theorem 23 is proved. $\qquad \square$

REMARK 9. *By (5.2) we see that if the norm of the initial data u_0 is sufficiently small, namely, $\|u_0\|_{\mathbf{H}_2^{1,\omega}} < \epsilon$, then for some time $T > 1$ there exists a unique solution u such that $\|u\|_{\mathbf{X}_T^{\varkappa,\omega}} < C\epsilon$.*

6. Large time asymptotics

We consider the initial-boundary value problem (1.1) with small initial data

(6.1) $$\|u_0\|_{\mathbf{H}_2^{1,\omega}} < \epsilon_1,$$

where

$$\omega > \frac{1}{2} + \mu, \mu \in \left(\frac{1}{2}, 1\right),$$

and $\epsilon_1 > 0$ is sufficiently small. Let us prove the estimate

(6.2) $$\sup_{t>0} \left(\sum_{n=0}^{2} t^{\alpha_n} \left\|u_x^{(n)}(t)\right\|_{\mathbf{L}^2} + \sum_{n=0}^{1} t^{\alpha_n - \frac{\delta_n}{2}} \left\|(\cdot)^{\delta_n} u_x^{(n)}(\cdot, t)\right\|_{\mathbf{L}^2}\right) < \varepsilon,$$

where

$$\alpha_n = \frac{3}{4} + \frac{n}{2} - \gamma, \delta_n = \frac{1}{2} + \mu n + \gamma \leq \omega,$$

and $\gamma, \varepsilon > 0$ are small enough. We prove this estimate by the contradiction. We assume that there exists some $T > 0$ such that

(6.3) $$\sup_{t\in[0,T]} \left(\sum_{n=0}^{1+\alpha} t^{\alpha_n} \left\|u_x^{(n)}(t)\right\|_{\mathbf{L}^2} + \sum_{n=0}^{1} t^{\alpha_n - \frac{\delta_n}{2}} \left\|(\cdot)^{\delta_n} u_x^{(n)}(\cdot, t)\right\|_{\mathbf{L}^2}\right) = \varepsilon.$$

From Lemma 22 we have

(6.4) $$\left\|(\cdot)^{\delta} \partial_x^n F_\alpha(\cdot, y, t)\right\|_{\mathbf{L}^2} \leq Cyt^{-\alpha_n + \frac{\delta}{2}}$$

all $y, t > 0$, where $\delta \geq 0$. Also by (4.4) and (4.6) we have

(6.5) $$F_\alpha(x, y, t) = H_\alpha(x, y, t) + G_\alpha(x - y, t),$$

where

$$\left\|(\cdot)^{\delta} H_x^{(n)}(\cdot, y, t)\right\|_{\mathbf{L}^2} \leq C\left(1 + y^{\gamma-1}\right) t^{-\frac{1}{6} - \frac{n-\delta+2\gamma}{3}},$$

$$\sup_{y>0} \left\|(\cdot)^{\delta} \partial_x^{(n)} G_\alpha(\cdot, t)\right\|_{\mathbf{L}^2} \leq Ct^{-\frac{1+2n}{4}}$$

for all $t > 0$. Therefore from (6.4) with $\delta_0 = 0, \frac{1}{2} - \gamma, \alpha_0 = \frac{3}{4} - \gamma$ we obtain

$$\|(\cdot)^{\delta_0} u(\cdot, t)\|_{\mathbf{L}^2}$$

$$\leq Ct^{-\alpha_0 + \frac{\delta_0}{2}} \|(\cdot)u_0(\cdot)\|_{\mathbf{L}^2}$$

$$+ \int_0^t d\tau \int_0^{+\infty} |uu_y(y, \tau)| \left\|(\cdot)^{\delta_0} F(\cdot, y, t - \tau)\right\|_{\mathbf{L}^2} dy$$

(6.6) $$\leq \epsilon_1 t^{-\alpha_0 + \frac{\delta_0}{2}} + \int_0^t \left\|(\cdot)^{\frac{1}{2}-\gamma} u(\cdot, \tau)\right\|_{\mathbf{L}^2}$$

$$\times \left\|(\cdot)^{\frac{1}{2}+\gamma} u_x(\cdot, \tau)\right\|_{\mathbf{L}^2} (t - \tau)^{-\alpha_0 + \frac{\delta_0}{2}} d\tau.$$

From (6.5) we have for $\delta_1 = 0$, $\frac{1+2\mu}{2} - \gamma$, $\alpha_1 = \alpha_n = \frac{5}{4} - \gamma$

$$\left\| (\cdot)^{\delta_1} u_x(\cdot, t) \right\|_{\mathbf{L}^2}$$

$$\leq Ct^{-\alpha_1 + \frac{\delta_1}{2}} \left\| (\cdot) u_0(\cdot) \right\|_{\mathbf{L}^2} + \int_0^t d\tau \int_0^{+\infty} |uu_y(y, \tau)| \left\| F_x^{(1)}(\cdot, y, t - \tau) \right\|_{\mathbf{L}^2} dy$$

$$\leq \epsilon_1 t^{-\alpha_1 + \frac{\delta_1}{2}} + \int_0^{\frac{t}{2}} \left\| (\cdot)^{\frac{1}{2} - \gamma} u(\cdot, \tau) \right\|_{\mathbf{L}^2} \left\| (\cdot)^{\frac{1}{2} + \gamma} u_x(\cdot, \tau) \right\|_{\mathbf{L}^2} (t - \tau)^{-\alpha_1 + \frac{\delta_1}{2}} d\tau$$

$$+ \int_{\frac{t}{2}}^t \left\| u(\cdot, \tau) \right\|_{\mathbf{L}^2} \left\| u_x(\cdot, \tau) \right\|_{\mathbf{L}^2} (t - \tau)^{-\frac{3}{4}} d\tau$$

(6.7) $$+ \int_{\frac{t}{2}}^t (t - \tau)^{-\frac{3}{4}} \left\| u(\cdot, \tau) \right\|_{\mathbf{L}^2}^{\frac{1}{2}} \left\| u_x(\cdot, \tau) \right\|_{\mathbf{L}^2}^{\frac{3}{2}} d\tau$$

and integrating by parts and using the boundary conditions $u(0, t) = u_x(0, t) = 0$ we get for $\alpha_2 = \frac{7}{4} - \gamma$

$$\left\| u_{xx}(t) \right\|_{\mathbf{L}^2}$$

$$\leq Ct^{-\alpha_2} \left\| (\cdot) u_0(\cdot) \right\|_{\mathbf{L}^2} + \int_0^t d\tau \int_0^{+\infty} |uu_y(y, \tau)| \left\| F_{xx}(\cdot, y, t - \tau) \right\|_{\mathbf{L}^2} dy$$

$$\leq \epsilon_1 t^{-\alpha_2}$$

$$+ \int_0^{\frac{t}{2}} \left\| (\cdot)^{\frac{1}{2} - \gamma} u(\cdot, \tau) \right\|_{\mathbf{L}^2} \left\| (\cdot)^{\frac{1}{2} + \gamma} u_x(\cdot, \tau) \right\|_{\mathbf{L}^2} (t - \tau)^{-\alpha_2} d\tau$$

$$+ \int_{\frac{t}{2}}^t \left\| u_x^2(\cdot, \tau) \right\|_{\mathbf{L}^2} (t - \tau)^{-\frac{3}{4}} d\tau$$

$$+ \int_{\frac{t}{2}}^t \left\| u_{xx}(\tau) \right\|_{\mathbf{L}^2} \left\| u(\tau) \right\|_{\mathbf{L}^2} (t - \tau)^{-\frac{3}{4}} d\tau$$

(6.8) $$+ \int_{\frac{t}{2}}^t (t - \tau)^{-\frac{5}{6}} \left\| u(\tau) \right\|_{\mathbf{L}^2} \left\| u_x(\tau) \right\|_{\mathbf{L}^2} d\tau.$$

Therefore substituting (6.3) into (6.6), (6.7) and (6.8) we have

$$\sup_{t \in [0, T]} \left(\sum_{n=0}^2 t^{\alpha_n} \left\| u_x^{(n)}(t) \right\|_{\mathbf{L}^2} + \sum_{n=0}^1 t^{\alpha_n - \frac{\delta_n}{2}} \left\| (\cdot)^{\delta_n} u_x^{(n)}(\cdot, t) \right\|_{\mathbf{L}^2} \right) < \varepsilon.$$

The contradiction obtained proves (6.2).

Now using estimate (6.2) and Lemma 22 we prove that if $\left\| (\cdot)^{1+\mu} u_0 \right\|_{\mathbf{L}^1} \leq C$, then the solution has the following asymptotics for $t \to +\infty$ uniformly with respect to $x > 0$

(6.9) $$u(x, t) = \frac{1}{t} A_\alpha \Phi_\alpha \left(\frac{x}{2\sqrt{t}}, t \right) + O\left(\min\left(\frac{x}{\sqrt{t}}, 1 \right) t^{-1 - \frac{\mu}{2}} \right),$$

where $\mu \in (0, 1)$ and

$$A_\alpha = \int_0^{+\infty} \Lambda_\alpha (y) \, u_0(y) dy + \int_0^{+\infty} d\tau \int_0^{+\infty} \Lambda_\alpha (y) \, u(y, \tau) u_y(y, \tau) dy < +\infty.$$

In fact, via (6.2), Lemma 22 and (4.1) we have

(6.10)
$$u(x, t) = \frac{1}{t} A_\alpha \Phi_\alpha \left(\frac{x}{2\sqrt{t}}, t \right) + R_\alpha(x, t),$$

where

$$
\begin{aligned}
|R_\alpha(x, t)| \quad \leq \quad & C \min \left(\frac{x}{\sqrt{t}}, 1 \right) t^{-1-\frac{\mu}{2}} \left(\|(\cdot)^{1+\mu} u_0\|_{\mathbf{L}^1} \right. \\
& + \int_0^t d\tau \|(\cdot)^{1+\mu} u(\cdot, \tau) u_y(\cdot, \tau)\|_{\mathbf{L}^1} \Big) \\
& + \int_0^t d\tau \int_0^{+\infty} dy |uu_y| \, |F_\alpha(x, y, t - \tau) - F_\alpha(x, y, t)| \\
& + \int_t^{+\infty} d\tau \int_0^{+\infty} |\Lambda_\alpha (y) \, u(y, \tau) u_y(y, \tau)| \, dy
\end{aligned}
$$

Using estimate (6.2) and Theorem 23 we see that

(6.11)
$$\int_0^t \tau \|(\cdot)^{\frac{1}{2}-\gamma} u(\cdot, \tau)\|_{\mathbf{L}^2} \|(\cdot)^{\frac{1}{2}+\gamma} u_x(\cdot, \tau)\|_{\mathbf{L}^2} d\tau \leq C + \int_1^t \tau^{-\frac{1}{2}} d\tau \leq C t^{\frac{1}{2}}$$

for $t > 1$.

Now we prove the estimate

$$\|\partial_t F_\alpha(\cdot, y, t)\|_{\mathbf{L}^\infty} \leq C y \min \left(\frac{x}{\sqrt{t}}, 1 \right) t^{-2+\gamma},$$

in the case $\alpha = 0$ (the case $\alpha = 1$ is considered by analogy). Since

$$\int_{-i\infty}^{i\infty} d\xi \xi e^{\xi t} e^{\phi_3(\xi)x} \sum_{j=1}^2 \left(1 - \phi_j(\xi) y \right) \phi_j'(\xi)$$

$$= -\int_{-i\infty}^{i\infty} dp K_0 (p) \, e^{px - K_0(p)t} (1 - py)$$

via (3.17) we have

$$\partial_t F_0(x, y, t)$$

$$= \frac{1}{2\pi i} \left(\int_{-i\infty}^{i\infty} d\xi \xi e^{\xi t} \left(e^{\phi_3(\xi)x} - 1 \right) \sum_{j=1}^2 \left(e^{-\phi_j(\xi)y} - 1 + \phi_j (\xi) y \right) \phi_j'(\xi) \right.$$

$$+ \int_{-i\infty}^{i\infty} dp K_0 (p) \, e^{-K_0(p)t} \left(e^{-py} - 1 + py \right) \left(e^{px} - 1 \right) \right).$$

By virtue of inequalities

$$\left| \left(e^{\phi_3(\xi)x} - 1 \right) \xi \sum_{j=1}^{2} \left(e^{-\phi_j(\xi)y} - 1 + \phi_j(\xi)y \right) \phi_j'(\xi) \right|$$

$$\leq \min\left(\frac{x}{\sqrt{t}}, 1 \right) \begin{cases} C|\xi|^{1-\frac{\mu}{2}} yx^{-\mu}, & \text{if } |\xi| \leq 1, \\ C|\xi|^{1-\frac{\mu+1}{3}} yx^{-\mu}, & \text{if } |\xi| > 1 \end{cases}$$

and

$$\left\| \int_{-i\infty}^{i\infty} dp K_0(p) e^{-K_0(p)t} \left(e^{-py} - 1 + py \right) \left(e^{p \cdot} - 1 \right) \right\|_{\mathbf{L}^\infty}$$

$$\leq Cy \min\left(\frac{x}{\sqrt{t}}, 1 \right) \int_{-i\infty}^{i\infty} e^{-\operatorname{Re} K_0(p)t} |K_0(p)| \, |p| \, |dp|$$

$$\leq Cy \min\left(\frac{x}{\sqrt{t}}, 1 \right) t^{-2},$$

we easily get

$$\|\partial_t F(\cdot, y, t)\|_{\mathbf{L}^\infty} \leq Cy \min\left(\frac{x}{\sqrt{t}}, 1 \right) \int_{\xi \in \mathcal{C}} |\xi|^{-\frac{1}{4} - \frac{\mu}{2} - \gamma} e^{\operatorname{Re} \xi t} d\xi$$

$$+ Cy \min\left(\frac{x}{\sqrt{t}}, 1 \right) t^{-\frac{7}{4} + \frac{\delta}{2}} \leq Cy \min\left(\frac{x}{\sqrt{t}}, 1 \right) t^{-2+\gamma},$$

whence

$$\|F_\alpha(\cdot, y, t - \tau) - F_\alpha(\cdot, y, t)\|_{\mathbf{L}^\infty}$$

$$\leq Cy\tau \min\left(\frac{x}{\sqrt{t}}, 1 \right) t^{-2+\gamma}.$$

Therefore using (6.11) we have

$$\int_0^t d\tau \left| \int_0^{+\infty} u(y, \tau) u_y(y, \tau) \right| |F_\alpha(x, y, t - \tau) - F_\alpha(x, y, t)|$$

$$\leq Ct^{\gamma-2} \int_0^t \tau \|(\cdot)^{\frac{1}{2}-\gamma} u(\cdot, \tau)\|_{\mathbf{L}^2} \|(\cdot)^{\frac{1}{2}+\gamma} u_y(\cdot, \tau)\|_{\mathbf{L}^2} d\tau$$

$$(6.12) \qquad \leq C \min\left(\frac{x}{\sqrt{t}}, 1 \right) t^{-1-\frac{\mu}{2}}$$

with some $\mu \in (0, 1)$. Since $\|(\cdot)^{1+\mu} u_0\|_{\mathbf{L}^1} \leq C$ using (6.2) and (6.12) we get

$$(6.13) \qquad |R_\alpha(x, t)| \leq C \min\left(\frac{x}{\sqrt{t}}, 1 \right) t^{-1-\frac{\mu}{2}},$$

where $\mu \in (0, 1)$. From (6.10) through (6.13) we obtain the asymptotics (6.9) for the solution. Theorem 21 is proved.

CHAPTER 7

Large Initial Data

1. Introduction

In this chapter we study the case of large initial data by considering the Korteweg-de Vries-Burgers (KdVB) equation

$$(1.1) \qquad \begin{cases} u_t + uu_x - u_{xx} + u_{xxx} = 0, \ t > 0, x > 0, \\ u(x,0) = u_0(x), \ x > 0, \ u(0,t) = 0, \ t > 0. \end{cases}$$

In the previous chapter we considered the large time asymptotics of solutions to (1.1) with small initial data $u_0 \in \mathbf{H}_2^{0,\omega}(\mathbf{R}^+) \cap u_0 \in \mathbf{H}_2^{1,0}(\mathbf{R}^+)$, where $\omega \in \left(1, \frac{3}{2}\right)$ and showed that there exists a unique solution $u \in \mathbf{C}([0, +\infty), \mathbf{H}_2^{0,\varkappa}(\mathbf{R}^+)) \cap \mathbf{C}\left((\mathbf{R}^+), \mathbf{H}_2^{1,\omega}(\mathbf{R}^+)\right) \cap \mathbf{C}\left((\mathbf{R}^+), \mathbf{H}_2^{2,0}(\mathbf{R}^+)\right)$, where $\varkappa \in \left(0, \frac{1}{2}\right)$. Since we considered the general case of any sign $\pm u_{xxx}$ in the KdVB equation the conditions on the initial data are not optimal in the case $+u_{xxx}$. For this case we can prove global existence of solutions with more general not small initial data $u_0 \in \mathbf{H}_2^{0,\omega}(\mathbf{R}^+) \cap \mathbf{H}_2^{1,0}(\mathbf{R}^+)$. There are a few results on the asymptotic behavior of solutions with large initial data. The Cauchy problem for nonlinear nonlocal Schrödinger equation was considered in [76], where the large time asymptotics of solutions was obtained. For the Ott-Sudan-Ostrovskiy equation the large time asymptotics was found in [103].

The aim of the present chapter is to consider problem (1.1) without a smallness condition on the data. We know (see [60]) that the solutions of (1.1) have more rapid time decay than that of solutions to the Cauchy problem due to the zero boundary value. Therefore the use of the symmetry property of the nonlinearity allows us to estimate the $\mathbf{H}_2^{1,0}(\mathbf{R}^+)$ - norm of solutions without any assumption on the size of the initial data. Furthermore by the energy type estimates and the estimates of the Green function of (1.1) it follows that for some time $T > 0$ the $\mathbf{L}^\infty$-norm of the solution becomes small. Our result below will be obtained by combining these estimates of solutions. Note that in the case $-u_{xxx}$ in the KdVB equation we do not know if we can estimate the $\mathbf{H}_2^{1,0}$-norm of solutions without smallness assumption on the initial data.

By the same letter C we denote different positive constants. We now state the main result of this chapter (see paper [59]).

THEOREM 24. *Suppose that the initial data* $u_0 \in \mathbf{H}_2^{0,\omega}(\mathbf{R}^+) \cap \mathbf{H}_2^{1,0}(\mathbf{R}^+)$, *where* $\omega \in \left(1, \frac{3}{2}\right)$. *Then there exists a unique solution*

$$u \in \mathbf{C}\left([0, +\infty), \mathbf{H}_2^{0,\omega}(\mathbf{R}^+) \cap \mathbf{H}_2^{1,0}(\mathbf{R}^+)\right)$$

of the initial-boundary value problem (1.1). Moreover if the initial data are such that $x^{1+\mu} u_0(x) \in \mathbf{L}^1(\mathbf{R}^+)$, $\mu = \omega - \frac{1}{2}$, *then there exists a constant A such that the solution has the following asymptotics*

$$u(x, t) = \frac{A}{t} \frac{x}{2\sqrt{\pi t}} e^{-\frac{x^2}{4t}} + O\left(\min\left(\frac{x}{\sqrt{t}}, 1\right) t^{-1-\frac{\mu}{2}}\right)$$

for $t \to +\infty$ *uniformly with respect to* $x > 0$.

We organize this chapter as follows. In Section 2 we consider the linear initial-boundary value problem corresponding to (1.1). We construct the Green function for the solution of the linear problem and prove the existence and uniqueness of the solution. Section 3 is devoted to the proof of the local existence of solutions to the nonlinear problem (1.1). In Section 4 we prove some preliminary estimates in Lemma 25 and energy estimate in Lemma 26. Theorem 24 is proved in Section 5. This Chapter was written in collaboration with Prof. H.F. Ruiz Paredes.

2. Linear problem

We consider the linear initial-boundary value problem corresponding to (1.1)

$$(2.1) \qquad \begin{cases} u_t - u_{xx} + u_{xxx} = f(x, t), & t > 0, x > 0, \\ u(x, 0) = u_0(x), & x > 0, \\ u(0, t) = 0, & t > 0. \end{cases}$$

From the results of the previous chapter we have the explicit formula for the solution of the linear problem (2.1) under the condition $u_0 \in \mathbf{L}^1(\mathbf{R}^+)$, $f \in \mathbf{L}_{loc}^q\left(0, +\infty; \mathbf{L}^1(\mathbf{R}^+)\right)$ with $q > 2$

$$(2.2) \qquad u(x, t) = \int_0^{+\infty} u_0(y) F(x, y, t) dy + \int_0^t d\tau \int_0^{+\infty} f(y, \tau) F(x, y, t - \tau) dy,$$

where

$$F(x, y, t) = -\frac{1}{4\pi^2} \int_{-i\infty}^{+i\infty} dp\, e^{-py} \int_{-i\infty}^{i\infty} e^{\xi t} \frac{e^{\phi_3(\xi) x} + e^{px}}{K(p) + \xi} d\xi,$$

$\phi_3(\xi)$ is the root of the equation $K(p) = -\xi$, such that $\operatorname{Re} \phi_3(\xi) < 0$ for $\operatorname{Re} \xi > 0$.

Now we prove the following local existence theorem on an arbitrary time interval $[0, T]$, $T > 0$.

THEOREM 25. *Let initial data* $u_0 \in \mathbf{H}_2^{0,\omega}(\mathbf{R}^+) \cap \mathbf{H}_2^{1,0}(\mathbf{R}^+)$, *where* $\omega \in \left(\frac{1}{2}, \frac{3}{2}\right)$. *Suppose that a force* $f \in \mathbf{L}^\infty\left(0, +\infty; \mathbf{L}^1(\mathbf{R}^+) \cap \mathbf{H}_2^{0,\omega}(\mathbf{R}^+)\right)$ *and* $f(0, t) = 0$. *Then there exists a unique solution*

$$u \in \mathbf{C}\left([0, T]; \mathbf{H}_2^{0,\omega}(\mathbf{R}^+) \cap \mathbf{H}_2^{1,0}(\mathbf{R}^+)\right)$$

of the initial-boundary value problem (2.1) such that

$$\sup_{t\in[0,T]} \|u\|_{\mathbf{H}_2^{0,\omega}} + \sup_{t\in[0,T]} \|u\|_{\mathbf{H}_2^{1,0}} \leq C\lambda,$$

where

$$\lambda = \|u_0\|_{\mathbf{H}_2^{0,\omega}} + \|u_0\|_{\mathbf{H}_2^{1,0}} + T^{\frac{1}{4}} \sup_{t\in[0,T]} \left(\|f(t)\|_{\mathbf{L}^1} + \|\langle\cdot\rangle^\omega f(\cdot,t)\|_{\mathbf{L}^2} \right).$$

PROOF. We prove the energy type estimates formally; the complete justification is carried out by introducing in a standard way test functions with compact support and then by taking a limit to extend the support for all $\mathbf{R}^+$. Multiplying equation (2.1) by $u(x,t)$, integrating the result with respect to $x > 0$ and using identities

$$\int_0^{+\infty} uu_{xx}dx = -\int_0^{+\infty} u_x^2 dx$$

and

$$\int_0^{+\infty} uu_{xxx}dx = \frac{1}{2}u_x^2(0,t)$$

we have

$$\frac{1}{2}\frac{d}{dt}\|u(t)\|_{\mathbf{L}^2}^2 + \int_0^{+\infty} u_x^2 dx + \frac{1}{2}u_x^2(0,t) = \int_0^{+\infty} uf dx.$$

Integrating the above equation with respect to t we get

$$(2.3) \qquad \|u(t)\|_{\mathbf{L}^2} \leq \|u_0(t)\|_{\mathbf{L}^2} + 2T \sup_{t\in[0,T]} \|f(t)\|_{\mathbf{L}^2}.$$

Differentiating equation (2.1) with respect to x, multiplying the result by $u_x(x,t)$ and integrating with respect to x, via the relation $u_{xx}(0,t) = u_{xxx}(0,t)$ (which follows from equation and boundary condition) and identities

$$\int_0^{+\infty} u_x f_x dx = -\int_0^{+\infty} fu_{xx}dx, \quad \int_0^{+\infty} u_x u_{xxx}dx$$

$$= u_x u_{xx}\big|_0^{+\infty} - \int_0^{+\infty} u_{xx}^2 dx,$$

$$\int_0^{+\infty} u_x u_{xxxx}dx = u_x u_{xx}\big|_0^{+\infty} + \frac{1}{2}u_{xx}^2(0,t),$$

we get

$$(2.4) \qquad \frac{1}{2}\frac{d}{dt}\|u_x(t)\|_{\mathbf{L}^2}^2 + \int_0^{+\infty} u_{xx}^2 dx + \frac{1}{2}u_{xx}^2(0,t) = -\int_0^{+\infty} fu_{xx}dx.$$

Since by the Schwarz inequality

$$\left|\int_0^{+\infty} fu_{xx}dx\right| \leq \|f(t)\|_{\mathbf{L}^2}\|u_{xx}(t)\|_{\mathbf{L}^2}$$

$$\leq \frac{1}{2}\|f(t)\|_{\mathbf{L}^2}^2 + \frac{1}{2}\|u_{xx}(t)\|_{\mathbf{L}^2}^2,$$

integrating (2.4) with respect to $t > 0$ we have

$$(2.5) \qquad \|u_x(t)\|_{\mathbf{L}^2} \ \leq \ \int_0^t \|f(t)\|_{\mathbf{L}^2}\, dt + \|u_{0x}\|_{\mathbf{L}^2}$$

$$\leq \ T \sup_{t \in [0,T]} \|f(t)\|_{\mathbf{L}^2} + \|u_{0x}\|_{\mathbf{L}^2}^2 \, .$$

From (2.3) and (2.5) we obtain

$$(2.6) \qquad \|u\|_{\mathbf{H}_2^{1,0}} \leq \|u_0\|_{\mathbf{H}_2^{1,0}} + CT \sup_{t \in [0,T]} \|f(t)\|_{\mathbf{L}^2} \, .$$

Multiplying equation (2.1) by $(1+x)^{2\omega} u(x,t)$ and integrating with respect to $x > 0$ we get

$$\frac{1}{2} \frac{d}{dt} \|(1+x)^{\omega} u(t)\|_{\mathbf{L}^2}^2$$

$$- \int_0^{+\infty} (1+x)^{2\omega} u u_{xx} dx + \int_0^{+\infty} (1+x)^{2\omega} u u_{xxx} dx$$

$$(2.7) \qquad = \ \int_0^{+\infty} (1+x)^{2\omega} u f \, dx.$$

Integrating by parts we have

$$\int_0^{+\infty} (1+x)^{2\omega} u u_{xx} dx \ = \ - \int_0^{+\infty} (1+x)^{2\omega} u_x^2 dx$$

$$+ \omega \left(2\omega - 1 \right) \int_0^{+\infty} (1+x)^{2\omega-2} u^2 dx,$$

and

$$\int_0^{+\infty} (1+x)^{2\omega} u u_{xxx} dx$$

$$= \ \frac{1}{2} (1+x)^{2\omega} u_x^2 (0,t) + 3\omega \int_0^{+\infty} (1+x)^{2\omega-1} u_x^2 dx$$

$$- \omega \left(2\omega - 1 \right) \left(2\omega - 2 \right) \int_0^{+\infty} (1+x)^{2\omega-3} u^2 dx.$$

Therefore we integrate (2.7) with respect to t to get

$$\|(\cdot)^{\omega} u(\cdot,t)\|_{\mathbf{L}^2}^2 \ \leq \ C \int_0^T dt \int_0^{+\infty} \left(x^{2\omega-2} u^2 + x^{2\omega} |f| \, |u| \right) dx$$

$$+ \|(\cdot)^{\omega} u_0(\cdot)\|_{\mathbf{L}^2}$$

$$\leq \ C \|(\cdot)^{\omega} u(t)\|_{\mathbf{L}^2} \int_0^T \left(\|(\cdot)^{\omega} f(\cdot,t)\|_{\mathbf{L}^2} + \|u(t)\|_{\mathbf{H}_2^{1,0}} \right) dt$$

$$+ \|(\cdot)^{\omega} u_0(\cdot)\|_{\mathbf{L}^2} \, ,$$

where $\omega \in \left(\frac{1}{2}, \frac{3}{2}\right)$. Using (2.6) we obtain

$$(2.8) \quad \|(\cdot)^\omega u(\cdot, t)\|_{\mathbf{L}^2} \leq C\left(\|u_0\|_{\mathbf{H}_2^{0,\omega}} + \|u_0\|_{\mathbf{H}_2^{1,0}} + T \sup_{t \in (0,T]} \|(\cdot)^\omega f(\cdot, t)\|_{\mathbf{L}^2}\right).$$

Recall (2.2); the solution of problem (2.1) can be represented in the following manner:

$$(2.9) \quad u_x(x, t) = \int_0^{+\infty} u_0(y) F_x(x, y, t) dy + \int_0^t d\tau \int_0^{+\infty} dy f(y, \tau) F_x(x, y, t - \tau).$$

As in the previous chapter we rewrite the function $F_x(x, y, t)$ as

$$F_x(x, y, t) = \mathcal{F}_x(x, y, t) + G_x(x - y, t),$$

where

$$\mathcal{F}_x(x, y, t) = \frac{1}{2\pi i} \int_{\operatorname{Re} \xi = 0} e^{\xi t} \phi_3(\xi) e^{\phi_3(\xi)x} \sum_{j=1}^{2} e^{-\phi_j(\xi)y} \phi_j'(\xi) d\xi$$

and

$$G_x(x - y, t) = \frac{1}{2\pi i} \int_{\operatorname{Re} p = 0} dp e^{px - K(p)t} e^{-py} p \, dp.$$

Due to estimates (4.4)-(4.6) we have

$$\sup_{t \in [0,T]} \left(\sup_{y \geq 0} (1 + y)^{1-\gamma} \|(\cdot)^\omega \mathcal{F}_x(\cdot, y, t)\|_{\mathbf{L}^2} + \|(\cdot)^\omega G_x(\cdot, t)\|_{\mathbf{L}^1}\right.$$
$$\left. + t^{\frac{3}{4}} \|(\cdot)^\omega \mathcal{F}_x(\cdot, t)\|_{\mathbf{L}^2}\right)$$
$$\leq \quad C$$

for all $y > 0$, $t \in [0, T]$, ($\gamma > 0$ is small enough). Applying the Young inequality

$$\left\|\int_0^{+\infty} g(x - y) v(x) dx\right\|_{\mathbf{L}^2} \leq C \|g\|_{\mathbf{L}^2} \|v\|_{\mathbf{L}^1}$$

we have

$$\| (\cdot)^\omega \, \partial_x u(\cdot, t) \|_{\mathbf{L}^2} \le C \left\| \int_0^{+\infty} u_0(y) \, (\cdot)^\omega \, F_x(\cdot, y, t) dy \right\|_{\mathbf{L}^2}$$

$$+ C \int_0^T |f(y, \tau)| \, \| (\cdot)^\omega \, F_x(\cdot, y, t - \tau) \|_{\mathbf{L}^2} \, dy$$

$$\le \ C \int_0^{+\infty} |u_0(y)| \| (\cdot)^\omega \, H_x(\cdot, y, t) \|_{\mathbf{L}^2} dy$$

$$+ C \left\| \int_0^{+\infty} (\cdot)^\omega \, u_0(y) G_x(\cdot - y, t) dy \right\|_{\mathbf{L}^2}$$

$$+ C \sup_{t \in [0, T]} \| f(t) \|_{\mathbf{L}^1} \int_0^T \frac{d\tau}{(t - \tau)^{\frac{3}{4}}}$$

$$\le \ C \left(\int_0^{+\infty} |u_0(y)| \left(1 + y^{\gamma - 1} \right) dy + \| u_0 \|_{\mathbf{L}^2} \| (\cdot)^\omega \, G_x(\cdot, t) \|_{L^1} \right.$$

$$(2.10) \qquad \left. + \| (\cdot)^\omega \, u_0(\cdot) \|_{\mathbf{L}^2} \| G_x(t) \|_{L^1} + T^{\frac{1}{4}} \sup_{t \in [0, T]} \| f(t) \|_{\mathbf{L}^1} \right)$$

$$\le \ C\lambda.$$

By virtue of (2.6), (2.8) and (2.10) we obtain the a priori estimate of Theorem 25. By virtue of the explicit representation (2.2) we have the existence of the solution to the initial-boundary value problem (2.1). The uniqueness follows from the estimates in the conclusion of the theorem and the fact that equation (2.1) is linear. Theorem 25 is proved. $\square$

REMARK 10. *Via estimate (2.6) we see that if the initial data $u_0 \in \mathbf{H}_2^{1,0}(\mathbf{R}^+)$,
then for each $T > 0$ there exists a unique solution $u \in \mathbf{C}\left([0, T]; \mathbf{H}_2^{1,0}(\mathbf{R}^+)\right)$ of
the initial-boundary value problem (2.1) such that*

$$\sup_{t \in [0, T]} \| u(t) \|_{\mathbf{H}_2^{1,0}} \le C\lambda,$$

where

$$\lambda = \| u_0 \|_{\mathbf{H}_2^{1,0}} + T \sup_{t \in [0, T]} \| f(t) \|_{\mathbf{L}^2}.$$

3. Local existence

In this section using result of Theorem 25 we prove the theorem on the local existence of solutions to the nonlinear problem (1.1).

THEOREM 26. *Suppose that the initial data $u_0(x) \in \mathbf{H}_2^{0,\omega}(\mathbf{R}^+) \cap \mathbf{H}_2^{1,0}(\mathbf{R}^+)$,
where $\omega \in \left(\frac{1}{2}, \frac{3}{2} \right)$. Then for some sufficiently small $T > 0$ there exists a unique
solution $u(x, t) \in \mathbf{C}\left([0, T], \mathbf{H}_2^{0,\omega}(\mathbf{R}^+) \cap \mathbf{H}_2^{1,0}(\mathbf{R}^+)\right)$ of the initial-boundary value
problem (1.1).*

PROOF. We prove the local existence of solutions by the contraction mapping principle in the space

$$\mathbf{H}^{1,\omega}_{T,\rho} = \left\{ \phi(t) \in \mathbf{C}\left([0,T], \mathbf{H}^{0,\omega}_2(\mathbf{R}^+) \cap \mathbf{H}^{1,0}_2(\mathbf{R}^+)\right) : \right.$$

$$\left. \sup_{t \in [0,T]} \left(\|\phi(t)\|_{\mathbf{H}^{0,\omega}_2} + \|\phi(t)\|_{\mathbf{H}^{1,0}_2} \right) \leq \rho \right\},$$

where

$$T > 0, \omega \in \left(\frac{1}{2}, \frac{3}{2}\right), \rho > 0.$$

Let $u(x,t)$ be a solution of the following linear problem

(3.1)
$$\begin{cases} u_t + N(w) - u_{xx} + u_{xxx} = 0, & t > 0, x > 0, \\ u(x,0) = u_0(x), & x > 0, \ u(0,t) = 0, \ t > 0, \end{cases}$$

where $N(w) = ww_x$ is well-defined since $w \in \mathbf{H}^{1,\omega}_{T,\rho}(\mathbf{R}^+)$. Note that the initial-boundary value problem (3.1) defines a mapping $\mathbb{M}$ by $u = \mathbb{M}w$, and we will show that $\mathbb{M}$ is the contraction mapping from $\mathbf{H}^{1,\omega}_{T,\rho}(\mathbf{R}^+)$ into itself for a sufficiently small $T > 0$. Since $w \in \mathbf{H}^{1,\omega}_{T,\rho}(\mathbf{R}^+)$ we have

$$\sup_{t \in [0,T]} \|N(w)(t)\|_{\mathbf{L}^1} \leq C\rho^2 \text{ and } \sup_{t \in [0,T]} \|\langle \cdot \rangle^\omega N(w)(\cdot,t)\|_{\mathbf{L}^2} \leq C\rho^2.$$

Via Theorem 25 problem (3.1) has a unique solution $u(x,t) \in \mathbf{H}^{1,\omega}_{T,\rho}(\mathbf{R}^+)$ with the norm

$$\|u\|_{\mathbf{H}^{1,\omega}_{T,\rho}} \leq C\lambda,$$

where

$$\lambda = \|u_0\|_{\mathbf{H}^{0,\omega}_2} + \|u_0\|_{\mathbf{H}^{1,0}_2}$$
$$+ T^{\frac{1}{4}} \sup_{t \in [0,T]} \left(\|N(w)(t)\|_{\mathbf{L}^1} + \|\langle \cdot \rangle^\omega N(w)(\cdot,t)\|_{\mathbf{L}^2} \right).$$

Since

$$\|u\|_{\mathbf{H}^{1,\omega}_{T,\rho}} \leq C\|u_0\|_{\mathbf{H}^{0,\omega}_2} + C\|u_0\|_{\mathbf{H}^{1,0}_2} + CT^{\frac{1}{4}}\rho^2,$$

we get $\|u\|_{\mathbf{H}^{1,\omega}_{T,\rho}} \leq \rho$ if $T > 0$ is sufficiently small. Thus the mapping $\mathbb{M}$ transforms $\mathbf{H}^{1,\omega}_{T,\rho}$ into itself. Analogously we can prove the estimate

$$\sup_{t \in [0,T]} \|u - \tilde{u}\|_{\mathbf{L}^2} < \sup_{t \in [0,T]} \|w - \tilde{w}\|_{\mathbf{L}^2}$$

for small T. Therefore $\mathbb{M}$ is a contraction mapping in $\mathbf{H}^{1,\omega}_{T,\rho}$ and there exists a unique solution $u(x,t) \in \mathbf{H}^{1,\omega}_{T,\rho}$ of the initial-value problem (1.1). Theorem 26 is proved. $\square$

REMARK 11. *Via Remark 10 we can obtain that if the initial data* $u_0 \in$ $\mathbf{H}_2^{1,0}(\mathbf{R}^+)$, *then there exists a* $T > 0$ *and a unique solution* $u \in \mathbf{C}\left([0,T], \mathbf{H}_2^{1,0}(\mathbf{R}^+)\right)$ *of the initial-boundary value problem (1.1).*

4. Preliminary estimates

As in Section 2 we denote

$$(4.1) \qquad F(x,y,t) = -\frac{1}{4\pi^2} \int_{-i\infty}^{i\infty} dp e^{-py} \int_{-i\infty}^{i\infty} d\xi e^{\xi t} \frac{e^{\phi_3(\xi)x} + e^{px}}{K(p) + \xi},$$

for all $x, y, t \geq 0$, where $K(p) = -p^2 + p^3$.

In the next lemma we prove some preliminary estimates of the functions F for large t.

LEMMA 25. *The estimates*

$$(4.2) \qquad \sup_{t \geq 0} t^{\frac{3}{4} - \frac{\delta}{2}} \sup_{y \geq 0} \left(\|y^\varkappa F_y(\cdot, y, t)\|_{\mathbf{L}^2} + \left\| (\cdot)^\delta F_y(\cdot, y, t) \right\|_{\mathbf{L}^2} \right) \leq C$$

$$(4.3) \qquad \|F(\cdot, y, t)\|_{\mathbf{L}^2} \leq C y^\mu t^{-\frac{1+2\mu}{4}}$$

and

$$(4.4) \qquad \left\| (\cdot)^\delta \partial_x F(\cdot, y, t) \right\|_{\mathbf{L}^2} \leq C y t^{-\frac{5}{4} + \frac{\delta}{2}}$$

are true for $\delta \in (0, \frac{3}{2})$, $\mu \in (0, 1)$, $t > 0$, *where* $\gamma > 0$ *is small.*

PROOF. Now we prove the estimate (4.2). From (4.1) we have

$$F_y(x, y, t) = \frac{1}{4\pi^2} \int_{-i\infty}^{i\infty} dp p e^{-py} \int_{-i\infty}^{i\infty} d\xi e^{\xi t} \frac{e^{\phi_3(\xi)x} + e^{px}}{K(p) + \xi}.$$

Then since

$$\int_{-i\infty}^{i\infty} \frac{p e^{-py}}{\xi + K(p)} dp = -2\pi i \sum_{j=1}^{2} res_{p=\phi_j(\xi)} \frac{p e^{-py}}{\xi + K(p)}$$

$$= 2\pi i \sum_{j=1}^{2} e^{-\phi_j(\xi)y} \phi_j(\xi) \phi_j'(\xi)$$

we find

$$F_y(x, y, t) = -\frac{1}{2\pi i} \left(\int_{-i\infty}^{i\infty} d\xi e^{\xi t} e^{\phi_3(\xi)x} \sum_{j=1}^{2} e^{-\phi_j(\xi)y} \phi_j(\xi) \phi_j'(\xi) \right.$$

$$(4.5) \qquad\qquad \left. + \int_{-i\infty}^{i\infty} dp p e^{p(x-y) - K(p)t} \right).$$

Applying estimates

$$\left| e^{\phi_3(\xi)x} \sum_{j=1}^{2} e^{-\phi_j(\xi)y} \phi_j(\xi) \phi_j'(\xi) \right| \leq C\xi^{-\frac{\varkappa+\delta}{2}} x^{-\delta} y^{-\varkappa}$$

for $\xi \in \mathcal{C}, |\xi| < 1$ and

$$\left| e^{\phi_3(\xi)x} \sum_{j=1}^{2} e^{-\phi_j(\xi)y} \phi_j(\xi) \phi_j'(\xi) \right| \leq C\xi^{-\frac{\varkappa+\delta}{3}} x^{-\delta} y^{-\varkappa}$$

for $\xi \in \mathcal{C}, |\xi| > 1$, where $\varkappa, \delta > 0$, by changing contour of integration into $\mathcal{C}$ in the first integral of (4.5) and by using the estimates

$$\sup_{y \geq 0} \left\| y^\varkappa \int_{-i\infty}^{i\infty} dp p e^{p \cdot -K(p)t} e^{-py} \right\|_{\mathbf{L}^2}$$

$$= \sup_{y \geq 0} \left\| y^\varkappa p e^{-py-K(p)t} \right\|_{\mathbf{L}^2(\operatorname{Re} p=0)}$$

$$\leq C \sup_{y \geq 0} \left(y^\varkappa \left| \int_{-\infty}^{+\infty} p^2 e^{-2py-C|p|^2 t} dp \right|^{\frac{1}{2}} \right)$$

$$\leq C \left| \int_{-\infty}^{+\infty} p^{2-2\varkappa} e^{-C|p|^2 t} dp \right|^{\frac{1}{2}} \leq Ct^{-\frac{3}{4}+\frac{\varkappa}{2}}$$

we get

$$\sup_{y \geq 0} \| y^\varkappa F_y(\cdot, y, t) \|_{\mathbf{L}^2} \leq Ct^{-\frac{3}{4}+\frac{\varkappa}{2}},$$

where $\varkappa \in [0, \frac{3}{2})$. By analogy using

$$\sup_{y \geq 0} \left\| (\cdot)^\delta \int_{-i\infty}^{i\infty} dp p e^{p \cdot -K(p)t} e^{-py} \right\|_{\mathbf{L}^2}$$

$$\leq C \left| \int_{-\infty}^{+\infty} p^{2-2\delta} e^{-C|p|^2 t} dp \right|^{\frac{1}{2}} \leq Ct^{-\frac{3}{4}+\frac{\delta}{2}},$$

we obtain

$$\sup_{y \geq 0} \left\| (\cdot)^\delta F_y(\cdot, y, t) \right\|_{\mathbf{L}^2} \leq Ct^{-\frac{3}{4}+\frac{\delta}{2}},$$

where $\delta \in [0, \frac{3}{2})$. Now we prove estimate (4.3). We rewrite the function $F(x, y, t)$ as

$$F(x, y, t) = \frac{1}{2\pi i} \left(\int_{-i\infty}^{i\infty} d\xi e^{\xi t} e^{\phi_3(\xi)x} \sum_{j=1}^{2} \left(e^{-\phi_j(\xi)y} - 1 \right) \phi_j'(\xi) \right.$$

$$\left. + \int_{-i\infty}^{i\infty} dp e^{px - K(p)t} \left(e^{-py} - 1 \right) \right).$$

Since for $\mu, \delta \in [0, 1]$

$$\left| e^{\phi_3(\xi)x} \sum_{j=1}^{2} \left(e^{-\phi_j(\xi)y} - 1 + \phi_j(\xi)y \right) \phi'_j(\xi) \right|$$

$$\leq \begin{cases} C\xi^{-\frac{\delta-\mu+1}{2}} x^{-\delta} y^{\mu}, |\xi| < 1 \\ C\xi^{-\frac{\delta+1-\mu}{3}} x^{-\delta} y^{\mu}, |\xi| > 1 \end{cases}$$

and

$$\left| e^{px} \left(e^{-py} - 1 \right) \right| \leq C p^{\mu-\delta} x^{-\delta} y^{\mu}$$

we get

$$\|F(\cdot, y, t)\|_{\mathbf{L}^2} \leq C y^{\mu} \left(\int_C e^{\operatorname{Re}\xi t} |\xi|^{-\frac{1}{4}-\gamma+\frac{\mu}{2}} |d\xi| \right.$$

$$\left. + \int_{-i\infty}^{i\infty} e^{-\operatorname{Re} K(p)t} |p|^{\frac{\mu-1}{2}-\gamma} |dp| \right)$$

$$\leq C y^{\mu} t^{-\frac{1+2\mu}{4}}.$$

Estimate (4.4) is proved in the same way. Lemma 25 is proved. $\square$

In the next lemma we prove the energy type estimates.

LEMMA 26. *We have*

(4.6)
$$\|u(t)\|_{\mathbf{H}_2^{1,0}} + \int_0^t \|u_x(t)\|_{\mathbf{L}^2}^2 \, dt \leq C \|u_0\|_{\mathbf{H}_2^{1,0}}$$

and

(4.7)
$$\|u_x(t)\|_{\mathbf{L}^2}^2 \leq C \|u_0\|_{\mathbf{L}^2}^2 \int_0^t \|u_x(t)\|_{\mathbf{L}^2}^2 \, dt + \|u_{0x}\|_{\mathbf{L}^2}^2$$

for all $t > 0$.

PROOF. Multiplying equation (1.1) by $u(x, t)$ and integrating with respect to $x > 0$ we get

$$\frac{d}{dt} \|u(t)\|_{\mathbf{L}^2}^2 + \int_0^{+\infty} u^2 u_x dx$$

$$- \int_0^{+\infty} u u_{xx} dx + \int_0^{+\infty} u u_{xxx} dx = 0.$$

Since

$$\int_0^{+\infty} u^2 u_x dx = 0, \int_0^{+\infty} u u_{xx} dx = - \int_0^{+\infty} u_x^2 dx$$

and

$$\int_0^{+\infty} u u_{xxx} dx = \frac{1}{2} u_x^2(0, t),$$

we have

(4.8)
$$\frac{1}{2}\frac{d}{dt}\|u(t)\|_{\mathbf{L}^2}^2 + \int_0^{+\infty} u_x^2 dx + \frac{1}{2}u_x^2(0,t) = 0.$$

Integrating this estimate with respect to $t > 0$ we get

(4.9)
$$\|u(t)\|_{\mathbf{L}^2}^2 + 2\int_0^{+\infty}\|u_x(t)\|_{\mathbf{L}^2}^2\,dt \le \|u_0\|_{\mathbf{L}^2}^2.$$

Differentiating equation (1.1) with respect to x, multiplying the result by $u_x(x,t)$ and integrating with respect to $x > 0$ we obtain

$$\frac{1}{2}\frac{d}{dt}\|u_x(t)\|_{\mathbf{L}^2}^2 + \int_0^{+\infty} u_x duu_x$$

$$-\int_0^{+\infty} u_x u_{xxx} dx + \int_0^{+\infty} u_x u_{xxxx} dx = 0.$$

Using identities

$$\int_0^{+\infty} u_x duu_x = -\int_0^{+\infty} uu_x u_{xx} dx,$$

$$\int_0^{+\infty} u_x u_{xxx} dx = u_x u_{xx}\big|_0^{+\infty} - \int_0^{+\infty} u_{xx}^2 dx,$$

$$\int_0^{+\infty} u_x u_{xxxx} dx = u_x u_{xxx}\big|_0^{+\infty} + \frac{1}{2}u_{xx}^2(0,t)$$

and relation $u_{xx}(0,t) = u_{xxx}(0,t)$ we get

(4.10)
$$\frac{1}{2}\frac{d}{dt}\|u_x(t)\|_{\mathbf{L}^2}^2 - \int_0^{+\infty} uu_x u_{xx} dx + \int_0^{+\infty} u_{xx}^2 dx + \frac{1}{2}u_{xx}^2(0,t) = 0.$$

Since

$$\|u(t)\|_{\mathbf{L}^\infty} \le C\|u(t)\|_{\mathbf{L}^2}^{\frac{1}{2}}\|u_x(t)\|_{\mathbf{L}^2}^{\frac{1}{2}}$$

and

$$\|u_x(t)\|_{\mathbf{L}^2} \le C\|u(t)\|_{\mathbf{L}^2}^{\frac{1}{2}}\|u_{xx}(t)\|_{\mathbf{L}^2}^{\frac{1}{2}}$$

we get

$$\left|\int_0^{+\infty} uu_x u_{xx} dx\right| \le C\|u(t)\|_{\mathbf{L}^\infty}\|u_x(t)\|_{\mathbf{L}^2}\|u_{xx}(t)\|_{\mathbf{L}^2}$$

$$\le C\|u(t)\|_{\mathbf{L}^2}^{\frac{1}{2}}\|u_x(t)\|_{\mathbf{L}^2}^{\frac{3}{2}}\|u_{xx}(t)\|_{\mathbf{L}^2}$$

$$\le C\|u_0\|_{\mathbf{L}^2}\|u_x(t)\|_{\mathbf{L}^2}^{\frac{1}{2}}\|u_{xx}(t)\|_{\mathbf{L}^2}^{\frac{3}{2}}.$$

Therefore by inequality

$$\|u_x(t)\|_{\mathbf{L}^2}^{\frac{1}{2}}\|u_{xx}(t)\|_{\mathbf{L}^2}^{\frac{3}{2}} \le \frac{1}{\varepsilon}\|u_x(t)\|_{\mathbf{L}^2}^2 + \varepsilon\|u_{xx}(t)\|_{\mathbf{L}^2}^2$$

where $\varepsilon > 0$, and using (4.9) we obtain

$$\left|\int_0^{+\infty} uu_x u_{xx} dx\right| \le C\|u_0\|_{\mathbf{L}^2}\left(\frac{1}{\varepsilon}\|u_x(t)\|_{\mathbf{L}^2}^2 + \varepsilon\|u_{xx}(t)\|_{\mathbf{L}^2}^2\right).$$

Choosing $\varepsilon = (C \left\| u_0 \right\|_{\mathbf{L}^2})^{-1}$ and integrating (4.10) with respect to t we get

(4.11) $$\left\| u_x(t) \right\|_{\mathbf{L}^2}^2 \leq C \left\| u_0 \right\|_{\mathbf{L}^2}^2 \int_0^t \left\| u_x(t) \right\|_{\mathbf{L}^2}^2 dt + \left\| u_{0x} \right\|_{\mathbf{L}^2}^2 .$$

From (4.9) we have

$$\int_0^{+\infty} \left\| u_x(t) \right\|_{\mathbf{L}^2}^2 dt \leq C \left\| u_0 \right\|_{\mathbf{L}^2}^2$$

and

$$\left\| u_x(t) \right\|_{\mathbf{L}^2} \leq C \left\| u_0 \right\|_{\mathbf{H}_2^{1,0}} .$$

Lemma 26 is proved.   □

5. Global existence

We now prove Theorem 24. Consider the initial-boundary value problem (1.1) with initial data

(5.1) $$\left\| u_0 \right\|_{\mathbf{H}_2^{0,\omega}} + \left\| u_0 \right\|_{\mathbf{H}_2^{1,0}} < C_1,$$

where $\omega > \frac{1}{2} + \mu$, $\mu \in \left(\frac{1}{2}, 1 \right)$. By Lemma 26 we get for all $t > 0$

$$\left\| u(t) \right\|_{\mathbf{H}_2^{1,0}} + \int_0^{+\infty} \left\| u_x(t) \right\|_{\mathbf{L}^2} dt \leq C \left\| u_0 \right\|_{\mathbf{H}_2^{1,0}}.$$

Taking into account Remark 11 by the standard continuation argument we ascertain that there exists a unique global solution $u \in \mathbf{C}\left((0, +\infty) ; \mathbf{H}_2^{1,0}\left(\mathbf{R}^+ \right) \right)$. Integral $\int_0^{+\infty} \left\| u_x(\tau) \right\|_{\mathbf{L}^2}^2 d\tau$ is convergent; hence there exists a positive time T such that

$$\int_T^{+\infty} \left\| u_x(\tau) \right\|_{\mathbf{L}^2}^2 d\tau < \varepsilon.$$

Therefore there exists a time $T_1 > T$ such that

$$\left\| u_x(T_1) \right\|_{\mathbf{L}^2}^2 < \varepsilon$$

(if $\left\| u_x(t) \right\|_{\mathbf{L}^2}^2 > \varepsilon$ for all $t > T$, then the integral $\int_T^{+\infty} \left\| u_x(\tau) \right\|_{\mathbf{L}^2}^2 d\tau = +\infty$). Also

$$\int_{T_1}^{+\infty} \left\| u_x(\tau) \right\|_{\mathbf{L}^2}^2 d\tau \leq \int_T^{+\infty} \left\| u_x(\tau) \right\|_{\mathbf{L}^2}^2 d\tau < \varepsilon,$$

thus we have

$$\left\| u_x(T_1) \right\|_{\mathbf{L}^2}^2 + \int_{T_1}^{+\infty} \left\| u_x(\tau) \right\|_{\mathbf{L}^2}^2 d\tau < 2\varepsilon.$$

To prove $\left\| u_x(t) \right\|_{\mathbf{L}^2}^2 < C\varepsilon$ for all $t > T_1$, we consider the initial-boundary value problem with the initial time at $t = T_1$. By Lemma 26 we have

$$\left\| u_x(t) \right\|_{\mathbf{L}^2}^2 \leq C \left\| u(T_1) \right\|_{\mathbf{L}^2}^2 \int_{T_1}^t \left\| u_x(t) \right\|_{\mathbf{L}^2}^2 dt + \left\| u_x(T_1) \right\|_{\mathbf{L}^2}^2$$

for all $t > T$. Here $\|u(T_1)\|_{\mathbf{L}^2}^2$ is bounded, $\int_{T_1}^t \|u_x(t)\|_{\mathbf{L}^2}^2 \, dt \le \int_T^{+\infty} \|u_x(t)\|_{\mathbf{L}^2}^2 \, dt < \varepsilon$ and $\|u_x(T_1)\|_{\mathbf{L}^2}^2 < \varepsilon$; hence $\|u_x(t)\|_{\mathbf{L}^2}^2 < C\varepsilon$ for all $t > T_1$. Thus we find that there exists a positive time T such that

$$\int_T^{+\infty} \|u_x(\tau)\|_{\mathbf{L}^2}^2 \, d\tau < \varepsilon, \quad \sup_{t \ge T} \|u_x(t)\|_{\mathbf{L}^2}^2 < C\varepsilon.$$

By the Sobolev inequality we see that

$$(5.2) \qquad \sup_{t \ge T} \|u(t)\|_{\mathbf{L}^\infty} \le C\varepsilon.$$

Now let us prove the estimate

$$(5.3) \qquad \sup_{t \ge T} t^{\frac{5}{8}} \|u(t)\|_{\mathbf{L}^2} < \widetilde{C}.$$

From (2.9) with $f(x,t) = u u_x$ we have

$$(5.4) \qquad u(x,t) \;=\; \int_0^{+\infty} u_0(y) F(x,y,t) dy$$
$$+ \int_0^t d\tau \int_0^{+\infty} dy\, u(y,\tau) u_y(y,\tau) F(x,y,t-\tau).$$

Integration by parts yields

$$(5.5) \qquad u(x,t) \;=\; \int_0^{+\infty} u_0(y) F(x,y,t) dy$$
$$- \frac{1}{2} \int_0^t d\tau \int_0^{+\infty} dy\, u^2(y,\tau) F_y(x,y,t-\tau).$$

From Lemma 25 we have for $\varkappa \in (0, \frac{1}{2})$

$$(5.6) \qquad \sup_{y \ge 0} \left(\|\langle y \rangle^\varkappa F_y(\cdot, y, t)\|_{\mathbf{L}^2} + y^{\varkappa - 1} \|F(\cdot, y, t)\|_{\mathbf{L}^2} \right) < Ct^{-\frac{3}{4} + \frac{\varkappa}{2}}.$$

Therefore from (5.5) we obtain

$$\|u(t)\|_{\mathbf{L}^2}$$
$$\le \; Ct^{-\frac{3}{4} + \frac{\varkappa}{2}} \|(\cdot)^{1-\varkappa} u_0(\cdot)\|_{\mathbf{L}^2}$$
$$+ C \int_0^t d\tau \int_0^{+\infty} \left| \frac{u(y,\tau)}{\langle y \rangle^\varkappa} \right|^2 \|\langle y \rangle^\varkappa F_y(\cdot, y, t)\|_{\mathbf{L}^2} \, dy$$
$$\le \; C \left(C_1 t^{-\frac{3}{4}} + \int_0^t \|\langle \cdot \rangle^{-\varkappa} u(\cdot, \tau)\|_{\mathbf{L}^2}^2 (t-\tau)^{-\frac{3}{4} + \frac{\varkappa}{2}} d\tau \right).$$

By the estimate

$$\left\|\langle\cdot\rangle^{-\varkappa}u(\cdot,t)\right\|_{\mathbf{L}^2}^2$$

$$= \int_0^{+\infty}\frac{u^2(x,t)}{\langle x\rangle^{2\varkappa}}dx$$

$$\leq C\left\|u(t)\right\|_{\mathbf{L}^\infty}^\varkappa\left(\int_0^{+\infty}u^2dx\right)^{\frac{2-\varkappa}{2}}\left(\int_0^{+\infty}\langle x\rangle^{-4}dx\right)^{\frac{\varkappa}{2}}$$

$$\leq C\left\|u(t)\right\|_{\mathbf{L}^\infty}^\varkappa\left\|u(t)\right\|_{\mathbf{L}^2}^{2-\varkappa}$$

we get

(5.7) $$\qquad \left\|u(t)\right\|_{\mathbf{L}^2} \leq CC_1 t^{-\frac{3}{4}+\frac{\varkappa}{2}}$$

$$+C\left\|u(t)\right\|_{\mathbf{L}^\infty}^\varkappa\int_0^t\left\|u(t)\right\|_{\mathbf{L}^2}^{2-\varkappa}(t-\tau)^{-\frac{3}{4}+\frac{\varkappa}{2}}d\tau.$$

We prove (5.3) by the contradiction. We assume that there exists some $T_1 > T$ such that

$$\sup_{t\in(0,T_1]}t^{\frac{5}{8}}\left\|u(t)\right\|_{\mathbf{L}^2} = \widetilde{C}.$$

Then from (5.7) and (5.2) with $\varkappa = \frac{1}{4}$ we obtain

$$\widetilde{C} < CC_1 + C\varepsilon^{\frac{1}{4}}\widetilde{C}^{\frac{15}{8}} < \widetilde{C}.$$

The contradiction obtained proves (5.3). Using (5.3) we obtain

(5.8) $$\qquad \left\|u(t)\right\|_{\mathbf{L}^2}^2 < C\langle t\rangle^{-1-\gamma}.$$

Also by virtue of Lemma 25 we have

$$\sup_{y\geq0}\left\|(\cdot)^\delta F_y(\cdot,y,t)\right\|_{\mathbf{L}^2} < Ct^{-\frac{3}{4}+\frac{\delta}{2}}$$

for $0 < \delta < \frac{3}{2}$. Therefore from (5.5) we get

$$\left\|(\cdot)^\delta u(\cdot,t)\right\|_{\mathbf{L}^2} \leq Ct^{-\frac{3}{4}+\frac{\delta}{2}}\left\|(\cdot)u_0\right\|_{\mathbf{L}^2}$$

$$+\int_0^t d\tau\int_0^{+\infty}|u(y,\tau)|^2\left\|(\cdot)^\delta F_y(\cdot,y,t-\tau)\right\|_{\mathbf{L}^2}$$

(5.9) $$\qquad \leq Ct^{-\frac{3}{4}+\frac{\delta}{2}}+\int_0^t\left\|u(\tau)\right\|_{\mathbf{L}^2}^2(t-\tau)^{-\frac{3}{4}+\frac{\delta}{2}}d\tau < Ct^{-\frac{3}{4}+\frac{\delta}{2}}.$$

We see that the norm

$$\left\|u(t)\right\|_{\mathbf{H}_2^{0,\omega}} + \left\|u(t)\right\|_{\mathbf{H}_2^{1,0}} \leq \left\|u_0\right\|_{\mathbf{H}_2^{0,\omega}} + \left\|u_0\right\|_{\mathbf{H}_2^{1,0}}$$

for all $t \geq 0$. By the standard continuation argument applying Theorem 26 we obtain that there exists a unique global solution $u \in \mathbf{C}\left([0,+\infty);\mathbf{H}_2^{1,\omega}\left(\mathbf{R}^+\right)\right)$.

Let us prove the estimate

(5.10) $$\qquad \sup_{t>T}t^{1-\frac{\gamma}{2}}\left\|(\cdot)^{\frac{1}{2}+\gamma}u_x(\cdot,t)\right\|_{\mathbf{L}^2} < C.$$

We prove (5.10) by the contradiction. We assume that there exists some $T_1 > T$ such that

(5.11) $$T_1^{1-\frac{\gamma}{2}} \left\| (\cdot)^{\frac{1}{2}+\gamma} u_x(\cdot, T_1) \right\|_{\mathbf{L}^2} = C.$$

We have from (5.4)

$$\| (\cdot)^{\frac{1}{2}-\gamma} u_x(\cdot, t) \|_{\mathbf{L}^2}$$
$$\leq C_1 \left(\int_0^{+\infty} |u_0(y)| \left\| (\cdot)^{\frac{1}{2}-\gamma} F_x(\cdot, y, t) \right\|_{\mathbf{L}^2} dy \right.$$
$$\left. + \int_0^t d\tau \int_0^{+\infty} \| u u_y(y, \tau) \|_{\mathbf{L}^1} \left\| (\cdot)^{\frac{1}{2}-\gamma} F_x(\cdot, y, t-\tau) \right\|_{\mathbf{L}^2} \right).$$

Clearly

$$\| y u(\cdot, \tau) u_y(\cdot, \tau) \|_{\mathbf{L}^1}$$
$$\leq C_1 \| u(\cdot, t) \|_{\mathbf{L}^\infty}^{\frac{1}{8}} \int_0^{+\infty} \left| y^{\frac{1}{2}-\gamma} u^{\frac{7}{8}} \right| \left| y^{\frac{1}{2}+\gamma} u_y \right| dx$$
$$\leq C_1 \| u(t) \|_{\mathbf{L}^\infty}^{\frac{1}{8}} \left\| (\cdot)^{\frac{1}{2}+\gamma} u_x(\cdot, t) \right\|_{\mathbf{L}^2}$$
$$\times \left(\int_0^{+\infty} \left(\langle y \rangle^{\frac{5}{4}-2\gamma} u^{\frac{7}{8}} \right)^{\frac{16}{7}} dy \right)^{\frac{7}{16}} \left(\int_0^{+\infty} \langle y \rangle^{-2} dy \right)^{\frac{1}{16}}$$
$$\leq C_1 \| u(t) \|_{\mathbf{L}^\infty}^{\frac{1}{8}} \left\| (\cdot)^{\frac{1}{2}+\gamma} u_x(\cdot, t) \right\|_{\mathbf{L}^2} \left\| (\cdot)^{\frac{5}{4}-2\gamma} u(\cdot, t) \right\|_{\mathbf{L}^2}^{\frac{7}{8}},$$

where $\gamma > 0$ is small. From Lemma 25

$$\left\| (\cdot)^{\frac{1}{2}+\gamma} \partial_x F(\cdot, y, t) \right\|_{\mathbf{L}^2} \leq C_1 y t^{-1+\frac{\gamma}{2}}$$

using (5.9) we obtain

$$\| (\cdot)^{\frac{1}{2}+\gamma} u_x(\cdot, t) \|_{\mathbf{L}^2}$$
$$\leq C_1 t^{-1+\frac{\gamma}{2}} + C_1 \| u(t) \|_{\mathbf{L}^\infty}^{\frac{1}{8}} \int_0^t \left\| (\cdot)^{\frac{1}{2}+\gamma} u_x(\cdot, t) \right\|_{\mathbf{L}^2}$$
$$\times \left\| (\cdot)^{\frac{5}{4}-2\gamma} u(\cdot, t) \right\|_{\mathbf{L}^2}^{\frac{7}{8}} (t-\tau)^{-1+\frac{\gamma}{2}} d\tau$$
$$\leq C_1 t^{-1+\frac{\gamma}{2}} + C_1 \| u(t) \|_{\mathbf{L}^\infty}^{\frac{1}{8}} \int_0^t \left\| (\cdot)^{\frac{1}{2}+\gamma} u_x(\cdot, t) \right\|_{\mathbf{L}^2}$$
$$\times (1+t)^{-\frac{1}{8}-\gamma} (t-\tau)^{-1+\frac{\gamma}{2}} d\tau.$$

Using (5.11), (5.9) we have

$$T^{1-\frac{\gamma}{2}} \left\| (\cdot)^{\frac{1}{2}+\gamma} u_x(\cdot, T) \right\|_{\mathbf{L}^2} = C < C_1 + C_1 \varepsilon^{\frac{1}{8}} C < C.$$

The contradiction obtained proves (5.10). From estimates (5.10), (5.9) we have

$$\int_0^t d\tau \|(\cdot)^{1+\mu} u(\cdot,\tau) u_y(\cdot,\tau)\|_{\mathbf{L}^1}$$

$$\le C \int_0^{+\infty} d\tau \|(\cdot)^{\frac{1}{2}+\mu-\gamma} u(\cdot,\tau)\|_{\mathbf{L}^2} \|(\cdot)^{\frac{1}{2}+\gamma} u_y(\cdot,\tau)\|_{\mathbf{L}^2}$$

$$(5.12) \qquad \le C \int_0^{+\infty} (1+t)^{-1-\frac{\gamma}{2}} dt \le C$$

for $\mu \in (0,1)$ and

$$(5.13) \qquad \|(\cdot) u(\cdot,\tau) u_x(\cdot,\tau)\|_{\mathbf{L}^1}$$
$$\le C \|(\cdot)^{\frac{1}{2}-\gamma} u(\cdot,\tau)\|_{\mathbf{L}^2} \|(\cdot)^{\frac{1}{2}+\gamma} u_x(\cdot,\tau)\|_{\mathbf{L}^2} \le C t^{-\frac{3}{2}}.$$

Now we prove that if $\|(\cdot)^{1+\mu} u_0\|_{\mathbf{L}^1} \le C$, then the solution has the following asymptotics for $t \to +\infty$ uniformly with respect to $x > 0$

$$(5.14) \qquad u(x,t) = \frac{1}{t} A \Phi\left(\frac{x}{2\sqrt{t}}, t\right) + O\left(\min\left(\frac{x}{\sqrt{t}}, 1\right) t^{-1-\frac{\mu}{2}}\right),$$

where $\mu \in (0,1)$, $\Lambda(y) = e^{-y} - 1 + y$, $\Phi(q,t) = \frac{q}{\sqrt{\pi}} e^{-q^2}$

$$A = \int_0^{+\infty} \Lambda(y) u_0(y) dy + \int_0^t d\tau \int_0^{+\infty} \Lambda(y) u(y,\tau) u_y(y,\tau) dy < +\infty.$$

From Lemma 22 we have

$$(5.15) \qquad F(x,y,t) = t^{-1} \Lambda(y) \Phi\left(\frac{x}{2\sqrt{t}}, t\right) + R(x,y,t)$$

for any $x,y,t > 0$, where

$$R(x,y,t) = y^{1+\mu} O\left(t^{-1-\frac{\mu}{2}} \min\left(\frac{x}{\sqrt{t}}, 1\right)\right).$$

Therefore, via (5.4) we get

$$(5.16) \qquad u(x,t) = \frac{1}{t} A \Phi\left(\frac{x}{2\sqrt{t}}, t\right) + R(x,t),$$

where by virtue of (5.13) $A < +\infty$ and

$$|R(x,t)|$$
$$\le C \min\left(\frac{x}{\sqrt{t}}, 1\right) t^{-1-\frac{\mu}{2}} \left(\|(\cdot)^{1+\mu} u_0\|_{\mathbf{L}^1} + \int_0^t d\tau \|(\cdot)^{1+\mu} u(\cdot,\tau) u_y(\cdot,\tau)\|_{\mathbf{L}^1}\right)$$
$$+ \int_0^t d\tau \int_0^{+\infty} dy |u u_y| |F(x,y,t-\tau) - F(x,y,t)|.$$

Using estimate (5.10) we see that

$$(5.17) \qquad \int_0^t \tau \|(\cdot)^{\frac{1}{2}-\gamma} u(\cdot,\tau)\|_{\mathbf{L}^2} \|(\cdot)^{\frac{1}{2}+\gamma} u_x(\cdot,\tau)\|_{\mathbf{L}^2} d\tau \le C + \int_1^t \tau^{-\frac{1}{2}} d\tau \le C t^{\frac{1}{2}}$$

for $t > 1$. As in the proof of Lemma 22 we get

$$\|\partial_t F(\cdot, y, t)\|_{\mathbf{L}^\infty} \le Cy \min\left(\frac{x}{\sqrt{t}}, 1\right) t^{-2+\gamma},$$

whence

$$\|F(\cdot, y, t - \tau) - F(\cdot, y, t)\|_{\mathbf{L}^\infty} \le Cy\tau \min\left(\frac{x}{\sqrt{t}}, 1\right) t^{-2+\gamma}.$$

Therefore using (5.17) we have

$$\int_0^t d\tau \left| \int_0^{+\infty} u(y, \tau) u_y(y, \tau) \right| |F(x, y, t - \tau) - F(x, y, t)|$$

$$\le \; Ct^{\gamma-2} \int_0^t \tau \|(\cdot)^{\frac{1}{2}-\gamma} u(\cdot, \tau)\|_{\mathbf{L}^2} \|(\cdot)^{\frac{1}{2}+\gamma} u_x(\cdot, \tau)\|_{\mathbf{L}^2} d\tau$$

(5.18) $\qquad \le \; C \min\left(\frac{x}{\sqrt{t}}, 1\right) t^{-1-\frac{\mu}{2}}$

with some $\mu \in (0, 1)$. Since $\|(\cdot)^{1+\mu} u_0\|_{\mathbf{L}^1} \le C$ using (5.10) and (5.18) we get

(5.19) $\qquad\qquad |R(x, t)| \le C \min\left(\frac{x}{\sqrt{t}}, 1\right) t^{-1-\frac{\mu}{2}},$

where $\mu \in (0, 1)$. From (5.16) and (5.19) we obtain the asymptotics (5.14) for the solution. Theorem 24 is proved.

CHAPTER 8

KdV-B Type Equation

1. Setting of the problem

In this chapter we want to show how to apply our methods developed in previous sections to another type of the nonlinearity, for example, of the form u_x^2. We consider the following initial-boundary value problem for the nonlinear Korteweg-de Vries-Burgers equation:

$$(1.1) \qquad \begin{cases} u_t + u_x^2 - u_{xx} + u_{xxx} = 0, & t > 0, x > 0, \\ u(x,0) = u_0(x), & x > 0, \\ u(0,t) = 0, & t > 0. \end{cases}$$

Note that because of the Dirichlet boundary condition the initial-boundary value problem (1.1) cannot be reduced to the initial-boundary value problem for equation (1.1); on the contrary in the case of the Neumann boundary condition the initial-boundary value problem for equation (1.1) can be reduced to problem (1.1) by virtue of the change $u_x \to u$.

We introduce the following function space

$$\mathbf{Z}_{T,\gamma} = \{\phi(x,t) \in \mathbf{C}([0,T]; \mathbf{L}^2(\mathbf{R}^+)) \cap \mathbf{C}((0,T]; \mathbf{H}_2^{1,0}(\mathbf{R}^+)); \|\phi\|_{\mathbf{Z}_{T,\gamma}} < +\infty\}$$

with the norm

$$\|\phi\|_{\mathbf{Z}_{T,\gamma}} = \sup_{t \in [0,T]} \left(\|\phi(t)\|_{\mathbf{L}^2} + t^{\frac{1}{6}+\gamma} \|\phi_x(t)\|_{\mathbf{L}^2} \right)$$

and

$$\mathbf{X} = \{\varphi(x) \in \mathbf{L}^1(\mathbf{R}^+)) \cap \mathbf{H}_2^{1,0}(\mathbf{R}^+)); \|\varphi\|_{\mathbf{X}} < +\infty\}$$

with the norm

$$\|\varphi\|_{\mathbf{X}} = \left(\|\varphi\|_{\mathbf{L}^1} + \|\varphi\|_{\mathbf{H}_2^1} \right).$$

Now we state result of this chapter (see paper [58]).

We give some sufficient conditions for global existence and asymptotic behavior in time of solutions.

THEOREM 27. *Suppose that the initial data satisfy $\|u_0\|_{\mathbf{X}} \le \epsilon_1$ and $u_0(0) = 0$, where $\epsilon_1 > 0$ is small enough. Then there exists a unique solution u of (1.1) such that $u \in \mathbf{C}([0,+\infty), \mathbf{H}_2^{1,0}(\mathbf{R}^+))$. Furthermore if $\|x^{1+\delta} u_0\|_{\mathbf{L}^1} + \|x^{1+\delta} u_0\|_{\mathbf{L}^2} \le C$, then the solution u has the following asymptotics for large time*

$$u(x,t) = \frac{A}{t} e^{-\frac{x^2}{4t}} \frac{x}{2\sqrt{t}} + O\left(\min\left(\frac{x}{\sqrt{t}}, 1 \right) t^{-1-\frac{\delta}{2}} \right),$$

where $\delta \in (0, \frac{1}{2})$ and

$$A = \frac{1}{\sqrt{\pi}} \int_0^{+\infty} (e^{-x} - 1 + x) u_0(x) dx$$

$$+ \int_0^{+\infty} d\tau \int_0^{+\infty} (e^{-x} - 1 + x) u_x^2 dx.$$

REMARK 12. *Comparing to the result of Chapter 6 in Theorem 27 we obtain the continuity of the solution at $t = 0$ in the $\mathbf{H}_2^{1,0}$-norm due to the compatibility condition for the initial and boundary data $u_0(0) = 0$.*

REMARK 13. *We cannot use here the method developed in Chapter 7 in order to exclude the smallness of the initial data, since due to the Dirichlet boundary conditions the contribution of the nonlinear term does not vanish in the energy type estimates.*

We organize this chapter as follows. In Section 2 we consider the linear initial-boundary value problem corresponding to (1.1), and we discuss the question of the number of the boundary data which are necessary to show existence and uniqueness of solutions to the problem (1.1). In Section 2 we also prove some estimates of the Green function of problem (1.1), and in Section 3 we show $\mathbf{H}_2^{1,0}(\mathbf{R}^+)$ estimates of solutions to (1.1) which are needed to prove that the solution belongs to $\mathbf{C}([0, +\infty), \mathbf{H}_2^{1,0}(\mathbf{R}^+))$. Sections 4-5 are devoted to the proof of Theorem 29. We prove Theorem 27 in Section 5. This Chapter was written in collaboration with Prof. H.F. Ruiz Paredes.

2. Linear problem

In this section we translate the Cauchy problem of the linear boundary value problem to the corresponding integral equation by finding the Green function. We now consider the following linear initial-boundary value problem

(2.1)
$$\begin{cases} u_t + \mathbb{K}u = f(x,t), & t > 0, x > 0, \\ u(x,0) = u_0(x), & x \in R^+, \\ \partial_x^{(j-1)} u(0,t) = 0, & \text{for} \quad j = 1, 2, ..., n, \end{cases}$$

with the integer number $n \le 3$ will be defined below. Here $\mathbb{K}u$ is the differential operator defined as follows

$$\mathbb{K}u = \frac{1}{2\pi i} \left(-\int_{-i\infty}^{i\infty} e^{px} p^2 \left(\hat{u}(p,t) - \sum_{j=1}^{2} \frac{\partial_x^{j-1} u(0,t)}{p^j} \right) dp \right.$$

$$\left. + \int_{-i\infty}^{i\infty} e^{px} p^3 \left(\hat{u}(p,t) - \sum_{j=1}^{3} \frac{\partial_x^{j-1} u(0,t)}{p^j} \right) dp \right),$$

where the symbol the operator $\mathbb{K}u$ has the form $K(p) = -p^2 + p^3$.

From result of the previous chapter we have the explicit formula for the solution of the linear problem (2.1) under the condition $u_0 \in \mathbf{L}^1(\mathbf{R}^+)$, $f \in \mathbf{L}^q_{loc}(0, +\infty; \mathbf{L}^1(\mathbf{R}^+))$ with $q > 2$

$$(2.2) \quad u(x,t) = \int_0^{+\infty} u_0(y) F(x,y,t) dy + \int_0^t d\tau \int_0^{+\infty} f(y,\tau) F(x,y,t-\tau) dy,$$

where

$$F(x,y,t) = -\frac{1}{4\pi^2} \int_{-i\infty}^{+i\infty} dp e^{-py} \int_{-i\infty}^{i\infty} e^{\xi t} \frac{e^{\phi_3(\xi)x} + e^{px}}{K(p) + \xi} d\xi,$$

$\phi_3(\xi)$ is the root of the equation $K(p) = -\xi$, such that $\operatorname{Re} \phi_3(\xi) < 0$ for $\operatorname{Re} \xi > 0$. Thus we have the following result.

THEOREM 28. *For some $T > 0$ there exists a unique solution $u \in \mathbf{Z}_{T,\gamma}$ of the initial-boundary value problem (2.1) such that*

$$\|u\|_{\mathbf{Z}_{T,\gamma}} \leq C\lambda,$$

provided that $\lambda < +\infty$, where $\lambda = \|u_0\|_{\mathbf{X}} + T^\mu \sup_{t \in [0,T]} t^\nu \|f\|_{\mathbf{L}^1}$, $\nu \in [0, \frac{5}{12})$, $\mu = \frac{5}{12} - \nu$ and $\gamma > 0$.

PROOF. As in the previous chapter we rewrite the function $F_x(x,y,t)$ as

$$F(x,y,t) = \mathcal{F}(x,y,t) + G(x-y,t),$$

where

$$\mathcal{F}(x,y,t) = \frac{1}{2\pi i} \int_{\operatorname{Re}\xi=0} e^{\xi t} e^{\phi_3(\xi)x} \sum_{j=1}^2 e^{-\phi_j(\xi)y} \phi_j'(\xi) d\xi$$

and

$$G(q,t) = \frac{1}{2\pi i} \int_{\operatorname{Re} p=0} dp e^{pq - K(p)t} dp.$$

Let us prove the following estimate

$$(2.3) \quad \|u\|_{\mathbf{Z}_{T,\gamma}} = \sup_{t \in [0,T]} \left(\|u(t)\|_{\mathbf{L}^2} + t^{\frac{1}{6}+\gamma} \|u_x(t)\|_{\mathbf{L}^2} \right) \leq C\lambda,$$

where $\lambda = \|u_0\|_{\mathbf{X}} + T^\mu \sup_{t \in [0,T]} t^\nu \|f(\cdot,t)\|_{\mathbf{L}^1}$, $T > 0$, $\mu = \frac{5}{12} - \nu$, $\gamma > 0$, $\nu < \frac{5}{12}$. From (2.2) we have

$$\|\partial_x^n u\|_{\mathbf{L}^2} \leq C \left\| \int_0^{+\infty} u_0(y) \partial_x^n G(\cdot - y, t) dy \right\|_{\mathbf{L}^2}$$

$$+ C \int_0^{+\infty} |u_0(y)| \|\partial_x^n \mathcal{F}(\cdot, y, t)\|_{\mathbf{L}^2} dy$$

$$+ C \int_0^t d\tau \left\| \int_0^{+\infty} f(y,\tau) \partial_x^n G(\cdot - y, t-\tau) dy \right\|_{\mathbf{L}^2}$$

$$(2.4) \quad + C \int_0^t d\tau \int_0^{+\infty} |f(y,\tau)| \|\partial_x^n \mathcal{F}(\cdot, y, t-\tau)\|_{\mathbf{L}^2} dy.$$

By virtue of Lemma 24 we have $\|G(\cdot, t)\|_{\mathbf{L}^1} \leq C$. Therefore by the Young inequality $\left\|\int_0^{+\infty} f(x - y)\phi(y)dy\right\|_{\mathbf{L}^2} \leq C\|f\|_{\mathbf{L}^2}\|\phi\|_{\mathbf{L}^1}$ we get

$$(2.5) \qquad \left\|\int_0^{+\infty} u_0(y)G(x - y, t)dy\right\|_{\mathbf{L}^2} \leq \|u_0\|_{\mathbf{L}^2}\|G(t)\|_{\mathbf{L}^1} \leq C\lambda$$

Integrating by parts in the case $n = 1$ and using the condition $u_0(0) = 0$ we get

$$\left\|\int_0^{+\infty} u_0(y)G_x(\cdot - y, t)dy\right\|_{\mathbf{L}^2}$$
$$\leq \quad C|u_0(0)|\|G(\cdot, t)\|_{\mathbf{L}^2} + C\|\partial_x u_0\|_{\mathbf{L}^2}\|G(t)\|_{\mathbf{L}^1}$$
$$(2.6) \qquad \leq \quad C\|\partial_x u_0\|_{\mathbf{L}^2} \leq C\lambda.$$

Let $\delta(0) = 0$ and $\delta(1) = \frac{1}{6} + \gamma$. By virtue of Lemma 23 we get for $n = 0, 1$

$$\|\partial_x^n \mathcal{F}(\cdot, y, t)\|_{\mathbf{L}^2} \leq Ct^{-\delta(n)}y^{-1+\gamma}.$$

Therefore we see that

$$\int_0^{+\infty} |u_0(y)| \, \|\partial_x^n \mathcal{F}(\cdot, y, t)\|_{\mathbf{L}^2} dy$$
$$\leq \quad Ct^{-\delta(n)} \left(\|u_0\|_{\mathbf{L}^\infty} \int_0^1 y^{-1+\gamma}dy + \int_1^{+\infty} |u_0(y)|dy \right)$$
$$(2.7) \qquad \leq \quad Ct^{-\delta(n)} \left(\|u_0\|_{\mathbf{H}^1} + \|u_0\|_{\mathbf{L}^1} \right) \leq Ct^{-\delta(n)}\lambda.$$

By Lemmas 23 through 24 we have

$$\|\partial_x^n G(t)\|_{\mathbf{L}^2} + \sup_{y \geq 0} \|\partial_x^n \mathcal{F}(\cdot, y, t)\|_{\mathbf{L}^2} \leq Ct^{-\frac{2n+1}{4}+\gamma}.$$

Hence we obtain the following estimate

$$\int_0^t d\tau \left(\left\|\int_0^{+\infty} f(y, \tau)\partial_x^n G(\cdot - y, t - \tau)dy\right\|_{\mathbf{L}^2} \right.$$
$$\left. + \int_0^{+\infty} |f(y, \tau)| \, \|\partial_x^n \mathcal{F}(\cdot, y, t - \tau)\|_{\mathbf{L}^2} \, dy \right)$$
$$\leq \quad C \int_0^t d\tau \|f(\cdot, \tau)\|_{\mathbf{L}^1} \left(\|\partial_x^n G(t - \tau)\|_{\mathbf{L}^2} + \sup_{y \geq 0} \|\partial_x^n \mathcal{F}(\cdot, y, t - \tau)\|_{\mathbf{L}^2} \right) dy$$
$$\leq \quad C \sup_{t \in [0,T]} t^\nu \|f(\cdot, t)\|_{\mathbf{L}^1} \int_0^t \frac{d\tau}{\tau^\nu (t - \tau)^{\frac{2n+1}{4} - \gamma}}$$
$$(2.8) \quad \leq \quad Ct^{-\frac{n}{6} - \gamma_1(n)}T^\mu \sup_{t \in [0,T]} t^\nu \|f(\cdot, t)\|_{\mathbf{L}^1} \leq Ct^{-\delta(n)}\lambda,$$

where $\mu = \frac{5}{12} - \nu$, $\nu < 1$. Thanks to (2.4)-(2.8) we obtain $\|\partial_x^n u(t)\|_{\mathbf{L}^2} \leq Ct^{-\delta(n)}\lambda$. Theorem 28 is proved. $\qquad\qquad\square$

3. Energy estimate

In this section we show some estimates of solutions to (1.1) by using the energy estimate. We prove

$$
\|u(\cdot,t)\|_{\mathbf{H}_2^{1,0}}^2 + \int_0^t \|u_x(\cdot,\tau)\|_{\mathbf{H}_2^{1,0}}^2 d\tau
$$

$$
(3.1) \qquad \leq 6 \int_0^t \|u(\cdot,\tau)\|_{\mathbf{H}_2^{1,0}} \|u_x(\cdot,\tau)\|_{\mathbf{H}_2^{1,0}}^2 d\tau + \|u_0\|_{\mathbf{H}_2^{1,0}}^2
$$

and for any $\varepsilon > 0$

$$
\|u_x(\cdot,t)\|_{\mathbf{L}^2}^2 + \int_0^t \|u_{xx}(\cdot,\tau)\|_{\mathbf{L}^2}^2 d\tau
$$

$$
\leq \varepsilon^{-1} \int_0^t \|u_x(\cdot,\tau)\|_{\mathbf{L}^2}^3 d\tau
$$

$$
(3.2) \qquad + \varepsilon \|u_x(\cdot,\tau)\|_{\mathbf{L}^2} \int_0^t \|u_{xx}(\cdot,\tau)\|_{\mathbf{L}^2}^2 d\tau + \|\partial_x u_0\|_{\mathbf{L}^2}^2 .
$$

We have from equation (1.1)

$$
\frac{d}{dt}\|u(\cdot,t)\|_{\mathbf{L}^2}^2 = 2\left(-\int_0^{+\infty} u(u_x)^2 dx + \int_0^{+\infty} uu_{xx} dx - \int_0^{+\infty} uu_{xxx} dx\right).
$$

Therefore by the identities $uu_{xx} = \partial_x uu_x - u_x^2$ and $uu_{xxx} = \partial_x uu_{xx} - \frac{1}{2}\partial_x u_x^2$ we get

$$
\frac{d}{dt}\|u(\cdot,t)\|_{\mathbf{L}^2}^2 + 2\|u_x(\cdot,t)\|_{\mathbf{L}^2}^2
$$

$$
(3.3) \qquad = -2\int_0^{+\infty} u(u_x)^2 dx + 2u(0,t)\left(u_x(0,t) - u_{xx}(0,t)\right)
$$

$$
-2\left(u_x(0,t)\right)^2 .
$$

Since

$$
\left|\int_0^{+\infty} uu_x^2 dx\right| \leq \|u(\cdot,t)\|_{\mathbf{L}^\infty} \|u_x(\cdot,t)\|_{\mathbf{L}^2}^2
$$

so integrating (3.3) with respect to time and using the condition $u(0,t) = 0$ we obtain

$$
\|u(\cdot,t)\|_{\mathbf{L}^2}^2 + 2\int_0^t \|u_x(\cdot,\tau)\|_{\mathbf{L}^2}^2 d\tau + 2\int_0^t \left(u_x(0,\tau)\right)^2 d\tau
$$

$$
(3.4) \qquad \leq 2\int_0^t \|u(\cdot,\tau)\|_{\mathbf{L}^\infty} \|u_x(\cdot,\tau)\|_{\mathbf{L}^2}^2 d\tau + \|u_0\|_{\mathbf{L}^2}^2 .
$$

In the same manner, by the fact that $u_t(0,t) = 0$ we get from (1.1)

$$\frac{d}{dt}\|u_x(\cdot,t)\|_{\mathbf{L}^2}^2 = -2\int_0^{+\infty} u_{xx}u_t dx$$

(3.5)
$$= -2\int_0^{+\infty} u_{xx}\left(-u_x^2 + u_{xx} - u_{xxx}\right)dx.$$

Using the identity $u_{xx}u_{xxx} = \frac{1}{2}\partial_x\left(u_{xx}^2\right)$, then applying the estimates

$$\left|\int_0^{+\infty} u_{xx}u_x^2 dx\right| \leq \|u_x(\cdot,t)\|_{\mathbf{L}^4}^2\|u_{xx}(\cdot,t)\|_{\mathbf{L}^2}$$

and

$$\|u_x(\cdot,t)\|_{\mathbf{L}^4}^2 \leq 3\|u(\cdot,t)\|_{\mathbf{L}^\infty}\|u_{xx}(\cdot,t)\|_{\mathbf{L}^2}$$

we get integrating (3.5) with respect to time

$$\|u_x(\cdot,t)\|_{\mathbf{L}^2}^2 + 2\int_0^t \|u_{xx}(\cdot,\tau)\|_{\mathbf{L}^2}^2 d\tau + 2\int_0^t u_{xx}(0,\tau)^2 d\tau$$

(3.6)
$$\leq 6\int_0^t \|u(\cdot,\tau)\|_{\mathbf{L}^\infty}\|u_{xx}(\cdot,\tau)\|_{\mathbf{L}^2}^2 d\tau + \|\partial_x u_0\|_{\mathbf{L}^2}^2.$$

Then using the estimate $\|u(\cdot,t)\|_{\mathbf{L}^\infty} \leq \|u(\cdot,t)\|_{\mathbf{H}_2^{1,0}}$ we obtain (3.1) from (3.4) and (3.6). Also using the inequalities

$$\|u_x(\cdot,t)\|_{\mathbf{L}^\infty} \leq C\left(|u_x(0,t)| + \|u_x(\cdot,t)\|_{\mathbf{L}^2}^{1/2}\|u_{xx}(\cdot,t)\|_{\mathbf{L}^2}^{1/2}\right)$$

and

$$|f(0)| \leq C\|f\|_{\mathbf{L}^2}^{1/2}\|f_x\|_{\mathbf{L}^2}^{1/2}$$

we find

$$\left|\int_0^{+\infty} u_{xx}u_x^2 dx\right|$$
$$\leq C\|u_x(\cdot,t)\|_{\mathbf{L}^\infty}\|u_x(\cdot,t)\|_{\mathbf{L}^2}\|u_{xx}(\cdot,t)\|_{\mathbf{L}^2}$$
$$\leq \|u_x(\cdot,t)\|_{\mathbf{L}^2}(\varepsilon^{-1}\|u_x(\cdot,t)\|_{\mathbf{L}^2}^2 + C\varepsilon\|u_{xx}(\cdot,t)\|_{\mathbf{L}^2}^2).$$

Therefore from (3.5) we have (3.2) .

4. Local existence

We formulate the local existence of the solutions of the initial-boundary value problem (1.1).

THEOREM 29. *Suppose that the initial data $u_0 \in \mathbf{X}$ and the boundary data $u_0(0) = 0$. Then there exists a unique solution $u(x,t) \in \mathbf{C}([0,T];\mathbf{H}_2^{1,0}(\mathbf{R}^+))$ for some $T > 0$.*

REMARK 14. *If the initial data u_0 are small, i.e. the norm $\|u_0\|_{\mathbf{X}} \leq \epsilon$, where $\epsilon > 0$ is sufficiently small, then there exists $T > 1$, such that there exists a unique solution u, which is also small: $\|u(t)\|_{\mathbf{H}_2^{1,0}} < C\epsilon$.*

PROOF. We prove the local existence of solutions by the contraction mapping principle. We let $u(x,t)$ be a solution of the following linear problem

(4.1)
$$\begin{cases} u_t - u_{xx} + u_{xxx} = \mathbb{N}(w), & t > 0, x > 0, \\ u(x,0) = u_0(x), & x > 0, \\ u(0,t) = 0, & t > 0, \end{cases}$$

where $\mathbb{N}(w) = -w_x^2$ and $w \in \mathbf{Z}_{T,\gamma,\rho}$, with $\mathbf{Z}_{T,\gamma,\rho} = \{w \in \mathbf{Z}_{T,\gamma}; \|w\|_{\mathbf{Z}_{T,\gamma}} \le \rho\}$. Note that initial value problem (4.1) defines a mapping $\mathbb{M}$ by $u = \mathbb{M}(w)$, and we will show that $\mathbb{M}$ is the contraction mapping from $\mathbf{Z}_{T,\gamma,\rho}$ into itself for a sufficiently small $T > 0$. Since $w \in \mathbf{Z}_{T,\gamma,\rho}$ we have

$$\sup_{t\in[0,T]} t^{\frac{1}{3}+2\gamma}\|\mathbb{N}(w)(\cdot,t)\|_{\mathbf{L}^1} \le C \sup_{t\in[0,T]} t^{\frac{1}{3}+2\gamma}\|w_x(\cdot,t)\|_{\mathbf{L}^2}^2 \le C\rho^2.$$

Via Theorem 28, problem (2.2) has a unique solution $u(x,t) \in \mathbf{Z}_{T,\gamma}$ with the norm

$$\|u\|_{\mathbf{Z}_{T,\gamma}} \le C\lambda,$$

where $\lambda = \|u_0\|_{\mathbf{X}} + T^\mu \sup_{t\ge[0,T]} t^\nu \|\mathbb{N}(w)(\cdot,t)\|_{\mathbf{L}^1}$ with $\mu = \frac{5}{12} - \nu$. Therefore we obtain

(4.2)
$$\|u\|_{\mathbf{Z}_{T,\gamma}} \le C\|u_0\|_{\mathbf{X}} + CT^\mu\rho^2,$$

for $\nu = \frac{1}{3}+2\gamma$, where $\mu = \frac{1}{12}-2\gamma > 0$ if $\gamma \in (0,\frac{1}{24})$. Therefore we get $\|u(\cdot,t)\|_{\mathbf{Z}_{T,\gamma}} \le \rho$ if $T = (2C\rho^2)^{-\mu}$ and $\|u_0\|_{\mathbf{X}} \le \frac{\mathcal{C}}{2}$. Thus the mapping $\mathbb{M}$ transforms the closed ball $\mathbf{Z}_{T,\gamma,\rho}$ in $\mathbf{Z}_{T,\gamma}$ with a center at the origin and a radius ρ into itself. Analogously we can prove the estimate $\|u - \tilde{u}\|_{\mathbf{Z}_{T,\gamma}} < \|w - \tilde{w}\|_{\mathbf{Z}_{T,\gamma}}$ for small T. Therefore the mapping $\mathbb{M}$ is a contraction mapping in $\mathbf{Z}_{T,\gamma}$ and there exists a unique solution $u(x,t) \in \mathbf{Z}_{T,\gamma,\rho}$ of the initial-value problem (1.1). Moreover using the energy estimate (3.2) we obtain that there exists some $T > 0$ such that

$$\sup_{t\in[0,T]} \|u_x(\cdot,t)\|_{\mathbf{L}^2} < C.$$

Theorem 29 is proved. □

REMARK 15. *By virtue of estimates (4.2) and (3.1) we see that if the norm of the initial data u_0 is sufficiently small $\|u_0\|_{\mathbf{X}} < \epsilon$, then for some time $T > 1$ there exists a unique solution u which is also small $\|u(\cdot,t)\|_{\mathbf{H}^1} < C\epsilon$.*

5. Large time asymptotics

We consider the initial-boundary value problem (1.1) with small initial data

(5.1)
$$\|u_0\|_{\mathbf{X}} < \epsilon_1,$$

where $\epsilon_1 > 0$ is sufficiently small. We are now in a position to prove Theorem 27.

Proof of Theorem 27. Let us prove that for all $t \ge 0$

(5.2)
$$\|u\|_{\mathbf{H}^1} \le \epsilon.$$

Via Remark 15 there exists a time $T > 0$ such that $\|u\|_{\mathbf{H}^1} \leq \epsilon$. Hence using the energy estimate (4.1) we obtain (4.2), and moreover, we get

$$(5.3) \qquad \int_0^t \|u_x(\cdot, \tau)\|_{\mathbf{L}^2}^2 d\tau < \epsilon.$$

Using the estimate (5.3) and Lemma 25 we can prove that if $\|x^{1+\delta}u_0\|_{\mathbf{L}^2} + \|x^{1+\delta}u_0\|_{\mathbf{L}^1} \leq C$, then the solution has the following asymptotics for $t \to +\infty$ uniformly with respect to $x > 0$

$$u(x, t) = \frac{A}{t} e^{-\frac{x^2}{4t}} \frac{x}{2\sqrt{t}} + O\left(\min\left(\frac{x}{2\sqrt{t}}, 1\right) t^{-1-\frac{\delta}{2}}\right),$$

where

$$A = \frac{1}{\sqrt{\pi}} \int_0^{+\infty} (e^{-y} - 1 + y)u_0(y)dy$$
$$+ \int_0^t d\tau \int_0^{+\infty} (e^{-y} - 1 + y)u_y^2(y, \tau)dy$$

and $\delta \in (0, 1)$. In fact, by virtue of Lemma 25 we have

$$G(x - y, t) + F(x, y, t) = \frac{1}{\sqrt{\pi t}} e^{-\frac{x^2}{4t}} \frac{x}{2\sqrt{t}} (e^{-y} - 1 + y)$$
$$+ y^{1+\delta} O\left(\min\left(\frac{x}{\sqrt{t}}, 1\right) t^{-1-\frac{\delta}{2}}\right),$$

where $\delta \in [0, 1)$. We also have the integral equation

$$u(x, t) = \int_0^{+\infty} u_0(y)(G(x - y, t) + F(x, y, t))dy$$
$$+ \int_0^t d\tau \int_0^{+\infty} u_y^2(y, \tau)(G(x - y, t - \tau) + F(x, y, t - \tau))dy.$$

Therefore we obtain

$$(5.4) \qquad u(x, t) = \frac{A}{t} e^{-\frac{x^2}{4t}} \frac{x}{2\sqrt{t}} + R(x, t),$$

where

$$|R(x, t)| \leq C \min\left(\frac{x}{\sqrt{t}}, 1\right) t^{-1-\frac{\delta}{2}} \left(\|y^{1+\delta}u_0\|_{\mathbf{L}^1} + \|u_0\|_{\mathbf{L}^1}\right)$$
$$+ \int_0^t d\tau (\|y^{1+\delta}u_y^2(\cdot, \tau)\|_{\mathbf{L}^1} + \|u_y^2(\cdot, \tau)\|_{\mathbf{L}^1})$$
$$+ \int_0^t d\tau \int_0^{+\infty} dy u_y^2 |F(x, y, (t - \tau)) - F(x, y, t)$$
$$+ G(x - y, t - \tau) - G(x - y, t)|.$$

To estimate $R(x,t)$ we need to estimate the integral

$$\int_0^t \|x^{(1+\delta)/2} u_x(\cdot, \tau)\|_{\mathbf{L}^2}^2 d\tau.$$

From equation (1.1) via integrating by parts we have

$$\| \langle x \rangle^{(1+\delta)/2} u\|_{\mathbf{L}^2}^2 + 2 \int_0^t \| \langle x \rangle^{(1+\delta)/2} u_x(\cdot, \tau)\|_{\mathbf{L}^2}^2 d\tau$$

$$\leq C \int_0^t \|u(\cdot, \tau)\|_{\mathbf{H}^1} \| \langle x \rangle^{(1+\delta)/2} u_x(\cdot, \tau)\|_{\mathbf{L}^2}^2 d\tau$$

$$+ C \int_0^t \|u(\cdot, \tau)\|_{\mathbf{L}^\infty} \|u_x(\cdot, \tau)\|_{\mathbf{L}^2} d\tau$$

$$+ C \int_0^t \|u_x(\cdot, \tau)\|_{\mathbf{L}^2}^2 d\tau + C\| \langle x \rangle^{(1+\delta)/2} u_0\|_{\mathbf{L}^2}^2$$

for $\delta \in (0,1)$. Then from (5.2) and (5.3) we have

$$(5.5) \qquad \int_0^t \| \langle x \rangle^{(1+\delta)/2} u_x(\cdot, \tau)\|_{\mathbf{L}^2}^2 d\tau$$

$$\leq C \int_0^t \|u(\cdot, \tau)\|_{\mathbf{L}^\infty} \|u_x(\cdot, \tau)\|_{\mathbf{L}^2} d\tau + C\epsilon.$$

Now we prove the estimate

$$(5.6) \qquad \sup_{t>0} t^{\frac{1}{2}-\gamma} \|u(\cdot, t)\|_{\mathbf{L}^\infty} + \sup_{t>0} t^{\frac{3}{4}-\gamma} \|u_x(\cdot, t)\|_{\mathbf{L}^2} \leq C.$$

From Lemmas 23 through 24 we easily get

$$\|F(\cdot, \cdot, t)\|_{\mathbf{L}^\infty} + \|G(\cdot, t)\|_{\mathbf{L}^\infty} \leq Ct^{-\frac{1}{2}+\gamma}$$

and

$$\sup_{y \geq 0} (\|\partial_x F(\cdot, y, t)\|_{\mathbf{L}^2} + \|\partial_x G(\cdot, t)\|_{\mathbf{L}^2}) \leq Ct^{-\frac{3}{4}+\gamma}.$$

Therefore we obtain

$$\|u_x(\cdot, t)\|_{\mathbf{L}^2} \leq Ct^{-\frac{3}{4}+\gamma} \|u_0\|_{\mathbf{L}^1}$$

$$+ \int_0^t \|u_x(\cdot, \tau)\|_{\mathbf{L}^2}^2 \sup_{y \geq 0} (\|\partial_x F(\cdot, y, t)\|_{\mathbf{L}^2} + \|\partial_x G(\cdot, t)\|_{\mathbf{L}^2}) d\tau$$

$$(5.7) \qquad \leq \epsilon_1 t^{-\frac{3}{4}+\gamma} + \int_0^t \|u_x(\cdot, \tau)\|_{\mathbf{L}^2}^2 (t-\tau)^{-\frac{3}{4}+\gamma} d\tau.$$

By virtue of Theorem 29 there exists a $T > 0$ such that $\sup_{t \in [0,T]} t^{\frac{3}{4}-\gamma} \|u_x(\cdot, t)\|_{\mathbf{L}^2} = \epsilon$. Therefore from (5.7) we get

$$\frac{\epsilon}{t^{\frac{3}{4}-\gamma}} \leq \frac{(C\epsilon_1 + C\epsilon^2)}{t^{\frac{3}{4}-\gamma}} < \frac{\epsilon}{t^{\frac{3}{4}-\gamma}}.$$

The contradiction obtained proves the second part of estimate (5.6). Analogously we can prove the first part of estimate (5.6).

Putting (5.6) into (5.5) for $t > 1$ and using Theorem 29 we obtain for $\delta \in (0, 1)$

(5.8)
$$\int_0^t \|x^{(1+\delta)/2} u_x(\cdot, \tau)\|_{\mathbf{L}^2}^2 d\tau \le C\epsilon.$$

Also using (5.6) we can see that for $\mu \in (0, \frac{1}{2})$

(5.9)
$$\int_0^t \tau^\mu \|u_x(\cdot, \tau)\|_{\mathbf{L}^2}^2 d\tau \le C + \int_1^t \tau^{\mu - \frac{3}{2}} d\tau \le C.$$

As in the proof of Lemma 25 we get

$$\|F_t(x, \cdot, t) + G_t(x - \cdot, t)\|_{\mathbf{L}^\infty} \le C \min\left(\frac{x}{\sqrt{t}}, 1\right) t^{-2+\gamma},$$

whence we obtain for $\mu \in (0, 1)$

$$\|F(\cdot, \cdot, t - \tau) - F(x, \cdot, t) + G(\cdot, t - \tau) - G(\cdot, t)\|_{\mathbf{L}^\infty}$$
$$\le \quad C \min\left(\frac{x}{\sqrt{t}}, 1\right) \tau^\mu t^{-2+2\mu+\gamma}.$$

Therefore using (5.9) with $\mu = \frac{1-2\delta}{4} < \frac{1}{2}$ we have

$$\int_0^t d\tau \int_0^{+\infty} u_y^2 |F(x, y, (t - \tau)) - F(x, y, t)$$
$$+ G(x - y, t - \tau) - G(x - y, t)|$$

(5.10)
$$\le \quad C \min\left(\frac{x}{\sqrt{t}}, 1\right) t^{-2+2\mu+\gamma} \int_0^t \tau^\mu \|u_x(\cdot, \tau)\|_{\mathbf{L}^2}^2 d\tau$$
$$\le \quad C \min\left(\frac{x}{\sqrt{t}}, 1\right) t^{-1-\frac{\delta}{2}}.$$

Now by virtue of (5.8), (5.10) we have the desired estimate for $R(x, t)$. In fact, since $\|y^{1+\delta} u_0\|_{\mathbf{L}^1} + \|u_0\|_{\mathbf{L}^1} \le C$ we get from (5.4)

$$|R(x, t)| \le C \min\left(\frac{x}{\sqrt{t}}, 1\right) t^{-1-\frac{\delta}{2}},$$

where $\delta \in (0, 1)$. Theorem 27 is proved.

CHAPTER 9

Dirichlet Problem for KdV Equation

1. Introduction

We consider the initial-boundary value problem on a half-line for the Korteweg-de Vries (KdV) equation

(1.1)
$$\begin{cases} u_t + uu_x + u_{xxx} = 0, & t > 0, x > 0, \\ u(x,0) = u_0(x), & x > 0, \\ u(0,t) = 0, & t > 0. \end{cases}$$

The KdV equation (1.1) is a simple universal model appearing in many fields of physics and Technology as the first approximation in the description of dispersive nonlinear waves [24], [103], [136]. Note that the KdV equation (1.1) is conservative in the case of the Cauchy problem. In the case of the Cauchy problem on the line, the KdV equation was solved via the inverse scattering transform method [1] and the well-posedness on the line and the periodic domain was studied extensively by many authors, see for example, [14], [82], [83] and references cited therein. The asymptotic behavior of small solutions was studied in the case of the generalized KdV equations in [19], [63], [64], [65], [112]. However large time asymptotics of small solutions to the Cauchy problem of the KdV on the line has not been achieved until now. From the heuristic point of view the quadratic nonlinearity of the shallow water type uu_x is sub critical for large time: the nonlinear term decays more slowly than the linear part of the equation. In the case of the initial-boundary value problem on the positive line it is expected by the linear problem discussed in Section 3 that the solution has an additional time-decay and as a result the nonlinear term in the boundary-value problem (1.1) appears to be super critical contrary to the corresponding Cauchy problem. Our main goal in the present chapter is to obtain the large time asymptotics of solutions to problem (1.1). The local and global existence of solutions in Sobolev spaces for the initial-boundary value problem

$$u_t + uu_x + u_{xxx} + u_x = f$$

on the positive line were studied in paper [20], [13]. Their results are also valid for the equation without the term u_x. In [36] the integral formula was used to get the formal long time asymptotics of solutions and in [13] the integral formula which is different from the one used in [36] was employed to derive the various smoothing properties of solutions. Our approach is also based on the estimates of the integral equation in the weighted Sobolev spaces. The integral formula is obtained by using the Laplace transform with respect to space variable. The Laplace transform requires the boundary data $u(0,t)$, $u_x(0,t)$, $u_{xx}(0,t)$ and so $u_x(0,t)$, $u_{xx}(0,t)$ should be determined by the given data. In order to do this we use the method developed in the previous works [58], [60]. Our method to derive the integral formula is different from that [13] or [36], but the representation of the integral formula of ours (2.3) stated below is the same as one obtained in [36].

Now we state the results of this chapter (see paper [52]).

THEOREM 30. *Suppose that the initial data* $u_0 \in \mathbf{H}_1^{0,2}(\mathbf{R}^+) \cap \mathbf{H}_2^{0,\frac{3}{2}}(\mathbf{R}^+)$ *and the norm*

$$\|u_0\|_{\mathbf{H}_1^{0,2}} + \|u_0\|_{\mathbf{H}_2^{0,\frac{3}{2}}} \le \varepsilon_1,$$

where $\varepsilon_1 > 0$ *is small enough. Then there exists a unique solution*

$$u \in \mathbf{C}([0,+\infty),\mathbf{H}_2^{0,1}(\mathbf{R}^+)) \cap \mathbf{L}^\infty(0,+\infty,\mathbf{H}_2^{0,\frac{3}{2}}(\mathbf{R}^+))$$
$$\cap \mathbf{C}\left((0,+\infty),\mathbf{H}_2^{2,0}(\mathbf{R}^+)\right) \cap \mathbf{L}^\infty(0,+\infty,\mathbf{H}_2^{3,0}(\mathbf{R}^+))$$

of the initial-boundary value problem (1.1). Moreover there exists a constant B *such that the solution has the following asymptotics*

$$u(x,t) = t^{-1}B\frac{x}{\sqrt[3]{t}}Ai\left(\frac{x}{\sqrt[3]{t}}\right) + O\left(t^{-1-\frac{\delta}{3}}\max\left(1,\frac{x}{\sqrt[3]{t}}\right)\right)$$

for $t \to +\infty$ *uniformly with respect to* $x > 0$, *where* $\delta \in \left(0,\frac{1}{2}\right)$, $Ai(q)$ *is the Airy function defined by*

$$Ai(q) = \int_{-i\infty}^{i\infty} e^{-z^3+zq}dz.$$

We organize this chapter as follows. In Section 2 we construct the integral formula of the solution of the linear problem corresponding to (1.1). Section 3 is devoted to the study of the asymptotic behavior of solutions to the linear problem by using the integral formula made in Section 2. Theorem 30 is proved in Section 4.

2. Linear problem

We consider the linear initial-boundary value problem corresponding to (1.1)

$$(2.1) \qquad \begin{cases} u_t + u_{xxx} = f(x,t), & t > 0, x > 0, \\ u(x,0) = u_0(x), & x > 0, \\ u(0,t) = h(t), & t > 0. \end{cases}$$

In this section we obtain the explicit formula for the solution of the linear problem (2.1) under the condition

$$u_0 \in \mathbf{L}^1(\mathbf{R}^+), f \in \mathbf{L}^q(0,T;\mathbf{L}^1(\mathbf{R}^+))$$

with $q > 2$.

In Chapter 3 we defined the Green operator as follows

$$(2.2) \qquad \mathcal{F}\left[v(x,\tau)\right](x,t) = \int_0^{+\infty} F(x,y,t)v(y,\tau)dy.$$

Due to the results of Theorem 13 we have (since $h = 0$)

$$u(x,t) = \mathcal{F}\left[u_0\right](x,t) + \int_0^t \mathcal{F}\left[f(\tau)\right](x,t-\tau)\,d\tau,$$

where $F(x,y,t)$ has the following form

$$(2.3) \qquad F(x,y,t) = -\frac{1}{2\pi i}\int_{-i\infty}^{i\infty} e^{\xi t + \phi_3(\xi)x}\sum_{j=1}^{3} e^{-\phi_j(\xi)y}\phi_j'(\xi)d\xi.$$

Indeed from Example 2 and formula (2.25) we have for the Green function in the general case

$$(2.4) \qquad F(x,y,t) = -\frac{1}{2\pi i}\sum_{j=1}^{m}\sum_{l=1}^{N}\int_{-i\infty}^{i\infty} e^{\xi t + \phi_{m+l}(\xi)x}\phi'_{m+l}(\xi)$$

$$\times \left(\theta_j(\xi)\phi_{m+l}^{-N-j}(\xi) + \frac{1}{N}e^{-\phi_{m+l}(\xi)y}\right)d\xi,$$

where

$$\theta_j(\xi) = \begin{pmatrix} \phi_1^{-N-1}(\xi) & \cdots & \phi_1^{-[\alpha]}(\xi) \\ \cdots & \cdots & \cdots \\ \phi_m^{-N-1}(\xi) & \cdots & \phi_m^{-[\alpha]}(\xi) \end{pmatrix}^{-1} \begin{pmatrix} e^{-\phi_1(\xi)y} \\ \cdots \\ e^{-\phi_m(\xi)y} \end{pmatrix}_j,$$

and also $\phi_i(\xi), i = 1, ..., [\alpha]$ are inverse functions $\phi_i(\xi) = K^{-1}(-\xi)$. In our case $K(p) = p^3$ the equation $-p^3 = \xi$ has three roots $\phi_1(\xi)$, $\phi_2(\xi)$ and $\phi_3(\xi)$ such that

$$p = \phi_1(\xi) = (\xi\exp(i\pi))^{\frac{1}{3}}, p = \phi_2(\xi) = (\xi\exp(-i\pi))^{\frac{1}{3}},$$

$$p = \phi_3(\xi) = (\xi\exp(3i\pi))^{\frac{1}{3}}$$

which transform the half-complex plane $\mathrm{Re}\,\xi > 0$ to domains, where

$$\mathrm{Re}\,\phi_1(\xi) > 0, \mathrm{Re}\,\phi_2(\xi) > 0, \mathrm{Re}\,\phi_3(\xi) < 0.$$

Therefore we get $(m = 2, N = 1)$ for $F(x,y,t)$

$$(2.5) \qquad F(x,y,t) = -\frac{1}{2\pi i}\sum_{j=1}^{2}\int_{-i\infty}^{i\infty} e^{\xi t + \phi_3(\xi)x}\phi_3'(\xi)$$

$$\times \left(\theta_j(\xi)\phi_3^{-1-j}(\xi) + e^{-\phi_3(\xi)y}\right)d\xi,$$

where using $\phi_j^3(\xi) = -\xi$

$$(2.6) \qquad \theta_j(\xi) = \begin{pmatrix} \phi_1^{-2}(\xi) & \phi_1^{-3}(\xi) \\ \phi_2^{-2}(\xi) & \phi_2^{-3}(\xi) \end{pmatrix}^{-1} \begin{pmatrix} e^{-\phi_1(\xi)y} \\ e^{-\phi_2(\xi)y} \end{pmatrix}_j$$

$$= \frac{\phi_1^3 \phi_2^3}{\phi_2 - \phi_1} \begin{pmatrix} e^{-\phi_1(\xi)y}\phi_2^{-3} - e^{-\phi_2(\xi)y}\phi_1^{-3} \\ -e^{-\phi_1(\xi)y}\phi_2^{-2} + e^{-\phi_2(\xi)y}\phi_1^{-2} \end{pmatrix}_j$$

$$= \frac{-\xi}{\phi_2 - \phi_1} \begin{pmatrix} e^{-\phi_1(\xi)y} - e^{-\phi_2(\xi)y} \\ -e^{-\phi_1(\xi)y}\phi_2 + e^{-\phi_2(\xi)y}\phi_1 \end{pmatrix}_j.$$

We have

$$\phi_1'(\xi) = \frac{1}{(\phi_2 - \phi_1)(\phi_3 - \phi_1)}, \phi_2'(\xi) = \frac{1}{(\phi_3 - \phi_2)(\phi_1 - \phi_2)}$$

$$\phi_3'(\xi) = \frac{1}{(\phi_2 - \phi_3)(\phi_1 - \phi_3)}.$$

Therefore from (2.6) we obtain

$$-\frac{1}{2\pi i}\sum_{j=1}^{2}\int_{-i\infty}^{i\infty} e^{\xi t + \phi_3(\xi)x}\phi_3'(\xi)\left(\theta_j(\xi)\phi_3^{1-j}(\xi)\right)d\xi$$

$$= -\frac{1}{2\pi i}\int_{-i\infty}^{i\infty} e^{\xi t + \phi_3(\xi)x}\phi_3'(\xi)\frac{\phi_3\left(e^{-\phi_1(\xi)y} - e^{-\phi_2(\xi)y}\right)}{(\phi_2 - \phi_1)}d\xi$$

$$-\frac{1}{2\pi i}\int_{-i\infty}^{i\infty} e^{\xi t + \phi_3(\xi)x}\phi_3'(\xi)\frac{\left(-e^{-\phi_1(\xi)y}\phi_2 + e^{-\phi_2(\xi)y}\phi_1\right)}{(\phi_2 - \phi_1)}d\xi$$

$$= -\frac{1}{2\pi i}\int_{-i\infty}^{i\infty} e^{\xi t + \phi_3(\xi)x}\phi_3'(\xi)\frac{e^{-\phi_1(\xi)y}\left(\phi_3 - \phi_2\right)\left(\phi_3 - \phi_1\right)}{(\phi_2 - \phi_1)\left(\phi_3 - \phi_1\right)}d\xi$$

$$-\frac{1}{2\pi i}\int_{-i\infty}^{i\infty} e^{\xi t + \phi_3(\xi)x}\phi_3'(\xi)\frac{e^{-\phi_2(\xi)y}\left(\phi_1 - \phi_3\right)\left(\phi_2 - \phi_3\right)}{(\phi_2 - \phi_1)\left(\phi_2 - \phi_3\right)}d\xi$$

$$= -\frac{1}{2\pi i}\int_{-i\infty}^{i\infty} e^{\xi t + \phi_3(\xi)x}\sum_{j=1}^{2}e^{-\phi_j(\xi)y}\phi_j(\xi)d\xi$$

and thus the representation (2.3).

Let us now prove that the representation (2.3) satisfies the problem (2.1) with conditions $u(x,0) = u_0(x)$ and $u(0,t) = 0$. Since in Re $\xi > 0$ ϕ_1, ϕ_2 and ϕ_3 are analytic functions, and Re $\phi_j > 0, j = 1,2$ and Re $\phi_3 < 0$, using Cauchy's theorem we have

$$\int_{-i\infty}^{i\infty} e^{\phi_3(\xi)x}\sum_{j=1}^{2}e^{-\phi_j(\xi)y}\phi_j'(\xi)d\xi = 0$$

and so

$$(2.7) \qquad F(x,y,0) = \frac{1}{2\pi i}\int_{-i\infty}^{i\infty} e^{p(x-y)}dp = \delta(x-y).$$

We let

$$\Gamma_1 = \left\{ p \in (-i\infty, 0) \cup \left(0, e^{-i\frac{\pi}{6}}\infty\right) \right\},$$
$$\Gamma_2 = \left\{ p \in \left(e^{i\frac{\pi}{6}}\infty, 0\right) \cup (0, i\infty) \right\},$$

and

$$\Gamma_3 = \left\{ p \in \left(e^{i\frac{5\pi}{6}}\infty, 0\right) \cup \left(0, e^{i\frac{7\pi}{6}}\infty\right) \right\}.$$

Changing the variable

$$\phi_1(\xi) = (\xi \exp(i\pi))^{\frac{1}{3}} = p$$

in the first integral in (2.3) with $x = 0$ and

$$\phi_2(\xi) = (\xi \exp(-i\pi))^{\frac{1}{3}} = p$$

in the second integral we get in (2.3) with $x = 0$

$$\int_{-i\infty}^{+i\infty} e^{\xi t} \sum_{j=1}^{2} e^{-\phi_j(\xi)y} \phi_j'(\xi) d\xi$$

$$= \int_{-i\infty}^{+i\infty} e^{\xi t} e^{-\phi_1(\xi)y} \phi_1'(\xi) d\xi + \int_{-i\infty}^{+i\infty} e^{\xi t} e^{-\phi_2(\xi)y} \phi_2'(\xi) d\xi$$

$$= \int_{\Gamma_2} e^{-p^3 t - py} dp + \int_{\Gamma_1} e^{-p^3 t - py} dp = \int_{-i\infty}^{+i\infty} e^{-p^3 t - py} dp;$$

also we have

$$\int_{-i\infty}^{+i\infty} e^{\xi t} e^{-\phi_3(\xi)y} \phi_3'(\xi) d\xi = \int_{\Gamma_3} e^{-p^3 t - py} dp = -\int_{-i\infty}^{+i\infty} e^{-p^3 t - py} dp.$$

Therefore

(2.8) $$F(0, y, t) = 0.$$

From (2.7) and (2.8) the desired conclusion follows.

3. Preliminaries

In this section we prepare a useful lemma which is needed to obtain our results. As in Section 2 we denote by

(3.1) $$\phi_1(\xi) = (\xi \exp(i\pi))^{\frac{1}{3}}, \phi_2(\xi) = (\xi \exp(-i\pi))^{\frac{1}{3}}, \phi_3(\xi) = (\xi \exp(3i\pi))^{\frac{1}{3}}$$

the roots of the equation $p^3 = -\xi$, such that $\operatorname{Re} \phi_l(\xi) > 0$, $l = 1, 2$, $\operatorname{Re} \phi_3(\xi) < 0$, for all $\operatorname{Re} \xi > 0$. We introduce the operator

$$\Phi(t)v(\tau) = \int_0^{+\infty} F(x, y, t)v(y, \tau)dy,$$

where

(3.2) $$F(x, y, t) = -\frac{1}{2\pi i} \int_{-i\infty}^{+i\infty} e^{\xi t + \phi_3(\xi)x} \sum_{j=1}^{3} e^{-\phi_j(\xi)y} \phi_j'(\xi) d\xi.$$

LEMMA 27. *We have for $m \geq 0$, $\beta \in [0, n+5]$ and $\gamma > 0, t > \frac{1}{4}, n = 0, 1, 2, 3$*

$$(3.3) \qquad \|\partial_x^n \Phi(t)v(\tau)\|_{\mathbf{H}_2^{0,m}} \leq Ct^{-\frac{5-2m+2n}{6}+\gamma} \|v(\tau)\|_{\mathbf{H}_1^{0,2}}$$

$$+Ct^{-\frac{4+5n+\beta}{12}} \|v(\tau)\|_{\mathbf{H}_1^{0,m+\frac{2+n+\beta}{4}+\gamma}}$$

and for $0 < t < \frac{1}{4}$

$$(3.4) \qquad \|\partial_x \Phi(t)v(\tau)\|_{\mathbf{H}_2^{0,m}} \leq Ct^{-\frac{3}{4}} \|v(\tau)\|_{\mathbf{H}_1^{0,m+\frac{3}{4}}}.$$

Moreover for any $x, y, t > 0$, $\delta > 0$

$$(3.5) \quad \Phi(t)v(\tau) = \frac{B}{3\pi t} \frac{x}{\sqrt[3]{t}} Ai\left(\frac{x}{\sqrt[3]{t}}\right) + O\left(\min\left(1, \frac{x}{\sqrt[3]{t}}\right) t^{-1-\frac{\delta}{3}} \|v(\tau)\|_{\mathbf{H}_1^{0,2+\delta}}\right),$$

where $Ai(q)$ is the Airy function

$$Ai(q) = \int_{-\infty}^{+\infty} e^{-iz^3+izq} dz,$$

and constant $B > 0$

$$B = \int_0^{+\infty} y^2 v(y) dy.$$

PROOF. Using

$$\sum_{j=1}^3 \phi_j(\xi) = 0, \sum_{j=1}^3 \phi_j^2(\xi) = 0$$

which implies

$$\sum_{j=1}^3 \phi_j'(\xi) = 0, \sum_{j=1}^3 \phi_j'(\xi)\phi_j(\xi) = 0$$

and

$$\sum_{j=1}^3 \phi_j^2(\xi)\phi_j'(\xi) = -1,$$

we have

$$F(x, y, t) = -\frac{1}{2\pi i} \int_{-i\infty}^{i\infty} d\xi e^{\xi t + \phi_3(\xi)x}$$

$$\times \left(\sum_{j=1}^3 \left(e^{-\phi_j(\xi)y} - \sum_{k=0}^2 \frac{(-1)^k (\phi_j(\xi)y)^k}{k!}\right) \phi_j'(\xi) - \frac{y^2}{2}\right)$$

$$(3.6) \qquad = \frac{y^2}{4\pi i} \int_{-i\infty}^{i\infty} e^{\xi t + \phi_3(\xi)x} d\xi + R(x, y, t),$$

where

$$R(x, y, t) \;=\; -\frac{1}{2\pi i} \sum_{j=1}^{3} \int_{-i\infty}^{i\infty} e^{\xi t + \phi_3(\xi)x}$$

$$\times \left(e^{-\phi_j(\xi)y} - \sum_{k=0}^{2} \frac{(-1)^k \left(\phi_j(\xi)y\right)^k}{k!} \right) \phi'_j(\xi) d\xi.$$

Let the notation Γ_j $(j = 1, 2, 3)$ be the same ones defined in the previous section. We have

$$\sum_{j=1}^{2} \int_{\Gamma_j} e^{-p^3 t - py} dp \;=\; \int_{-i\infty}^{i\infty} e^{-p^3 t - py} dp,$$

$$\int_{\Gamma_3} e^{-p^3 t - py} dp \;=\; -\int_{-i\infty}^{i\infty} e^{-p^3 t - py} dp.$$

Therefore by changing the variables $\phi_j(\xi) = p$ we obtain

$$\int_{-i\infty}^{i\infty} e^{\xi t} \sum_{j=1}^{3} \left(e^{-\phi_j(\xi)y} - \sum_{k=0}^{2} \frac{(-1)^k \left(\phi_j(\xi)y\right)^k}{k!} \right) \phi'_j(\xi) d\xi$$

$$= \sum_{j=1}^{3} \int_{-i\infty}^{i\infty} e^{\xi t} e^{-\phi_j(\xi)y} \phi'_j(\xi) d\xi = \sum_{j=1}^{3} \int_{\Gamma_j} e^{-p^3 t - py} dp = 0.$$

For simplicity we state that

$$q = \frac{x}{\sqrt[3]{t}}, \theta = \frac{y}{\sqrt[3]{t}}.$$

Then we may write the remainder term $R(x, y, t)$ in the form

$$R(x, y, t) \;=\; \left(\widetilde{F}(q, \theta) - \widetilde{F}(0, \theta) \right.$$

$$\left. - \sum_{k=0}^{2} \frac{\left(\partial_\theta^k \widetilde{F}(q, 0) - \partial_\theta^k \widetilde{F}(0, 0) \right) \theta^k}{k!} \right)$$

$$= - \int_0^\theta \left(\partial_\xi^3 \widetilde{F}(q, \theta - \xi) - \partial_\xi^3 \widetilde{F}(0, \theta - \xi) \right) \xi^2 d\xi,$$

where

$$\widetilde{F}(q, \theta) = -\frac{1}{2\pi i} \sum_{j=1}^{3} \int_{-i\infty}^{i\infty} e^{\xi t} e^{\phi_3(\xi) \sqrt[3]{t} q} e^{-\phi_j(\xi) \sqrt[3]{t}\theta} \phi'_j(\xi) d\xi.$$

Changing the variable $z = \phi_j t^{\frac{1}{3}}$, by using the identities

$$\phi_3(\xi) = r_j \phi_j(\xi), \quad r_j = -\exp\left((-1)^j i\frac{\pi}{3} \right), \quad j = 1, 2 \text{ and } r_3 = 1$$

we rewrite the function $\widetilde{F}$ as

$$
\begin{aligned}
\widetilde{F}(q,\theta) &= -\frac{1}{2\pi i}\sum_{j=1}^{3}\int_{-i\infty}^{+i\infty} e^{\xi t}e^{\phi_3(\xi)x}e^{-\phi_j(\xi)y}\phi_j'(\xi)d\xi \\
&= -\frac{1}{2\pi i}\frac{1}{\sqrt[3]{t}}\sum_{j=1}^{3}\int_{-i\infty}^{+i\infty} e^{-z^3}e^{r_j zq}e^{-z\theta}dz \\
(3.7)\qquad &= -\frac{1}{2\pi i}\frac{1}{\sqrt[3]{t}}\left(\sum_{j=1}^{2}H_j(q,\theta)+Ai(q-\theta)\right),
\end{aligned}
$$

where

$$
H_j(q,\theta)=\int_{-i\infty}^{i\infty} e^{-z^3}e^{r_j zq}e^{-z\theta}dz, j=1,2.
$$

Now we prove that

$$
\begin{aligned}
(3.8)\qquad \left|\partial_\theta^n \widetilde{F}(q,\theta)\right| &\leq C\left|\partial_q^n \widetilde{F}(q,\theta)\right|\leq \frac{C}{\sqrt[3]{t}}\langle\theta\rangle^{\frac{2n-1}{4}}, 2\theta>q \\
\left|\partial_\theta^n \widetilde{F}(q,\theta)\right| &\leq C\left|\partial_q^n \widetilde{F}(q,\theta)\right|\leq \frac{C}{\sqrt[3]{t}}\langle|q|\rangle^{-\alpha}, 2\theta<q.
\end{aligned}
$$

It is well-known that for any $\alpha>0$

$$
(3.9)\qquad \left|Ai^{(n)}(-|\zeta|)\right|\leq \langle|\zeta|\rangle^{\frac{2n-1}{4}}, \left|Ai^{(n)}(|\zeta|)\right|\leq C\langle|\zeta|\rangle^{-\alpha}.
$$

We prove for $j=1,2$, $\alpha>0$

$$
\begin{aligned}
(3.10)\qquad |\partial_\theta^n H_j(q,\theta)| &= |\partial_q^n H_j(q,\theta)|\leq C\langle\theta\rangle^{\frac{2n-1}{4}}, 2\theta>q \\
|\partial_\theta^n H_j(q,\theta)| &= |\partial_q^n H_j(q,\theta)|\leq C\langle|q|\rangle^{-\alpha}, 2\theta<q.
\end{aligned}
$$

Using the identity

$$
e^{-z^3-z\theta}=\left(\sqrt{3}+\left(\sqrt{3}z\pm i\sqrt{\theta}\right)(3z^2+\theta)\right)^{-1}\frac{d}{dz}\left(\sqrt{3}z\pm i\sqrt{\theta}\right)e^{-z^3-z\theta}
$$

and integrating by parts we get in the case $2\theta>q$

$$
\begin{aligned}
(3.11)\qquad |\partial_\theta^n H_j(q,\theta)| &= \left|\int_{-i\infty}^{+i\infty} e^{-z^3}e^{r_j zq}e^{-z\theta}z^n dz\right| \\
&\leq C\left|\int_{\mp 2i\sqrt{\theta}}^0 e^{-z^3}e^{-z\theta}f_\pm(z,q,\theta)dz\right| \\
&\quad +C\left|\int_{|z|>2\sqrt{\theta},\mathrm{Re}\,z=0} e^{-z^3}e^{r_j zq}e^{-z\theta}z^n dz\right|,
\end{aligned}
$$

where

$$
\begin{aligned}
f_\pm&(z,q,\theta)\\
&= \left(e^{r_j z q}\left(r_j q z^n + n z^{n-1}\right)\left(\sqrt{3}+\left(\sqrt{3}z \pm i\sqrt{\theta}\right)(3z^2+\theta)\right)\right.\\
&\quad\left.-e^{r_j z q}z^n\left(\sqrt{3}\left(3z^2+\theta\right)+6z\left(\sqrt{3}z\pm i\sqrt{\theta}\right)\right)\right)\\
&\quad\times\left(\sqrt{3}z+i\sqrt{\theta}\right)\left(\sqrt{3}+\left(\sqrt{3}z+i\sqrt{\theta}\right)(3z^2+\theta)\right)^{-2}.
\end{aligned}
$$

By virtue of $\mathrm{Re}\,r_j q z < 0$ we see that for $|z| < 2\sqrt{\theta}$

$$
|f_\pm(z,q,\theta)| \le C\theta^{\frac{2n+3}{4}}\left|\sqrt{3}z\pm i\sqrt{\theta}\right|\left(\sqrt{\frac{3}{\theta}}+\left|\sqrt{3}z\pm i\sqrt{\theta}\right|^2\right)^{-2}.
$$

Therefore integrating with respect to z, we obtain

$$
\left|\int_{\mp 2i\sqrt{\theta}}^{0} e^{-z^3}e^{r_j z q}e^{-z\theta}\left(\sqrt{3}z\pm i\sqrt{\theta}\right)f_\pm(z,q,\theta)\left(\sqrt{3}+3z^2+\theta\right)^{-2}dz\right|
$$

$$
\le\ C\langle\theta\rangle^{\frac{2n+3}{4}}\left|\int_{\mp 2i\sqrt{\theta}}^{0}\left|\sqrt{3}z+i\sqrt{\theta}\right|\left(\sqrt{\frac{3}{\theta}}+\left|\sqrt{3}z+i\sqrt{\theta}\right|^2\right)^{-2}dz\right|
$$

$$
(3.12)\le\ C\langle\theta\rangle^{\frac{2n-1}{4}}.
$$

Also we change the contour

$$
\left\{\mathrm{Re}\,z = 0, |z| > 2\sqrt{\theta}\right\}
$$

into

$$
\Omega = \left\{z = |z|\,e^{\pm i\left(\frac{\pi}{2}\pm\varepsilon\right)}, |z| > 2\sqrt{\theta}, \varepsilon\in\left(0,\frac{\pi}{3}\right)\right\}.
$$

Then for all $z\in\Omega$,

$$
\left|e^{\mathrm{Re}(-z^3-z\theta)}\right| < e^{-C|z|^3}\ \text{ and }\ \left|e^{r_j z q}\right|\le e^{-C\,\mathrm{Re}|z|\theta},
$$

and we get for $\alpha > 0$

$$
(3.13)\qquad\left|\int_{|z|>2\sqrt{\theta}, z\in\Omega} e^{-z^3}e^{r_j z q}e^{-z\theta}z^n dz\right|\le C\langle\theta\rangle^{-\alpha}.
$$

Therefore we have the first inequality in (3.10). In the same manner we can prove that in the case $2\theta < q$

$$
(3.14)\qquad\left|\int_{-i\infty}^{+i\infty}e^{-z^3}e^{r_j z q}e^{-z\theta}z^n dz\right|\le C\langle|q|\rangle^{-\alpha}.
$$

From (3.11)-(3.14) we get (3.10) and then (3.8). Using (3.8) we obtain for $\delta \in (0,1)$ and $\theta \in (0,1)$

$$
|R(x,y,t)| = \left| \int_0^\theta \left(\widetilde{F}^{(3)}(q, \theta - \xi) - \widetilde{F}^{(3)}(0, \theta - \xi) \right) \xi^2 d\xi \right|
$$

$$
\leq C \min \left(q \int_0^\theta \widetilde{F}^{(4)}(q^*, \theta - \xi) \xi^2 d\xi \, , \right.
$$

$$
\left. \int_0^\theta \left| \widetilde{F}^{(3)}(q, \theta - \xi) \right| \xi^2 d\xi + \int_0^\theta \left| \widetilde{F}^{(3)}(0, \theta - \xi) \right| \xi^2 d\xi \right)
$$

$$
(3.15) \qquad \leq C \frac{1}{\sqrt[3]{t}} \min(1, q) \theta^{2+\delta},
$$

where $q^* \in (0, q)$. For $\theta > 1$ we have

$$
|R(x, y, t)|
$$

$$
\leq C \left(\left| \widetilde{F}(q, \theta) - \widetilde{F}(0, \theta) \right| + \sum_{j=0}^2 \left| \widetilde{F}^{(k)}(q, 0) - \widetilde{F}^{(k)}(0, 0) \right| \theta^k \right)
$$

$$
(3.16) \qquad \leq C \min \frac{1}{\sqrt[3]{t}} (1, q) \theta^{2+\delta}.
$$

Via (3.15) and (3.16) we get

$$
(3.17) \qquad\qquad R = y^{2+\delta} O \left(\min \left(1, \frac{x}{\sqrt[3]{t}} \right) t^{-1-\frac{\delta}{3}} \right).
$$

From (3.6), (3.17) we have

$$
F(x,y,t)
$$

$$
= \frac{y^2}{\pi i t} \int_{-i\infty}^{+i\infty} e^{-z^3 + z \frac{x}{\sqrt[3]{t}}} z^2 dz + y^{2+\delta} O(t^{-1-\frac{\delta}{3}})
$$

$$
= \frac{y^2}{3\pi t} \frac{x}{\sqrt[3]{t}} Ai \left(\frac{x}{\sqrt[3]{t}} \right) + y^{2+\delta} O \left(\min \left(1, \frac{x}{\sqrt[3]{t}} \right) t^{-1-\frac{\delta}{3}} \right),
$$

whence we get asymptotics (3.5). Now we prove (3.4). We have for $\alpha = m+1, t \in [0, \frac{1}{4}]$

$$
\begin{aligned}
|\partial_x F(x,y,t)| &\leq Ct^{-\frac{3}{4}} \langle y \rangle^{\frac{1}{4}}, 2y > x \\
|\partial_x F(x,y,t)| &\leq Ct^{-\frac{2}{3}} \langle x \rangle^{-\alpha}, 2y < x.
\end{aligned}
$$

Therefore we get

$$
\begin{aligned}
\|\partial_x \Phi \nu\|_{\mathbf{H}_2^{0,m}} \;\leq\; & \left\| \int_0^{\frac{z}{2}} \partial_x F(\cdot, y, t)\nu(y)dy \right\|_{\mathbf{H}_2^{0,m}} \\
& + \left\| \int_{\frac{z}{2}}^{+\infty} \partial_x F(\cdot, y, t)\nu(y)dy \right\|_{\mathbf{H}_2^{0,m}} \\
\leq\; & t^{-\frac{2}{3}} \left\| \langle \cdot \rangle^{-\alpha} \int_0^{+\infty} |\nu(y)| \, dy \right\|_{\mathbf{H}_2^{0,m}} \\
& + t^{-\frac{3}{4}} \left\| \int_{\frac{z}{2}}^{+\infty} \langle y \rangle^{\frac{1}{4}} |\nu(y)| \, dy \right\|_{\mathbf{H}_2^{0,m}} \\
\leq\; & Ct^{-\frac{3}{4}} \|v\|_{\mathbf{H}_2^{0,m+\frac{3}{4}}} .
\end{aligned}
$$

To prove (3.3) by analogue to (3.6) we rewrite the function $F(x, y, t)$ as

$$
(3.18) \qquad F(x, y, t) = \left(\widetilde{F}(q, \theta) - \sum_{j=0}^{1} \frac{\partial_\theta^j \widetilde{F}(q, 0)}{j!} \theta^j \right).
$$

In the case $x - y > \frac{x}{2}$ we rewrite F as

$$
\begin{aligned}
\partial_x^n F(x, y, t) \;=\; & t^{-\frac{n}{3}} \partial_q^n \left(\widetilde{F}(q, \theta) - \sum_{j=0}^{1} \frac{\partial_\theta^j \widetilde{F}(q, 0)}{j!} \theta^j \right) \\
=\; & -t^{-\frac{n}{3}} \left(\partial_q^n \int_0^\theta \partial_\xi^2 \widetilde{F}(q, \theta - \xi) \xi d\xi \right).
\end{aligned}
$$

Using (3.8) in the case $q > 2\theta$ with $\alpha = \frac{1}{2} + 3\gamma + m > 0$ we get for $t > \frac{1}{4}$

$$
\begin{aligned}
|\partial_x^n F| \;\leq\; & Ct^{-\frac{1+n}{3}} \langle q \rangle^{-\alpha} \int_0^\theta \xi d\xi \\
\leq\; & Ct^{\frac{\alpha - 3 - n}{3}} \langle x \rangle^{-\alpha} y^2 = Ct^{-\frac{5 + 2n - 2m}{6} + \gamma} \langle y \rangle^2 \frac{1}{\langle x \rangle^{m + \frac{1}{2} + \gamma}} .
\end{aligned}
$$

Therefore we obtain

$$
(3.19) \qquad \left\| \int_0^{\frac{z}{2}} F(\cdot, y, t)v(y, t)dy \right\|_{\mathbf{H}_2^{0,m}} \leq Ct^{-\frac{5 - 2m + 2n}{6} + \gamma} \|v(t)\|_{\mathbf{H}_1^{0,2}} .
$$

Using (3.8) in the case $2\theta > q$ we have for $\beta \in [0, n+5]$

$$
\begin{aligned}
&|\partial_x^n F(x,y,t)| \\
&\leq Ct^{-\frac{n}{3}} \min\left(\left|\partial_q^n \widetilde{F}(q,\theta) - \partial_q^n \widetilde{F}(q,0)\right| + \left|\partial_q^n \partial_\theta \widetilde{F}(q,0)\right|\theta, \right.\\
&\qquad \left|\partial_q^n \widetilde{F}(q,\theta)\right| + \left|\partial_q^n \widetilde{F}(q,0)\right| + \left|\partial_q^n \partial_\theta \widetilde{F}(q,0)\right|\theta, \\
&\qquad \left. \left|\partial_q^n \widetilde{F}(q,\theta) - \sum_{k=0}^{1} \frac{\partial_\theta^j \partial_q^n \widetilde{F}(q,0)}{j!}\theta^j\right| \right) \\
&\leq Ct^{-\frac{1+n}{3}} \min\left(\theta\langle\theta\rangle^{\frac{1+2n}{4}} + \langle|q|\rangle^{-\alpha}\theta \,, \langle\theta\rangle^{\frac{n}{4}} + \langle|q|\rangle^{-\alpha}\theta, \theta^2\langle\theta\rangle^{\frac{4+2n}{4}} \right)
\end{aligned}
$$

$$
(3.20) \quad \leq Ct^{-\frac{1+n}{3}} \min\left(\langle|q|\rangle^{-\alpha}\theta + \langle\theta\rangle^{\frac{n+\beta}{4}}, \theta^2\langle\theta\rangle^{\frac{4+2n}{4}} \right).
$$

In the case $y > \sqrt[3]{t}$

$$
\langle\theta\rangle^{\frac{n+\beta}{4}} \leq Ct^{-\frac{n+\beta}{12}} \langle\sqrt[3]{t} + y\rangle^{\frac{n+\beta}{4}} < t^{-\frac{n+\beta}{12}} \langle y\rangle^{\frac{n+\beta}{4}}.
$$

Therefore choosing $\alpha = m + \frac{1}{2} + \gamma$ and taking into account that $\theta > 1$ we get from (3.20)

$$
\begin{aligned}
&|\partial_x^n F(x,y,t)| \\
&\leq Ct^{-\frac{1+n}{3}} \left(\langle|q|\rangle^{-\alpha}\theta^2 + \langle\theta\rangle^{\frac{n+\beta}{4}} \right) \\
&\leq C\left(t^{-\frac{5+2n-2m}{6}+\gamma}\langle y\rangle^2 \langle x\rangle^{-m-\frac{1}{2}-\gamma} + t^{-\frac{4+5n+\beta}{12}}\langle y\rangle^{\frac{n+\beta}{4}+m+\frac{1}{2}+\gamma} \right),
\end{aligned}
$$

and in the case $y < \sqrt[3]{t}$ using $\langle\theta\rangle^{\frac{4+2n}{4}} < C$ and $x < \sqrt[3]{t}$ we have from (3.20)

$$
\begin{aligned}
|\partial_x^n F(x,y,t)| &\leq Ct^{-\frac{1+n}{3}}\theta^2\langle\theta\rangle^{\frac{4+2n}{4}} \leq Ct^{-\frac{3+n}{3}}\langle y\rangle^2 \\
&\leq Ct^{-\frac{5+2n-2m}{6}+\gamma}\langle y\rangle^2\langle x\rangle^{-m-\frac{1}{2}-\gamma}.
\end{aligned}
$$

Therefore we get

$$
(3.21) \quad \left\| \int_{\frac{\cdot}{2}}^{+\infty} \partial_x^n F(\cdot,y,t)v(y)dy \right\|_{\mathbf{H}_2^{0,m}} \leq C\left(t^{-\frac{5-2m+2n}{6}+\gamma} \|v\|_{\mathbf{H}_1^{0,2}} + t^{-\frac{4+5n+\beta}{12}} \|v\|_{\mathbf{H}_1^{0,m+\frac{2+n+\beta}{4}+\gamma}} \right).
$$

From (3.19)-(3.21) we obtain for $\beta \in [0, n+5]$, $t > \frac{1}{4}$

$$\|\partial_x^n \Phi v\|_{\mathbf{H}_2^{0,m}}$$

$$\leq C \left\| \int_0^{\frac{x}{2}} \partial_x^n F(\cdot, y, t) v(y) dy \right\|_{\mathbf{H}_2^{0,m}}$$

$$+ C \left\| \int_{\frac{x}{2}}^{+\infty} \partial_x^n F(\cdot, y, t) v(y) dy \right\|_{\mathbf{H}_2^{0,m}}$$

$$\leq Ct^{-\frac{5-2m+2n}{6}+\gamma} \|v\|_{\mathbf{H}_1^{0,2}} + Ct^{-\frac{4+5n+\beta}{12}} \|v\|_{\mathbf{H}_1^{0,m+\frac{2+n+\beta}{4}+\gamma}}.$$

Lemma 27 is proved. □

4. Global existence

We now prove Theorem 30. Consider the linearized problem of the (1.1)

(4.1)
$$\begin{cases} u_t + u_{xxx} = -vu_x, & t > 0, x > 0, \\ u(x, 0) = u_0(x), & x > 0, \\ u(0, t) = 0, & t > 0. \end{cases}$$

We suppose that

$$\|u_0\|_{\mathbf{H}_1^{0,2}} + \|u_0\|_{\mathbf{H}_2^{0,\frac{3}{2}}} \leq \varepsilon_1$$

and $v \in X_\varepsilon$, where $\varepsilon_1 > 0$ is small enough, $\varepsilon = 100C\varepsilon_1$ with a constant C as in (3.3), and

$$X_\varepsilon = \{v \in X; \||v|\|_X \leq \varepsilon, v(0, t) = 0\},$$

where

$$X = \left\{ v \in C\left([0, +\infty); \mathbf{L}^2\left(\mathbf{R}^+\right)\right); \||v|\|_X = \sum_{m=1}^3 \sup_{t>0} (1+t)^{\frac{5-m}{6}-\gamma} \|v(t)\|_{\mathbf{H}_2^{0,\frac{m}{2}}} \right.$$

$$+ \sum_{m=0}^3 \sup_{t>0} t^{\frac{1}{3}} (1+t)^{\frac{5-m}{6}-\gamma} \|\partial_x v(t)\|_{\mathbf{H}_2^{0,\frac{m}{2}}} + \sup_{t>0} (1+t)^{\frac{3}{4}} \|v(t)\|_{\mathbf{L}^2}$$

$$+ \sum_{m=0}^3 \sup_{t>0} t^{\frac{m}{2}} \left\| x^{\frac{3}{2}-\frac{m}{2}} \partial_x^m v(t) \right\|_{\mathbf{L}^2} + \sup_{t>0} t^{\frac{3}{2}} \|\partial_t v(t)\|_{\mathbf{L}^2}$$

$$< +\infty \right\}.$$

We define the mapping $\mathbb{M}$ by $u = \mathbb{M}v$.

By Lemma 27 (3.3) with $\beta = 5 - 2m$ we get for $t \geq \frac{1}{4}, \gamma < \frac{1}{12}, m \geq \frac{1}{2}$

$$\|u(t)\|_{\mathbf{H}_2^{0,\frac{m}{2}}}$$

$$\leq Ct^{-\frac{5-m}{6}+\gamma} \|u_0\|_{\mathbf{H}_1^{0,2}} + C \int_0^t (t-\tau)^{-\frac{5-m}{6}+\gamma} \|vu_x(\tau)\|_{\mathbf{H}_1^{0,2}} d\tau$$

$$\leq Ct^{-\frac{5-m}{6}+\gamma} \|u_0\|_{\mathbf{H}_1^{0,2}}$$

$$+C \int_0^t (t-\tau)^{-\frac{5-m}{6}+\gamma} \|v(\tau)\|_{\mathbf{H}_2^{0,1}} \|u_x(\tau)\|_{\mathbf{H}_2^{0,1}} d\tau$$

$$\leq Ct^{-\frac{5-m}{6}+\gamma} \|u_0\|_{\mathbf{H}_1^{0,2}}$$

$$+C\varepsilon \int_0^t (t-\tau)^{-\frac{5-m}{6}+\gamma} (1+\tau)^{-1+2\gamma} \tau^{-\frac{1}{3}} d\tau$$

$$\times \sup_t t^{\frac{1}{3}} (1+t)^{\frac{1}{2}-\gamma} \|u_x(t)\|_{\mathbf{H}_2^{0,1}}$$

(4.2) $$\leq Ct^{-\frac{5-m}{6}+\gamma} \left(\|u_0\|_{\mathbf{H}_1^{0,2}} + \varepsilon \sup_t t^{\frac{1}{3}} (1+t)^{\frac{1}{2}-\gamma} \|u_x(t)\|_{\mathbf{H}_2^{0,1}} \right).$$

Using the estimate (3.3) of Lemma 27 with $\beta = 5$ we get

$$\|u(t)\|_{\mathbf{L}^2}$$

$$\leq Ct^{-\frac{3}{4}} \|u_0\|_{\mathbf{H}_1^{0,2}} + C \int_0^t (t-\tau)^{-\frac{3}{4}} \|vu_x(\tau)\|_{\mathbf{H}_1^{0,2}} d\tau$$

$$\leq Ct^{-\frac{3}{4}} \|u_0\|_{\mathbf{H}_1^{0,2}} + C \int_0^t (t-\tau)^{-\frac{3}{4}} \|v(\tau)\|_{\mathbf{H}_2^{0,1}} \|u_x(\tau)\|_{\mathbf{H}_2^{0,1}} d\tau$$

$$\leq Ct^{-\frac{3}{4}} \|u_0\|_{\mathbf{H}_1^{0,2}}$$

$$+C\varepsilon \int_0^t (t-\tau)^{-\frac{3}{4}} (1+\tau)^{-1+2\gamma} \tau^{-\frac{1}{3}} d\tau$$

$$\times \sup_t t^{\frac{1}{3}} (1+t)^{\frac{1}{2}-\gamma} \|u_x(t)\|_{\mathbf{H}_2^{0,1}}$$

(4.3) $$\leq Ct^{-\frac{3}{4}} \left(\|u_0\|_{\mathbf{H}_1^{0,2}} + \varepsilon \sup_t t^{\frac{1}{3}} (1+t)^{\frac{1}{2}-\gamma} \|u_x(t)\|_{\mathbf{H}_2^{0,1}} \right).$$

We again use Lemma 27, formula (3.3) with $n = 1$, $\beta = 6 - 2m$, and estimate (3.4), to get for $t \geq \frac{1}{4}$, $0 \leq m \leq 3$

$$\|u_x(t)\|_{\mathbf{H}_2^{0,\frac{m}{2}}} \leq Ct^{-\frac{7-m}{6}+\gamma} \|u_0\|_{\mathbf{H}_1^{0,2}}$$

$$+C\int_0^{\frac{t}{2}} (t-\tau)^{-\frac{7-m}{6}+\gamma} \|vu_x(\tau)\|_{\mathbf{H}_1^{0,2}} d\tau$$

$$+\int_{\frac{t}{2}}^t (t-\tau)^{-\frac{3}{4}} \|vu_x(\tau)\|_{\mathbf{H}_1^{0,\frac{2m+3}{4}}} d\tau$$

$$\leq Ct^{-\frac{7-m}{6}+\gamma} \|u_0\|_{\mathbf{H}_1^{0,2}} + C\int_0^{\frac{t}{2}} (t-\tau)^{-\frac{7-m}{6}+\gamma} (1+\tau)^{-1+2\gamma} \tau^{-\frac{1}{3}} d\tau$$

$$\times \varepsilon \sup_t t^{\frac{1}{3}} (1+t)^{\frac{1}{2}-\gamma} \|u_x(t)\|_{\mathbf{H}_2^{0,1}}$$

$$+C\int_{\frac{t}{2}}^t (t-\tau)^{-\frac{3}{4}} (1+\tau)^{-\frac{17-2m}{24}+2\gamma} \tau^{-\frac{1}{3}} d\tau$$

$$\times \varepsilon \sup_t t^{\frac{1}{3}} (1+t)^{\frac{17-2m}{24}-\gamma} \|u_x(t)\|_{\mathbf{H}_2^{0,\frac{2m+3}{4}}}$$

$$\leq Ct^{-\frac{7-m}{6}+\gamma} \left(\|u_0\|_{\mathbf{H}_1^{0,2}} + \varepsilon \sup_{t>0} t^{\frac{1}{3}} (1+t)^{\frac{1}{2}-\gamma} \|u_x(t)\|_{\mathbf{H}_2^{0,1}} \right.$$

$$(4.4) \qquad \left. +\varepsilon \sup_{t>0} t^{\frac{1}{3}} (1+t)^{\frac{17-2m}{24}-\gamma} \|u_x(t)\|_{\mathbf{H}_2^{0,\frac{2m+3}{4}}} \right).$$

Hence we have by (4.2)-(4.4) for $t > \frac{1}{4}$

$$\sum_{m=1}^3 \sup_{t>0} t^{\frac{5-m}{6}-\gamma} \|u(t)\|_{\mathbf{H}_2^{0,\frac{m}{2}}}$$

$$+\sum_{m=0}^3 \sup_{t>0} t^{\frac{7-m}{6}-\gamma} \|u_x(t)\|_{\mathbf{H}_2^{0,\frac{m}{2}}} + \sup_{t>0} t^{\frac{3}{4}} \|u_x(t)\|_{\mathbf{L}^2}$$

$$(4.5) \qquad \leq \frac{\varepsilon}{10}.$$

By the energy method we first consider the case $t \in [0, \frac{1}{4}]$. Multiplying both sides of (4.1) by u, tu_x, t^3u_t, respectively, and integrating the resulting equations in space variable, we find

$$\frac{d}{dt} \left(\|u(t)\|_{\mathbf{L}^2}^2 + t\|u_x(t)\|_{\mathbf{L}^2}^2 + t^3 \|u_t(t)\|_{\mathbf{L}^2}^2 \right)$$

$$+ \left(u_x(0,t)^2 + tu_{xx}(0,t)^2 + t^3u_{xt}(0,t)^2 \right)$$

$$- \|u_x(t)\|_{\mathbf{L}^2}^2 - 3t^2 \|u_t(t)\|_{\mathbf{L}^2}^2$$

$$\leq C \left(\|v_x(t)\|_{\mathbf{L}^\infty} \left(\|u(t)\|_{\mathbf{L}^2}^2 + t\|u_x(t)\|_{\mathbf{L}^2}^2 + t^3 \|u_t(t)\|_{\mathbf{L}^2}^2 \right) \right.$$

$$(4.6) \qquad \left. + \|u_x(t)\|_{\mathbf{L}^\infty} t^3 \|u_t(t)\|_{\mathbf{L}^2} \|v_t(t)\|_{\mathbf{L}^2} \right).$$

As in the proof of (4.6), multiplying both sides of (4.1) by $xu, t^2 x u_{xxxx}, x^2 u, tx^2 u_{xx}, x^3 u$ we obtain

$$\frac{d}{dt}\left(\left\|(\cdot)^{\frac{1}{2}}u\left(\cdot,t\right)\right\|_{\mathbf{L}^2}^2 + t^2\left\|(\cdot)^{\frac{1}{2}}u_{xx}\left(\cdot,t\right)\right\|_{\mathbf{L}^2}^2\right)$$

$$+3\left\|u_x\left(t\right)\right\|_{\mathbf{L}^2}^2 + 3t^2\left\|u_{xxx}\left(t\right)\right\|_{\mathbf{L}^2}^2 - 2t\left\|(\cdot)^{\frac{1}{2}}u_{xx}\left(\cdot,t\right)\right\|_{\mathbf{L}^2}^2$$

$$\leq\ C\left\|v_x\left(t\right)\right\|_{\mathbf{L}^\infty}\left(\left\|(\cdot)^{\frac{1}{2}}u\left(\cdot,t\right)\right\|_{\mathbf{L}^2}^2 + t^2\left\|(\cdot)^{\frac{1}{2}}u_{xx}\left(\cdot,t\right)\right\|_{\mathbf{L}^2}^2\right)$$

$$+C\left\|v\left(t\right)\right\|_{\mathbf{L}^\infty}\left\|u\left(t\right)\right\|_{\mathbf{L}^2}^2$$

$$+Ct^2\left\|(\cdot)^{\frac{1}{2}}v(\cdot,t)\right\|_{\mathbf{L}^\infty}\left\|u_{xxx}(t)\right\|_{\mathbf{L}^2}\left\|(\cdot)^{\frac{1}{2}}u_{xx}(\cdot,t)\right\|_{\mathbf{L}^2}$$

(4.7)
$$+Ct^2\left\|u_x(t)\right\|_{\mathbf{L}^\infty}\left\|(\cdot)^{\frac{1}{2}}v_{xx}(\cdot,t)\right\|_{\mathbf{L}^2}\left\|(\cdot)^{\frac{1}{2}}u_{xx}(\cdot,t)\right\|_{\mathbf{L}^2},$$

$$\frac{d}{dt}\left(\left\|(\cdot)u\left(\cdot,t\right)\right\|_{\mathbf{L}^2}^2 + t\left\|(\cdot)u_x\left(\cdot,t\right)\right\|_{\mathbf{L}^2}^2\right)$$

$$+6\left\|(\cdot)^{\frac{1}{2}}u_x\left(\cdot,t\right)\right\|_{\mathbf{L}^2}^2 + 6t\left\|(\cdot)^{\frac{1}{2}}u_{xx}\left(\cdot,t\right)\right\|_{\mathbf{L}^2}^2$$

$$-\left\|(\cdot)u_x\left(\cdot,t\right)\right\|_{\mathbf{L}^2}^2 - 2tu_x\left(0,t\right)^2$$

$$\leq\ C\left\|v_x\left(t\right)\right\|_{\mathbf{L}^\infty}\left\|(\cdot)u\left(\cdot,t\right)\right\|_{\mathbf{L}^2}^2$$

(4.8)
$$+C\left\|v\left(t\right)\right\|_{\mathbf{L}^\infty}\left(\left\|(\cdot)^{\frac{1}{2}}u\left(t\right)\right\|_{\mathbf{L}^2}^2 + t\left\|(\cdot)^{\frac{1}{2}}u_x\left(\cdot,t\right)\right\|_{\mathbf{L}^2}^2\right)$$

and

$$\frac{d}{dt}\left\|x^{\frac{3}{2}}u\left(\cdot,t\right)\right\|_{\mathbf{L}^2}^2 + 9\left\|(\cdot)u_x\left(\cdot,t\right)\right\|_{\mathbf{L}^2}^2 - 3\left\|u\left(t\right)\right\|_{\mathbf{L}^2}^2$$

(4.9)
$$\leq\ C\left(\left\|v_x\left(t\right)\right\|_{\mathbf{L}^\infty}\left\|(\cdot)^{\frac{3}{2}}u\left(\cdot,t\right)\right\|_{\mathbf{L}^2}^2 + \left\|v\left(t\right)\right\|_{\mathbf{L}^\infty}\left\|(\cdot)u\left(\cdot,t\right)\right\|_{\mathbf{L}^2}^2\right).$$

We have by a direct calculation

$$tu_x\left(0,t\right)^2 = -t\int_0^{+\infty}\partial_x u_x\left(x,t\right)^2 dx$$

$$=\ -2t\int_0^{+\infty}u_{xx}\left(x,t\right)u_x\left(x,t\right)dx$$

$$=\ 2t\int_0^{+\infty}u_{xxx}\left(x,t\right)u\left(x,t\right)dx$$

$$=\ 2t\int_0^{+\infty}\left(-u_t\left(x,t\right)-v\left(x,t\right)u_x\left(x,t\right)\right)u\left(x,t\right)dx$$

(4.10)
$$=\ -t\frac{d}{dt}\left\|u\left(t\right)\right\|_{\mathbf{L}^2}^2 - 2t\int_0^{+\infty}v\left(x,t\right)u_x\left(x,t\right)u\left(x,t\right)dx.$$

Collecting (4.6)-(4.10) and using the notation

$$\|u(t)\|_{\mathbf{Y}}^2 = \|u(t)\|_{\mathbf{H}_2^{0,\frac{3}{2}}}^2 + t\|u_x(t)\|_{\mathbf{L}^2}^2$$

$$+ \sum_{j=0}^{2} t^j \left\|x^{\frac{3}{2}-\frac{j}{2}}\partial_x^j u(t)\right\|_{\mathbf{L}^2}^2 + t^3\|\partial_t u(t)\|_{\mathbf{L}^2}^2$$

we get

$$\frac{d}{dt}\left(\|u(t)\|_{\mathbf{Y}}^2 + t\|u(t)\|_{\mathbf{L}^2}^2\right)$$
$$+ \left(u_x(0,t)^2 + tu_{xx}(0,t)^2 + t^3 u_{xt}(0,t)^2\right) - 2\|u(t)\|_{\mathbf{L}^2}^2$$
$$\leq 2t\|u_x(t)\|_{\mathbf{L}^\infty}\|v(t)\|_{\mathbf{L}^2}\|u(t)\|_{\mathbf{L}^2}$$
$$+ Ct^2\left\|(\cdot)^{\frac{1}{2}}v(\cdot,t)\right\|_{\mathbf{L}^\infty}\|u_{xxx}(t)\|_{\mathbf{L}^2}\left\|(\cdot)^{\frac{1}{2}}u_{xx}(\cdot,t)\right\|_{\mathbf{L}^2}$$
$$+ C\left(\|v_x(t)\|_{\mathbf{L}^\infty}\|u(t)\|_{\mathbf{Y}}^2 + \|u_x(t)\|_{\mathbf{L}^\infty}\|u(t)\|_{\mathbf{Y}}\|v(t)\|_{\mathbf{Y}}\right)$$
$$+ C\|v(t)\|_{\mathbf{L}^\infty}\left(\|u(t)\|_{\mathbf{L}^2}^2 + \left\|(\cdot)^{\frac{1}{2}}u(\cdot,t)\right\|_{\mathbf{L}^2}^2\right)$$

$$(4.11) \qquad + C\|v(t)\|_{\mathbf{L}^\infty}\left(t\left\|(\cdot)^{\frac{1}{2}}u_x(\cdot,t)\right\|_{\mathbf{L}^2}^2 + \|(\cdot)u(\cdot,t)\|_{\mathbf{L}^2}^2\right).$$

We first consider the case $0 < t \leq \frac{1}{4}$. We have by the integration by parts

$$f(x)^2 = -\int_x^{+\infty} \partial_x f(x)^2\, dx \leq 2\|f_x\|_{\mathbf{L}^2}\|f\|_{\mathbf{L}^2}$$

which implies

$$(4.12) \qquad \|f\|_{\mathbf{L}^\infty} \leq \sqrt{2}\|f_x\|_{\mathbf{L}^2}^{\frac{1}{2}}\|f\|_{\mathbf{L}^2}^{\frac{1}{2}}$$

from which it follows

$$\int_x^{+\infty} f_x(x)^2\, dx = \int_x^{+\infty} \partial_x(f_x(x)f(x))\, dx - \int_x^{+\infty} f_{xx}(x)f(x)\, dx$$
$$\leq -f_x(x)f(x) + \|f_{xx}\|\|f\|$$
$$\leq 2\|f_{xx}\|_{\mathbf{L}^2}^{\frac{1}{2}}\|f_x\|_{\mathbf{L}^2}\|f\|_{\mathbf{L}^2}^{\frac{1}{2}} + \|f_{xx}\|_{\mathbf{L}^2}\|f\|_{\mathbf{L}^2}.$$

This gives us

$$(4.13) \qquad \|f_x\|_{\mathbf{L}^2} \leq \sqrt{6}\|f_{xx}\|_{\mathbf{L}^2}^{\frac{1}{2}}\|f\|_{\mathbf{L}^2}^{\frac{1}{2}}.$$

Also we have

$$(4.14) \qquad \left\|(\cdot)^{\frac{1}{2}}f(\cdot)\right\|_{\mathbf{L}^\infty}^2 \leq C\left(\|f\|_{\mathbf{L}^2}^2 + \|f\|_{\mathbf{L}^2}\|(\cdot)f_x(\cdot)\|_{\mathbf{L}^2}\right).$$

From equation (4.1) we get

$$
\begin{aligned}
\|u_{xxx}(t)\|_{\mathbf{L}^2} &\leq C(\|u_t(t)\|_{\mathbf{L}^2} + \|u_x(t)\|_{\mathbf{L}^2}\|v(t)\|_{\mathbf{L}^\infty}) \\
&\leq C\left(t^{-\frac{3}{2}}\|u\|_{\mathbf{Y}} + t^{-\frac{1}{2}}\|u\|_{\mathbf{Y}}\|v(t)\|_{\mathbf{L}^2}^{\frac{1}{2}}\|v_x(t)\|_{\mathbf{L}^2}^{\frac{1}{2}}\right) \\
&\leq Ct^{-\frac{3}{2}}\|u(t)\|_{\mathbf{Y}}.
\end{aligned}
$$

(4.15)

Applying (4.12), (4.13), (4.14) and (4.15) to the right-hand side of (4.11), we get

$$
\begin{aligned}
&\frac{d}{dt}\left(\|u(t)\|_{\mathbf{Y}}^2 + t\|u(t)\|_{\mathbf{L}^2}^2\right) \\
\leq\ & 2\|u(t)\|_{\mathbf{L}^2}^2 + 2t\|u_x(t)\|_{\mathbf{L}^\infty}\|v(t)\|_{\mathbf{L}^2}\|u(t)\|_{\mathbf{L}^2} \\
&+Ct\|v(t)\|_{\mathbf{L}^2}\|u_{xxx}(t)\|_{\mathbf{L}^2}\|u(t)\|_{\mathbf{Y}} \\
&+Ct\left\|(\cdot)^{\frac{1}{2}}v_x(\cdot,t)\right\|_{\mathbf{L}^2}^{\frac{1}{2}}\|v(t)\|_{\mathbf{L}^2}^{\frac{1}{2}}\|u_{xxx}(t)\|_{\mathbf{L}^2}\|u(t)\|_{\mathbf{Y}} \\
&+C\|v_{xx}(t)\|_{\mathbf{L}^2}^{\frac{1}{2}}\|v_x(t)\|_{\mathbf{L}^2}^{\frac{1}{2}}\|u(t)\|_{\mathbf{Y}}^2 \\
&+C\|u_{xx}(t)\|_{\mathbf{L}^2}^{\frac{1}{2}}\|u_x(t)\|_{\mathbf{L}^2}^{\frac{1}{2}}\|u(t)\|_{\mathbf{Y}}\|v(t)\|_{\mathbf{Y}} \\
&+C\|v_x(t)\|_{\mathbf{L}^2}^{\frac{1}{2}}\|v(t)\|_{\mathbf{L}^2}^{\frac{1}{2}}\|u(t)\|_{\mathbf{Y}}^2 \\
\leq\ & 2\|u(t)\|_{\mathbf{L}^2}^2 + Ct\|u_{xxx}(t)\|_{\mathbf{L}^2}^{\frac{1}{4}}\|u_x(t)\|_{\mathbf{L}^2}^{\frac{3}{4}}\|v(t)\|_{\mathbf{L}^2}\|u(t)\|_{\mathbf{L}^2} \\
&+Ct\|v(t)\|_{\mathbf{L}^2}\|u_{xxx}(t)\|_{\mathbf{L}^2}\|u(t)\|_{\mathbf{Y}} \\
&+Ct\|(\cdot)v(\cdot,t)\|_{\mathbf{L}^2}^{\frac{1}{2}}\|v(t)\|_{\mathbf{L}^2}^{\frac{1}{2}}\|u_{xxx}(t)\|_{\mathbf{L}^2}\|u(t)\|_{\mathbf{Y}} \\
&+C\|v_{xxx}(t)\|_{\mathbf{L}^2}^{\frac{1}{4}}\|v_x(t)\|_{\mathbf{L}^2}^{\frac{3}{4}}\|u(t)\|_{\mathbf{Y}}^2 \\
&+C\|u_{xxx}(t)\|_{\mathbf{L}^2}^{\frac{1}{4}}\|u_x(t)\|_{\mathbf{L}^2}^{\frac{3}{4}}\|u(t)\|_{\mathbf{Y}}\|v(t)\|_{\mathbf{Y}} \\
&+C\|v_x(t)\|_{\mathbf{L}^2}^{\frac{1}{2}}\|v(t)\|_{\mathbf{L}^2}^{\frac{1}{2}}\|u(t)\|_{\mathbf{Y}}^2 \\
\leq\ & 2\|u(t)\|_{\mathbf{Y}}^2 + C\max\left(t^{-\frac{3}{4}},1\right)\|v(t)\|_{\mathbf{Y}}\|u(t)\|_{\mathbf{Y}}^2.
\end{aligned}
$$

Integrating the inequality in time, we obtain

$$
\begin{aligned}
\|u(t)\|_{Y}^2 &\leq \varepsilon_1^2 + 2T\sup_{t\in[0,\frac{1}{4}]}\|u(t)\|_{Y}^2 + C\varepsilon T^{\frac{1}{4}}\sup_{t\in[0,\frac{1}{4}]}\|u(t)\|_{Y}^2 \\
&\leq \varepsilon_1^2 + \frac{1}{2}\sup_{t\in[0,\frac{1}{4}]}\|u(t)\|_{Y}^2 + C\varepsilon\sup_{t\in[0,\frac{1}{4}]}\|u(t)\|_{Y}^2\ ;
\end{aligned}
$$

hence we have

(4.16)

$$
\sup_{t\in[0,\frac{1}{4}]}\|u(t)\|_{Y} \leq \sqrt{2}\varepsilon_1.
$$

We consider the case $t > \frac{1}{4}$. By (4.5) we find that

$$\|u(t)\|_{\mathbf{L}^2} \leq \varepsilon t^{-\frac{3}{4}}, \|u(t)\|_{\mathbf{H}_2^{0,\frac{m}{2}}} \leq \varepsilon t^{-\frac{5-m}{6}+\gamma} \text{ for } m = 1, 2, 3$$

$$\|u_x(t)\|_{\mathbf{H}_2^{0,\frac{m}{2}}} \leq \varepsilon t^{-\frac{7-m}{6}+\gamma} \text{ for } m = 0, 1, 2, 3.$$

From the equation (4.1), (4.12) and (4.13) it follows that

$$t^{\frac{3}{2}} \|u_{xxx}(t)\|_{\mathbf{L}^2} \leq t^{\frac{3}{2}} \|u_t(t)\|_{\mathbf{L}^2} + C t^{\frac{3}{2}} \|v(t)\|_{\mathbf{L}^\infty} \|u_x(t)\|_{\mathbf{L}^2}$$

(4.17)
$$\leq \|u(t)\|_{\mathbf{Y}} + C\varepsilon^2 t^{\frac{3}{2}} t^{-1+\gamma} t^{-\frac{7}{6}+\gamma} \leq 2 \|u\|_{\mathbf{Y}}.$$

Therefore we get

$$\|u_x(t)\|_{\mathbf{L}^\infty} \leq C \|u_{xxx}(t)\|_{\mathbf{L}^2}^{\frac{1}{4}} \|u_x(t)\|_{\mathbf{L}^2}^{\frac{3}{4}}$$

$$\leq C\varepsilon^{\frac{3}{4}} t^{-\frac{3}{8}} \left(t^{-\frac{7}{6}+\gamma}\right)^{\frac{3}{4}} \|u(t)\|_{\mathbf{Y}} \leq C\varepsilon^{\frac{3}{4}} t^{-\frac{5}{4}+\frac{3}{4}\gamma} \|u(t)\|_{\mathbf{Y}}.$$

Hence we have by (4.11),

$$\frac{d}{dt} \left(\|u(t)\|_{\mathbf{Y}}^2 + 2t \|u(t)\|_{\mathbf{L}^2}^2\right)$$

$$\leq C\varepsilon^2 t^{-\frac{5}{4}+\gamma} + C\varepsilon t^{-\frac{5}{4}+\gamma} \|u(t)\|_{\mathbf{Y}}^2.$$

Integrating with respect to t, we get

$$\|u(t)\|_{\mathbf{Y}}^2 + 2t \|u(t)\|_{\mathbf{L}^2}^2$$

$$\leq \|u(T)\|_{\mathbf{Y}}^2 + 2T \|u(T)\|_{\mathbf{L}^2}^2 + C\varepsilon^3 T^{-\frac{1}{4}+\gamma}$$

$$+ C\varepsilon T^{-\frac{1}{4}+\gamma} \sup_{t \in [T,+\infty)} \|u(t)\|_{\mathbf{Y}}^2.$$

Thus we have

(4.18)
$$\sup_{t \in [\frac{1}{4},+\infty)} \|u(t)\|_{\mathbf{Y}} \leq 2\varepsilon_1.$$

Therefore by (4.11), (4.16)-(4.17)

(4.19)
$$\||u|\|_X \leq \varepsilon.$$

We introduce the metric in X such that

$$d(f, g) = \sum_{m=1}^{2} \sup_t (1+t)^{\frac{5-m}{6}-\gamma} \|f(t) - g(t)\|_{\mathbf{H}_2^{0,\frac{m}{2}}}$$

$$+ \sum_{m=0}^{2} \sup_t t^{\frac{1}{3}} (1+t)^{\frac{5-m}{6}-\gamma} \|\partial_x (f(t) - g(t))\|_{\mathbf{H}_2^{0,\frac{m}{2}}}$$

$$+ \sup_t (1+t)^{\frac{3}{4}} \|f(t) - g(t)\|_{\mathbf{L}^2}.$$

Then as in the proof of (4.19) we have

(4.20)
$$d(u_1, u_2) = d(\mathbb{M}v_1, \mathbb{M}v_2) \leq \frac{1}{2} d(v_1, v_2),$$

where

$$\begin{cases} (u_j)_t + (u_j)_{xxx} = -(v_j)(u_j)_x, & t > 0, x > 0, \\ u_j(x,0) = u_0(x), & x > 0, \\ u_j(0,t) = 0, & t > 0. \end{cases}$$

The estimates (4.19) and (4.20) show that $\mathbb{M}$ is a contraction mapping from X into itself. Therefore there exists a unique solution u satisfying the estimate (4.19). This completes the proof of the first part of Theorem 30. Now using estimate (4.19) we prove that solution has the following asymptotics for $t \to +\infty$ uniformly with respect to $x > 0$

$$(4.21) \qquad u(x,t) = \frac{B(t)}{3\pi t} \frac{x}{\sqrt[3]{t}} Ai\left(\frac{x}{\sqrt[3]{t}}\right) + O\left(\min\left(1, \frac{x}{\sqrt[3]{t}}\right) t^{-1-\frac{\delta}{3}}\right),$$

where $\delta \in (0, \frac{1}{2})$ and

$$\begin{aligned} B(t) &= \int_0^{+\infty} x^2 u_0(x) dx + \int_0^t d\tau \int_0^{+\infty} x^2 u(x,\tau) u_x(x,\tau) dx \\ &= B(+\infty) - \int_t^{+\infty} d\tau \int_0^{+\infty} x^2 u(x,\tau) u_x(x,\tau) dx \\ (4.22) \qquad &\leq B(+\infty) + \int_t^{+\infty} \|u(\tau)\|_{\mathbf{H}_2^{0,1}} \|u_x(\tau)\|_{\mathbf{H}_2^{0,1}} d\tau \\ &\leq 2\varepsilon_1 + C\varepsilon^2 t^{-\frac{1}{6}+2\gamma}. \end{aligned}$$

In fact, via Lemma 27 we have

$$(4.23) \qquad u(x,t) = \frac{B(t)}{3\pi t} \frac{x}{\sqrt[3]{t}} Ai\left(\frac{x}{\sqrt[3]{t}}\right) + R(x,t),$$

where

$$\begin{aligned} &|R(x,t)| \\ &\leq C \min\left(1, \frac{x}{\sqrt[3]{t}}\right) t^{-1-\frac{\delta}{3}} \left(\|u_0\|_{\mathbf{H}_1^{0,2+\delta}} + \int_0^t \|u(\tau)u_y(\tau)\|_{\mathbf{H}_1^{0,2+\delta}} d\tau\right) \\ (4.24) \qquad &+ \int_0^t d\tau \int_0^{+\infty} dy |uu_y| \, |F(x,y,t-\tau) - F(x,y,t)|. \end{aligned}$$

Using estimate (4.19) we see that for $\delta \in (0, \frac{1}{2} - 6\gamma)$

$$\begin{aligned} \int_0^t \|u(\tau)u_y(\tau)\|_{\mathbf{H}_1^{0,2+\delta}} d\tau &\leq \int_0^t \|u(\tau)\|_{\mathbf{H}_2^{0,1+\frac{\delta}{2}}} \|u_y(\tau)\|_{\mathbf{H}_2^{0,1+\frac{\delta}{2}}} d\tau \\ (4.25) \qquad &\leq C\varepsilon^2 \int_0^t \tau^{-\frac{1}{3}} \langle\tau\rangle^{-\frac{5-2\delta}{6}+2\gamma} d\tau \leq C\varepsilon^2. \end{aligned}$$

We consider the last term of the right-hand side of (4.24). In the same way as in the proof of formula (3.3) of Lemma 27 we have

$$(4.26) \qquad |F_t(x,y,t)| < C\min\left(1, \frac{x}{\sqrt[3]{t}}\right) y^2 t^{-2},$$

since the asymptotics of $|F_t(x, y, t)|$ is the same as $\left|\partial_x^3 F(x, y, t)\right|$. By (4.26)

$$\|F(\cdot, y, t - \tau) - F(\cdot, y, t)\|_{\mathbf{L}^\infty} \leq C \min\left(1, \frac{x}{\sqrt[3]{t}}\right) y^2 t^{-2} \tau$$

and by (4.19)

$$\|u(\tau)\|_{\mathbf{H}_2^{0,1}} \|u_x(\tau)\|_{\mathbf{H}_2^{0,1}} < C\tau^{-\frac{7}{6} + 2\gamma}.$$

Hence we get

$$\int_0^t d\tau \left| \int_0^{+\infty} u(y, \tau) u_y(y, \tau) \right| |F(x, y, t - \tau) - F(x, y, t)|$$

$$\leq Ct^{-2} \min\left(1, \frac{x}{\sqrt[3]{t}}\right) \int_0^t \tau \|u(\tau)\|_{\mathbf{H}_2^{0,1}} \|u_y(\tau)\|_{\mathbf{H}_2^{0,1}} d\tau$$

$$\leq Ct^{-2} \min\left(1, \frac{x}{\sqrt[3]{t}}\right) \int_0^t \tau^{-\frac{1}{6} + 2\gamma} d\tau$$

$$\text{(4.27)} \qquad \leq C \min\left(1, \frac{x}{\sqrt[3]{t}}\right) t^{-1 - \frac{1}{6} + 2\gamma}.$$

Applying (4.25) and (4.27) to (4.24), we get

$$\text{(4.28)} \qquad |R(x, t)| \leq C\varepsilon_1 \min\left(1, \frac{x}{\sqrt[3]{t}}\right) t^{-1 - \frac{\delta}{3}},$$

where $\delta \in (0, \frac{1}{2} - 3\gamma]$. From (4.21), (4.22) and (4.28) we obtain the asymptotics (3.5) for the solution. Theorem 30 is proved.

CHAPTER 10

Neumann Problem for KdV Equation

1. Introduction

We consider the Neumann initial-boundary value problem on a half-line for the Korteweg-de Vries (KdV) equation

$$(1.1) \qquad \begin{cases} u_t + \lambda u u_x + u_{xxx} = 0, & t > 0, x > 0, \\ u(x,0) = u_0(x), & x > 0, \\ u_x(0,t) = 0, & t > 0. \end{cases}$$

The simple universal Korteweg-de Vries equation (1.1) appears in many fields of physics and technology as the first approximation in the description of the dispersive nonlinear waves [103], [136]. In the case of the Cauchy problem on the line, the KdV equation was solved via the inverse scattering transform method [1], and the well-posedness on the line as well as periodic domains was studied extensively by many authors, see for example,[14], [82], [83] and references cited therein. The asymptotic behavior of small solutions was studied in the case of the generalized KdV equations in [19], [63], [64], [65], [112]. However large time asymptotics of small solutions to the Cauchy problem for the KdV equation on the line has not been reached completely even now. From the heuristic point of view the quadratic nonlinearity of the shallow water type $u u_x$ is sub critical for large time: the nonlinear term decays more slowly than the linear part of the equation. The case of the Dirichlet initial-boundary value problem for the KdV equation on the positive line was considered recently in paper [52]. It was proved that the solution has an additional time-decay and as a result the nonlinear term of KdV equation behaves as super critical in the contrary to the corresponding Cauchy problem. In the case of the Neumann boundary-value problem (1.1) we expect that the time decay rate of the solutions will be slower, so that the nonlinear term in the problem (1.1) behaves as a critical one. Our main goal in the present chapter is to obtain the large time asymptotics of solutions to problem (1.1). Our approach is also based on the estimates of the integral equation in the weighted Sobolev spaces and weighted $\mathbf{L}^2$ space is used to get smooth solutions.

We also use the methods of our previous chapters. The integral formula is obtained by using the Laplace transform with respect to the space variable. The Laplace transform requires the boundary data $u(0,t), u_x(0,t), u_{xx}(0,t)$ and so $u(0,t), u_{xx}(0,t)$ should be determined by the given data. To achieve this we use the method developed in the papers [58], [60]. Our method to derive an integral formula is different from that [13] or [36], but the representation of the integral formula of ours (2.9) stated below is the same as one obtained in [36]. In [36] integral formulas for various boundary value problems for linear equation including Dirichlet and Neumann problems were constructed. However estimates of solutions applied to nonlinear problem were not given. In [13] an integral formula for the Dirichlet problem which is different from one used in [36] was constructed and used to derive the various smoothing properties of solutions. Now we state our result.

THEOREM 31. *Suppose that the initial data* $u_0 \in \mathbf{H}_1^{0,\frac{21}{4}}(\mathbf{R}^+) \cap \mathbf{H}_2^{1,\frac{7}{2}}(\mathbf{R}^+)$ *and the norm*

$$\|u_0\|_{\mathbf{H}_1^{0,\frac{21}{4}}} + \|u_0\|_{\mathbf{H}_2^{1,\frac{7}{2}}} \le \varepsilon,$$

where $\varepsilon > 0$ *is small enough and*

$$\lambda \int_0^{+\infty} x u_0(x) dx = \lambda \theta < 0.$$

Then there exists a unique solution

$$u \in \mathbf{C}([0, +\infty), \mathbf{H}_2^{1,\frac{7}{2}}(\mathbf{R}^+)) \cap \mathbf{L}^2\left(0, T; \mathbf{H}_2^{2,3}(\mathbf{R}^+)\right)$$

of the initial-boundary value problem (1.1) satisfying the boundary condition such that $u_x(0,t) = 0$ *for* $t > 0$. *Moreover the solution has the following asymptotics*

$$u(x,t) = 3\theta \left(1 + \eta \log t\right)^{-1} t^{-\frac{2}{3}} Ai'\left(\frac{x}{\sqrt[3]{t}}\right) + O\left(\varepsilon^2 t^{-\frac{2}{3}}(1 + \eta \log t)^{-\frac{6}{5}}\right)$$

for $t \to +\infty$ *uniformly with respect to* $x > 0$, *where*

$$\eta = -9\theta\lambda \int_0^{+\infty} Ai'^2(x) dx > 0$$

and $Ai(q)$ *is Airy function*

$$Ai(q) = \frac{1}{2\pi i} \int_{-i\infty}^{i\infty} e^{-z^3 + zq} dz.$$

We note here that solutions of our problem (1.1) decay faster than those of the linear problem and $Ai^{(2)}(0) = 0$, then we see that the main term satisfies the Neumann boundary condition for $t > 0$.

For the convenience of the reader we now state of our strategy of the proof. In Section 2 we construct the integral formula of the solution of the linear problem corresponding to (1.1)

$$u(x,t) = \Phi(x,t) u_0 = \int_0^{+\infty} F(x,y,t) u_0(y) dy,$$

where

$$F(x, y, t) = \frac{1}{2\pi i} \int_{-i\infty}^{i\infty} e^{-p^3 t + p(x-y)} dp$$

$$- \frac{1}{2\pi i} \int_{-i\infty}^{i\infty} e^{\xi t + \phi_3(\xi) x} \phi_3^{-1}(\xi) \sum_{j=1}^{2} e^{-\phi_j(\xi) y} \phi_j(\xi) \phi_j'(\xi) d\xi$$

and $\phi_j(\xi)$ are the roots of equation $p^3 = -\xi$ such that

$$p = \phi_1(\xi) = (\xi \exp(i\pi))^{\frac{1}{3}}, p = \phi_2(\xi) = (\xi \exp(-i\pi))^{\frac{1}{3}},$$

$$p = \phi_3(\xi) = (\xi \exp(3i\pi))^{\frac{1}{3}}.$$

Section 3 is devoted to the study of the asymptotic behavior of solutions to the linear problem by using the integral formula constructed in Section 2. We will show that for $t > 1, 0 < \delta < 1$

$$\Phi(t)u_0 = 3\theta t^{-\frac{2}{3}} Ai'\left(\frac{x}{\sqrt[3]{t}}\right) + O(t^{-\frac{2+\delta}{3}}),$$

where $\theta = \int_0^{+\infty} x u_0 dx$. Therefore the nonlinear term in the equation $u_t + \lambda u u_x + u_{xxx} = 0$ has the same decay rate as linear terms. Section 5 attempts to prove a global result and to establish asymptotic formulas of solutions. Here as in paper [54] we change variable $u = e^{-\phi(t)}v$. Then for the new function v we get the following problem

$$\begin{cases} v_t - \phi_t v + \lambda e^{-\phi(t)} v v_x + v_{xxx} = 0, & t > 0, x > 0, \\ v(x, 0) = e^{\phi(0)} u_0(x), & x > 0, \\ v_x(0, t) = 0, & t > 0. \end{cases}$$

In order to obtain an additional time decay rate we assume that the real-valued function $\phi(t)$ satisfies the following condition

$$\int_0^{+\infty} x\left(-\phi_t v + \lambda e^{-\phi(t)} v v_x\right) dx = 0, \phi(0) = 0,$$

which implies that

$$e^{\phi(t)} = g(t) = 1 + \lambda \theta^{-1} \int_0^t d\tau \int_0^{+\infty} x v v_x dx$$

$$= 1 - \lambda \theta^{-1} \int_0^t \|v(\tau)\|^2 d\tau.$$

We look for the solution v in the neighborhood of the first approximation

$$\Phi(t)u_0 \approx 3\theta t^{-\frac{2}{3}} Ai'\left(\frac{x}{\sqrt[3]{t}}\right) = \theta G.$$

We put $r = v - \Phi(t)u_0$ then we get the following integral formula

$$\begin{cases} r = \int_0^t g(\tau)^{-1} \Phi(x, t - \tau)(\lambda G\theta^2 \|G\|^2 + GG_x)d\tau + R_1 \\ g = 1 - \lambda \theta \int_0^t \|G\|^2 d\tau + R_2, \end{cases}$$

where R_1, R_2 are considered as remainder terms in our function space defined later. By Lemmas 32, 33 in Section 3 we obtain

$$g(t) = \frac{C}{1 + \eta \log(1+t)} \quad \text{and} \quad \|r\|_{L^\infty} \leq Ct^{-\frac{2}{3}} g(t)^{-\frac{1}{5}},$$

where under condition $\lambda\theta < 0$

$$\eta = -9\theta\lambda \int_0^{+\infty} Ai'^2(z)dz > 0.$$

Hence from the representation $u = \frac{v}{g} = \frac{r+\Phi}{g}$ we get the result of Theorem 31. We prove a local existence result of solutions in the special Section 4 by using energy method. However it seems that the energy method is not sufficient to get a global result and a sharp asymptotics of solutions. Therefore we show various estimates of solutions of linear problem with the Neumann boundary condition in Section 3. This Chapter was written in collaboration with Prof. L. Guardado Zavala.

2.   Linear problem

We consider the linear initial-boundary value problem corresponding to (1.1)

(2.1)
$$\begin{cases} u_t + u_{xxx} = f(x,t), \ t > 0, x > 0, \\ u(x,0) = u_0(x), \ x > 0, \\ u_x(0,t) = h(t), \ t > 0. \end{cases}$$

Taking the Laplace transformation of the problem (2.1) with respect to the space variable x we have

$$\hat{u}_t(p,t) + p^3\hat{u}(p,t) = f_1(p,t),$$

where

(2.2) $$f_1(p,t) = u(0,t)p^2 + u_{xx}(0,t) + \hat{f}(p,t) + ph(t),$$

$$\hat{u}(p,t) = \int_0^{+\infty} e^{-px} u(x,t)\, dx,$$

and so we have the following representation for the Laplace transform of the solution

(2.3) $$\hat{u}(p,t) = e^{-p^3 t}\hat{u}_0(p) + \int_0^t e^{-p^3(t-\tau)} f_1(p,\tau)d\tau.$$

In order to get the integral formula associated with (2.1), we find $u(0,t)$ and $u_{xx}(0,t)$ by using the given data. The condition

(2.4) $$|\hat{u}(p,t)| \leq M(1+|p|)^\beta, \quad \text{for all Re } p \geq 0,$$

with some M; $\beta > 0$ is necessary and sufficient for the existence of the inverse Laplace transformation. Clearly condition (2.4) is fulfilled in domains $\operatorname{Re} p^3 > 0$

of the right half-complex plane $\operatorname{Re} p \geq 0$. In domains $\operatorname{Re} p^3 < 0$ of the right-half complex plane $\operatorname{Re} p \geq 0$, we rewrite formula (2.3) as

$$
\begin{aligned}
\widehat{u}(p,t) \;=\;& e^{-p^3 t}\left(\widehat{u}_0(p) + \int_0^{+\infty} e^{p^3 \tau} f_1(p,\tau)d\tau\right) \\
& - \int_t^{+\infty} e^{p^3(t-\tau)} f_1(p,\tau)d\tau.
\end{aligned}
$$

It is clear that the last integral

$$
\int_t^{+\infty} e^{p^3(t-\tau)} f_1(p,\tau)d\tau
$$

satisfies the condition (2.4) for all $\operatorname{Re} p \geq 0$, $\operatorname{Re} p^3 < 0$. Therefore in order to satisfy the condition (2.4) we must assume that

$$
(2.5) \qquad\qquad \widehat{u}_0(p) + \int_0^{+\infty} e^{p^3 \tau} f_1(p,\tau)d\tau = 0
$$

for all $\operatorname{Re} p \geq 0$, $\operatorname{Re} p^3 < 0$. We use (2.5) to find the boundary functions $u(0,t)$ and $u_{xx}(0,t)$ involved in (2.2). Making the change of variable $-p^3 = \xi$ we transform domains $\operatorname{Re} p^3 < 0$ of the right half-complex plane $\operatorname{Re} p \geq 0$ to the half-complex plane $\operatorname{Re} \xi > 0$. The equation $-p^3 = \xi$ has three roots $\phi_1(\xi)$, $\phi_2(\xi)$ and $\phi_3(\xi)$ such that

$$
p = \phi_1(\xi) = (\xi \exp(i\pi))^{\frac{1}{3}}, p = \phi_2(\xi) = (\xi \exp(-i\pi))^{\frac{1}{3}},
$$

$$
p = \phi_3(\xi) = (\xi \exp(3i\pi))^{\frac{1}{3}}
$$

which transform the half-complex plane $\operatorname{Re} \xi > 0$ to domains where

$$
\operatorname{Re} \phi_1(\xi) > 0, \operatorname{Re} \phi_2(\xi) > 0 \text{ and } \operatorname{Re} \phi_3(\xi) < 0.
$$

The condition (2.5) can be written as a system of equations in the half-complex plane $\operatorname{Re} \xi > 0$

$$
\begin{aligned}
0 \;=\;& \int_0^{+\infty} e^{-\phi_l(\xi)x} u_0(x)dx \\
& + \int_0^{+\infty} e^{-\xi\tau} u(0,\tau)d\tau \phi_l(\xi)^2 + \int_0^{+\infty} e^{-\xi\tau} \partial_{xx} u(0,\tau)d\tau \\
& + \int_0^{+\infty}\int_0^{+\infty} e^{-(\phi_l(\xi)x+\xi\tau)} f(x,\tau)dx d\tau + \left(\int_0^{+\infty} e^{-\xi\tau} h(\tau)d\tau\right)\phi_l(\xi)
\end{aligned}
$$

for $l = 1, 2$. From which it follows that

$$
\begin{pmatrix} \phi_1^2(\xi) & 1 \\ \phi_2^2(\xi) & 1 \end{pmatrix} \begin{pmatrix} \int_0^{+\infty} e^{-\xi t} u(0, t) dt \\ \int_0^{+\infty} e^{-\xi t} u_{xx}(0, t) dt \end{pmatrix}
$$

$$
= -\begin{pmatrix} \int_0^{+\infty} e^{-\phi_1(\xi) x} u_0(x) dx \\ \int_0^{+\infty} e^{-\phi_2(\xi) x} u_0(x) dx \end{pmatrix}
$$

(2.6)
$$
-\begin{pmatrix} \int_0^{+\infty} \int_0^{+\infty} e^{-(\phi_1(\xi) x + \xi t)} f(x, t) dx dt \\ + \int_0^{+\infty} \int_0^{+\infty} e^{-(\phi_2(\xi) x + \xi t)} f(x, t) dx dt \end{pmatrix}
$$

$$
-\begin{pmatrix} \left(\int_0^{+\infty} e^{-\xi t} h(t) dt \right) \phi_1(\xi) \\ \left(\int_0^{+\infty} e^{-\xi t} h(t) dt \right) \phi_2(\xi) \end{pmatrix}.
$$

Solving (2.6) we find

$$
\begin{pmatrix} \int_0^{+\infty} e^{-\xi t} u(0, t) dt \\ \int_0^{+\infty} e^{-\xi t} u_{xx}(0, t) dt \end{pmatrix}
$$

$$
= \begin{pmatrix} -\frac{1}{\phi_1^2(\xi) - \phi_2^2(\xi)} \left(\int_0^{+\infty} \left(e^{-\phi_1(\xi) x} - e^{-\phi_2(\xi) x} \right) u_0(x) dx \right) \\ -\frac{1}{\phi_1^2(\xi) - \phi_2^2(\xi)} \int_0^{+\infty} \left(\phi_1^2(\xi) e^{-\phi_2(\xi) x} - \phi_2^2(\xi) e^{-\phi_1(\xi) x} \right) u_0(x) dx \end{pmatrix}
$$

$$
- \frac{1}{\phi_1(\xi) - \phi_2(\xi)}
$$

$$
\times \int_0^{+\infty} \int_0^{+\infty} \begin{pmatrix} e^{-(\phi_1(\xi) x + \xi t)} - e^{-(\phi_2(\xi) x + \xi t)} \\ \phi_1^2(\xi) e^{-(\phi_2(\xi) x + \xi t)} - \phi_2^2(\xi) e^{-(\phi_1(\xi) x + \xi t)} \end{pmatrix} f(x, t) dx dt
$$

$$
+ \begin{pmatrix} - \left(\int_0^{+\infty} e^{-\xi t} h(t) dt \right) (\phi_1(\xi) + \phi_2(\xi)) \\ - \left(\int_0^{+\infty} e^{-\xi t} h(t) dt \right) \phi_1(\xi) \phi_2(\xi) (\phi_1(\xi) + \phi_2(\xi))^{-1} \end{pmatrix}.
$$

By the inverse Laplace transform with respect to the time variable we get

$$
\begin{pmatrix} u(0, t) \\ u_{xx}(0, t) \end{pmatrix}
$$

(2.7)
$$
= -\frac{1}{2\pi i (\phi_1(\xi) - \phi_2(\xi))}
$$

$$
\times \int_{-i\infty}^{i\infty} e^{\xi t} \begin{pmatrix} (g(\phi_1(\xi)) - g(\phi_2(\xi))) \\ (\phi_1^2(\xi) g(\phi_2(\xi)) - \phi_2^2(\xi) g(\phi_1(\xi))) \end{pmatrix} d\xi,
$$

where

$$
g(\phi_l) = \widehat{u}_0(\phi_l) + \widehat{\widehat{f}}(\phi_l(\xi), \xi) + \phi_l \widehat{h}(\xi)
$$

and

$$\widehat{u_0}(\phi_l(\xi)) = \int_0^{+\infty} e^{-\phi_l(\xi)y} u_0(y) dy,$$

$$\widehat{f}(\phi_l(\xi), \xi) = \int_0^{+\infty} \int_0^{+\infty} e^{-(\phi_l(\xi)y+\xi t)} f(y, t) dy dt$$

$$\widehat{h}(\xi) = \int_0^{+\infty} e^{-\xi t} h(t) dt.$$

By (2.3) we have

$$u(x, t) = \frac{1}{2\pi i} \int_{-i\infty}^{i\infty} e^{-p^3 t + px} \widehat{u_0}(p) dp$$

$$+ \frac{1}{2\pi i} \int_{-i\infty}^{i\infty} e^{px} \int_0^t e^{-p^3(t-\tau)} \left(\widehat{f}(p, \tau) + h(\tau) p \right) d\tau dp$$

(2.8)

$$+ \frac{1}{2\pi i} \int_{-i\infty}^{i\infty} e^{px} e^{-p^3 t} \int_0^t e^{p^3 \tau} \left(u(0, \tau) p^2 + u_{xx}(0, \tau) \right) d\tau dp.$$

We consider the last term on the right-hand side of (2.8). We put

$$H(p, \xi) = -\frac{1}{\phi_1^2(\xi) - \phi_2^2(\xi)} \left(p^2 \left(g(\phi_1(\xi)) - g(\phi_2(\xi)) \right) \right.$$
$$\left. + \phi_1^2(\xi) g(\phi_2(\xi)) - \phi_2^2(\xi) g(\phi_1(\xi)) \right)$$

and by the fact that

$$\phi_1'(\xi) = -\frac{1}{(\phi_1(\xi) - \phi_2(\xi))(\phi_1(\xi) - \phi_3(\xi))},$$

$$\phi_2'(\xi) = -\frac{1}{(\phi_2(\xi) - \phi_1(\xi))(\phi_2(\xi) - \phi_3(\xi))},$$

$$\sum_{j=1}^3 \phi_j(\xi) = 0$$

we get

$$\int_{-i\infty}^{i\infty} \frac{e^{px}}{p^3 + \xi} H(p, \xi) dp = \int_{-i\infty}^{i\infty} \frac{e^{px}}{\prod_{j=1}^3 (p - \phi_j(\xi))} H(p, \xi) dp$$

$$= -2\pi i e^{\phi_3(\xi)x} \frac{g(\phi_1)(\phi_3^2 - \phi_2^2) + g(\phi_2)(\phi_1^2 - \phi_3^2)}{(\phi_1^2 - \phi_2^2)(\phi_3 - \phi_1)(\phi_3 - \phi_2)}$$

$$= -2\pi i e^{\phi_3(\xi)x} \left(\frac{g(\phi_1)(\phi_3 + \phi_2)}{(\phi_1^2 - \phi_2^2)(\phi_3 - \phi_1)} - \frac{g(\phi_2)(\phi_1 + \phi_3)}{(\phi_1^2 - \phi_2^2)(\phi_3 - \phi_2)} \right)$$

$$= -2\pi i e^{\phi_3(\xi)x} \phi_3^{-1} \left(g(\phi_1)\phi_1\phi_1' + g(\phi_2)\phi_2\phi_2' \right).$$

Therefore we obtain

$$\frac{1}{2\pi i} \int_{-i\infty}^{i\infty} e^{px} e^{-p^3 t} \int_0^t e^{p^3 \tau} \left(u_x(0,\tau)p + u_{xx}(0,\tau) \right) d\tau dp$$

$$= -\frac{1}{4\pi^2} \int_{-i\infty}^{i\infty} e^{px} \int_{-i\infty}^{i\infty} \frac{e^{\xi t} - e^{-p^3 t}}{p^3 + \xi} H(p,\xi) \, d\xi dp$$

$$= -\frac{1}{4\pi^2} \int_{-i\infty}^{i\infty} e^{px} \int_{-i\infty}^{i\infty} \frac{e^{\xi t}}{p^3 + \xi} H(p,\xi) \, d\xi dp$$

$$= -\frac{1}{4\pi^2} \int_{-i\infty}^{i\infty} e^{\xi t} \int_{-i\infty}^{i\infty} \frac{e^{px}}{p^3 + \xi} H(p,\xi) \, d\xi dp$$

$$= -\frac{1}{2\pi i} \int_0^{+\infty} \sum_{j=1}^{2} \int_{-i\infty}^{i\infty} e^{\xi t} e^{\phi_3(\xi)x} \phi_3^{-1}(\xi)\phi_j(\xi)\phi_j'(\xi) e^{-y\phi_j(\xi)}$$

$$\times \left(u_0(y) + \int_0^{+\infty} e^{-\xi \tau} f(y,\tau) \, d\tau \right) d\xi dy$$

$$- \frac{1}{2\pi i} \int_{-i\infty}^{i\infty} e^{\xi t} e^{\phi_3(\xi)x} \phi_3^{-1}(\xi) \sum_{j=1}^{2} \phi_j^2(\xi)\phi_j(\xi) \int_0^{+\infty} e^{-\xi \tau} h(\tau) \, d\tau d\xi$$

$$= -\frac{1}{2\pi i} \int_0^{+\infty} dy u_0(y)$$

$$\times \left(\sum_{j=1}^{2} \int_{-i\infty}^{i\infty} e^{\xi t} e^{\phi_3(\xi)x} \phi_3^{-1}(\xi)\phi_j(\xi)\phi_j'(\xi) e^{-y\phi_j(\xi)} d\xi \right)$$

$$- \frac{1}{2\pi i} \int_0^t d\tau \int_0^{+\infty} dy f(y,\tau)$$

$$\times \left(\sum_{j=1}^{2} \int_{-i\infty}^{i\infty} e^{\xi(t-\tau)} e^{\phi_3(\xi)x} \phi_3^{-1}(\xi)\phi_j(\xi)\phi_j'(\xi) e^{-y\phi_j(\xi)} d\xi \right)$$

$$- \frac{1}{\pi i} \int_0^t \int_{-i\infty}^{i\infty} e^{\xi(t-\tau)} e^{\phi_3(\xi)x} \phi_3(\xi)\phi_3'(\xi) d\xi h(\tau) \, d\tau.$$

Hence (2.7) and (2.8) give the following integral representation of the solution $u(x,t)$ to the problem (2.1) with $h(t) \equiv 0$

$$(2.9) \quad u(x,t) = \int_0^{+\infty} u_0(y) F(x,y,t) dy + \int_0^t d\tau \int_0^{+\infty} f(x,y,\tau) F(x,y,t-\tau) dy$$

where

$$F(x,y,t) = \frac{1}{2\pi i}\int_{-i\infty}^{i\infty} e^{-p^3 t + p(x-y)}\,dp$$

$$-\frac{1}{2\pi i}\int_{-i\infty}^{i\infty} e^{\xi t + \phi_3(\xi)x}\phi_3^{-1}(\xi)\sum_{j=1}^{2} e^{-\phi_j(\xi)y}\phi_j(\xi)\phi_j'(\xi)\,d\xi.$$

3. Preliminaries

In this section we provide a useful and necessary lemma for our results. As in Section 2 we denote by

$$\phi_1(\xi) = (\xi\exp(i\pi))^{\frac{1}{3}}, \phi_2(\xi) = (\xi\exp(-i\pi))^{\frac{1}{3}},$$

(3.1) $$\phi_3(\xi) = (\xi\exp(3i\pi))^{\frac{1}{3}},$$

the roots of the equation $p^3 = -\xi$, such that $\operatorname{Re}\phi_l(\xi) > 0$, $l = 1,2$, $\operatorname{Re}\phi_3(\xi) < 0$, for all $\operatorname{Re}\xi > 0$. We introduce the following functions

(3.2) $$\mathcal{G}_k(x,y,t) = \frac{1}{2\pi i}\int_{-i\infty}^{i\infty} e^{-p^3 t + px}\Big(e^{-py} - \sum_{j=0}^{k}(-1)^j\frac{p^j y^j}{j!}\Big)dp$$

and

$$\mathcal{F}_k(x,y,t) = -\frac{1}{2\pi i}\sum_{j=1}^{2}\int_{-i\infty}^{i\infty} e^{\xi t + \phi_3(\xi)x}\phi_3^{-1}(\xi)$$

(3.3)

$$\times\Big(e^{-\phi_j(\xi)y} - \sum_{l=0}^{k}(-1)^l\frac{\phi_j^l(\xi)y^l}{l!}\Big)\phi_j\phi_j'(\xi)\,d\xi.$$

LEMMA 28. *We have for* $m \geq 0, t > 0$

$$\left\|(\cdot)^{\frac{m}{2}}\partial_x^{(n)}\mathcal{G}_0(\cdot,y,t)\right\|_{\mathbf{L}^2}$$

$$\leq C\begin{cases} t^{-\frac{2n+1-m}{6}}\langle t\rangle^{-\frac{1}{3}}\langle y\rangle + t^{-\frac{n+1}{3}}\langle t\rangle^{-\frac{1}{3}}\langle y\rangle^{\frac{m+3}{2}} \\ \quad + t^{-\frac{3n+2}{6}+\frac{1}{12}}\langle t\rangle^{-\frac{1}{4}}\langle y\rangle^{\frac{m+2+n}{2}}, \\ t^{-\frac{2n+1-m}{6}} + t^{-\frac{n+1}{3}}\langle y\rangle^{\frac{m+1}{2}} + t^{-\frac{3n+2}{6}+\frac{1}{12}}\langle y\rangle^{\frac{m+2+n}{2}-\frac{3}{4}}, \end{cases}$$

for $n \geq 1$ *and*

$$\left\|(\cdot)^{\frac{m}{2}}\mathcal{G}_0(\cdot,y,t)\right\|_{\mathbf{L}^2}$$

$$\leq C\begin{cases} t^{-\frac{1-m}{6}}\langle t\rangle^{-\frac{1}{3}}\langle y\rangle + t^{-\frac{1}{3}}\langle t\rangle^{-\frac{1}{3}}\langle y\rangle^{\frac{m+3}{2}} + t^{-\frac{1}{2}}\langle t\rangle^{-\frac{1}{4}}\langle y\rangle^{\frac{m+2}{2}+\frac{3}{4}} \\ t^{-\frac{1-m}{6}} + t^{-\frac{1}{3}}\langle y\rangle^{\frac{m+1}{2}} + t^{-\frac{1}{2}}\langle y\rangle^{\frac{m+2}{2}}. \end{cases}$$

PROOF. We let $p^3 t = z^3$, then we have

$$\partial_x^{(n)} \mathcal{G}_0(x, y, t) = \frac{1}{2\pi i} \int_{-i\infty}^{i\infty} e^{-p^3 t} e^{px} \left(e^{-py} - 1 \right) p^n dp$$

$$= \frac{1}{2\pi i} t^{-\frac{n+1}{3}} \int_{-i\infty}^{i\infty} e^{-z^3} z^n e^{z\tilde{x}} \left(e^{-z\tilde{y}} - 1 \right) dz$$

$$= t^{-\frac{n+1}{3}} \left(Ai^{(n)}(\tilde{x} - \tilde{y}) - Ai^{(n)}(\tilde{x}) \right),$$

where $Ai(q)$ is the Airy function

$$Ai(q) = \frac{1}{2\pi i} \int_{-i\infty}^{i\infty} e^{-z^3 + zq} dz.$$

It is known that we have

$$\left| Ai^{(n)}(-|q|) \right| \le C(1 + |q|)^{\frac{2n-1}{4}}$$

and for any $\beta > 0$

$$\left| Ai^{(n)}(|q|) \right| \le C(1 + |q|)^{-\beta},$$

(see, for example, [1]). Therefore we get for $\xi \in (\tilde{x} - \tilde{y}, \tilde{x})$, $\tilde{x} = \frac{x}{\sqrt[3]{t}}$ and $\tilde{y} = \frac{y}{\sqrt[3]{t}}$, $\mu \in [0, 1]$, $x < 2y$

$$\left| \partial_x^{(n)} \mathcal{G}_0(x, y, t) \right|$$

$$\le Ct^{-\frac{n+1+\mu}{3}} y^\mu \left| Ai^{(n+1)}(\xi) \right|^\mu$$

$$\times \left(\left| Ai^{(n)}(\tilde{x} - \tilde{y}) \right|^{1-\mu} + \left| Ai^{(n)}(\tilde{x}) \right|^{1-\mu} \right)$$

$$\le Ct^{-\frac{n+1+\mu}{3}} y^\mu (1 + |\tilde{y} - \tilde{x}|)^{\frac{(2n+1)\mu}{4}}$$

$$\times \left((1 + |\tilde{y} - \tilde{x}|)^{\frac{(2n-1)}{4}(1-\mu)} + (1 + \tilde{x})^{-\beta} \right).$$

Therefore we get for $y > \sqrt[3]{t}$ and $\frac{(2n-1)}{2} + \mu \ge 0$

$$\left(\int_0^{2y} x^m \left| \partial_x^{(n)} \mathcal{G}_0(x, y, t) \right|^2 dx \right)^{\frac{1}{2}}$$

$$\le Ct^{-\frac{n+1+\mu}{3}} y^\mu \left(\int_0^{2y} x^m (1 + |\tilde{y} - \tilde{x}|)^{\frac{(2n-1)}{2} + \mu} dx \right)^{\frac{1}{2}}$$

$$\le Ct^{-\frac{n+1+\mu}{3} - \frac{1}{3}\left(\frac{(2n-1)}{4} + \frac{1}{2}\mu \right)} y^{\frac{(2n-1)}{4} + \frac{3}{2}\mu} \left(\int_0^{2y} x^m dx \right)^{\frac{1}{2}}$$

$$\le Ct^{-\frac{n+1+\mu}{3} - \frac{1}{3}\left(\frac{(2n-1)}{4} + \frac{1}{2}\mu \right)} y^{\frac{(2n-1)}{4} + \frac{3}{2}\mu + \frac{m+1}{2}}$$

and for $y < \sqrt[3]{t}$

$$\left(\int_0^{2y} x^m \left|\mathcal{G}_0(x,y,t)\right|^2 dx\right)^{\frac{1}{2}}$$

$$\leq \ Ct^{-\frac{n+1+\mu}{3}} y^\mu \left(\int_0^{2y} x^m dx\right)^{\frac{1}{2}} \leq Ct^{-\frac{n+1+\mu}{3}} y^{\mu+\frac{m+1}{2}}.$$

Since $Ai^{(n)}(|\xi|) \leq C(1+|\xi|)^{-10}$ we get for $x > 2y$

$$\left|\partial_x^{(n)} \mathcal{G}_0(x,y,t)\right|$$

$$\leq \ Ct^{-\frac{n+1+\mu}{3}} y^\mu \left|Ai^{(n+1)}(\xi)\right|^\mu \left(\left|Ai^{(n)}(\tilde{x}-\tilde{y})\right|^{1-\mu} + \left|Ai^{(n)}(\tilde{x})\right|^{1-\mu}\right)$$

$$\leq \ Ct^{-\frac{n+1+\mu}{3}} y^\mu (1+\tilde{x})^{-10}.$$

Thus letting $x = \sqrt[3]{tz}$ we obtain

$$\left(\int_y^{+\infty} x^m \left|\partial_x^{(n)} \mathcal{G}_0(x,y,t)\right|^2 dx\right)^{\frac{1}{2}}$$

$$\leq \ Ct^{-\frac{n+1+\mu}{3}} y^\mu \left(\int_y^{+\infty} x^m (1+\tilde{x})^{-20} dx\right)^{\frac{1}{2}}$$

$$\leq \ Ct^{-\frac{n+1+\mu}{3}} y^\mu t^{-\frac{m}{6}+\frac{1}{6}} \left(\int_0^{+\infty} z^{2m}(2+z)^{-20} dz\right)^{\frac{1}{2}}$$

$$\leq \ Ct^{-\frac{2n+1+2\mu-m}{6}} y^\mu.$$

Therefore we have for $t > 0$ and $\mu_j \in [0,1]$, $j = 1,2,3$, $\frac{(2n-1)}{2} + \mu_1 \geq 0$

$$\left\|(\cdot)^{\frac{m}{2}} \partial_x^{(n)} \mathcal{G}_0(\cdot,y,t)\right\|_{\mathbf{L}^2} \leq \ Ct^{-\frac{n+1+\mu_1}{3} - \frac{1}{3}\left(\frac{(2n-1)}{4}+\frac{1}{2}\mu_1\right)} y^{\frac{(2n-1)}{4}+\frac{3}{2}\mu_1+\frac{m+1}{2}}$$

$$+Ct^{-\frac{n+1+\mu_2}{3}} y^{\mu_2+\frac{m+1}{2}} + Ct^{-\frac{2n+1+2\mu_3-m}{6}} y^{\mu_3}.$$

Thus if we take $\mu_2, \mu_3 = 0, 1$ and $\mu_1 = \frac{1}{2}, 1$ for $n = 0$ and $\mu_1 = 0, \frac{1}{2}$ for $n = 1, 2$, we have the desired estimates. Lemma 28 is proved. $\qquad\square$

LEMMA 29. *We have $m \geq 0, t > 0$*

$$\left\|(\cdot)^{\frac{m}{2}} \partial_x^{(n)} \mathcal{G}_1(\cdot,y,t)\right\|_{\mathbf{L}^2}$$

$$\leq \begin{cases} Ct^{-\frac{n+1}{3}} \langle y\rangle^{\frac{m+1}{2}} + Ct^{-\frac{2n+1-m}{6}} + Ct^{-\frac{3n+2}{6}+\frac{1}{12}} \langle y\rangle^{\frac{m+1+n}{2}-\frac{1}{4}} \\ Ct^{-\frac{n+1}{3}} \langle t\rangle^{-\frac{1}{3}} \langle y\rangle^{\frac{m+3}{2}} + Ct^{-\frac{2n+1-m}{6}} \langle t\rangle^{-\frac{2}{3}} \langle y\rangle^2 \\ \qquad + Ct^{-\frac{3n+2}{6}+\frac{1}{12}} \langle t\rangle^{-\frac{1}{4}} \langle y\rangle^{\frac{m+2+n}{2}} \end{cases}$$

for $n = 1, 2$ and

$$\left\|(\cdot)^{\frac{m}{2}} \mathcal{G}_1(\cdot,y,t)\right\|_{\mathbf{L}^2}$$

$$\leq \ Ct^{-\frac{1}{3}} \langle t\rangle^{-\frac{1}{3}} \langle y\rangle^{\frac{m+3}{2}} + Ct^{-\frac{1-m}{6}} \langle t\rangle^{-\frac{2}{3}} \langle y\rangle^2 + Ct^{-\frac{1}{2}} \langle t\rangle^{-\frac{1}{3}} \langle y\rangle^{\frac{m+3}{2}}.$$

PROOF. Changing the variable of integration $p^3 t = z^3$ we have

$$\partial_x^{(n)} \mathcal{G}_1(x, y, t)$$

$$= \frac{1}{2\pi i} \int_{-i\infty}^{i\infty} e^{-p^3 t} e^{px} \left(e^{-py} - \sum_{k=0}^{1} (-1)^k (py)^k \right) p^n dp$$

$$= \frac{1}{2\pi i} t^{-\frac{n+1}{3}} \int_{-i\infty}^{i\infty} e^{-z^3} z^n e^{z\widetilde{x}} \left(e^{-z\widetilde{y}} - \sum_{k=0}^{1} (-1)^k (z\widetilde{y})^k \right) dz$$

$$= t^{-\frac{n+1}{3}} \left(Ai^{(n)}(\widetilde{x} - \widetilde{y}) - \sum_{j=0}^{1} (-1)^j \frac{\widetilde{y}^j}{j!} Ai^{(n+j)}(\widetilde{x}) \right).$$

We again use the asymptotic properties of the Airy function such that

$$\left| Ai^{(n)}(-|q|) \right| \leq C(1 + |q|)^{\frac{2n-1}{4}}$$

and for any $\beta > 0$

$$\left| Ai^{(n)}(|q|) \right| \leq C(1 + |q|)^{-\beta}.$$

Then we get for $\xi \in (\widetilde{x} - \widetilde{y}, \widetilde{x}), \widetilde{x} = \frac{x}{\sqrt[3]{t}}$ and $\widetilde{y} = \frac{y}{\sqrt[3]{t}}, \mu \in [0, 1], x < 2y$

$$\left| \partial_x^{(n)} \mathcal{G}_1(x, y, t) \right| \leq C t^{-\frac{n+1+2\mu}{3}} y^{2\mu} \left| Ai^{(n+2)}(\xi) \right|^{\mu}$$

$$\times \left(\left| Ai^{(n)}(\widetilde{x} - \widetilde{y}) \right|^{1-\mu} + \left| Ai^{(n)}(\widetilde{x}) \right|^{1-\mu} + \left| Ai^{(n+1)}(\widetilde{x}) \right|^{1-\mu} \widetilde{y}^{1-\mu} \right)$$

$$\leq t^{-\frac{n+1+2\mu}{3}} y^{2\mu} (1 + |\widetilde{y} - \widetilde{x}|)^{\frac{(2n+3)\mu}{4}}$$

$$\times \left((1 + |\widetilde{y} - \widetilde{x}|)^{\frac{2n-1}{4}(1-\mu)} + (1 + \widetilde{x})^{-\beta} + (1 + \widetilde{x})^{-\beta} \widetilde{y}^{1-\mu} \right).$$

Therefore we get for $y > \sqrt[3]{t}$ and $\frac{(2n-1)}{2} + 2\mu \geq 0$

$$\left(\int_0^{2y} x^m \left| \partial_x^{(n)} \mathcal{G}_1(x, y, t) \right|^2 dx \right)^{\frac{1}{2}}$$

$$\leq C t^{-\frac{n+1+2\mu}{3}} y^{2\mu} \left(\int_0^{2y} x^m (1 + |\widetilde{y} - \widetilde{x}|)^{\frac{(2n-1)}{2} + 2\mu} dx \right)^{\frac{1}{2}}$$

$$\leq C t^{-\frac{n+1+2\mu}{3} - \frac{1}{3} \left(\frac{(2n-1)}{4} + \mu \right)} y^{\frac{(2n-1)}{4} + 3\mu} \left(\int_0^{2y} x^m dx \right)^{\frac{1}{2}}$$

$$\leq C t^{-\frac{n+1+2\mu}{3} - \frac{1}{3} \left(\frac{(2n-1)}{4} + \mu \right)} \langle y \rangle^{\frac{(2n-1)}{4} + 3\mu + \frac{m+1}{2}}.$$

In the case $y < \sqrt[3]{t}$ we get

$$\left(\int_0^{2y} x^m \left| \mathcal{G}_1(x, y, t) \right|^2 dx \right)^{\frac{1}{2}}$$

$$\leq Ct^{-\frac{1+2\mu}{3}} y^{2\mu} \left(\int_0^{2y} x^m dx \right)^{\frac{1}{2}} \leq Ct^{-\frac{1+2\mu}{3}} \langle y \rangle^{2\mu + \frac{1+m}{2}}.$$

Since

$$\left| Ai^{(n)}(|\xi|) \right| \leq C(1 + |\xi|)^{-10}$$

we have for $x > 2y$

$$\left| \partial_x^{(n)} \mathcal{G}_1(x, y, t) \right| \leq Ct^{-\frac{n+1}{3}} \widetilde{y}^{2\mu} \left| Ai^{(n+2)}(\xi) \right|^\mu$$

$$\times \left(\left| Ai^{(n)}(\widetilde{x} - \widetilde{y}) \right|^{1-\mu} + \left| Ai^{(n)}(\widetilde{x}) \right|^{1-\mu} + \left| Ai^{(n+1)}(\widetilde{x}) \right|^{1-\mu} \widetilde{y}^{1-\mu} \right)$$

$$\leq Ct^{-\frac{n+1+2\mu}{3}} y^{2\mu} (1 + \widetilde{y})^{1-\mu} (1 + \widetilde{x})^{-10} \leq Ct^{-\frac{n+1+2\mu}{3}} y^{2\mu} (1 + \widetilde{x})^{-9},$$

which gives us with $x = \sqrt[3]{t}z$

$$\left(\int_{2y}^{+\infty} x^m \left| \partial_x^{(n)} \mathcal{G}_1(x, y, t) \right|^2 dx \right)^{\frac{1}{2}}$$

$$\leq Ct^{-\frac{n+1+2\mu}{3}} y^{2\mu} \left(\int_{2y}^{+\infty} x^m (2 + \widetilde{x})^{-18} dx \right)^{\frac{1}{2}}$$

$$\leq Ct^{-\frac{n+1+2\mu}{3}} y^{2\mu} t^{-\frac{m}{6} + \frac{1}{6}} \left(\int_0^{+\infty} z^{2m} (2 + z)^{-18} dz \right)^{\frac{1}{2}}$$

$$\leq Ct^{-\frac{2n+1-m+4\mu}{6}} y^{2\mu} \leq Ct^{-\frac{2n+1-m+4\mu}{6}} \langle y \rangle^{2\mu}.$$

Therefore we obtain for $t > 0$,

$$\left\| x^{\frac{m}{2}} \partial_x^{(n)} \mathcal{G}_1(x, y, t) \right\|_{L^2}$$

$$\leq Ct^{-\frac{n+1+2\mu_1}{3} - \frac{1}{3}\left(\frac{(2n-1)}{4} + \mu_1\right)} \langle y \rangle^{\frac{(2n-1)}{4} + 3\mu_1 + \frac{m+1}{2}}$$

$$+ Ct^{-\frac{n+1+2\mu_2}{3}} \langle y \rangle^{2\mu_2 + \frac{1+m}{2}} + Ct^{-\frac{2n+1-m+4\mu_3}{6}} \langle y \rangle^{2\mu_3},$$

where $\mu_j \in [0, 1]$, $j = 1, 2, 3$, $\frac{(2n-1)}{2} + 2\mu_1 \geq 0$. If we take $\mu_2 = 0, \frac{1}{2}$; $\mu_3 = 0, 1$ and $\mu_1 = 0, \frac{1}{4}$ for $n = 1, 2$, then we have the result for $n = 1, 2$ and if we take $\mu_1 = \frac{1}{4}, \frac{5}{12}$ for $n = 0$, then we have the last estimate of the lemma. Lemma 29 is proved. $\qquad \square$

LEMMA 30. *We have for $t > 1$, n, $m \geq 0$, $k = 0, 1$, $\delta \in [0, 1]$*

$$\left\| (\cdot)^{\frac{m}{2}} \partial_x^{(n)} \mathcal{F}_k(\cdot, y, t) \right\|_{L^2} \leq Ct^{-\frac{1-m+2n+2k+2\delta}{6}} \langle y \rangle^{k+\delta}.$$

PROOF. Using $\phi_j'(\xi) = -\frac{1}{3\phi_j^2}$ we get for the Laplace transform of function $\mathcal{F}_k(x,y,t)$

$$\mathcal{L}\left\{\partial_x^{(n)}\mathcal{F}_k(x,y,t)\right\} = \widehat{\partial_x^{(n)}\mathcal{F}_k}(p,y,t)$$

$$= -\frac{1}{6\pi i}\sum_{j=1}^{2}\int_{-i\infty}^{i\infty}e^{\xi t}\frac{\phi_3^{n-1}(\xi)}{(p-\phi_3(\xi))}\left(e^{-\phi_j(\xi)y} - \sum_{l=0}^{k}(-1)^l\frac{\phi_j^l(\xi)y^l}{l!}\right)\phi_j^{-1}(\xi)d\xi.$$

Using the Plancherel theorem we get for $t > 1$, and an integer $m \geq 0$

$$\left\|x^m\partial_x^{(n)}\mathcal{F}_k(x,y,t)\right\|_{\mathbf{L}^2} \leq \left\|\partial_p^{(m)}\widehat{\partial_x^{(n)}\mathcal{F}_k}(p,y,t)\right\|_{\mathbf{L}^2(\mathrm{Re}\,p=0)}$$

$$\leq C\left\|\sum_{j=1}^{2}\int_{-i\infty}^{i\infty}e^{\xi t}\frac{\phi_3^{n-1}(\xi)}{(p-\phi_3(\xi))^{m+1}}\right.$$

$$\left.\times\left(e^{-\phi_j(\xi)y} - \sum_{l=0}^{k}(-1)^l\frac{\phi_j^l(\xi)y^l}{l!}\right)\phi_j^{-1}(\xi)d\xi\right\|_{\mathbf{L}^2}.$$

We have for $\mathrm{Re}\,\xi = 0$

$$\left|\int_{-i\infty}^{i\infty}\frac{1}{(p-\phi_3(\xi))^{m+1}}\frac{1}{(p-\phi_3(\xi_1))^{m+1}}|dp|\right|$$

$$\leq C\left|\int_{-i\infty}^{i\infty}\frac{1}{\left(\left|p+\xi^{\frac{1}{3}}\right|+|\xi|^{\frac{1}{3}}\right)^{m+1}}\frac{1}{\left(\left|p+\xi_1^{\frac{1}{3}}\right|+|\xi_1|^{\frac{1}{3}}\right)^{m+1}}|dp|\right|$$

$$\leq C\left(\left|\int_{-i\infty}^{i\infty}\frac{1}{\left(\left|p+\xi^{\frac{1}{3}}\right|^2+|\xi|^{\frac{2}{3}}\right)^{m+1}}|dp|\right|\right)^{\frac{1}{2}}$$

$$\times\left(\left|\int_{-i\infty}^{i\infty}\frac{1}{\left(\left|p+\xi_1^{\frac{1}{3}}\right|^2+|\xi_1|^{\frac{2}{3}}\right)^{m+1}}|dp|\right|\right)^{\frac{1}{2}}$$

$$\leq C|\xi|^{-\frac{m+1}{3}}|\xi_1|^{-\frac{m+1}{3}}.$$

Since $e^{\xi t} = \frac{1}{1+\xi t}\partial_\xi \xi e^{\xi t}$ we integrate by parts and use the estimate for $\delta \in [0,1]$

$$\sum_{j=1}^{2} \left| e^{-\phi_j(\xi)y} - \sum_{l=0}^{k}(-1)^l \frac{\phi_j^l(\xi)y^l}{l!} \right|$$

$$\leq C\sum_{j=1}^{2} |\phi_j(\xi)y|^{k+\delta} \leq Cy^{k+\delta} |\xi|^{\frac{k+\delta}{3}}$$

to get for $t > 1$

$$\sum_{j=1}^{2} \int_{-i\infty}^{i\infty} e^{\xi t} \frac{\phi_3^{n-1}(\xi)}{(p-\phi_3(\xi))^{m+1}}$$

$$\times \left(e^{-\phi_j(\xi)y} - \sum_{l=0}^{k}(-1)^l \frac{\phi_j^l(\xi)y^l}{l!} \right) \phi_j^{-1}(\xi)d\xi$$

$$= \int_{-i\infty}^{i\infty} \frac{1}{1+\xi t} \left(\xi M_\xi(\xi,p) + \frac{M(\xi,p)\xi}{1+\xi t} \right) d\xi,$$

where for $k = 0,1$

$$M(\xi,p)$$

$$= \phi_3^{n-1}(\xi)\frac{1}{(p-\phi_3(\xi))^{m+1}}$$

$$\times \sum_{j=1}^{2} \left(e^{-\phi_j(\xi)y} - \sum_{l=0}^{k}(-1)^l \frac{\phi_j^l(\xi)y^l}{l!} \right) \phi_j^{-1}(\xi)$$

$$= \frac{y^{k+\delta}}{(p-\phi_3(\xi))^{m+1}}O(|\xi|^{\frac{n-2+k+\delta}{3}}),$$

$$\xi M_\xi(\xi,p)$$

$$= \xi\phi_3^{n-4}(\xi)\frac{1}{(p-\phi_3(\xi))^{m+1}}\sum_{j=1}^{2}\left(e^{-\phi_j(\xi)y} - \sum_{l=0}^{k}(-1)^l \frac{\phi_j^l(\xi)y^l}{l!} \right)\phi_j^{-1}(\xi)$$

$$- \frac{\xi y\phi_3^{n-1}(\xi)}{3}\frac{1}{(p-\phi_3(\xi))^{m+1}}\sum_{j=1}^{2}\left(e^{-\phi_j(\xi)y} - \sum_{l=0}^{k-1}(-1)^l \frac{\phi_j^l(\xi)y^l}{l!} \right)\phi_j^{-3}(\xi)$$

$$+ \frac{\xi\phi_3^{n-1}(\xi)}{3}\frac{1}{(p-\phi_3(\xi))^{m+1}}\sum_{j=1}^{2}\left(e^{-\phi_j(\xi)y} - \sum_{l=0}^{k}(-1)^l \frac{\phi_j^l(\xi)y^l}{l!} \right)\phi_j^{-4}(\xi)$$

$$+ \xi\frac{\phi_3^{n-3}(\xi)}{(p-\phi_3(\xi))^{m+2}}\sum_{j=1}^{2}\left(e^{-\phi_j(\xi)y} - \sum_{l=0}^{k}(-1)^l \frac{\phi_j^l(\xi)y^l}{l!} \right)\phi_j^{-1}(\xi)$$

$$= \frac{y^{k+\delta}}{(p-\phi_3(\xi))^{m+1}}O(|\xi|^{\frac{n-2+k+\delta}{3}}).$$

Note that when we integrate by parts the substitution in $\xi = 0$ vanishes for $\operatorname{Re} p = 0$ since the integrand is estimated as

$$
\left| \frac{\xi e^{\xi t}}{1 + \xi t} \frac{\phi_3^{n-1}(\xi)}{(p - \phi_3(\xi))^{m+1}} \sum_{j=1}^{2} \left(e^{-\phi_j(\xi) y} - \sum_{l=0}^{k} (-1)^l \frac{\phi_j^l(\xi) y^l}{l!} \right) \phi_j^{-1}(\xi) \right|
$$

$$
\leq C |\xi|^{\frac{n+1}{3}} .
$$

Thus we get

$$
\left\| (\cdot)^m \partial_x^{(n)} \mathcal{F}_k(\cdot, y, t) \right\|_{\mathbf{L}^2}
$$

$$
= \left(\int_{-i\infty}^{i\infty} |dp| \left| \int_{-i\infty}^{i\infty} e^{\xi t} \frac{\phi_3^{n-1}(\xi)}{(p - \phi_3(\xi))^{m+1}} \right. \right.
$$

$$
\times \sum_{j=1}^{2} \left(e^{-\phi_j(\xi) y} - \sum_{l=0}^{k} (-1)^l \frac{\phi_j^l(\xi) y^l}{l!} \right) \phi_j^{-1}(\xi) d\xi \Bigg|^2 \Bigg)^{\frac{1}{2}}
$$

$$
\left(\int_{-i\infty}^{i\infty} |dp| \int_{-i\infty}^{i\infty} \frac{1}{1 + \xi t} \left(\xi M_\xi(\xi, p) + \frac{M(\xi, p) \xi t}{1 + \xi t} \right) d\xi \right.
$$

$$
\times \int_{-i\infty}^{i\infty} \frac{1}{1 + \xi_1 t} \left(\xi_1 M_{\xi_1}(\xi_1, p) + \frac{M(\xi_1, p) \xi_1 t}{1 + \xi_1 t} \right) d\xi_1 \Bigg)^{\frac{1}{2}}
$$

$$
= y^{1+k} \left(\int_{-i\infty}^{i\infty} \frac{1}{1 + \xi t} \left(O(|\xi|^{\frac{n-2+k+\delta}{3}}) + \frac{O(|\xi|^{\frac{n-2+k+\delta}{3}}) \xi t}{1 + \xi t} \right) d\xi \right.
$$

$$
\overline{\int_{-i\infty}^{i\infty} \frac{1}{1 + \xi_1 t} \left(O(|\xi_1|^{\frac{n-2+k+\delta}{3}}) + \frac{O(|\xi_1|^{\frac{n-2+k+\delta}{3}}) \xi_1 t}{1 + \xi_1 t} \right) d\xi_1}
$$

$$
\times \int_{-i\infty}^{i\infty} \frac{1}{(p - \phi_3(\xi))^{m+1} (p - \phi_3(\xi_1))^{m+1}} |dp| \Bigg)^{\frac{1}{2}}
$$

$$
\leq C y^{k+\delta} \int_{-i\infty}^{i\infty} \frac{1}{1 + |\xi| t} |\xi|^{\frac{n-2+k+\delta}{3} - \frac{m+1}{6}} |d\xi| .
$$

Therefore we have for an integer $m \geq 0$

(3.4) $\left\| (\cdot)^m \partial_x^{(n)} \mathcal{F}_k(\cdot, y, t) \right\|_{\mathbf{L}^2} \leq C y^{k+\delta} t^{-\frac{1-2m+2n+2k+2\delta}{6}} .$

For any fractional $m \geq 0$ estimate (3.4) is obtained by interpolation. Lemma 30 is proved.   □

We introduce the following operator

(3.5) $\Phi(x, t) \psi(\tau) = \int_0^{+\infty} F(x, y, t) \psi(y, \tau) dy.$

The function $F(x, y, t)$ is the Green function for lineal problem (2.1)

$$F(x, y, t) = \frac{1}{2\pi i} \int_{-i\infty}^{i\infty} e^{-p^3 t + p(x-y)} dp$$

$$- \frac{1}{2\pi i} \int_{-i\infty}^{i\infty} e^{\xi t + \phi_3(\xi)x} \phi_3^{-1}(\xi) \sum_{j=1}^{2} e^{-\phi_j(\xi)y} \phi_j(\xi) \phi_j'(\xi) d\xi.$$

LEMMA 31. *The following estimates are valid for $m \geq 0, t > 0$*

$$\left\| (\cdot)^{\frac{m}{2}} \partial_x^{(n)} \Phi(t) f \right\|_{L^2}$$

$$\leq C \begin{cases} t^{-\frac{4+2n}{6}} \|f\|_{H_1^{0,\frac{m+3}{2}}} + t^{-\frac{3+2n-m}{6}} \|f\|_{H_1^{0,1}} + t^{-\frac{3+3n}{6}} \|f\|_{H_1^{0,\frac{m+2+n}{2}}}, \\ t^{-\frac{n+1}{3}} \|f\|_{H_1^{0,\frac{m+1}{2}}} + t^{-\frac{1+2n-m}{6}} \|f\|_{L^1} \\ +t^{-\frac{3n+2}{6}+\frac{1}{12}} \|f\|_{H_1^{0,\frac{m+2+n}{2}-\frac{3}{4}}} + t^{-\frac{3+2n-m}{6}} \|f\|_{H_1^{0,1}} \end{cases}$$

if $n = 1, 2$ and

$$\left\| x^{\frac{m}{2}} \Phi(t) f \right\|_{L^2}$$

$$\leq C \begin{cases} t^{-\frac{2}{3}} \|f\|_{H_1^{0,\frac{m+3}{2}}} + t^{-\frac{3-m}{6}} \|f\|_{H_1^{0,1}} + t^{-\frac{3}{4}} \|f\|_{H_1^{0,\frac{m+2}{2}+\frac{3}{4}}}, \\ t^{-\frac{1}{3}} \|f\|_{H_1^{0,\frac{m+1}{2}}} + t^{-\frac{1-m}{6}} \|f\|_{L^1} \\ +t^{-\frac{1}{2}} \|f\|_{H_1^{0,\frac{m+2}{2}}} + t^{-\frac{3-m}{6}} \|f\|_{H_1^{0,1}} \end{cases}$$

provided that the right-hand sides are finite.

PROOF. Using the identity

$$\sum_{j=1}^{3} \phi_j'(\xi) \phi_j(\xi) = 0,$$

we obtain the following representation for the Green function $F(x, y, t)$

$$F(x, y, t) = \frac{1}{2\pi i} \int_{-i\infty}^{i\infty} e^{-p^3 t + px} (e^{-py} - 1) dp$$

$$- \frac{1}{2\pi i} \int_{-i\infty}^{i\infty} e^{\xi t + \phi_3(\xi)x} \phi_3^{-1}(\xi) \sum_{j=1}^{2} \left(e^{-\phi_j(\xi)y} - 1 \right) \phi_j(\xi) \phi_j'(\xi) d\xi$$

$$= \mathcal{G}_0(x, y, t) + \mathcal{F}_0(x, y, t).$$

The functions $\mathcal{G}_0(x, y, t)$, $\mathcal{F}_0(x, y, t)$ were defined in (3.3) and (3.2) with $k = 0$. We obtain the following representation

$$\Phi(x, t) f(x) = \int_0^{+\infty} F(x, y, t) f(y) dy$$

$$= \int_0^{+\infty} \mathcal{G}_0(x, y, t) f(y) dy + \int_0^{+\infty} \mathcal{F}_0(x, y, t) f(y) dy.$$

From Lemma 28 we have for $t > 0$

$$\left\| (\cdot)^{\frac{m}{2}} \partial_x^{(n)} \mathcal{G}_0(\cdot, y, t) \right\|_{\mathbf{L}^2}$$

$$\leq C \begin{cases} t^{-\frac{2n+1-m}{6}} \langle t \rangle^{-\frac{1}{3}} \langle y \rangle + t^{-\frac{n+1}{3}} \langle t \rangle^{-\frac{1}{3}} \langle y \rangle^{\frac{m+3}{2}} \\ \quad + t^{-\frac{3n+2}{6}+\frac{1}{12}} \langle t \rangle^{-\frac{1}{4}} \langle y \rangle^{\frac{m+2+n}{2}} , \\ t^{-\frac{2n+1-m}{6}} + t^{-\frac{n+1}{3}} \langle y \rangle^{\frac{m+1}{2}} + t^{-\frac{3n+2}{6}+\frac{1}{12}} \langle y \rangle^{\frac{m+2+n}{2}-\frac{3}{4}} \end{cases}$$

if $n = 1, 2$ and

$$\left\| (\cdot)^{\frac{m}{2}} \mathcal{G}_0(\cdot, y, t) \right\|_{\mathbf{L}^2}$$

$$\leq C \begin{cases} t^{-\frac{1-m}{6}} \langle t \rangle^{-\frac{1}{3}} \langle y \rangle + t^{-\frac{1}{3}} \langle t \rangle^{-\frac{1}{3}} \langle y \rangle^{\frac{m+3}{2}} \\ \quad + t^{-\frac{1}{2}} \langle t \rangle^{-\frac{1}{4}} \langle y \rangle^{\frac{m+2}{2}+\frac{3}{4}} , \\ t^{-\frac{1-m}{6}} + t^{-\frac{1}{3}} \langle y \rangle^{\frac{m+1}{2}} + t^{-\frac{1}{2}} \langle y \rangle^{\frac{m+2}{2}} . \end{cases}$$

From Lemma 30 we have for $t > 1$, $n \geq 0$, $k = 0$, $\delta = 1$

$$\left\| (\cdot)^{\frac{m}{2}} \partial_x^{(n)} \mathcal{F}_0(\cdot, y, t) \right\|_{\mathbf{L}^2} \leq C t^{-\frac{3+2n-m}{6}} \langle y \rangle .$$

Therefore

$$\left\| (\cdot)^{\frac{m}{2}} \partial_x^{(n)} \Phi(t) f \right\|_{\mathbf{L}^2}$$

$$\leq C \begin{cases} t^{-\frac{4+2n}{6}} \|f\|_{\mathbf{H}_1^{0,\frac{m+3}{2}}} + t^{-\frac{3+2n-m}{6}} \|f\|_{\mathbf{H}_1^{0,1}} + t^{-\frac{3+3n}{6}} \|f\|_{\mathbf{H}_1^{0,\frac{m+2+n}{2}}} , \\ t^{-\frac{n+1}{3}} \|f\|_{\mathbf{H}_1^{0,\frac{m+1}{2}}} + t^{-\frac{1+2n-m}{6}} \|f\|_{\mathbf{L}^1} \\ \quad + t^{-\frac{3n+2}{6}+\frac{1}{12}} \|f\|_{\mathbf{H}_1^{0,\frac{m+2+n}{2}-\frac{3}{4}}} + t^{-\frac{3+2n-m}{6}} \|f\|_{\mathbf{H}_1^{0,1}} \end{cases}$$

for $n = 1, 2$ and

$$\left\| (\cdot)^{\frac{m}{2}} \Phi(t) f \right\|_{\mathbf{L}^2}$$

$$\leq C \begin{cases} t^{-\frac{2}{3}} \|f\|_{\mathbf{H}_1^{0,\frac{m+3}{2}}} + t^{-\frac{3-m}{6}} \|f\|_{\mathbf{H}_1^{0,1}} + t^{-\frac{3}{4}} \|f\|_{\mathbf{H}_1^{0,\frac{m+2}{2}+\frac{3}{4}}} , \\ t^{-\frac{1}{3}} \|f\|_{\mathbf{H}_1^{0,\frac{m+1}{2}}} + t^{-\frac{1-m}{6}} \|f\|_{\mathbf{L}^1} \\ \quad + t^{-\frac{1}{2}} \|f\|_{\mathbf{H}_1^{0,\frac{m+2}{2}}} + t^{-\frac{3-m}{6}} \|f\|_{\mathbf{H}_1^{0,1}} . \end{cases}$$

Lemma 31 is proved. □

Using identities

$$\sum_{j=1}^{3} \phi_j(\xi) = 0, \quad \sum_{j=1}^{3} \phi_j^2(\xi) = 0$$

which imply

$$\sum_{j=1}^{3} \phi_j'(\xi) = 0, \quad \sum_{j=1}^{3} \phi_j'(\xi) \phi_j(\xi) = 0 \text{ and } \sum_{j=1}^{3} \phi_j^2(\xi) \phi_j'(\xi) = -1,$$

we obtain the following representation for the function $F(x, y, t)$

$$F(x, y, t) = \frac{3y}{2\pi i} \int_{-i\infty}^{i\infty} e^{-p^3 t + px} p\, dp + G(x, y, t)$$

$$(3.6) \qquad = 3yt^{-\frac{2}{3}} Ai'\left(\frac{x}{\sqrt[3]{t}}\right) + G(x, y, t),$$

where $G(x, y, t) = \mathcal{G}_1(x, y, t) + \mathcal{F}_1(x, y, t)$. The functions $\mathcal{G}_1(x, y, t), \mathcal{F}_1(x, y, t)$ were defined in (3.2) and (3.3) with $k = 1$.

LEMMA 32. *Let the function $f(x, t)$ such that $\int_0^{+\infty} x f(x, t)dx = 0$ and satisfy the estimate*

$$\left\|(\cdot)^{\frac{l}{2}} f(\cdot, t)\right\|_{L^1} \leq C\varepsilon^\beta (1+t)^{-\frac{8-l}{6}} g(t)^\alpha, \, 0 \leq l \leq 5,$$

where $\alpha = \frac{4}{5}, 0 < \beta$. We also assume that the function $g(t)$ satisfies the inequalities for $\eta > 0$

$$\frac{1}{2}(1 + \eta \log(1+t)) < g(t) < 2(1 + \eta \log(1+t)).$$

Then the following estimates are valid for $n = 0, 1, t > 0$

$$\left\|(\cdot)^{\frac{m}{2}} \int_0^t g^{-1}(\tau) \partial_x^{(n)} \Phi(t-\tau) f(\tau) d\tau\right\|_{L^2} \leq C\varepsilon^\beta g^{-1+\alpha}(t) t^{-\frac{3-m+2n}{6}},$$

where $0 \leq m \leq 2$.

PROOF. Since

$$\int_0^{+\infty} x f(x, t)dx = 0,$$

using (3.6) we obtain the following representation

$$\Phi(x, t-\tau)f(\tau) = \int_0^{+\infty} F(x, y, t-\tau)f(y, \tau)dy$$

$$= \int_0^{+\infty} G(x, y, t-\tau)f(y, \tau)dy.$$

From Lemma 29 we have for $t > 0$, $m \geq 0$

$$(3.7) \qquad \left\|(\cdot)^{\frac{m}{2}} \partial_x \mathcal{G}_1(\cdot, y, t)\right\|_{L^2}$$

$$\leq C \begin{cases} t^{-\frac{2}{3}} \langle y\rangle^{\frac{m+1}{2}} + t^{-\frac{3-m}{6}} + Ct^{-\frac{5}{6}+\frac{1}{12}} \langle y\rangle^{\frac{m+2}{2}-\frac{1}{4}} \\ t^{-\frac{2}{3}} \langle t\rangle^{-\frac{1}{3}} \langle y\rangle^{\frac{m+3}{2}} + t^{-\frac{3-m}{6}} \langle t\rangle^{-\frac{2}{3}} \langle y\rangle^2 \\ + t^{-\frac{5}{6}+\frac{1}{12}} \langle t\rangle^{-\frac{1}{4}} \langle y\rangle^{\frac{m+3}{2}} \end{cases}$$

and

$$(3.8) \qquad \left\|(\cdot)^{\frac{m}{2}} \mathcal{G}_1(\cdot, y, t)\right\|_{L^2} \leq Ct^{-\frac{2}{3}} \langle y\rangle^{\frac{m+3}{2}} + Ct^{-\frac{5-m}{6}} \langle y\rangle^2.$$

From Lemma 30 we have $k = 1, \delta \in [0, 1]$

$$(3.9) \qquad \left\|(\cdot)^{\frac{m}{2}} \partial_x^{(n)} \mathcal{F}_1(\cdot, y, t)\right\|_{L^2} \leq Ct^{-\frac{3-m+2n+2\delta}{6}} \langle y\rangle^{1+\delta}.$$

Therefore

$$\left\|(\cdot)^{\frac{m}{2}}\Phi(t-\tau)f(\tau)\right\|_{\mathbf{L}^2} = \left\|\int_0^{+\infty}(\cdot)^{\frac{m}{2}}G(\cdot,y,t-\tau)f(y,\tau)dy\right\|_{\mathbf{L}^2}$$

$$(3.10)\quad \leq\ C(t-\tau)^{-\frac{2}{3}}\|f(\tau)\|_{\mathbf{H}_1^{0,\frac{m+3}{2}}}+C(t-\tau)^{-\frac{5-m}{6}}\|f(\tau)\|_{\mathbf{H}_1^{0,2}}\,.$$

Furthermore we have

$$\left\|(\cdot)^{\frac{m}{2}}\partial_x\Phi(t-\tau)f(\tau)\right\|_{\mathbf{L}^2} = \left\|\int_0^{+\infty}(\cdot)^{\frac{m}{2}}\partial_x G(\cdot,y,t-\tau)f(y,\tau)dy\right\|_{\mathbf{L}^2}$$

$$\leq\ C(t-\tau)^{-1}\|f(\tau)\|_{\mathbf{H}_1^{0,\frac{m+3}{2}}}+C(t-\tau)^{-\frac{7-m}{6}}\|f(\tau)\|_{\mathbf{H}_1^{0,2}}\,,$$

$$\left\|(\cdot)^{\frac{m}{2}}\partial_x\Phi(t-\tau)f(\tau)\right\|_{\mathbf{L}^2} = \left\|\int_0^{+\infty}(\cdot)^{\frac{m}{2}}\partial_x G(\cdot,y,t-\tau)f(y,\tau)dy\right\|$$

$$\leq\ C(t-\tau)^{-\frac{2}{3}}\|f(\tau)\|_{\mathbf{H}_1^{0,\frac{m+1}{2}}}+C(t-\tau)^{-\frac{3-m}{6}}\|f(\tau)\|_{\mathbf{L}^1}$$

$$(3.11)\qquad +C\,(t-\tau)^{-\frac{3}{4}}\|f(\tau)\|_{\mathbf{H}_1^{0,\frac{m}{2}+\frac{3}{4}}}+C\,(t-\tau)^{-\frac{5-m}{6}}\|f(\tau)\|_{\mathbf{H}_1^{0,1}}\,.$$

Also from the condition of the lemma for the function $g(t)$ we have for $t>1$

$$t^{-\alpha}<\frac{1}{g(t)},\ \alpha>0$$

and

$$\sup_{\tau\in[\sqrt{t},t]}\frac{1}{g(\tau)}<\frac{C}{g(t)}.$$

Hence we get for $t>4$

$$\left\|\int_0^t(\cdot)^{\frac{m}{2}}g^{-1}(\tau)\partial_x\Phi(t-\tau)f(\tau)d\tau\right\|_{\mathbf{L}^2}$$

$$\leq\ C\int_0^{\sqrt{t}}t^{-1}\|f(\tau)\|_{\mathbf{H}_1^{0,\frac{m+3}{2}}}+t^{-\frac{7-m}{6}}\|f(\tau)\|_{\mathbf{H}_1^{0,2}}\,d\tau$$

$$+Cg^{-1}(t)\int_{\sqrt{t}}^{\frac{t}{2}}t^{-1}\|f(\tau)\|_{\mathbf{H}_1^{0,\frac{m+3}{2}}}+t^{-\frac{7-m}{6}}\|f(\tau)\|_{\mathbf{H}_1^{0,2}}\,d\tau$$

$$+C\sup_{\frac{t}{2}<\tau<t}g^{-1}(\tau)\|f(\tau)\|_{\mathbf{H}_1^{0,\frac{m+1}{2}}}\int_{\frac{t}{2}}^t(t-\tau)^{-\frac{2}{3}}d\tau$$

$$+C\sup_{\frac{t}{2}<\tau<t}g^{-1}(\tau)\|f(\tau)\|_{\mathbf{H}_1^{0,\frac{m}{2}+\frac{3}{4}}}\int_{\frac{t}{2}}^t(t-\tau)^{-\frac{3}{4}}d\tau$$

$$+C\sup_{\frac{t}{2}<\tau<t}g^{-1}(\tau)\|f(\tau)\|_{\mathbf{L}^1}\int_{\frac{t}{2}}^t(t-\tau)^{-\frac{3-m}{6}}d\tau$$

and

$$\left\| \int_0^t (\cdot)^{\frac{m}{2}} g^{-1}(\tau) \Phi(t-\tau) f(\tau) d\tau \right\|_{\mathbf{L}^2}$$

$$\leq C \int_0^{\sqrt{t}} t^{-\frac{2}{3}} \|f(\tau)\|_{\mathbf{H}_1^{0,\frac{m+3}{2}}} + t^{-\frac{5-m}{6}} \|f(\tau)\|_{\mathbf{H}_1^{0,2}} d\tau$$

$$+ Cg^{-1}(t) \int_{\sqrt{t}}^{\frac{t}{2}} t^{-\frac{2}{3}} \|f(\tau)\|_{\mathbf{H}_1^{0,\frac{m+3}{2}}} + t^{-\frac{5-m}{6}} \|f(\tau)\|_{\mathbf{H}_1^{0,2}} d\tau$$

$$+ C \sup_{\frac{t}{2} < \tau < t} g^{-1}(\tau) \|f(\tau)\|_{\mathbf{H}_1^{0,\frac{m+3}{2}}} \int_{\frac{t}{2}}^t (t-\tau)^{-\frac{2}{3}} d\tau$$

$$+ C \sup_{\frac{t}{2} < \tau < t} g^{-1}(\tau) \|f(\tau)\|_{\mathbf{H}_1^{0,2}} \int_{\frac{t}{2}}^t (t-\tau)^{-\frac{5-m}{6}} d\tau.$$

By the previous assumptions we have

$$\left\| x^{\frac{m}{2}} f(t) \right\|_{\mathbf{L}^1} \leq C\varepsilon^\beta (1+t)^{-\frac{5}{6}} t^{\frac{m-3}{6}} g(t)^\alpha, \ 0 \leq m \leq 5;$$

hence

$$\left\| \int_0^t (\cdot)^{\frac{m}{2}} g^{-1}(\tau) \partial_x \Phi(t-\tau) f(\tau) d\tau \right\|_{\mathbf{L}^2}$$

$$\leq Cg^\alpha(t) \int_0^{\sqrt{t}} t^{-1} (1+\tau)^{-\frac{5}{6}} \tau^{\frac{m}{6}} + t^{-\frac{7-m}{6}} (1+\tau)^{-\frac{5}{6}} \tau^{\frac{1}{6}} d\tau$$

$$+ Cg^{-1+\alpha}(t) \int_{\sqrt{t}}^{\frac{t}{2}} t^{-1} (1+\tau)^{-\frac{5}{6}} \tau^{\frac{m}{6}} + t^{-\frac{7-m}{6}} (1+\tau)^{-\frac{5}{6}} \tau^{\frac{1}{6}} d\tau$$

$$+ Ct^{\frac{1}{3}} \sup_{\frac{t}{2} < \tau < t} g^{-1}(\tau) \|f(\tau)\|_{\mathbf{H}_1^{0,\frac{m+1}{2}}}$$

$$+ Ct^{\frac{1}{4}} \sup_{\frac{t}{2} < \tau < t} g^{-1}(\tau) \|f(\tau)\|_{\mathbf{H}_1^{0,\frac{m+1}{2}+\frac{1}{4}}}$$

$$+ Ct^{\frac{3+m}{6}} \sup_{\frac{t}{2} < \tau < t} g^{-1}(\tau) \|f(\tau)\|_{\mathbf{L}^1}$$

$$\leq C\varepsilon^\beta g^{-1+\alpha}(t) t^{-\frac{5-m}{6}}$$

and

$$\left\| (\cdot)^{\frac{m}{2}} \int_0^t g^{-1}(\tau)\Phi(t-\tau)f(\tau)d\tau \right\|_{\mathbf{L}^2}$$

$$\leq C \int_0^{\sqrt{t}} t^{-\frac{2}{3}} (1+\tau)^{-\frac{5}{6}} \tau^{\frac{m}{6}} + t^{-\frac{5-m}{6}} (1+\tau)^{-\frac{5}{6}} \tau^{\frac{1}{6}} d\tau$$

$$+C \int_{\sqrt{t}}^{\frac{t}{2}} \left(t^{-\frac{2}{3}} g^{-1+\alpha}(t)(1+\tau)^{-\frac{5}{6}} \tau^{\frac{m}{6}} \right.$$

$$+ t^{-\frac{5-m}{6}} g^{-1+\alpha}(t)(1+\tau)^{-\frac{5}{6}} \tau^{\frac{1}{6}} \Big) d\tau$$

$$+C t^{\frac{1}{3}} \sup_{\frac{t}{2}<\tau<t} g^{-1}(\tau) \|f(\tau)\|_{\mathbf{H}_1^{0,\frac{m+3}{2}}}$$

$$+C t^{\frac{m+1}{6}} \sup_{\frac{t}{2}<\tau<t} g^{-1}(\tau) \|f(\tau)\|_{\mathbf{H}_1^{0,2}}$$

$$\leq \beta^\beta C g^{-1+\alpha}(t) t^{-\frac{3-m}{6}}.$$

Lemma 32 is proved.   □

LEMMA 33. *We assume that $u_0 \in \mathbf{H}_1^{0,\frac{m+3}{2}}(\mathbf{R}^+)$ and $\|u_0\|_{\mathbf{H}_1^{0,\frac{m+3}{2}}} = \varepsilon$ is small enough and*

$$\lambda \int_0^{+\infty} x u_0(x)dx = \lambda\theta < 0;$$

in addition a function $v(x,t)$ satisfies the estimate for all $t > 0$

$$\|xvv_x\|_{\mathbf{L}^1} \leq C\varepsilon^\beta (1+t)^{-1}$$

and has the asymptotic representation for all $t > 0$, $m \geq 0$

$$\left\| \partial_x^{(n)} (v(t) - \Phi(t)u_0) \right\|_{\mathbf{H}_2^{0,\frac{m}{2}}} \leq C\varepsilon^\beta g^{-1+\alpha}(t)(1+t)^{-\frac{3+2n-m}{6}},$$

where $1 < \beta < 2$. We also assume that the function $g(t)$ is such that for all $t > 0$

$$\frac{1}{2}(1+\eta\log(1+t)) < g(t) < 2(1+\eta\log(1+t)),$$

where

$$0 < \eta = -9\theta\lambda \int_0^{+\infty} Ai'^2(z)dz \leq C\varepsilon.$$

Then the following inequality is valid for all $t > 0$

$$\frac{1}{2}(1+\eta\log(1+t)) < 1 + \frac{\lambda}{\theta}\int_0^t d\tau \int_0^{+\infty} xvv_x dx$$

$$(3.12) \qquad\qquad\qquad < 2(1+\eta\log(1+t)).$$

PROOF. For $t < 1$ the estimate (3.12) is trivial. Using (3.6) we have

$$\Phi(t)u_0 = 3\theta t^{-\frac{2}{3}} Ai'\left(\frac{x}{\sqrt[3]{t}}\right) + R(x,t),$$

where

$$R(x,t) = \int_0^{+\infty} u_0(y)\left(\mathcal{G}_1(x,y,t) + \mathcal{F}_1(x,y,t)\right)dy.$$

We again use the estimates (3.7), (3.8) and (3.9) to find that for $t > 1$ and $n = 0,1$

$$\left\| x^{\frac{m}{2}}\partial_x^{(n)}R(t)\right\|_{\mathbf{L}^2} \leq C\left(t^{-\frac{5-m+2n}{6}}\|u_0\|_{\mathbf{H}_1^{0,2}} + t^{-\frac{4+2n}{6}}\|u_0\|_{\mathbf{H}_1^{0,\frac{m+3}{2}}}\right)$$

$$\leq C\varepsilon t^{-\frac{3-m+2n}{6}}\max\left\{t^{-\frac{1}{3}}, t^{-\frac{m+1}{6}}\right\}.$$

We have

$$\left\| Ai^{(n)}\left(\frac{\cdot}{\sqrt[3]{t}}\right)\right\|_{\mathbf{H}_2^{0,\frac{1}{2}}}^2 = \int_0^{+\infty} x\left| Ai^{(n)}\left(\frac{x}{\sqrt[3]{t}}\right)\right|^2 dx \leq Ct^{\frac{2}{3}}.$$

Therefore taking

$$r = v - \Phi(t)u_0 = v - \frac{3\theta}{2\pi}t^{-\frac{2}{3}} Ai'\left(\frac{x}{\sqrt[3]{t}}\right) - R(x,t)$$

by a direct computation we have

$$\left\|(\cdot)\left(vv_x - \frac{\theta^2}{4\pi^2}t^{-\frac{5}{3}} Ai'\left(\frac{\cdot}{\sqrt[3]{t}}\right) Ai''\left(\frac{\cdot}{\sqrt[3]{t}}\right)\right)\right\|_{\mathbf{H}_2^{0,1}}$$

$$= \left\|(\cdot)\left(\frac{\theta}{2\pi}t^{-\frac{2}{3}} Ai'\left(\frac{\cdot}{\sqrt[3]{t}}\right)(R_x(\cdot,t) + r_x(\cdot,t))\right.\right.$$

$$\left.\left. + \frac{\theta}{2\pi}t^{-1} Ai''\left(\frac{\cdot}{\sqrt[3]{t}}\right)(R+r) + RR_x(\cdot,t) + rr_x + Rr_x + R_xr\right)\right\|_{\mathbf{H}_2^{0,1}}$$

$$\leq C\theta t^{-\frac{1}{3}}\left(\|R_x(t)\|_{\mathbf{H}_2^{0,\frac{1}{2}}} + \|r_x(t)\|_{\mathbf{H}_2^{0,\frac{1}{2}}}\right)$$

$$+ C\theta t^{-\frac{2}{3}}\left(\|R(t)\|_{\mathbf{H}_2^{0,\frac{1}{2}}} + \|r(t)\|_{\mathbf{H}_2^{0,\frac{1}{2}}}\right)$$

$$+ \left(\|R(t)\|_{\mathbf{H}_2^{0,\frac{1}{2}}} + \|r(t)\|_{\mathbf{H}_2^{0,\frac{1}{2}}}\right)\left(\|r_x(t)\|_{\mathbf{H}_2^{0,\frac{1}{2}}} + \|R_x(t)\|_{\mathbf{H}_2^{0,\frac{1}{2}}}\right)$$

$$\leq C\varepsilon^2 t^{-\frac{4}{3}} + C\varepsilon^{1+\beta}t^{-1}g^{-1+\alpha}(t) + C\varepsilon^2 t^{-\frac{4}{3}} + C\varepsilon^{1+\beta}t^{-1}g^{-1+\alpha}(t)$$

$$+ C\varepsilon^{1+\beta}t^{-\frac{4}{3}}g^{-1+\alpha}(t) + C\varepsilon^{1+\beta}t^{-1}g^{-2+2\alpha}(t)$$

$$+ C\varepsilon^2 t^{-\frac{5}{3}} + C\varepsilon^{1+\beta}t^{-\frac{4}{3}}g^{-1+\alpha}(t)$$

$$\leq C\varepsilon^2 t^{-\frac{4}{3}} + C\varepsilon^{1+\beta}t^{-1}g^{-1+\alpha}(t)$$

and

$$9\theta^2 t^{-\frac{5}{3}} \int_0^{+\infty} x Ai'\left(\frac{x}{\sqrt[3]{t}}\right) Ai''\left(\frac{x}{\sqrt[3]{t}}\right) dx$$
$$= -9\theta^2 t^{-1} \int_0^{+\infty} Ai'^2(z)dz.$$

Therefore by virtue of the estimate

$$g^{-1}(t) < C(1 + \eta \log(1 + t))^{-1}$$

we get for $t > 1$

$$\left| \frac{\lambda}{\theta} \int_0^{+\infty} x v v_x dx + 9\theta \lambda (1 + t)^{-1} \int_0^{+\infty} Ai'^2(z)dz \right|$$
$$\leq C\varepsilon t^{-\frac{3+\delta}{3}} + C\varepsilon^2 t^{-1}(1 + \eta \log(1 + t))^{-1+\alpha}.$$

Thus integrating with respect to time we obtain

$$\left| \frac{\lambda}{\theta} \int_1^t d\tau \int_0^{+\infty} x v v_x dx - \eta \log(1 + t) \right|$$
$$\leq C\varepsilon \int_1^t \tau^{-\frac{3+\delta}{3}} d\tau + C\varepsilon^\beta \int_1^t \tau^{-1}(1 + \eta \log \tau)^{-1+\alpha} d\tau)$$
$$\leq C\varepsilon + C\varepsilon^\beta \eta^{-1}(1 + \eta \log(2 + t))^\alpha,$$

where

$$\eta = -9\theta\lambda \int_0^{+\infty} Ai'^2(z)dz.$$

Therefore we get

$$1 + \eta \log(1 + t) - C\varepsilon^\beta \eta^{-1}(1 + \eta \log t)^\alpha$$
$$\leq 1 + \frac{1}{\theta} \int_1^t d\tau \int_0^{+\infty} x v v_x dx$$
$$\leq 1 + \eta \log(1 + t) + C\varepsilon^\beta \eta^{-1}(1 + \eta \log t)^\alpha.$$

Since $C\varepsilon^\beta \eta^{-1} < \frac{1}{2}$, Lemma 33 is proved. □

4. Local existence

In this section we prove the following theorem.

THEOREM 32. *Let* $u_0(x) \in \mathbf{H}_2^{1,\frac{7}{2}}(\mathbf{R}^+)$. *Then there exist a positive time* T *and a unique solution of the problem* (1.1) *such that*

$$u(t) \in C([0, T]; \mathbf{H}_2^{1,\frac{7}{2}}(\mathbf{R}^+)) \cap L^2\left(0, T; \mathbf{H}_2^{2,3}(\mathbf{R}^+)\right).$$

PROOF. We consider the linearized equation of (1.1) such that

$$(4.1) \quad \begin{cases} u_t + u_{xxx} = -\lambda v u_x, & t > 0, x > 0, \\ u(x,0) = u_0(x), & x > 0, \\ u_x(0,t) = 0, & t > 0. \end{cases}$$

For simplicity, we let $\lambda = 1$. Let

$$\begin{aligned} \mathbf{Z}_T &= \left\{ f \in \mathbf{C}([0,T], \mathbf{H}_2^{1,\frac{7}{2}}(\mathbf{R}^+)); \|f\|_{\mathbf{Z}_T} \right. \\ &= \sup_{t \in [0,T]} \|f(t)\|_{\mathbf{H}_2^{1,\frac{7}{2}}} + \left(\int_0^T \|\partial_x f(t)\|_{\mathbf{H}_2^{1,3}}^2 \, dt \right)^{\frac{1}{2}} < +\infty \right\} \end{aligned}$$

and the closed ball

$$\mathbf{Z}_{T,\rho} = \left\{ f \in \mathbf{Z}_T; \|f\|_{\mathbf{Z}_T} \le \rho \right\}.$$

We define the mapping $\mathbb{M}$ by $u = \mathbb{M}v$, where $v \in \mathbf{Z}_{T,\rho}$. Applying both sides of equation (4.1) by ∂_x, multiplying the resulting equation by $x^{7-3j}\partial_x u$, and integrating over $\mathbf{R}^+$, we get

$$\frac{1}{2} \sum_{j=0}^{2} \frac{d}{dt} \left\| (\cdot)^{\frac{7-3j}{2}} \partial_x u(\cdot, t) \right\|_{\mathbf{L}^2}^2$$

$$- \sum_{j=0}^{1} \frac{(7-3j)(6-3j)(5-3j)}{2} \left\| (\cdot)^{\frac{4-3j}{2}} \partial_x u(\cdot, t) \right\|_{\mathbf{L}^2}^2$$

$$+ \sum_{j=0}^{2} \frac{3(7-3j)}{2} \left\| (\cdot)^{\frac{6-3j}{2}} \partial_x^2 u(\cdot, t) \right\|_{\mathbf{L}^2}^2$$

$$(4.2) \qquad = -\sum_{j=0}^{2} \left(x^{7-3j} \partial_x(v u_x), \partial_x u \right).$$

By the boundary condition

$$(4.3) \qquad \frac{1}{2}\frac{d}{dt} \|\partial_x u(t)\|_{\mathbf{L}^2}^2 + \frac{1}{2}\left(\partial_x^2 u(0,t)\right)^2 = -\left(\partial_x(v u_x), \partial_x u\right).$$

As in the proof of (4.2) through (4.3)

$$\frac{1}{2} \sum_{j=0}^{2} \frac{d}{dt} \left\| (\cdot)^{\frac{7-3j}{2}} u(\cdot, t) \right\|_{\mathbf{L}^2}^2$$

$$- \sum_{j=0}^{1} \frac{(7-3j)(6-3j)(5-3j)}{2} \left\| (\cdot)^{\frac{4-3j}{2}} u(\cdot, t) \right\|_{\mathbf{L}^2}^2$$

$$+ \sum_{j=0}^{2} \frac{3(7-3j)}{2} \left\| (\cdot)^{\frac{6-3j}{2}} \partial_x u(\cdot, t) \right\|_{\mathbf{L}^2}^2$$

$$(4.4) \qquad = \; - \sum_{j=0}^{2} \left(x^{7-3j}(vu_x), u \right)$$

and

$$(4.5) \qquad \frac{1}{2} \frac{d}{dt} \|u(t)\|_{\mathbf{L}^2}^2 - \left(\partial_x^2 u(0,t) \right) (u(0,t)) = - \left((vu_x), u \right).$$

Via (4.2)-(4.5) we find that there exists a positive constant C_1 such that

$$\frac{1}{2} \frac{d}{dt} \|u(t)\|_{\mathbf{H}_2^{1,\frac{7}{2}}}^2 + C_1 \|\partial_x u(t)\|_{\mathbf{H}_2^{1,3}}^2 \leq C \|u(t)\|_{\mathbf{L}^\infty}^2$$

$$+ C \|v(t)\|_{\mathbf{H}_2^{1,\frac{7}{2}}} \|u(t)\|_{\mathbf{H}_2^{1,\frac{7}{2}}} \left(\|\partial_x u(t)\|_{\mathbf{H}_2^{1,3}} + \|u(t)\|_{\mathbf{H}_2^{1,\frac{7}{2}}} \right)$$

from which it follows that

$$\frac{1}{2} \frac{d}{dt} \|u(t)\|_{\mathbf{H}_2^{1,\frac{7}{2}}}^2 + \frac{C_1}{2} \|\partial_x u(t)\|_{\mathbf{H}_2^{1,3}}^2$$

$$\leq \; C \|u(t)\|_{\mathbf{L}^2} \|\partial_x u(t)\|_{\mathbf{L}^2} + C\rho(1+\rho) \|u(t)\|_{\mathbf{H}_2^{1,\frac{7}{2}}}^2$$

$$(4.6) \qquad \leq \; C(1+\rho)^2 \|u(t)\|_{\mathbf{H}_2^{1,\frac{7}{2}}}^2 .$$

Integrating (4.6) with respect to time t, we have

$$\|u(t)\|_{\mathbf{H}_2^{1,\frac{7}{2}}}^2 + C_1 \int_0^t \|\partial_x u(\tau)\|_{\mathbf{H}_2^{1,3}}^2 \, d\tau$$

$$\leq \; C \|u_0\|_{\mathbf{H}_2^{1,\frac{7}{2}}}^2 + C(1+\rho)^2 \int_0^t \|u(\tau)\|_{\mathbf{H}_2^{1,\frac{7}{2}}}^2 \, d\tau \, ;$$

hence

$$\sup_{t \in [0,T]} \|u(t)\|_{\mathbf{H}_2^{1,\frac{7}{2}}} \leq C \|u_0\|_{\mathbf{H}_2^{1,\frac{7}{2}}} + C(1+\rho) T^{\frac{1}{2}} \sup_{t \in [0,T]} \|u(t)\|_{\mathbf{H}_2^{1,\frac{7}{2}}}$$

and

$$\int_0^T \|\partial_x u(t)\|_{\mathbf{H}_2^{1,3}}^2 \, dt \leq \frac{1}{C_1} \|u_0\|_{\mathbf{H}_2^{1,\frac{7}{2}}}^2$$

$$+ C\,(1+\rho)^2\,T\left(\sup_{t\in[0,T]} \|u(t)\|_{\mathbf{H}_2^{1,\frac{7}{2}}}^2\right)^2.$$

If we choose T such that $C\,(1+\rho)\,T^{\frac{1}{2}} \leq \frac{1}{2}$, then

(4.7)
$$\|u\|_{\mathbf{Z}_T} \leq C\,\|u_0\|_{\mathbf{H}_2^{1,\frac{7}{2}}}.$$

We let $u_j = \mathbb{M}v_j$, $j=1,2$. Then as in the proof of (4.7) we find that there exists a T such that

$$\|\mathbb{M}v_1 - \mathbb{M}v_2\|_{\mathbf{Z}_T} \leq \frac{1}{2}\|v_1 - v_2\|_{\mathbf{Z}_T}.$$

Therefore there exists a unique solution the problem (1.1) in the set $Z_{T,\rho}$. Theorem 32 is proved. □

The above theorem says that the solutions have a smoothing property if the data decay rapidly at infinity. Therefore we have the following theorem. We do not give a proof since we do not need the result of a smoothing property to prove asymptotic behavior of solutions.

THEOREM 33. *Let $u_0(x) \in \mathbf{H}_2^{1,\frac{7}{2}}(\mathbf{R}^+)$. Then there exist a positive time T and a unique solution of the problem (1.1) such that*

$$u(t) \in C([0,T]; \mathbf{H}_2^{1,\frac{7}{2}}(\mathbf{R}^+)),$$

and

$$t^{\frac{7-j}{2}}\partial_x^{7-j}u(t) \in \mathbf{C}((0,T]; \mathbf{H}_2^{0,\frac{i}{2}}(\mathbf{R}^+)), \ 0 \leq j \leq 6.$$

5. Proof of Theorem 31

We suppose that for a sufficiently small $\varepsilon > 0$

$$\|u_0\|_{\mathbf{H}_1^{0,\frac{21}{4}}} + \|u_0\|_{\mathbf{H}_2^{1,\frac{7}{2}}} \leq \varepsilon$$

and

$$\lambda \int_0^{+\infty} x u_0(x)\,dx = \lambda\theta < 0.$$

We denote

$$\eta = -9\theta\lambda \int_0^{+\infty} Ai'^2(z)\,dz > 0.$$

We let $u(x,t) = e^{-\phi(t)}v(x,t)$. Then we get from (1.1)

(5.1)
$$\begin{cases} v_t - \phi_t v + \lambda e^{-\phi(t)}vv_x + v_{xxx} = 0, & t > 0, x > 0, \\ v(x,0) = e^{\phi(0)}u_0(x), \ x > 0, \\ v_x(0,t) = 0, \ t > 0. \end{cases}$$

Now we assume that the real-valued function $\phi(t)$ satisfies the following condition

(5.2) $$\int_0^{+\infty} x\left(-\phi_t v + \lambda e^{-\phi(t)} v v_x\right) dx = 0.$$

Then via (5.1) we have for all $t > 0$

$$\frac{d}{dt}\int_0^{+\infty} xv(x,t)dx = -\int_0^{+\infty} xv_{xxx}(x,t)dx = v_x(0,t) = 0.$$

Therefore choosing $\phi(0) = 0$ we get by (5.2)

$$\phi_t e^{\phi(t)} = \theta^{-1}\lambda \int_0^{+\infty} xvv_x dx.$$

Integrating with respect to time we get

$$
\begin{aligned}
e^{\phi(t)} &= g(t) = 1 + \theta^{-1}\lambda \int_0^t d\tau \int_0^{+\infty} xvv_x dx \\
&= 1 - \frac{1}{2}\theta^{-1}\lambda \int_0^t \|v(\tau)\|_{\mathbf{L}^2}^2 d\tau.
\end{aligned}
$$

Thus we get the following problem

(5.3) $$\begin{cases} v_t + \lambda g(t)^{-1}(\frac{v}{2\theta}\|v(t)\|_{\mathbf{L}^2}^2 + vv_x) + v_{xxx} = 0, \quad t > 0, x > 0, \\ g(t) = 1 - \frac{\lambda}{2\theta}\int_0^t \|v(\tau)\|_{\mathbf{L}^2}^2 d\tau, \\ v(x,0) = u_0(x), \ x > 0, \\ v_x(0,t) = 0, \ g(0) = 1, t > 0. \end{cases}$$

We consider the integral equation associated with (5.3) which is written as

$$
\begin{aligned}
v(x,t) &= \Phi(x,t)u_0(x) \\
&\quad + \lambda \int_0^t g(\tau)^{-1}\Phi(x,t-\tau)(\frac{v}{2\theta}\|v(\tau)\|_{\mathbf{L}^2}^2 + vv_x)d\tau.
\end{aligned}
$$

Changing the variables

(5.4) $$v(x,t) = \Phi(x,t)u_0(x) + r(x,t)$$

we get the system of integral equations $(r,g) = (\mathbb{M}_1(r,g), \mathbb{M}_2(r,g))$ for the first approximation of perturbation theory

(5.5) $$\begin{cases} \mathbb{M}_1(r,g) = \lambda \int_0^t g(\tau)^{-1}\Phi(x,t-\tau)(\frac{v}{2\theta}\|v(\tau)\|_{\mathbf{L}^2}^2 + vv_x)d\tau. \\ \mathbb{M}_2(r,g) = 1 - \lambda\frac{1}{2\theta}\int_0^t \|v(\tau)\|_{\mathbf{L}^2}^2 d\tau. \end{cases}$$

We prove that $(\mathbb{M}_1(r,g), \mathbb{M}_2(r,g))$ is a contradiction mapping in the set

$$
\begin{aligned}
\mathbf{X}_\varepsilon = \Big\{ &r \in C([0,+\infty),\mathbf{X}); g \in C(0,+\infty), |||r|||_{\mathbf{X}_1} \leq \varepsilon, |||r|||_{\mathbf{X}_2} \leq \varepsilon^{\frac{3}{4}}, \\
&r_x(0,t) = 0, \frac{1}{2}(1 + \eta \log t) < g(t) < 2(1 + \eta \log t), t > 0\Big\}
\end{aligned}
$$

with the norm

$$|||r|||_{\mathbf{X}} = |||r|||_{\mathbf{X}_1} + |||r|||_{\mathbf{X}_2}$$

$$|||r|||_{\mathbf{X}_1} = \sum_{n=0}^{1}\sum_{m=0}^{2} \sup_{t>0} g(t)^{\frac{1}{5}} (1+t)^{\frac{3+2n-m}{6}} \left\| x^{\frac{m}{2}} \partial_x^{(n)} r(t) \right\|_{\mathbf{L}^2},$$

$$|||r|||_{\mathbf{X}_2} = \sum_{n=0}^{1}\sum_{m=3}^{7} \sup_{t>0} g(t)^{-1} (1+t)^{\frac{3+2n-m}{6}} \left\| x^{\frac{m}{2}} \partial_x^{(n)} (\Phi(t)u_0 + r(t)) \right\|_{\mathbf{L}^2}.$$

We assume that

(5.6) $$|||r|||_{\mathbf{X}_1} \le \varepsilon, |||r|||_{\mathbf{X}_2} \le \varepsilon^{\frac{3}{4}}.$$

First we prove that the mapping transforms the set X_ε into itself if $\varepsilon > 0$ is small. By the local existence theorem we have

$$|||\mathbb{M}_1(r,g)|||_{\mathbf{X}_1} \le \varepsilon, |||\mathbb{M}_1(r,g)|||_{\mathbf{X}_2} \le \varepsilon^{\frac{3}{4}}$$

for $t < 1$ since data are sufficiently small. Now we consider the case $t > 1$. From Lemma 31 we have for $t > 1$ and $0 \le m \le 7$

$$\left\| (\cdot)^{\frac{m}{2}} \partial_x^{(n)} \Phi(t)u_0(\cdot) \right\|_{\mathbf{L}^2} \le C\varepsilon t^{-\frac{3-m+2n}{6}}$$

since $u_0 \in \mathbf{H}_1^{0,\frac{21}{4}}$. Thus via (5.4) and (5.6) we have

$$\sum_{n=0}^{1}\sum_{m=0}^{2} \sup_{t>0} (1+t)^{\frac{3+2n-m}{6}} \left\| (\cdot)^{\frac{m}{2}} \partial_x^{(n)} v(t) \right\|_{\mathbf{L}^2}$$

$$\le \sum_{n=0}^{1}\sum_{m=0}^{2} \sup_{t>0} (1+t)^{\frac{3+2n-m}{6}} \left\| (\cdot)^{\frac{m}{2}} \partial_x^{(n)} \Phi(t)u_0(\cdot) \right\|_{\mathbf{L}^2}$$

$$+ \sum_{n=0}^{1}\sum_{m=0}^{2} \sup_{t>0} (1+t)^{\frac{3+2n-m}{6}} \left\| (\cdot)^{\frac{m}{2}} \partial_x^{(n)} r(\cdot,t) \right\|_{\mathbf{L}^2} \le C\varepsilon$$

and

(5.7) $$\sum_{n=0}^{1}\sum_{m=3}^{7} \sup_{t>0} g(t)^{-1} (1+t)^{\frac{3+2n-m}{6}} \left\| (\cdot)^{\frac{m}{2}} \partial_x^{(n)} v(\cdot,t) \right\|_{\mathbf{L}^2} \le C\varepsilon^{\frac{3}{4}}.$$

We have by the Schwarz inequality with $\rho = \|xf\|_{\mathbf{L}^2} / \|f\|_{\mathbf{L}^2}$

$$\|f\|_{\mathbf{L}^1} \le \left(\int (\rho^2 + x^2) |f|^2 \, dx \right)^{\frac{1}{2}} \left(\int (\rho^2 + x^2)^{-1} \, dx \right)^{\frac{1}{2}}$$

$$\le \rho^{-\frac{1}{2}} \left(\int (1+x^2)^{-1} \, dx \right)^{\frac{1}{2}} \left(\int (\rho^2 + x^2) |f|^2 \, dx \right)^{\frac{1}{2}}$$

$$\le \sqrt{2} \left(\int (1+x^2)^{-1} \, dx \right)^{\frac{1}{2}} \|f\|_{\mathbf{L}^2}^{\frac{1}{2}} \|xf\|_{\mathbf{L}^2}^{\frac{1}{2}}.$$

Therefore we get for $t > 0$

$$\left\| (\cdot)^{\frac{m+3}{2}} \left(-\frac{v(\cdot,t)}{\theta} \int_0^{+\infty} y v v_y dy + v v_x(\cdot,t) \right) \right\|_{\mathbf{L}^1}$$

$$\leq \ C\theta^{-1} \left\| (\cdot)^{\frac{m+5}{2}} v(\cdot,t) \right\|_{\mathbf{L}^2}^{\frac{1}{2}} \left\| (\cdot)^{\frac{m+3}{2}} v(\cdot,t) \right\|_{\mathbf{L}^2}^{\frac{1}{2}} \|v(t)\|_{\mathbf{L}^2}^2$$

$$+ C \left\| (\cdot)^{\frac{m+1}{2}} v(\cdot,t) \right\|_{\mathbf{L}^2} \|(\cdot) v_x(\cdot,t)\|_{\mathbf{L}^2}$$

$$\leq \ C\theta^{-1} \left\| \langle \cdot \rangle^{\frac{m+5}{2}} v(\cdot,t) \right\|_{\mathbf{L}^2}^{\frac{1}{2}+\frac{1}{2}\frac{m+1}{m+3}} \|\langle \cdot \rangle v(\cdot,t)\|_{\mathbf{L}^2}^{\frac{1}{2}\frac{2}{m+3}} \|v(t)\|_{\mathbf{L}^2}^2$$

$$+ C \left\| (\cdot)^{\frac{m+1}{2}} v(\cdot,t) \right\|_{\mathbf{L}^2} \|(\cdot) v_x(\cdot,t)\|_{\mathbf{L}^2}$$

$$\leq \ C\varepsilon^{\frac{9}{5}} (1+t)^{-\frac{5-m}{6}} (\log(1+t))^{\frac{4}{5}}$$

for $m = 0, 1, 2$. Also by the construction of v we see that

$$\int_0^{+\infty} x\left(-\frac{v}{\theta} \int_0^{+\infty} y v v_y dy + v v_x \right) dx = 0.$$

Thus using results of Lemma 32 we have for $t > 1$, $\alpha = \frac{4}{5}$, $m = 0, 1, 2$

$$\sum_{n=0}^{1} \sum_{m=0}^{2} \sup_{t>1} g(t)^{1-\alpha} t^{\frac{3-m+2n}{6}} \left\| (\cdot)^{\frac{m}{2}} \partial_x^{(n)} \mathbb{M}_1(r,g) \right\|_{\mathbf{L}^2}$$

$$\leq \ C \sum_{n=0}^{1} \sum_{m=0}^{2} \sup_{t>1} g(t)^{1-\alpha} t^{\frac{3-m+2n}{6}}$$

$$\times \left\| (\cdot)^{\frac{m}{2}} \int_0^t g(\tau)^{-1} \partial_x^{(n)} \Phi(\cdot, t-\tau)\left(-\frac{v}{\theta} \int_0^{+\infty} y v v_y dy + v v_x \right) d\tau \right\|_{\mathbf{L}^2}$$

$$(5.8) \quad \leq \ C\varepsilon^{\frac{9}{5}} \leq \varepsilon.$$

For higher order m, we turn to the original equation

$$u(x,t) = \Phi(x,t) u_0(x) + \lambda \int_0^t \Phi(x, t-\tau) u u_x d\tau.$$

By Lemma 31

$$\left\| (\cdot)^{\frac{m}{2}} \Phi(t) f \right\| \leq C \left(t^{-\frac{2}{3}} \|f\|_{\mathbf{H}_1^{0,\frac{m+3}{2}}} + t^{-\frac{3-m}{6}} \|f\|_{\mathbf{H}_1^{0,1}} + t^{-\frac{3}{4}} \|f\|_{\mathbf{H}_1^{0,\frac{m+2}{2}+\frac{3}{4}}} \right)$$

and

$$\left\| (\cdot)^{\frac{m}{2}} \partial_x \Phi(t) f \right\|$$

$$\leq \ C \begin{cases} t^{-1} \|f\|_{\mathbf{H}_1^{0,\frac{m+3}{2}}} + t^{-\frac{5-m}{6}} \|f\|_{\mathbf{H}_1^{0,1}}, \\ t^{-\frac{2}{3}} \|f\|_{\mathbf{H}_1^{0,\frac{m+1}{2}}} + t^{-\frac{3-m}{6}} \|f\|_{\mathbf{L}^1} \\ + t^{-\frac{5}{6}+\frac{1}{12}} \|f\|_{\mathbf{H}_1^{0,\frac{m+3}{2}-\frac{3}{4}}} + t^{-\frac{5-m}{6}} \|f\|_{\mathbf{H}_1^{0,1}} \end{cases}$$

Therefore by the local existence theorem we get for $3 \leq m \leq 7$

$$\left\|(\cdot)^{\frac{m}{2}} u(\cdot, t)\right\|_{\mathbf{L}^2}$$
$$\leq \quad C(1+t)^{-\frac{3-m}{6}} \|u_0\|_{\mathbf{H}_1^{0, \frac{m+2}{2}+\frac{3}{4}}}$$
$$+ C \int_0^t \left((t-\tau)^{-\frac{2}{3}} \|uu_x(t)\|_{\mathbf{H}_1^{0, \frac{m+3}{2}}} + (t-\tau)^{-\frac{3}{4}} \|uu_x(t)\|_{\mathbf{H}_1^{0, \frac{m+2}{2}+\frac{3}{4}}} \right.$$
$$\left. + (t-\tau)^{-\frac{3-m}{6}} \|uu_x(t)\|_{\mathbf{H}_1^{0,1}} \right) d\tau.$$

Hence

$$\left\|(\cdot)^{\frac{m}{2}} u(\cdot, t)\right\|_{\mathbf{L}^2}$$
$$\leq \quad C(1+t)^{-\frac{3-m}{6}} \|u_0\|_{\mathbf{H}_1^{0, \frac{m+2}{2}+\frac{3}{4}}}$$
$$+ C\varepsilon^{\frac{3}{2}} \int_0^t \left((t-\tau)^{-\frac{2}{3}} (1+\tau)^{-\frac{5-m}{6}} + (t-\tau)^{-\frac{3}{4}} (1+\tau)^{-\frac{5-m}{6}+\frac{1}{12}} \right) d\tau$$
$$+ C\varepsilon^2 \int_0^t (t-\tau)^{-\frac{3-m}{6}} (1+\tau)^{-1} (1+\eta \log(1+\tau))^{-2} d\tau$$
$$\leq \quad C(1+t)^{-\frac{3-m}{6}} \|u_0\|_{\mathbf{H}_1^{0, \frac{m+3}{2}}} + C\varepsilon (1+t)^{-\frac{3-m}{6}} \leq C\varepsilon (1+t)^{-\frac{3-m}{6}}.$$

In the same way

$$\left\| x^{\frac{m}{2}} \partial_x u(t) \right\|_{\mathbf{L}^2}$$
$$\leq \quad C(1+t)^{-\frac{5-m}{6}} \|u_0\|_{\mathbf{H}_1^{0, \frac{m+3}{2}}}$$
$$+ C \int_0^{\frac{t}{2}} \left((t-\tau)^{-1} \|uu_x\|_{\mathbf{H}_1^{0, \frac{m+3}{2}}} + (t-\tau)^{-\frac{5-m}{6}} \|uu_x\|_{\mathbf{H}_1^{0,1}} \right) d\tau$$
$$+ C \int_{\frac{t}{2}}^t \left((t-\tau)^{-\frac{2}{3}} \|uu_x\|_{\mathbf{H}_1^{0, \frac{m+1}{2}}} + (t-\tau)^{-\frac{3-m}{6}} \|uu_x\|_{\mathbf{L}^1} \right.$$
$$\left. + (t-\tau)^{-\frac{5}{6}+\frac{1}{12}} \|uu_x\|_{\mathbf{H}_1^{0, \frac{m+3}{2}-\frac{3}{4}}} + (t-\tau)^{-\frac{5-m}{6}} \|uu_x\|_{\mathbf{H}_1^{0,1}} \right) d\tau.$$

Therefore by a direct calculation

$$\left\| x^{\frac{m}{2}} \partial_x u(t) \right\|_{\mathbf{L}^2}$$

$$\leq C(1+t)^{-\frac{5-m}{6}} \|u_0\|_{\mathbf{H}_1^{0,\frac{m+3}{2}}} + C\varepsilon^2 \int_0^{\frac{t}{2}} (t-\tau)^{-1} (1+\tau)^{-\frac{5-m}{6}} d\tau$$

$$+ C\varepsilon^2 \int_0^{\frac{t}{2}} (t-\tau)^{-\frac{5-m}{6}} (1+\tau)^{-1} (1+\eta \log(1+\tau))^{-2} d\tau$$

$$+ C\varepsilon^{\frac{3}{2}} \int_{\frac{t}{2}}^t \left((t-\tau)^{-\frac{2}{3}} (1+\tau)^{-\frac{3-m}{6}} + (t-\tau)^{-\frac{3-m}{6}} (1+\tau)^{-\frac{4}{3}} \right.$$

$$\left. + (t-\tau)^{-\frac{5}{6}+\frac{1}{12}} (1+\tau)^{-1-\frac{1}{12}+\frac{m}{6}} \right) d\tau$$

$$+ \varepsilon^2 \int_{\frac{t}{2}}^t (t-\tau)^{-\frac{5-m}{6}} (1+\tau)^{-1} (1+\eta \log(1+\tau))^{-2} d\tau$$

$$\leq C(1+t)^{-\frac{5-m}{6}} \|u_0\|_{\mathbf{H}_1^{0,\frac{m+3}{2}}} + C\varepsilon(1+t)^{-\frac{5-m}{6}} \leq C\varepsilon(1+t)^{-\frac{5-m}{6}}.$$

Thus we get

$$\sum_{n=0}^1 \sum_{m=3}^7 \sup_{t>0} (1+t)^{\frac{3+2n-m}{6}} \left\| (\cdot)^{\frac{m}{2}} \partial_x^{(n)} u(\cdot,t) \right\|_{\mathbf{L}^2} \leq C\varepsilon$$

which implies

$$\sum_{n=0}^1 \sum_{m=3}^7 \sup_{t>0} g(t)^{-1} (1+t)^{\frac{3+2n-m}{6}} \left\| (\cdot)^{\frac{m}{2}} \partial_x^{(n)} \mathbb{M}_2(r,g) \right\|_{\mathbf{L}^2} \leq C\varepsilon \leq \varepsilon^{\frac{3}{4}}.$$

Then applying (5.8) to Lemma 33 we get for $t > 0$

$$\frac{1}{2}(1+\eta \log(1+t)) < g(t) < 2(1+\eta \log(1+t)).$$

Thus $(\mathbb{M}_1(r,g), \mathbb{M}_2(r,g))$ transform the set X_ε into itself. Similarly, we can prove that the transformation $(\mathbb{M}_1(r,g), \mathbb{M}_2(r,g))$ is the contraction mapping. Hence there exists a unique solution (r,g) of the system of integral equations (5.5) in the set X and for $t > 1$ and

$$\|r(t)\|_{\mathbf{L}^\infty} = \|v(t) - \Phi(t) u_0\|_{\mathbf{L}^\infty}$$

$$\leq C \|r(t)\|_{\mathbf{L}^2}^{\frac{1}{2}} \|r_x(t)\|_{\mathbf{L}^2}^{\frac{1}{2}} \leq Ct^{-\frac{2}{3}} g^{-1+\frac{4}{5}}(t).$$

From Lemma 31 we have

$$\Phi(t)u_0 = 3\theta t^{-\frac{2}{3}} Ai'\left(\frac{x}{\sqrt[3]{t}}\right) + R(x,t),$$

where for $t > 1, 0 < \delta < 1$

$$\left\| \partial_x^{(n)} R(t) \right\|_{\mathbf{H}_2^{0,\frac{m}{2}}} \leq C\varepsilon t^{-\frac{3-m+2n+2\delta}{6}}.$$

and therefore
$$\|R\|_{\mathbf{L}^\infty} \le C\,\|R\|_{\mathbf{L}^2}^{\frac{1}{2}}\,\|R_x\|_{\mathbf{L}^2}^{\frac{1}{2}} \le Ct^{-\frac{2+\delta}{3}}.$$
From the proof of Lemma 33 we have for $t > 1$
$$g^{-1}(t) = (\eta \log t)^{-1}.$$
Thus we obtain the following asymptotics of solutions for $t \to +\infty$ uniformly with respect to $x > 0$ such that
$$
\begin{aligned}
u(x,t) &= g^{-1}(t)v(x,t) = g^{-1}(t)\left(r\left(x,t\right) + \Phi(t)u_0\right) \\
&= 3\theta\,(\eta\log t)^{-1}\,t^{-\frac{2}{3}}\,Ai'\left(\frac{x}{\sqrt[3]{t}}\right) + g^{-1}(t)R(x,t) + g^{-1}(t)r(x,t) \\
&= 3\theta\,(\eta\log t)^{-1}\,t^{-\frac{2}{3}}\,Ai'\left(\frac{x}{\sqrt[3]{t}}\right) + O\left(t^{-\frac{2}{3}}(\eta\log t)^{-\frac{6}{5}}\right).
\end{aligned}
$$
Theorem 31 is now proved.

CHAPTER 11

Landau-Ginzburg Equations

1. Introduction

We consider the initial-boundary value problem for the nonlinear Landau-Ginzburg type equations with Dirichlet boundary conditions

(1.1)
$$\begin{cases} \mathcal{L}u + \beta|u|^\sigma u + \gamma|u|^\kappa u = 0, & x \in \mathbf{R}_+^n, \ t \in \mathbf{R}^+, \\ u(0,x) = u_0(x), & x \in \mathbf{R}_+^n, \\ u(t,x) = 0, & x \in \partial\mathbf{R}_+^n, \ t \in \mathbf{R}^+, \end{cases}$$

where $\mathcal{L} = \partial_t - \alpha\Delta$, $\alpha, \beta, \gamma \in \mathbf{C}$, $\operatorname{Re}\alpha > 0$, $0 < \sigma < \kappa$. Equation (1.1) with $\sigma = 2, \gamma = 0$ is known as the complex Landau-Ginzburg equation. The Cauchy problem for the complex Landau-Ginzburg equation

(1.2)
$$\begin{cases} u_t - \alpha\Delta u + \beta|u|^\sigma u + \gamma|u|^\kappa u = 0, & x \in \mathbf{R}^n, \ t \in \mathbf{R}^+, \\ u(0,x) = u_0(x), & x \in \mathbf{R}^n, \end{cases}$$

where $\mathcal{L} = \partial_t - \alpha\Delta$, $\alpha, \beta, \gamma \in \mathbf{C}$, $\operatorname{Re}\alpha > 0$ was studied extensively by many authors (see, for example, [44], [45], [54], [55], [107] and references cited therein). Blow up in finite time of positive solutions to the Cauchy problem for the heat equation $u_t - \Delta u = u^{1+\sigma}$ was proved in papers [39], [48], [85], [130]. Large time behavior of positive solutions for the nonlinear heat equation (which is a particular case of (1.2)) $u_t - \Delta u + u^{1+\sigma} = 0$ was studied for any $\sigma > 0$ (see paper [78] for the super critical case $\sigma > \frac{2}{n}$, [41] for the critical case $\sigma = \frac{2}{n}$ and papers [30], [31], [46], [81] for the sub critical case $\sigma \in (0, \frac{2}{n})$). Global in time existence of small solutions to (1.2) in the super critical case $\sigma > \frac{2}{n}$ was also shown in [39]. Large time asymptotic behavior of small solutions to the Cauchy problem (1.2) was investigated in papers [55], [56], [57] under the condition

$$\operatorname{Re}\beta\left((2+\sigma)|\alpha|^2 + \sigma\alpha^2\right)^{-\frac{n}{2}} > 0.$$

The critical case $\sigma = \frac{2}{n}$ was studied in paper [56] for small initial data in $\mathbf{L}^\infty \cap \mathbf{L}^{1,a}$, $a \in (0,1)$. Paper [57] considered the sub critical case $1 < \sigma < \frac{2}{n}$, when $\frac{2}{n} - \sigma$ is small.

In this chapter we are interested in the global existence and large time behavior of solutions to the initial-boundary value problem (1.1) in the critical $\sigma = \frac{1}{n}$ and sub critical $0 < \sigma < \frac{1}{n}$ cases. Our result below shows that solutions of (1.1) decay faster than those of the Cauchy problem (1.2) due to the zero Dirichlet boundary conditions. Thus the critical power $\sigma = \frac{1}{n}$ for the boundary-value problem (1.1) is different from that $\sigma = \frac{2}{n}$ corresponding to the case of the Cauchy problem (1.2). In what follows we use the Lebesgue space $\mathbf{L}^p$ on $\mathbf{R}^n_+$ and the weighted Lebesgue space $\mathbf{L}^{1,a}$, $a \geq 0$, with norm

$$\|\phi\|_{\mathbf{L}^{1,a}} = \|\langle\cdot\rangle^a \phi\|_{\mathbf{L}^1}, \langle x\rangle = \sqrt{1 + x^2}.$$

By $\mathbf{C}(\mathbf{I}; \mathbf{B})$ we denote the space of continuous functions from a time interval $\mathbf{I}$ to the Banach space $\mathbf{B}$. Different positive constants might be denoted by the same letter C.

We assume that the initial data $u_0 \in \mathbf{L}^\infty \cap \mathbf{L}^{1,n+1}$ and satisfy conditions

(1.3) $$\|u_0\|_{\mathbf{L}^\infty} + \|u_0\|_{\mathbf{L}^{1,n+1}} = \varepsilon, \quad |\theta| \geq C\varepsilon,$$

where

$$\theta = \int_{\mathbf{R}^n_+} \tilde{x} u_0(x)\, dx, \tilde{x} = \Pi^n_{j=1} x_j.$$

Also we suppose that the following angular conditions are fulfilled for $(\arg \alpha, \arg \beta)$

$$\arg \beta + \frac{\sigma n}{2} \arg \alpha - \frac{\sigma n + 3n}{2} \arctan \frac{\sigma \sin(2\arg\alpha)}{\sigma + 2 + \sigma \cos(2\arg\alpha)}$$

(1.4) $$\in \left(-\frac{\pi}{2}, \frac{\pi}{2}\right),$$

$\arg \alpha \in \left(-\frac{\pi}{2}, \frac{\pi}{2}\right)$ and the condition on the order σ

(1.5) $$0 \leq \frac{1}{n} - \sigma \leq \varepsilon,$$

where $\varepsilon > 0$ is sufficiently small. We denote

$$\eta = \operatorname{Re} \beta \delta(\alpha, \sigma) > 0,$$

where

$$\delta(\alpha, \sigma) = t^{\sigma n} \int_{\mathbf{R}^n_+} \tilde{x} |F(t, x)|^\sigma F(t, x)\, dx,$$

$$F(t, x) = \left(4\pi\alpha^2\right)^{-\frac{n}{2}} t^{-\frac{3}{2}n} \tilde{x} e^{-\frac{x^2}{4\alpha t}}.$$

We denote

$$\tilde{\eta} = -\operatorname{Im} \beta t^{\sigma n} \int_{\mathbf{R}^n_+} \tilde{x} |F(t, x)|^\sigma F(t, x)\, dx,$$

$$\chi_\sigma(t) = g(t) \text{ if } \sigma = \frac{1}{n} \text{ and}$$

$$\chi_\sigma(t) = 1 + \frac{\sigma |\theta|^\sigma \eta}{1 - \sigma n} t^{1-\sigma n} \text{ if } \sigma \in \left(0, \frac{1}{n}\right),$$

$g(t) = 1 + |\theta|^\sigma \sigma\eta \log(1 + t).$

Our purpose in this chapter is to prove the following theorem.

THEOREM 34. *We assume that the conditions (1.3) through (1.5) are valid. Then there exists a unique solution*

$$u(t, x) \in \mathbf{C}\left([0, +\infty); \mathbf{L}^{\infty}(\mathbf{R}_+^n) \cap \mathbf{L}^{1, n+1}(\mathbf{R}_+^n)\right)$$

of the initial-boundary value problem (1.1) satisfying the following time decay estimate

$$\left\| u(t) - \theta F(t) \chi_\sigma^{-\frac{1}{\sigma}}(t) e^{i\psi(t)} \right\|_{\mathbf{L}^{\infty}}$$

$$\leq \begin{cases} C\varepsilon^{1+\sigma} (1+t)^{-\frac{1}{\sigma}} (\log(1+t))^{-\frac{1}{\sigma}} & \text{if } \sigma = \frac{1}{n}, \\ C\varepsilon^{1+\sigma} (1+t)^{-\frac{1}{\sigma}} & \text{if } \sigma \in \left(\frac{1}{n} - \varepsilon, \frac{1}{n}\right), \end{cases}$$

and ψ obey the estimate

$$\left| \psi(t) - \arg\theta + |\theta|^\sigma \tilde{\eta} \int_0^t \chi_\sigma^{-\frac{1}{\sigma}}(\tau)(1+\tau)^{-\sigma n} \, d\tau \right|$$

$$\leq \begin{cases} C \int_0^t \chi_\sigma^{-1-\frac{1}{\sigma}}(\tau)(\log \chi_\sigma(t))(1+\tau)^{-1} \, d\tau & \text{if } \sigma = \frac{1}{n}, \\ C \int_0^t \chi_\sigma^{-\frac{1}{\sigma}}(\tau)(1+\tau)^{-\sigma n} \, d\tau & \text{if } \sigma \in \left(\frac{1}{n} - \varepsilon, \frac{1}{n}\right). \end{cases}$$

It is interesting to compare the results for problems (1.1) and (1.2). As we mentioned above the critical value $\sigma = \frac{1}{n}$ for problem (1.1) differs from that of the Cauchy problem since the solution of (1.1) possesses a faster time decay rate. Consider the one dimensional case $n = 1$. In the case of problem (1.1) the points $(\arg\alpha, \arg\beta)$ satisfy condition (1.4) with $\sigma = 1$ and $n = 1$ written as

$$-\frac{\pi}{2} < \arg\beta + \frac{1}{2}\arg\alpha - 2\arctan\frac{\sin(2\arg\alpha)}{3 + \cos(2\arg\alpha)} < \frac{\pi}{2}.$$

Whereas in the case of problem (1.2) with $\sigma = 2$ and $n = 1$ the points $(\arg\alpha, \arg\beta)$ satisfy

$$(1.6) \qquad -\frac{\pi}{2} < \arg\beta - \arctan\frac{\sin 2\arg\alpha}{2 + \cos 2\arg\alpha} < \frac{\pi}{2}$$

under the conditions (1.3), (1.5). Inequalities (1.4) and (1.6) come from the positivity of the values

$$\operatorname{Re}\beta\delta(\alpha, \sigma) \quad \text{and} \quad \operatorname{Re}\beta \int_{\mathbf{R}^n} |G(x)|^2 G(x) \, dx$$

respectively, where $G(x) = \mathcal{F}^{-1}\left(e^{-\alpha\xi^2}\right)$. By a direct computation we have

$$\operatorname{Re}\beta\delta(\alpha, \sigma)$$

$$= \frac{|\beta|}{2^{\frac{\sigma n}{2} - \frac{3n}{2}} \pi^{\frac{\sigma n}{2} + \frac{n}{2}} |\alpha|^{\sigma n}} \left((\sigma + 2 + \sigma\cos(2\arg\alpha))^2 + (\sigma\sin(2\arg\alpha))^2 \right)^{-\frac{\sigma+3}{4}n}$$

$$\times \left(2 \int_0^{+\infty} x^{\sigma+2} \exp\left(-x^2\right) dx \right)^n$$

$$\times \cos\left(\arg\beta + \frac{\sigma n}{2}\arg\alpha - \frac{\sigma n + 3n}{2}\arctan\frac{\sigma\sin(2\arg\alpha)}{\sigma + 2 + \sigma\cos(2\arg\alpha)} \right).$$

If $\operatorname{Im}\alpha = 0$, $\operatorname{Im}\beta = 0$, namely, if $(\arg\alpha, \arg\beta) = (0,0), (0,\pm\pi)$, then solutions of equations (1.1) and (1.2) satisfy the same property. More precisely, if $(\arg\alpha, \arg\beta) = (0,0), (\beta > 0)$, then these problems have solutions globally in time. See [41] for (1.2) and [99] for (1.1). In addition if $(\arg\alpha, \arg\beta) = (0,\pm\pi), (\beta < 0)$, then the solutions of problems (1.1) and (1.2) blow up in finite time; see [48], [85] for (1.2) and [100] for (1.1). We have another situation in the case of complex coefficients α, β, since there are points $(\arg\alpha, \arg\beta)$ which satisfy (1.4) but not (1.6). This fact implies that the properties of solutions of (1.1) and (1.2) in the same points $(\arg\alpha, \arg\beta)$ could be different. For example, when $\arg\alpha = \frac{\pi}{3}$ we find approximately that $\frac{1}{2}\arg\alpha - 2\arctan\frac{\sin(2\arg\alpha)}{3+\cos(2\arg\alpha)} \approx -0.1$ and $-\frac{1}{2}\arctan\frac{\sin(2\arg\alpha)}{2+\cos(2\arg\alpha)} \approx -0.3$. Hence the points $\left(\frac{\pi}{3}, \arg\beta\right)$ satisfy condition (1.4), corresponding to the case of a half-line, if

$$-\frac{\pi}{2} + 0.1 < \arg\beta < \frac{\pi}{2} + 0.1$$

and satisfy condition (1.6), corresponding to the case of the line, if

$$-\frac{\pi}{2} + 0.3 < \arg\beta < \frac{\pi}{2} + 0.3$$

respectively.

We organize the rest of this chapter as follows. We prove preliminary lemmas in Section 2. In Lemma 34 we obtain estimates of the Green operator in the Lebesgue spaces $\mathbf{L}^p$, $1 \le p \le \infty$ and $\mathbf{L}^{1,a}$. Then in Lemma 35 we estimate the Green operator in our basic norm

$$\|\phi\|_{\mathbf{X}} = \sup_{t>0}\left((1+t)^n \|\phi(t)\|_{\mathbf{L}^\infty} + (1+t)^{\frac{n}{2}}\|\phi(t)\|_{\mathbf{L}^1}\right.$$
$$\left. + (1+t)^{-\frac{1}{2}}\|\phi(t)\|_{\mathbf{L}^{1,n+1}}\right).$$

Large time behavior of the first moments of the nonlinearity $\beta|u|^\sigma u$ in equation (1.1) is evaluated in Lemma 36. Section 3 is devoted to the proof of Theorem 34.

2. Preliminaries

Consider the linear Dirichlet initial-boundary value problem

(2.1)
$$\begin{cases} \mathcal{L}u = f(t,x), & x \in \mathbf{R}^n_+, \ t \in \mathbf{R}^+, \\ u(0,x) = u_0(x), & x \in \mathbf{R}^n_+, \\ u(t,x) = 0, & x \in \partial\mathbf{R}^n_+, \ t \in \mathbf{R}^+, \end{cases}$$

where $\mathcal{L} = \partial_t - \alpha\Delta$, $\operatorname{Re}\alpha > 0$. We write the solution $u(t,x)$ of the problem (2.1) by virtue of the Duhamel formula

$$u(t) = \mathcal{G}(t)u_0 + \int_0^t \mathcal{G}(t-\tau)f(\tau)\,d\tau,$$

where the Green operator $\mathcal{G}(t)$ is given by

$$\mathcal{G}(t)\phi = \int_{\mathbf{R}^n_+} G(t,x,y)\,\phi(y)\,dy,$$

where the kernel

$$G(t,x,y) = (4\pi\alpha t)^{-\frac{n}{2}}\,\Pi^n_{j=1}\left(e^{-\frac{(x_j-y_j)^2}{4\alpha t}} - e^{-\frac{(x_j+y_j)^2}{4\alpha t}}\right).$$

We first prepare some preliminary estimates of the Green operator $\mathcal{G}(t)$ in the Lebesgue norms $\|\phi\|_{\mathbf{L}^p}$ and $\|\phi\|_{\mathbf{L}^{1,a}} = \|\langle\cdot\rangle^a\,\phi\|_{\mathbf{L}^1}$, where $a \geq 0$, $1 \leq p \leq \infty$. We denote

$$\begin{aligned} F(t,x) &= \Pi^n_{j=1}\partial_{y_j}G(t,x,y)\big|_{y=0} = (4\pi\alpha^2)^{-\frac{n}{2}}\,t^{-\frac{3}{2}n}\widetilde{x}e^{-\frac{x^2}{4\alpha t}},\\ \widetilde{x} &= \Pi^n_{j=1}x_j. \end{aligned}$$

LEMMA 34. *Suppose that $\phi \in \mathbf{L}^p$, then the estimate*

$$\|\mathcal{G}(t)\,\phi\|_{\mathbf{L}^p} \leq C\,\|\phi\|_{\mathbf{L}^p}\,,$$

is true for all $t > 0$, $1 \leq p \leq \infty$. Furthermore we assume that $\phi \in \mathbf{L}^{1,n+1}$, then the estimate

$$\||\cdot|^\omega\,(\mathcal{G}(t)\,\phi - \vartheta F(t))\|_{\mathbf{L}^p} \leq Ct^{-n+\frac{n}{2p}+\frac{\omega-1}{2}}\,\|\phi\|_{\mathbf{L}^{1,n+1}}$$

is valid for all $t > 0$, where $1 \leq p \leq \infty$, $\omega \in [0, n+1]$ and

$$\vartheta = \int_{\mathbf{R}^n_+}\widetilde{x}\phi(x)\,dx.$$

PROOF. Since

(2.2)
$$|G(t,x,y)| \leq Ct^{-\frac{n}{2}}e^{-\frac{C}{t}(x-y)^2}$$

for all $x,y \in \mathbf{R}^n_+$, by the Young inequality we have

$$\begin{aligned} \|\mathcal{G}(t)\,\phi\|_{\mathbf{L}^p} &\leq Ct^{-\frac{n}{2}}\left\|\int_{\mathbf{R}^n_+}e^{-\frac{C}{t}(x-y)^2}\phi(y)\,dy\right\|_{\mathbf{L}^p}\\ &\leq Ct^{-\frac{n}{2}}\left\|e^{-\frac{C}{t}x^2}\right\|_{\mathbf{L}^1}\|\phi\|_{\mathbf{L}^p} \leq C\,\|\phi\|_{\mathbf{L}^p} \end{aligned}$$

for all $t > 0$, where $1 \leq p \leq \infty$, whence the first estimate of the lemma follows. For the second estimate we write

$$|x|^\omega\,(\mathcal{G}(t)\,\phi - \vartheta F(t,x)) = \int_{\mathbf{R}^n_+}|x|^\omega\,(G(t,x,y) - F(t,x)\,\widetilde{y})\,\phi(y)\,dy$$

for any $\omega \in [0, n+1]$. Applying the Taylor's formula, we obtain

(2.3)
$$|G(t,x,y) - F(t,x)\,\widetilde{y}| \leq Ct^{-n-\frac{1}{2}}\,|y|^{n+1}\left(e^{-\frac{C}{t}(x-y)^2} + e^{-\frac{C}{t}x^2}\right)$$

for all $x, y \in \mathbf{R}_+^n$. Thus in the domain $|y| \leq \frac{|x|}{2}$

$$
\begin{aligned}
|x|^\omega \, |G\,(t, x, y) - F\,(t, x)\,\widetilde{y}| \; &\leq \; Ct^{-n-\frac{1}{2}} \, |y|^{n+1} \, |x|^\omega \, e^{-\frac{C}{t} x^2} \\
&\leq \; Ct^{-n+\frac{\omega-1}{2}} \, |y|^{n+1} \, e^{-\frac{C}{t} x^2}.
\end{aligned}
$$

By the Lagrange finite differences theorem we have

$$
|G\,(t, x, y)| \leq Ct^{-n} \, |y|^n \, e^{-\frac{C}{t} (x-y)^2},
$$

whence in view of (2.2) we find

$$
(2.4) \qquad |G\,(t, x, y)| \leq Ct^{-\frac{n+\nu}{2}} \, |y|^\nu \, e^{-\frac{C}{t} (x-y)^2}
$$

for all $x, y \in \mathbf{R}_+^n$, where $\nu \in [0, n]$. Taking (2.4) with $\nu = n + 1 - \omega$, in the case $\omega \in [1, n+1]$ we get for $|y| \geq \frac{|x|}{2}$

$$
\begin{aligned}
&|x|^\omega \, |G\,(t, x, y) - F\,(t, x)\,\widetilde{y}| \\
\leq \; &|x|^\omega \, (|G\,(t, x, y)| + |F\,(t, x)\,\widetilde{y}|) \\
\leq \; &Ct^{-n+\frac{\omega-1}{2}} \, |x|^\omega \, |y|^{n+1-\omega} \, e^{-\frac{C}{t}(x-y)^2} + Ct^{-\frac{3}{2}n} \, |x|^{\omega+n} \, |y|^n \, e^{-\frac{C}{t} x^2} \\
(2.5) \qquad \leq \; &Ct^{-n+\frac{\omega-1}{2}} \, |y|^{n+1} \left(e^{-\frac{C}{t}(x-y)^2} + e^{-\frac{C}{t} x^2} \right),
\end{aligned}
$$

and in the case $\omega \in [0, 1]$ we write by virtue of (2.3) and (2.5) with $\omega = 1$

$$
\begin{aligned}
&|x|^\omega \, |G\,(t, x, y) - F\,(t, x)\,\widetilde{y}| \\
\leq \; &|x|^\omega \, (|G\,(t, x, y)| + |F\,(t, x)\,\widetilde{y}|)^\omega \, |G\,(t, x, y) - F\,(t, x)\,\widetilde{y}|^{1-\omega} \\
\leq \; &Ct^{-n\omega} \, |y|^{(n+1)\omega} \, t^{-\left(n+\frac{1}{2}\right)(1-\omega)} \, |y|^{(n+1)(1-\omega)} \\
&\times \left(e^{-\frac{C}{t}(x-y)^2} + e^{-\frac{C}{t} x^2} \right) \\
\leq \; &Ct^{-n+\frac{\omega-1}{2}} \, |y|^{n+1} \left(e^{-\frac{C}{t}(x-y)^2} + e^{-\frac{C}{t} x^2} \right),
\end{aligned}
$$

for all $x, y \in \mathbf{R}_+^n$, $|y| \geq \frac{|x|}{2}$. Thus we obtain the estimate

$$
\begin{aligned}
&|x|^\omega \, |G\,(t, x, y) - F\,(t, x)\,\widetilde{y}| \\
\leq \; &Ct^{-n+\frac{\omega-1}{2}} \, |y|^{n+1} \left(e^{-\frac{C}{t}(x-y)^2} + e^{-\frac{C}{t} x^2} \right)
\end{aligned}
$$

for all $x, y \in \mathbf{R}_+^n$, and for any $\omega \in [0, n+1]$. Applying the above estimate with Young inequality we find

$$
\begin{aligned}
&\left\| |\cdot|^\omega \, (\mathcal{G}\,(t)\,\phi - \vartheta F\,(t)) \right\|_{\mathbf{L}^p} \\
= \; &\left\| \int_{\mathbf{R}_+^n} |x|^\omega \, (G\,(t, x, y) - F\,(t, x)\,\widetilde{y})\,\phi\,(y)\, dy \right\|_{\mathbf{L}_x^p} \\
\leq \; &Ct^{-n+\frac{\omega-1}{2}} \left\| \int_{\mathbf{R}_+^n} \left(e^{-\frac{C}{t}(x-y)^2} + e^{-\frac{C}{t} x^2} \right) |y|^{n+1} \, |\phi\,(y)|\, dy \right\|_{\mathbf{L}_x^p} \\
\leq \; &Ct^{-n+\frac{n}{2p}+\frac{\omega-1}{2}} \, \|\phi\|_{\mathbf{L}^{1, n+1}}.
\end{aligned}
$$

Thus the second estimate of the lemma follows. Lemma 34 is proved. □

We introduce the function space

$$\|\phi\|_{\mathbf{X}} = \sup_{t>0} \left((1+t)^n \|\phi(t)\|_{\mathbf{L}^\infty} + (1+t)^{\frac{n}{2}} \|\phi(t)\|_{\mathbf{L}^1} \right.$$
$$\left. + (1+t)^{-\frac{1}{2}} \|\phi(t)\|_{\mathbf{L}^{1,n+1}} \right).$$

Define the function $g(t)$

$$g(t) = 1 + \varsigma \log(1+t)$$

for some $\varsigma > 0$.

LEMMA 35. Let the function $f(t,x)$ satisfy $\int_{\mathbf{R}_+^n} \widetilde{y} f(t,x)\, dx = 0$. Then the following inequality

$$\left\| g^l(t) \int_0^t g^{-l}(\tau) \mathcal{G}(t-\tau) f(\tau)\, d\tau \right\|_{\mathbf{X}} \le C \|(1+t) f(t)\|_{\mathbf{X}}$$

is valid for $l = 0, 1$, provided that the right-hand side is finite.

PROOF. By the estimate $g^{-1}(\tau) \le C$ and Lemma 34 we get

$$\left\| \int_0^t g^{-l}(\tau) \mathcal{G}(t-\tau) f(\tau)\, d\tau \right\|_{\mathbf{L}^\infty}$$
$$+ \left\| \int_0^t g^{-l}(\tau) \mathcal{G}(t-\tau) f(\tau)\, d\tau \right\|_{\mathbf{L}^{1,n+1}}$$
$$\le C \|(1+t) f(t,x)\|_{\mathbf{X}} \int_0^4 (1+\tau)^{-1}\, d\tau$$
$$\le C \|(1+t) f(t,x)\|_{\mathbf{X}} g^{-l}(t)$$

for all $0 \le t \le 4$. We now consider $t > 4$. Via the condition of the lemma for the function $g(t)$ we have the estimate $(1+t)^{-\frac{1}{4}} \le C g^{-1}(t)$ and

$$\sup_{\tau \in [\sqrt{t}, t]} g^{-1}(\tau) \le C \left(1 + \varsigma \log\left(1 + \sqrt{t}\right) \right)^{-1}$$
$$\le C \left(1 + \frac{\varsigma}{2} \log(1+t) \right)^{-1} \le C g^{-1}(t);$$

hence by virtue of Lemma 34 with $\omega = 0$ we obtain

$$\left\| \int_0^t g^{-l}(\tau) \mathcal{G}(t-\tau) f(\tau) \, d\tau \right\|_{\mathbf{L}^p}$$

$$\leq C \int_0^{\sqrt{t}} (t-\tau)^{-n+\frac{n}{2p}-\frac{1}{2}} (1+\tau)^{-\frac{1}{2}} \, d\tau$$

$$\times \sup_{\tau>0} (1+\tau)^{-\frac{1}{2}} \|(1+\tau) f(\tau)\|_{\mathbf{L}^{1,n+1}}$$

$$+ C g^{-l}(t) \int_{\sqrt{t}}^{\frac{t}{2}} (t-\tau)^{-n+\frac{n}{2p}-\frac{1}{2}} (1+\tau)^{-\frac{1}{2}} \, d\tau$$

$$\times \sup_{\tau>0} (1+\tau)^{-\frac{1}{2}} \|(1+\tau) f(\tau)\|_{\mathbf{L}^{1,n+1}}$$

$$+ C g^{-l}(t) \int_{\frac{t}{2}}^t (1+\tau)^{-n+\frac{n}{2p}-1} \, d\tau$$

$$\times \sup_{\tau>0} (1+\tau)^{n-\frac{n}{2p}} \|(1+\tau) f(\tau)\|_{\mathbf{L}^p}$$

$$\leq C t^{-n+\frac{n}{2p}} g^{-l}(t) \|(1+t) f\|_{\mathbf{X}}$$

for $1 \leq p \leq \infty$, and using the second estimate of Lemma 34 with $\omega = n+1$ we get

$$\left\| \int_0^t g^{-l}(\tau) \mathcal{G}(t-\tau) f(\tau) \, d\tau \right\|_{\mathbf{L}^{1,n+1}}$$

$$\leq C \int_0^{\sqrt{t}} (1+\tau)^{-\frac{1}{2}} \, d\tau \sup_{\tau>0} (1+\tau)^{-\frac{1}{2}} \|(1+\tau) f(\tau)\|_{\mathbf{L}^{1,n+1}}$$

$$+ C g^{-l}(t) \int_{\sqrt{t}}^t (1+\tau)^{-\frac{1}{2}} \, d\tau \sup_{\tau>0} (1+\tau)^{-\frac{1}{2}} \|(1+\tau) f(\tau)\|_{\mathbf{L}^{1,n+1}}$$

$$\leq C \left(t^{\frac{1}{4}} + g^{-l}(t) t^{\frac{1}{2}} \right) \|(1+t) f\|_{\mathbf{X}}$$

$$\leq C g^{-l}(t) t^{\frac{1}{2}} \|(1+t) f\|_{\mathbf{X}}$$

for all $t > 4$. Thus the result of the lemma follows. Lemma 35 is proved. $\qquad\square$

The next lemma will be employed in the proof of the theorem to evaluate large time behavior of the mean value of the nonlinearity in equation (1.1). We denote $\eta = \operatorname{Re} \beta \delta (\alpha, \sigma) > 0$,

$$\delta (\alpha, \sigma) = t^{\sigma n} \int_{\mathbf{R}_+^n} \widetilde{x} |F(t,x)|^\sigma F(t,x) \, dx,$$

$$F(t,x) = \left(4\pi\alpha^2 \right)^{-\frac{n}{2}} t^{-\frac{3}{2}n} \widetilde{x} e^{-\frac{x^2}{4\alpha t}},$$

$$\chi_\sigma(t) = g(t) \text{ if } \sigma = \frac{1}{n} \text{ and}$$

$$\chi_\sigma(t) = 1 + \frac{\sigma|\theta|^\sigma \eta}{1 - \sigma n} t^{1-\sigma n} \text{ if } \sigma \in \left(0, \frac{1}{n}\right),$$

$g(t) = 1 + |\theta|^\sigma \sigma\eta \log(1+t).$

LEMMA 36. *Assume that* $v_0 \in \mathbf{L}^\infty \cap \mathbf{L}^{1,n+1}$, *the norm*

$$\|v_0\|_{\mathbf{L}^\infty} + \|v_0\|_{\mathbf{L}^{1,n+1}} = \varepsilon$$

is sufficiently small and

$$\int_{\mathbf{R}_+^n} \widetilde{x} v_0(x)\, dx = \left|\int_{\mathbf{R}_+^n} \widetilde{x} u_0(x)\, dx\right| = \theta.$$

Let function $v(t, x)$ *satisfy the estimates*

$$\|v\|_{\mathbf{L}^\infty} \le C\varepsilon(1+t)^{-n}, \quad \|v\|_{\mathbf{L}^{1,n}} \le C\varepsilon \text{ and}$$

$$\|v(t) - \mathcal{G}(t)v_0\|_{\mathbf{L}^{1,n}} \le C\varepsilon^{1+\sigma} g^{-l}(t),$$

where $l = 1$ *in the critical case* $\sigma = \frac{1}{n}$ *and* $l = 0$ *in the sub critical case* $0 < \sigma < \frac{1}{n}$. *Then the inequality*

$$\left|1 + \frac{\sigma}{\theta}\int_0^t d\tau \, \mathrm{Re} \int_{\mathbf{R}_+^n} \widetilde{x}\beta |v|^\sigma v(\tau, x) dx - \chi_\sigma(t)\right|$$

$$(2.6) \qquad \le \begin{cases} C\varepsilon \log \chi_\sigma(t) & \text{if } \sigma = \frac{1}{n}, \\ C\varepsilon^{1+\sigma}\chi_\sigma(t) & \text{if } \sigma \in \left(0, \frac{1}{n}\right) \end{cases}$$

is valid for all $t > 0$.

PROOF. In view of the condition

$$\|v\|_{\mathbf{L}^\infty} + \|v\|_{\mathbf{L}^{1,n}} \le C\varepsilon$$

we get

$$\left|\frac{\sigma}{\theta}\int_0^t d\tau \, \mathrm{Re} \int_{\mathbf{R}_+^n} \widetilde{x}\beta |v|^\sigma v(\tau, x) dx\right| \le C\varepsilon^\sigma t;$$

hence (2.6) follows for all $0 < t < 1$.

We now consider the case $t \ge 1$. By the second estimate of Lemma 34 we find

$$\|\mathcal{G}(t)v_0 - \theta F(t)\|_{\mathbf{L}^{1,n}} \le Ct^{-\frac{1}{2}}\|v_0\|_{\mathbf{L}^{1,n+1}} \le C\varepsilon t^{-\frac{1}{2}}.$$

Hence we find

$$\||v|^\sigma v - |\theta|^\sigma \theta |F(t)|^\sigma F(t)\|_{\mathbf{L}^{1,n}}$$
$$\le C\left(\|v(t) - \mathcal{G}(t)v_0\|_{\mathbf{L}^{1,n}} + \|\mathcal{G}(t)v_0 - \theta F(t)\|_{\mathbf{L}^{1,n}}\right)$$
$$\times\left(\|v\|_{\mathbf{L}^\infty}^\sigma + \|\mathcal{G}v_0\|_{\mathbf{L}^\infty}^\sigma + |\theta|^\sigma\|F(t)\|_{\mathbf{L}^\infty}^\sigma\right)$$
$$\le C\varepsilon^{1+2\sigma}t^{-\sigma n}g^{-l}(t) + C\varepsilon^{1+\sigma}t^{-\sigma n-\frac{1}{2}}$$

for all $t \geq 1$, where $l = 1$ if $\sigma = \frac{1}{n}$ and $l = 0$ if $0 < \sigma < \frac{1}{n}$. Since

$$\int_{\mathbf{R}^n_+} \tilde{x} \left| F(t, x) \right|^\sigma F(t, x) \, dx = t^{-n\sigma} \delta(\alpha, \sigma)$$

and $\operatorname{Re} \beta \delta(\alpha, \sigma) = \eta > 0$ we get

$$\left| \operatorname{Re} \int_{\mathbf{R}^n_+} \tilde{x} \beta \left| v \right|^\sigma v(t, x) \, dx - \left| \theta \right|^\sigma \theta t^{-\sigma n} \eta \right|$$
$$\leq \quad C \left\| \left| v \right|^\sigma v - \left| \theta \right|^\sigma \theta \left| F(t) \right|^\sigma F(t) \right\|_{\mathbf{L}^{1,n}}$$
$$\leq \quad C\varepsilon^{1+2\sigma} t^{-\sigma n} g^{-l}(t) + C\varepsilon^{1+\sigma} t^{-\sigma n - \frac{1}{2}}$$

for all $t \geq 1$, where $0 < \sigma \leq \frac{1}{n}$. Therefore

$$\left| \frac{\sigma}{\theta} \int_1^t d\tau \operatorname{Re} \int_{\mathbf{R}^n_+} \tilde{x} \beta \left| v \right|^\sigma v(\tau, x) dx - \left| \theta \right|^\sigma \sigma \eta \log t \right|$$
$$\leq \quad \int_1^t \frac{C\varepsilon^{1+2\sigma} d\tau}{\tau \left(1 + \left| \theta \right|^\sigma \eta \log(1 + \tau) \right)} + C\varepsilon^{1+\sigma} \int_1^t \tau^{-1-\frac{1}{2}} d\tau$$
(2.7)
$$\leq \quad C\varepsilon^{1+\sigma} \log \left(1 + \left| \theta \right|^\sigma \eta \log (1 + t) \right)$$

for all $t \geq 1$ if $\sigma = \frac{1}{n}$. Thus in view of (2.7) we obtain estimate (2.6) in the case $\sigma = 1$. As in the proof of (2.7) we have the inequality

$$\left| \frac{\sigma}{\theta} \int_0^t d\tau \operatorname{Re} \int_{\mathbf{R}^n_+} \tilde{x} \beta \left| v \right|^\sigma v(\tau, x) dx - \left| \theta \right|^\sigma \frac{\sigma \eta}{1 - \sigma n} t^{1-\sigma n} \right|$$
$$\leq \quad C\varepsilon^{1+2\sigma} t^{1-\sigma n} + C\varepsilon^{1+\sigma} t^{\frac{1}{2}-\sigma n}$$

for all $t \geq 1$, which implies (2.6) in the case $0 < \sigma < \frac{1}{n}$. Lemma 36 is proved. $\qquad \square$

3. Proof of Theorem 34

We change the dependent variable $u = v e^{-\varphi(t) + i\psi(t)}$ as in [54]. Then for the new function v we get the equation

$$\mathcal{L}v + \beta e^{-\sigma\varphi} \left| v \right|^\sigma v + \gamma e^{-\kappa\varphi} \left| v \right|^\kappa v - (\varphi' - i\psi') v = 0,$$

where $\mathcal{L} = \partial_t - \alpha\Delta$. We assume that

$$\int_{\mathbf{R}^n_+} \tilde{x} \left(\beta e^{-\sigma\varphi} \left| v \right|^\sigma v + \gamma e^{-\kappa\varphi} \left| v \right|^\kappa v - (\varphi' - i\psi') v \right) dx = 0$$

and

$$\int_{\mathbf{R}^n_+} \tilde{x} v_0(x) \, dx = \left| \int_{\mathbf{R}^n_+} \tilde{x} u_0(x) \, dx \right| = \theta > 0, \quad \varphi(0) = 0,$$

where $\tilde{x} = \Pi_{j=1}^{n} x_j$. Thus we consider the Cauchy problem for the new dependent variables (v, φ)

(3.1)
$$
\begin{cases}
\mathcal{L}v = -\beta e^{-\sigma\varphi}\left(|v|^\sigma - \frac{1}{\theta}\int_{\mathbf{R}_+^n}\tilde{x}|v|^\sigma v\,dx\right)v \\
\qquad -\gamma e^{-\kappa\varphi}\left(|v|^\kappa - \frac{1}{\theta}\int_{\mathbf{R}_+^n}\tilde{x}|v|^\kappa v\,dx\right)v, \\
\varphi' = \frac{1}{\theta}e^{-\sigma\varphi}\operatorname{Re}\beta\int_{\mathbf{R}_+^n}\tilde{x}|v|^\sigma v\,dx + \frac{1}{\theta}e^{-\kappa\varphi}\operatorname{Re}\gamma\int_{\mathbf{R}_+^n}\tilde{x}|v|^\kappa v\,dx, \\
v(0,x) = v_0(x), \quad \varphi(0) = 0,
\end{cases}
$$

and
$$
\psi'(t) = -\frac{1}{\theta}e^{-\sigma\varphi}\operatorname{Im}\beta\int_{\mathbf{R}_+^n}\tilde{x}|v|^\sigma v\,dx - \frac{1}{\theta}e^{-\kappa\varphi}\operatorname{Im}\gamma\int_{\mathbf{R}_+^n}\tilde{x}|v|^\kappa v\,dx.
$$

We write (3.1) as

(3.2)
$$
\begin{cases}
\mathcal{L}v = N(v,h), \\
h' = \frac{\sigma}{\theta}\left(\operatorname{Re}\beta\int_{\mathbf{R}_+^n}\tilde{x}|v|^\sigma v\,dx + h^{\frac{\sigma-\kappa}{\sigma}}\operatorname{Re}\gamma\int_{\mathbf{R}_+^n}\tilde{x}|v|^\kappa v\,dx\right), \\
v(0,x) = v_0(x), \quad h(0) = 1,
\end{cases}
$$

where we denote $h = e^{\sigma\varphi(t)}$ and

$$
\begin{aligned}
N(v,h) &= -\beta h^{-1}\left(|v|^\sigma - \frac{1}{\theta}\int_{\mathbf{R}_+^n}\tilde{x}|v|^\sigma v\,dx\right)v \\
&\quad -\gamma h^{-\frac{\kappa}{\sigma}}\left(|v|^\kappa - \frac{1}{\theta}\int_{\mathbf{R}_+^n}\tilde{x}|v|^\kappa v\,dx\right)v.
\end{aligned}
$$

We note here that $h^{\frac{\sigma-\kappa}{\sigma}}\operatorname{Re}\gamma\int_{\mathbf{R}_+^n}\tilde{x}|v|^\kappa v\,dx$ decays in time more rapidly than $\operatorname{Re}\beta\int_{\mathbf{R}_+^n}\tilde{x}|v|^\sigma v\,dx$ and $\int_{\mathbf{R}_+^n}\tilde{x}F(v,h)\,dx = 0$. We prove a global existence of solutions $(v(t,x), h(t))$ for the Cauchy problem (3.2) by the successive approximations $(v_m(t,x), h_m(t))$, $m = 1, 2, \ldots$, defined as follows

$$
\begin{cases}
\mathcal{L}v_m = N(v_{m-1}, h_{m-1}), \\
\partial_t h_m = \frac{\sigma}{\theta}\operatorname{Re}\beta\int_{\mathbf{R}_+^n}\tilde{x}|v_{m-1}|^\sigma v_{m-1}\,dx \\
\quad +\frac{\sigma}{\theta}h_{m-1}^{1-\frac{\kappa}{\sigma}}\operatorname{Re}\gamma\int_{\mathbf{R}_+^n}\tilde{x}|v_{m-1}|^\kappa v_{m-1}\,dx, \\
v_m(0,x) = v_0(x), \quad h_m(0) = 1,
\end{cases}
$$

for all $m \geq 2$, where $v_1 = \mathcal{G}(t)v_0$, $h_1(t) = \chi_\sigma(t)$,

$$
\chi_\sigma(t) = g(t) \text{ if } \sigma = \frac{1}{n} \text{ and}
$$
$$
\chi_\sigma(t) = 1 + \frac{\sigma|\theta|^\sigma\eta}{1-\sigma n}t^{1-\sigma n} \text{ if } \sigma \in \left(0, \frac{1}{n}\right),
$$

$g(t) = 1 + |\theta|^\sigma\eta\log(1+t)$. We now prove by induction the following estimates

$$
\|v_m\|_{\mathbf{X}} \leq C\varepsilon, \quad \|v_m(t) - \mathcal{G}(t)v_0\|_{\mathbf{L}^{1,n}} \leq C\varepsilon^{1+\sigma}g^{-l}(t),
$$

(3.3)
$$
|h_m - \chi_\sigma(t)| \leq \begin{cases} C\varepsilon^{1+\sigma}\log\chi_\sigma(t) & \text{if } \sigma = \frac{1}{n} \\ C\varepsilon^{1+\sigma}\chi_\sigma(t) & \text{if } \sigma \in \left(\frac{1}{n} - \varepsilon, \frac{1}{n}\right) \end{cases}
$$

for all $m \geq 1$, where $l = 1$ in the critical case $\sigma = \frac{1}{n}$ and $l = 0$ in the sub critical case $\frac{1}{n} - \varepsilon < \sigma < \frac{1}{n}$; the norm $\|\cdot\|_X$ is defined as above by

$$\|\phi\|_X \;=\; \sup_{t>0} \left((1+t)^n \|\phi(t)\|_{L^\infty} + (1+t)^{\frac{n}{2}} \|\phi(t)\|_{L^1} \right.$$
$$\left. + (1+t)^{-\frac{1}{2}} \|\phi(t)\|_{L^{1,n+1}} \right).$$

By virtue of Lemma 34 we have

$$\|\mathcal{G}(t)v_0\|_{L^\infty} \;\leq\; C\varepsilon(1+t)^{-n}, \;\; \|\mathcal{G}(t)v_0\|_{L^1} \leq C\varepsilon(1+t)^{-\frac{n}{2}} \text{ and}$$
$$\|\mathcal{G}(t)v_0\|_{L^{1,n+1}} \;\leq\; \|F(t)\|_{L^{1,n+1}} + C\|v_0\|_{L^{1,n+1}} \leq C\varepsilon(1+t)^{\frac{1}{2}};$$

therefore estimates (3.3) are valid for $m = 1$. We assume that estimates (3.3) are true with m replaced by $m - 1$. The integral equation associated with (3.2) is written as

$$\begin{cases} v_m(t) = \mathcal{G}(t)v_0 + \int_0^t \mathcal{G}(t-\tau)N(v_{m-1}(\tau), h_{m-1}(\tau))\,d\tau, \\ h_m(t) = 1 + \frac{\sigma}{\theta}\int_0^t d\tau \left(\operatorname{Re}\beta \int_{\mathbf{R}_+^n} \widetilde{x}\,|v_{m-1}|^\sigma v_{m-1}dx \right. \\ \left. + h_{m-1}^{1-\frac{\varkappa}{\sigma}} \operatorname{Re}\gamma \int_{\mathbf{R}_+^n} \widetilde{x}\,|v_{m-1}|^\varkappa v_{m-1}dx \right). \end{cases}$$

Note that in the critical case $\sigma = \frac{1}{n}$ we have

$$\|N(v_{m-1}(t), h_{m-1}(t))\|_{L^\infty}$$
$$\leq\; Ch_{m-1}^{-1}(t)\|v_{m-1}(t)\|_{L^\infty}^{1+\sigma}\left(1 + \frac{1}{\theta}\|v_{m-1}(t)\|_{L^1}\right)$$
$$\leq\; C\varepsilon^{1+\sigma}(1+t)^{-1-\frac{1}{\sigma}}g^{-1}(t),$$

$$\|N(v_{m-1}(t), h_{m-1}(t))\|_{L^1}$$
$$\leq\; Ch_{m-1}^{-1}(t)\|v_{m-1}(t)\|_{L^\infty}^\sigma\left(\|v_{m-1}(t)\|_{L^1} + \frac{1}{\theta}\|v_{m-1}(t)\|_{L^1}^2\right)$$
$$\leq\; C\varepsilon^{1+\sigma}(1+t)^{-1}g^{-1}(t)$$

and

$$\|N(v_{m-1}(t), h_{m-1}(t))\|_{L^{1,n+1}}$$
$$\leq\; Ch_{m-1}^{-1}(t)\|v_{m-1}(t)\|_{L^\infty}^\sigma\|v_{m-1}(t)\|_{L^{1,n+1}}\left(1 + \frac{1}{\theta}\|v_{m-1}(t)\|_{L^1}\right)$$
$$\leq\; C\varepsilon^{1+\sigma}(1+t)^{-1}g^{-1}(t)$$

for all $t > 0$, provided that $(v_{m-1}(t), h_{m-1}(t))$ satisfies (3.3). Similarly in the sub critical case $\sigma \in \left(0, \frac{1}{n}\right)$ we obtain

$$\|N(v_{m-1}(t), h_{m-1}(t))\|_{\mathbf{L}^\infty}$$
$$\leq Ch_{m-1}^{-1}(t) \|v_{m-1}(t)\|_{\mathbf{L}^\infty}^{1+\sigma} \left(1 + \frac{1}{\theta}\|v_{m-1}(t)\|_{\mathbf{L}^1}\right)$$
$$\leq C\varepsilon^{1+\sigma}(1+t)^{-n-\sigma n} \left(1 + \frac{n|\theta|^\sigma}{1-\sigma n} t^{1-\sigma n}\right)^{-1}$$
$$\leq C\varepsilon(1-\sigma n)(1+t)^{-1-n},$$

$$\|N(v_{m-1}(t), h_{m-1}(t))\|_{\mathbf{L}^1}$$
$$\leq Ch_{m-1}^{-1}(t) \|v_{m-1}(t)\|_{\mathbf{L}^\infty}^\sigma \left(\|v_{m-1}(t)\|_{\mathbf{L}^1} + \frac{1}{\theta}\|v_{m-1}(t)\|_{\mathbf{L}^1}^2\right)$$
$$\leq C\varepsilon(1-\sigma n)(1+t)^{-1}$$

and

$$\|N(v_{m-1}(t), h_{m-1}(t))\|_{\mathbf{L}^{1,n+1}}$$
$$\leq Ch_{m-1}^{-1}(t) \|v_{m-1}(t)\|_{\mathbf{L}^\infty}^\sigma \|v_{m-1}(t)\|_{\mathbf{L}^{1,n+1}} \left(1 + \frac{1}{\theta}\|v_{m-1}(t)\|_{\mathbf{L}^1}\right)$$
$$\leq C\varepsilon(1-\sigma n)(1+t)^{-\frac{1}{2}}$$

for all $t > 0$. This yields the estimate

$$(3.4) \qquad \left\|(1+t)g^l(t)N(v_{m-1}(t), h_{m-1}(t))\right\|_{\mathbf{X}} \leq C\varepsilon^{1+\sigma}$$

if we suppose that $\frac{1}{n} - \sigma \leq \varepsilon$. Since $N(v_{m-1}(\tau), h_{m-1}(\tau))$ satisfy condition $\int_{\mathbf{R}_+^n} \tilde{x}N(t,x)\,dx = 0$, we get via Lemma 35

$$\left\|g^l(t)\int_0^t g^{-l}(\tau)\mathcal{G}(t-\tau)N(v_{m-1}(\tau), h_{m-1}(\tau))\,d\tau\right\|_{\mathbf{X}} \leq C\varepsilon^{1+\sigma};$$

hence it follows that

$$\|v_m\|_{\mathbf{X}} \leq C\varepsilon, \quad \|v_m(t) - \mathcal{G}(t)v_0\|_{\mathbf{L}^{1,n}} \leq C\varepsilon^{1+\sigma}g^{-k}(t).$$

By virtue of Lemma 36 we find that

$$|h_m(t) - \chi_\sigma(t)| \leq \begin{cases} C\varepsilon \log \chi_\sigma(t) & \text{if } \sigma = \frac{1}{n} \\ C\varepsilon^{1+\sigma}\chi_\sigma(t) & \text{if } \sigma \in \left(\frac{1}{n} - \varepsilon, \frac{1}{n}\right) \end{cases}$$

for all $t > 0$. Thus by induction we see that estimates (3.3) are valid for all $m \geq 1$. Again by induction we can prove that

$$\|v_m - v_{m-1}\|_{\mathbf{X}} \leq \frac{1}{4} \|v_{m-1} - v_{m-2}\|_{\mathbf{X}},$$

$$\sup_{t>0} \chi_\sigma^{-1}(t) |h_m(t) - h_{m-1}(t)| \leq \frac{1}{4} \sup_{t>0} \chi_\sigma^{-1}(t) |h_{m-1}(t) - h_{m-2}(t)|$$
$$+ \frac{1}{4} \|v_{m-1} - v_{m-2}\|_{\mathbf{X}}$$

for all $m > 2$. Therefore taking limits $\lim_{m \to \infty} v_m(t, x) = v(t, x)$ and $\lim_{m \to \infty} h_m(t) = h(t)$ we obtain a unique solution $v(t, x) \in \mathbf{X}$, $h(t) \in \mathbf{C}(0, +\infty)$ satisfying the estimates

(3.5) $|h(t) - \chi_\sigma(t)| \leq \begin{cases} C\varepsilon \log \chi_\sigma(t) & \text{if } \sigma = \frac{1}{n} \\ C\varepsilon^{1+\sigma} \chi_\sigma(t) & \text{if } \sigma \in \left(\frac{1}{n} - \varepsilon, \frac{1}{n}\right), \end{cases}$

$$\|v\|_{\mathbf{X}} \leq C\varepsilon, \|x(v(t) - \mathcal{G}(t)v_0)\|_{\mathbf{L}^1} \leq C\varepsilon^{1+\sigma} g^{-k}(t),$$

and integral equations

$$\begin{cases} v(t) = \mathcal{G}(t)v_0 + \int_0^t \mathcal{G}(t-\tau)N(v(\tau), h(\tau)) \, d\tau, \\ h(t) = 1 + \frac{\sigma}{\theta} \int_0^t d\tau \left(\operatorname{Re}\beta \int_{\mathbf{R}_+^n} \tilde{x} |v|^\sigma v dx + h^{1-\frac{\varkappa}{\sigma}} \operatorname{Re}\gamma \int_{\mathbf{R}_+^n} \tilde{x} |v|^\varkappa v dx \right), \end{cases}$$

$$\psi(t) = \arg \int_{\mathbf{R}_+^n} \tilde{x} u_0 dx - \frac{1}{\theta} \int_0^t d\tau e^{-\sigma\varphi} \operatorname{Im}\beta \int_{\mathbf{R}_+^n} \tilde{x} |v|^\sigma v dx$$
$$- \frac{1}{\theta} e^{-\varkappa\varphi} \operatorname{Im}\gamma \int_{\mathbf{R}_+^n} \tilde{x} |v|^\varkappa v dx.$$

By (3.4) we see that

(3.6) $\|v(t) - \mathcal{G}(t)v_0\|_{\mathbf{L}^\infty} \leq \begin{cases} C\varepsilon^{1+\sigma}(1+t)^{-n} g^{-1}(t) & \text{if } \sigma = \frac{1}{n} \\ C\varepsilon^{1+\sigma}(1+t)^{-n} & \text{if } \sigma \in \left(\frac{1}{n} - \varepsilon, \frac{1}{n}\right), \end{cases}$

and by the time decay property of the solution v we have

$$\left| \psi(t) - \arg \int_{\mathbf{R}_+^n} \tilde{x} u_0 dx + |\theta|^\sigma \tilde{\eta} \int_0^t h^{-\frac{1}{\sigma}}(\tau)(1+\tau)^{-\sigma n} d\tau \right|$$
$$\leq C\varepsilon^\varkappa t^{-(\varkappa-\sigma)n},$$

where $\tilde{\eta} = -\operatorname{Im}\beta\delta(\alpha, \sigma)$,

$$\delta(\alpha, \sigma) = t^{n\sigma} \int_{\mathbf{R}_+^n} |F(t, x)|^\sigma F(t, x) \, dx.$$

Hence by (3.5)

$$\left| \psi\left(t\right) - \arg \int_{\mathbf{R}_+^n} \widetilde{x} u_0 dx + |\theta|^\sigma \widetilde{\eta} \int_0^t \chi_\sigma^{-\frac{1}{\sigma}}\left(\tau\right)\left(1+\tau\right)^{-\sigma n} d\tau \right|$$

$$\leq \quad C |\theta|^\sigma \widetilde{\eta} \int_0^t \chi_\sigma^{-1-\frac{1}{\sigma}}\left(\tau\right) |h\left(t\right) - \chi_\sigma\left(t\right)| \left(1+\tau\right)^{-\sigma n} d\tau$$

$$(3.7) \qquad \leq \quad \begin{cases} C \int_0^t \chi_\sigma^{-1-\frac{1}{\sigma}}\left(\tau\right)\left(\log \chi_\sigma\left(t\right)\right)\left(1+\tau\right)^{-1} d\tau \text{ if } \sigma = \frac{1}{n}, \\ C \int_0^t \chi_\sigma^{-\frac{1}{\sigma}}\left(\tau\right)\left(1+\tau\right)^{-\sigma n} d\tau \text{ if } \sigma \in \left(\frac{1}{n}-\varepsilon,\frac{1}{n}\right). \end{cases}$$

Then via formulas

$$u\left(t,x\right) = e^{-\varphi(t)+i\psi(t)} v\left(t,x\right) = h^{-\frac{1}{\sigma}}\left(t\right) e^{i\psi(t)} v\left(t,x\right)$$

we find the estimates

$$\left\| u\left(t\right) - \theta F\left(t\right) e^{-\varphi(t)+i\psi(t)} \right\|_{\mathbf{L}^\infty}$$

$$\leq \quad \left\| u\left(t\right) - \left(\mathcal{G}\left(t\right) v_0\right) e^{-\varphi(t)+i\psi(t)} \right\|_{\mathbf{L}^\infty}$$

$$+ \left\| \left(\mathcal{G}\left(t\right) v_0 - \theta F\left(t\right)\right) e^{-\varphi(t)+i\psi(t)} \right\|_{\mathbf{L}^\infty}$$

$$(3.8) \qquad \leq \quad \begin{cases} C\varepsilon^{1+\sigma}\left(1+t\right)^{-\frac{1}{\sigma}} g^{-1}\left(t\right) \text{ if } \sigma = \frac{1}{n}, \\ C\varepsilon^{1+\sigma}\left(1+t\right)^{-\frac{1}{\sigma}} \text{ if } \sigma \in \left(\frac{1}{n}-\varepsilon,\frac{1}{n}\right), \end{cases}$$

via (3.6) and the inequality

$$\left\| \left(\mathcal{G}\left(t\right) v_0 - \theta F\left(t\right)\right) e^{-\varphi(t)+i\psi(t)} \right\|_{\mathbf{L}^\infty} \leq C t^{-\frac{1}{\sigma}-\frac{1}{2}} \|\phi\|_{\mathbf{L}^{1,n+1}}.$$

We also have by (3.5)

$$\left\| \theta F\left(t\right) h^{-\frac{1}{\sigma}}\left(t\right) e^{i\psi(t)} - \theta F\left(t\right) \chi_\sigma^{-\frac{1}{\sigma}}\left(t\right) e^{i\psi(t)} \right\|_{\mathbf{L}^\infty}$$

$$\leq \quad C\varepsilon t^{-n} \chi_\sigma^{-1-\frac{1}{\sigma}}\left(t\right) |h\left(t\right) - \chi_\sigma\left(t\right)|$$

from which in view of (3.8) it follows that

$$\left\| u\left(t\right) - \theta F\left(t\right) \chi_\sigma^{-\frac{1}{\sigma}}\left(t\right) e^{i\psi(t)} \right\|_{\mathbf{L}^\infty}$$

$$\leq \quad \begin{cases} C\varepsilon^{1+\sigma}\left(1+t\right)^{-\frac{1}{\sigma}} g^{-\frac{1}{\sigma}}\left(t\right) \text{ if } \sigma = \frac{1}{n}, \\ C\varepsilon^{1+\sigma}\left(1+t\right)^{-\frac{1}{\sigma}} \text{ if } \sigma \in \left(\frac{1}{n}-\varepsilon,\frac{1}{n}\right). \end{cases}$$

This completes the proof of Theorem 34.

CHAPTER 12

Burgers Equation with Pumping

1. Introduction

We study the large time asymptotic behavior of solutions to the initial-boundary value problem

(1.1)
$$\begin{cases} u_t + e^t u \, u_x - u_{xx} = 0, \ x \in \mathbf{R}, \ t > 0, \\ u(0, x) = u_0(x), \ x \in \mathbf{R}, \\ u(t, x) \to a_{\pm}, \ x \to \pm\infty, \ t > 0. \end{cases}$$

Equation (1.1) is obtained from the Burgers equation with a pumping $u_t + uu_x - u - u_{xx} = 0$ via a change of the dependent variable $\tilde{u}(t, x) = e^t u(t, x)$. This equation is an interesting example of a simple model of nonlinear interaction of long wave pumping with short wave dissipation. Such types of models were considered earlier in [89] and [121]. By the change of dependent and independent variables

$$u(t, x) = \beta v \left(\beta^2 t, \beta x - \alpha\beta e^t\right) + \alpha,$$

where $\alpha = \frac{1}{2}(a_+ + a_-)$, $\beta = \frac{1}{2}(a_+ - a_-)$ if $a_+ \neq a_-$ and $\beta = 1$, otherwise, we can reduce problem (1.1) to the following three different cases: 1) $a_{\pm} = \pm 1$, 2) $a_{\pm} = \mp 1$ and 3) $a_{\pm} = 0$. We hope that the methods of investigation of the large time asymptotic behavior of solutions to (1.1) developed in the present chapter can be applied also to the Kuramoto and Sivashinskiy equation

$$u_t + uu_x + u_{xx} + u_{xxxx} = 0.$$

Now let us mention known results about large time asymptotic behavior of solutions for some problems similar to (1.1).

Solutions to the Burgers equation

$$u_t + uu_x - u_{xx} = 0$$

are found by the Hopf-Cole substitution $u = -2\partial_x \log \phi$ reducing it to the linear heat equation

$$\phi_t - \phi_{xx} = 0$$

with initial data

$$\phi(0, x) = \phi_0(x) \equiv \exp\left(-\frac{1}{2} \int_0^x u_0(y) \, dy\right).$$

Therefore analyzing the explicit representation of solutions we can find the large time asymptotics

$$u = At^{-\frac{1}{2}}e^{-x^2/4t} + O\left(t^{-\frac{1}{2}-\gamma}\right)$$

for $t \to \infty$ uniformly with respect to $x \in \mathbf{R}$, where $\gamma > 0$, in the case of initial data $u_0(x)$ which tend to zero sufficiently rapidly as $x \to \pm\infty$. If the initial data $u_0(x)$ have a form of a step $u_0(x) \to a_\pm = \mp 1$, $x \to \pm\infty$ then the solution to the Burgers equation converges to the stationary solution $\varphi_0(x) = -\tanh x$ for large time (the so-called shock wave). If $u_0(x) \to a_\pm = \pm 1$ as well, then the solution of the Burgers equation converge as $t \to \infty$ towards to the rarefaction wave.

Paper [96] shows that the solutions of the one-dimensional barotropic model system for compressible viscous gas tends toward the centered rarefaction waves, provided that the initial data are suitably close to rarefaction wave at initial time. In paper [97] the authors prove results on the existence, regularity, and asymptotic behavior of solutions of the initial-value problem for the Burgers type equation

$$u_t + f(u)_x = \mu(|u_x|^p u_x)_x$$

on $\mathbf{R} \times \mathbf{R}^+$, where $p > 0$ and the initial data $u(x, 0) = u_0(x)$ satisfies $u_0(x) \to u_\pm$ as $x \to \pm\infty$, under the assumption $u_- \leq u_+$. Using a priori energy estimates, they proved that a solution $u(t, x)$ converges to the rarefaction wave $\varphi(t, x)$ as $t \to \infty$ uniformly in $x \in \mathbf{R}$, which is the entropy solution of the inviscid equation (with $\mu = 0$) with initial data $\varphi(0, x) = u_-$ if $x < 0$, and $\varphi(0, x) = u_+$ if $x > 0$. In paper [98] the asymptotic stability of traveling wave solutions with shock profile were considered for scalar viscous conservation laws $u_t + f(u)_x = \mu u_{xx}$ with initial data u_0 which converges to constant states $u_\pm$ as $x \to \pm\infty$. In the case of nonconvex flux f, when the shock speed $s = f'(u_\pm)$, the stability results were obtained by applying an elementary weighted energy method to the integrated equation of the original one. Moreover, the rate of asymptotics in time is investigated. For the case $f'(u_+) < s < f'(u_-)$, if the integral of the initial disturbances over $(-\infty, x)$ is small and decays at an algebraic rate as $|x| \to \infty$, then the solution approaches a traveling wave at the corresponding rate as $t \to \infty$. Paper [94] considered the large time asymptotic behavior of solutions of the initial-boundary value problem for the generalized Burgers equation

$$u_t + f(u)_x = u_{xx}$$

on the half-line with the conditions $u(t, 0) = u_-$, $u(t, \infty) = u_+$, where the corresponding Cauchy problem admits a rarefaction wave as an asymptotic state. Because of the Dirichlet boundary condition, the asymptotic states in this problem are divided into five cases dependent on the signs of the characteristic speeds $f'(u_\pm)$ of the boundary state $u_- = u(0)$ and the far field state $u_+ = u(\infty)$. In the case $f'(u_-) < 0 < f'(u_+)$ the solution behaves as a sum of a viscous shock wave as a boundary layer and rarefaction wave propagating away from the boundary. Large time behavior of positive solutions for nonlinear heat equation

$$u_t - \Delta u = -u^{1+\sigma},$$

$\sigma > 0$ was studied extensively (see paper [78] for the super critical case $\sigma > \frac{2}{n}$, [41] for the critical case $\sigma = \frac{2}{n}$ and papers [30], [31], [46], [81] for the sub critical case $\sigma \in (0, \frac{2}{n})$). In paper [46] for the sub critical case $\sigma \in (0, \frac{2}{n})$ it was proved that if the initial data are nonnegative $u_0 \geq 0$, $u_0 \in \mathbf{L}^1$ and decay slowly at infinity such that $\lim_{x \to \pm\infty} |x|^{\frac{2}{\sigma}} u_0(x) = +\infty$, then the solution has the asymptotic representation

$$u(t, x) = t^{-\frac{1}{\sigma}} \sigma^{-\frac{1}{\sigma}} + o\left(t^{-\frac{1}{\sigma}}\right)$$

as $t \to \infty$ uniformly in domains $\{x \in \mathbf{R}^n;\ |x| \leq C\sqrt{t}\}$ with any $C > 0$. On the other hand in papers [30], [31] the initial data $u_0 \in \mathbf{L}^1$, $u_0 \neq 0$ were considered such that $\lim_{|x| \to \infty} |x|^{\frac{2}{\sigma}} u_0(x) = \varkappa \geq 0$. It then was shown that the main term of the asymptotic representation of solutions has a self-similar character

$$u(t, x) = t^{-\frac{1}{\sigma}} w_\varkappa \left(\frac{x}{\sqrt{t}}\right) + o\left(t^{-\frac{1}{\sigma}}\right)$$

as $t \to \infty$ uniformly with respect to $x \in \mathbf{R}^n$, where $w_\varkappa(\xi)$ is a positive solution of equation

$$-\Delta w - \frac{1}{2}\xi\nabla w + w^{1+\sigma} = \frac{1}{\sigma}w,$$

such that $\lim_{|\xi| \to \infty} |\xi|^{\frac{2}{\sigma}} w_\varkappa(\xi) = \varkappa$. Note that there was no restriction on the size of the initial data in these papers.

However the methods of these papers are not applicable to the case of the Cauchy problem (1.1). Note that the nonlinearity in equation (1.1) is sub critical for large time asymptotic behavior (due to the coefficient e^t growing with time), so we cannot find the large time asymptotic expansion of solutions by using the usual methods of treating the nonlinearity as a small perturbation of the linear theory. One should expect that the large time asymptotics differs essentially from the corresponding linearized case. In the case $a_\pm = 0$ there are global existence results (see [103]) and smoothing properties for solutions, and as far as we know the large time asymptotic behavior of solutions is an open question.

We organize the rest of this chapter as follows. In Section 2 we will show that if the initial data are monotonically increasing and have small higher order derivatives, then solutions converge to the rarefaction wave as $t \to \infty$. In Section 3 we consider the case of the shock wave $a_+ < a_-$ and we will show that solutions converge to the self-similar solution $-\tanh(xe^t)$ as $t \to \infty$. The most difficult and intriguing case $a_+ = a_- = 0$ is treated in Section 4.

Finally we would like to comment that the results of this chapter could be extended to a more general nonlinearity $e^t u^m u_x$ with $m \geq 1$, odd, instead of $e^t u u_x$ in equation (1.1). The nonconvex case of even $m \geq 1$ remains an open problem.

We denote the usual Lebesgue space $\mathbf{L}^p = \left\{\phi \in \mathbf{S}';\ \|\phi\|_p < \infty\right\}$, where the norm $\|\phi\|_p = \left(\int_{\mathbf{R}} |\phi(x)|^p\, dx\right)^{1/p}$ if $1 \leq p < \infty$ and $\|\phi\|_\infty = \text{ess.sup}_{x \in \mathbf{R}} |\phi(x)|$

if $p = \infty$. For simplicity we write $\|\cdot\| = \|\cdot\|_2 = \|\cdot\|_{\mathbf{L}^2}$. Weighted Sobolev space is $\mathbf{H}^{m,k} = \left\{ \phi \in \mathbf{S}' : \|\phi\|_{m,k} \equiv \left\| \langle x \rangle^k \langle i\partial \rangle^m \phi \right\| < \infty \right\}$, $m, k \in \mathbf{R}$, $\langle x \rangle = \sqrt{1 + x^2}$. Different positive constants we denote by the same letter C. The results of this chapter were published in paper [66].

2. Rarefaction wave

In this section we consider the case of the rarefaction wave $a_\pm = \pm 1$. Consider the Cauchy problem for the Hopf equation

$$(2.1) \qquad \begin{cases} \varphi_t + e^t \varphi \varphi_x = 0, \ x \in \mathbf{R}, \ t > 0, \\ \varphi(0, x) = \varphi_0(x), \ x \in \mathbf{R}, \end{cases}$$

where the initial data $\varphi_0(x) \in \mathbf{C}^2(\mathbf{R})$ are monotonically increasing $0 < \varphi_0'(x) < C$ for all $x \in \mathbf{R}$ and $\varphi_0(x) \to \pm 1$ as $x \to \pm\infty$ and $\varphi_0(0) = 0$. The solution to problem (2.1) is given by $\varphi(t, y(t, \xi)) = \varphi_0(\xi)$, where the characteristics $y(t, \xi) = \xi + \varphi_0(\xi)(e^t - 1)$ for $\xi \in \mathbf{R}$, $t > 0$. Note that $\varphi(t, 0) = 0$ and

$$\varphi_x(t, x) = \frac{\varphi_0'(\xi)}{1 + \varphi_0'(\xi)(e^t - 1)} > 0$$

for all $x \in \mathbf{R}$, $t > 0$. Therefore

$$\begin{aligned} \|\varphi_x(t)\|^2 &= \int \frac{(\varphi_0'(\xi))^2 \, d\xi}{(1 + \varphi_0'(\xi)(e^t - 1))^2} \\ &\leq \int \varphi_0'(\xi) \, d\xi \sup_{\xi \in \mathbf{R}} \frac{\varphi_0'(\xi)}{(1 + \varphi_0'(\xi)(e^t - 1))^2} \\ &= (\varphi_0'(+\infty) - \varphi_0'(-\infty)) \sup_{\xi \in \mathbf{R}} \frac{\varphi_0'(\xi)}{(1 + \varphi_0'(\xi)(e^t - 1))^2} \\ &\leq C e^{-t} \end{aligned}$$

for all $t > 0$. By a direct calculation we have

$$\varphi_{xx}(t) = \varphi_0'' \left(1 + \varphi_0'(e^t - 1) \right)^{-3},$$

and we also assume that

$$\int_t^\infty \|\varphi_{xx}(\tau)\|_\infty \, d\tau = \int_t^\infty \left\| \varphi_0'' \left(1 + \varphi_0'(e^\tau - 1) \right)^{-3} \right\|_\infty d\tau \to 0$$

as $t \to \infty$.

First we give a sufficiently general result about convergence as $t \to \infty$ of solutions $u(t, x)$ of problem (1.1) to the rarefaction wave $\varphi(t, x)$.

THEOREM 35. *Let* $u_0 - \varphi_0 \in \mathbf{L}^2$. *Then* $u(t, x) \to \varphi(t, x)$ *as* $t \to \infty$ *uniformly with respect to* $x \in \mathbf{R}$.

Proof. For the difference $w = u - \varphi$ we get the Cauchy problem

$$(2.2) \qquad \begin{cases} w_t + e^t w w_x + e^t (\varphi w)_x - w_{xx} - \varphi_{xx} = 0, \ x \in \mathbf{R}, \ t > 0, \\ w(x, 0) = w_0(x), \ x \in \mathbf{R}, \end{cases}$$

where $w_0(x) = u_0(x) - \varphi_0(x) \in \mathbf{L}^2$. By the methods of book [103] we can easily prove existence of a unique solution $w(t,x) \in \mathbf{C}^\infty((0,\infty); \mathbf{H}^{\infty,0}) \cap \mathbf{C}([0,\infty); \mathbf{L}^2)$ to the Cauchy problem (2.2). We now derive energy type a priori estimates for the solution w (multiplying equation (2.2) by w and integrating with respect to x over $\mathbf{R}$)

$$\frac{d}{dt} \|w\|^2 + e^t \int \varphi_x w^2 dx + 2 \|w_x\|^2 + 2 \int w_x \varphi_x dx = 0,$$

whence by the Cauchy inequality and estimate $\|\varphi_x(t)\|^2 \le Ce^{-t}$ we have

$$\frac{d}{dt} \|w\|^2 + e^t \int \varphi_x w^2 dx + \|w_x\|^2 \le Ce^{-t}.$$

Integration with respect to time $t > 0$ yields

$$(2.3) \qquad \|w(t)\|^2 + \int_0^t d\tau \left(e^\tau \int \varphi_x w^2(\tau,x) dx + \|w_x(\tau)\|^2 \right) \le C.$$

Estimate (2.3) shows that $\|w(t)\| \le C$ for all $t > 0$ and via inequalities $\|w\|_\infty^4 \le 2 \|w\|^2 \|w_x\|^2 \le C \|w_x\|^2$ we obtain $\int_0^\infty \|w(t)\|_\infty^4 dt < C$. Therefore $\|w(t_k)\|_\infty \to 0$ for some sequence $t_k \to \infty$. In order to prove that $\|w(t)\|_\infty \to 0$ as $t \to \infty$, let us estimate $m(t) = \inf_{x \in \mathbf{R}} w(t,x)$ and $M(t) = \sup_{x \in \mathbf{R}} w(t,x)$. Since $w \in \mathbf{C}((0,\infty); \mathbf{H}^{1,0})$ we see that $\lim_{|x| \to \infty} w(t,x) = 0$; hence we have $m(t) \le 0$ and $M(t) \ge 0$ for all $t \in (0,\infty)$. By the method of paper [22] we prove the following result.

LEMMA 37. *Let* $w \in \mathbf{C}^1((T_1, T_2); \mathbf{H}^{1,0})$ *and* $m(t) < 0$ *for all* $t \in (T_1, T_2)$. *Then there exists a point* $\zeta(t) \in \mathbf{R}$ *such that* $m(t) = w(t, \zeta(t))$; *moreover* $m'(t) = w_t(t, \zeta(t))$ *almost everywhere on* (T_1, T_2).

PROOF. Since $m(t) = \inf_{x \in \mathbf{R}} w(t,x) < 0$ and $w(t,x) \to 0$ as $|x| \to \infty$ we see that there exists a point $\zeta(t) \in \mathbf{R}$ such that $m(t) = w(t, \zeta(t))$ for all $t \in (T_1, T_2)$. By virtue of the estimate

$$\begin{aligned} m(s) - m(t) &\le w(s, \zeta(t)) - w(t, \zeta(t)) \le \|w(s) - w(t)\|_\infty \\ &\le \left\| \int_t^s w_t(\tau) d\tau \right\|_\infty \le |t - s| \sup_{\tau \in (s,t)} \|w_\tau(\tau)\|_{1,0} \end{aligned}$$

for all $s, t \in (T_1, T_2)$ we see that $m(t)$ has a bounded variation on (T_1, T_2) and, hence, is almost everywhere differentiable on (T_1, T_2) (see [87]). Then we have

$$m'(t) = \lim_{s \to t+0} \frac{m(s) - m(t)}{s - t} \le \lim_{s \to t+0} \frac{w(s, \zeta(t)) - w(t, \zeta(t))}{s - t} = w_t(t, \zeta(t))$$

and

$$m'(t) = \lim_{s \to t-0} \frac{m(t) - m(s)}{t - s} \ge \lim_{s \to t-0} \frac{w(t, \zeta(t)) - w(s, \zeta(t))}{t - s} = w_t(t, \zeta(t))$$

almost everywhere on (T_1, T_2), since by the Sobolev embedding theorem $w_t(t,x) = \lim_{s \to t} \frac{1}{t-s}(w(t,x) - w(s,x))$ uniformly with respect to $x \in \mathbf{R}$. Therefore $m'(t) = w_t(t, \zeta(t))$ almost everywhere on (T_1, T_2). Lemma 37 is proved. $\qquad \square$

We now prove that $m(t) \to 0$ as $t \to \infty$. Recall that $\|w(t_k)\|_\infty \to 0$ for some sequence $t_k \to \infty$. Consider the time interval $T_2 > T_1 \geq t_k$ such that $m(t) < 0$ for all $t \in (T_1, T_2)$. By virtue of Lemma 37 we get from equation (2.2)

$$m' + e^t \varphi_x m - w_{xx}(t, \zeta(t)) - \varphi_{xx}(t, \zeta(t)) = 0$$

for almost all $t \in (T_1, T_2)$, where we have used the condition $w_x(t, \zeta(t)) = 0$. Thus integrating with respect to $t \in (T_1, T_2)$ via $w_{xx}(t, \zeta(t)) \geq 0$ and $\varphi_x m(t) \geq 0$ we have

$$m(t) \geq m(T_1) + \int_{T_1}^t \varphi_{xx}(\tau, \zeta(\tau)) \, d\tau.$$

Since $m(T_1) = 0$ or $m(T_1) = m(t_k)$ and $m(t_k) \to 0$ as $t_k \to \infty$, we have $m(T_1) \to 0$ as $T_1 \to \infty$. Also by our assumption $\left| \int_{T_1}^t \varphi_{xx}(\tau, \zeta(\tau)) \, d\tau \right| \leq \int_{T_1}^\infty \|\varphi_{xx}(t)\|_\infty \, dt = o(1)$ as $T_1 \to \infty$. Therefore $m(t) \to 0$ as $t \to \infty$. Similarly we prove that $M(t) = \sup_{x \in \mathbf{R}} w(t, x) \to 0$ as $t \to \infty$. Hence $\|w(t)\|_\infty \to 0$ as $t \to \infty$. Theorem 35 is proved.

Now we suppose some more conditions to be fulfilled for the initial data $u_0(x)$, and we compute more precisely the asymptotic behavior of solution $u(t, x)$ to problem (1.1). We assume that the initial data $u_0(x)$ monotonically increase and slowly vary, so that the higher order derivatives are less than the first one, such that

$$(2.4) \qquad \sum_{k=1}^{3} \int \left(\left(\frac{1}{u_0'(x)} \frac{d}{dx} \right)^k u_0'(x) \right)^2 (u_0'(x))^{k-\frac{3}{2}} \, dx < \varepsilon,$$

where $\varepsilon > 0$ is sufficiently small. For example, we can take $u_0'(x) = \varepsilon C_0 \langle \varepsilon x \rangle^{-n}$ with $n < 2$, where $C_0 > 0$ can be chosen independently of $\varepsilon > 0$ such that $u_0(x) \to a_\pm = \pm 1$ as $x \to \pm \infty$.

By Theorem 35 we know that solutions of (1.1) are similar to those of the Hopf equation (2.1). Therefore the nonlinearity in equation (1.1) grows with time more rapidly than the term with the second derivative; hence the large time behavior of solutions should be determined by the first two terms in equation (1.1). Thus we try to solve equation (1.1) by the method of characteristics. We define characteristics $y(t, \xi)$ as solutions to the following Cauchy problem

$$\begin{cases} y_t = e^t u(t, y) - \frac{u_{yy}}{u_y}, \ t > 0, \xi \in \mathbf{R}, \\ y(0, \xi) = \xi, \ \xi \in \mathbf{R}. \end{cases}$$

Then from equation (1.1) we get a simple equation

$$\begin{aligned} w_t(t, \xi) &= u_t + u_y y_t = u_t + u_y \left(e^t u - \frac{u_{yy}}{u_y} \right) \\ &= u_t - u_{yy} + e^t u u_y = 0 \end{aligned}$$

for the new dependent variable $w(t, \xi) = u(t, y(t, \xi))$. Hence $w(t, \xi) = u_0(\xi)$ for all $t > 0$, $\xi \in \mathbf{R}$. By a straightforward calculation we have

$$\partial_y u = \frac{u_0'(\xi)}{y_\xi(t, \xi)}, \quad \partial_y^2 u = \frac{u_0''(\xi)}{y_\xi^2(t, \xi)} - \frac{u_0'(\xi) y_{\xi\xi}(t, \xi)}{y_\xi^3(t, \xi)},$$

whence

$$y_t = e^t u_0(\xi) - \frac{1}{u_0'(\xi)} \partial_\xi \left(\frac{u_0'(\xi)}{y_\xi(t, \xi)} \right).$$

We now change the independent variable $\eta = u_0(\xi)$, then the real axis $\xi \in \mathbf{R}$ biuniquelly is transformed to a segment $\Omega = (-1, 1)$ by the assumptions for $u_0(\xi)$. We denote $m(\eta) = u_0'(\xi)$, $\tau = e^t - 1 \geq 0$ and $Z(\tau, \eta) = u_y(t, y) = \frac{m(\eta)}{y_\xi(t, \xi)}$. Then for $Z(\tau, \eta)$ we get the following initial-boundary value problem

(2.5)
$$\begin{cases} Z_\tau = -Z^2(1 + A), \ \tau > 0, \eta \in \Omega, \\ Z(0, \eta) = m(\eta), \ \eta \in \Omega, \\ \partial_\eta^k Z \big|_{\eta = \pm 1} = 0, \ \tau > 0, \ k \geq 1, \end{cases}$$

where $A(\tau, \eta) = -(1 + \tau)^{-1} \partial_\eta^2 Z(\tau, \eta)$. In what follows we only use the cases $k = 1, 2, 3$. From the existence of a unique solution $u(t, x)$ to problem (1.1) it follows that there exists a unique global solution $Z(\tau, \eta) \in \mathbf{C}([0, \infty), \mathbf{C}^2(\Omega)) \cap \mathbf{C}^1((0, \infty), \mathbf{C}(\Omega))$ to the initial-boundary problem (2.5). Integrating equation (2.5) with respect to time $\tau > 0$ we get the representation

(2.6)
$$Z(\tau, \eta) = m(\eta) \left(1 + m(\eta) \left(\tau + \int_0^\tau A(\tau', \eta) \, d\tau' \right) \right)^{-1}.$$

We prove the following result.

THEOREM 36. *Let condition (2.4) for initial data $u_0(x)$ be fulfilled with sufficiently small $\varepsilon > 0$. Then the estimate*

$$\sup_{\eta \in \Omega} \left| \int_0^\tau A(\tau', \eta) \, d\tau' \right| \leq C \sqrt{\varepsilon} \log(1 + \tau)$$

is true for all $\tau \geq 0$.

By virtue of representation (2.6) and the estimate of Theorem 36 we get the asymptotics

$$Z(\tau, \eta) = \frac{m(\eta)}{1 + m(\eta) \tau (1 + O(\tau^{-1} \log \tau))} \quad \text{as } \tau \to \infty,$$

whence

(2.7)
$$u_y(t, y(t, \xi)) = \frac{u_0'(\xi)}{1 + u_0'(\xi) e^t (1 + O(te^{-t}))},$$

where

(2.8)
$$y(t, \xi) = \xi + u_0(\xi)(e^t - 1) - \frac{1}{u_0'(\xi)} \partial_\xi \int_0^t u_y(t', y(t', \xi)) \, dt'.$$

Thus we see that the solution $u(t, x)$ to problem (1.1) asymptotically behaves as a solution of the Hopf equation (2.1). Note that via formulas (2.7) and (2.8) we can obtain a higher-order asymptotic expansion of the solution $u(t, x)$.

PROOF. We make a regularization of problem (2.5). We define Y as a solution to the following problem

(2.9)
$$\begin{cases} Y_\tau = -Y^2 + aY^2 Y_{\eta\eta}, & \tau > 0, \eta \in \Omega, \\ Y(0, \eta) = m(\eta) + \delta, & \eta \in \Omega, \\ \partial_\eta^k Y\big|_{\eta=\pm 1} = 0, & \tau > 0, \ k \geq 1, \end{cases}$$

where $a = (1 + \tau)^{-1}$, $\delta > 0$. By the methods of [103] we easily can prove that there exists a solution

$$Y(\tau, \eta) \in \mathbf{C}\left([0, \infty), \mathbf{C}^2(\Omega)\right) \cap \mathbf{C}^1\left((0, \infty), \mathbf{C}(\Omega)\right)$$

to problem (2.9) such that

$$Y(\tau, \eta) \geq (m(\eta) + \delta)(1 + 2(m(\eta) + \delta)\tau)^{-1} > 0$$

for all $\tau \geq 0, \eta \in \Omega$, therefore the integrals

$$J_k(\tau) = \int_\Omega \left(\partial_\eta^k Y(\tau, \eta)\right)^2 (Y(\tau, \eta))^{k-\frac{5}{2}} d\eta, \quad k = 1, 2, 3$$

are convergent for all $\tau \geq 0$. We now prove that

(2.10)
$$J_k(\tau) < \varepsilon$$

for all $\tau \geq 0$, $k = 1, 2, 3$. For $\tau = 0$ estimate (2.10) follows from (2.4). By the contradiction in view of the continuity of $J_k(\tau)$, we suppose that there exists a time $T > 0$ such that $J_k(T) = \varepsilon$ and

(2.11)
$$J_k(\tau) \leq \varepsilon$$

for all $\tau \in [0, T]$. By the Cauchy inequality we have

$$\left\| (Y_\eta(\tau))^2 Y^{-1}(\tau) \right\|_\infty$$

$$\leq \int_\Omega \left| \partial_\eta \left((Y_\eta(\tau, \eta))^2 Y^{-1}(\tau, \eta) \right) \right| d\eta$$

$$\leq 2 \int_\Omega |Y_{\eta\eta}(\tau, \eta) Y_\eta(\tau, \eta)| Y^{-1}(\tau, \eta) d\eta$$

$$+ \int_\Omega |Y_\eta(\tau, \eta)|^3 Y^{-2}(\tau, \eta) d\eta$$

$$\leq 2(J_1 J_2)^{\frac{1}{2}} + J_1 \left\| (Y_\eta(\tau))^2 Y^{-1}(\tau) \right\|_\infty^{\frac{1}{2}},$$

whence by virtue of (2.11) we obtain

(2.12)
$$\left\| (Y_\eta(\tau))^2 Y^{-1}(\tau) \right\|_\infty \leq C\varepsilon$$

for all $\tau \in [0, T]$. In the same manner

$$(2.13) \qquad \|Y_{\eta\eta}(\tau)\|_\infty^2 \leq 2 \int_\Omega |Y_{\eta\eta}(\tau, \eta) Y_{\eta\eta\eta}(\tau, \eta)| \, d\eta \leq 2 (J_2 J_3)^{\frac{1}{2}} \leq 2\varepsilon.$$

Now differentiating $J_k(\tau)$ with respect to $\eta \in \Omega$, using equation (2.9) and integrating by parts we obtain

$$\frac{dJ_1}{d\tau} = \int_\Omega \left(-\frac{5}{2} Y^{-\frac{1}{2}} Y_\eta^2 - 2a Y^{\frac{1}{2}} Y_{\eta\eta}^2 + \frac{a}{4} Y^{-\frac{3}{2}} Y_\eta^4 \right) d\eta,$$

$$\frac{dJ_2}{d\tau} = \int_\Omega \left(-\frac{2}{3} Y^{-\frac{3}{2}} Y_\eta^4 - \frac{7}{2} Y^{\frac{1}{2}} Y_{\eta\eta}^2 - 2a Y^{\frac{3}{2}} Y_{\eta\eta\eta}^2 \right.$$
$$\left. + \frac{11}{4} a Y^{-\frac{1}{2}} Y_\eta^2 Y_{\eta\eta}^2 + a Y^{\frac{1}{2}} Y_{\eta\eta}^3 \right) d\eta$$

and

$$\frac{dJ_3}{d\tau} = \int_\Omega \left(-\frac{9}{2} Y^{\frac{3}{2}} Y_{\eta\eta\eta}^2 - 2a Y^{\frac{5}{2}} Y_{\eta\eta\eta\eta}^2 - 4a Y^{\frac{1}{2}} Y_{\eta\eta}^4 + 3 Y^{-\frac{1}{2}} Y_\eta^2 Y_{\eta\eta}^2 \right.$$
$$\left. + 6 Y^{\frac{1}{2}} Y_{\eta\eta}^3 - 2a Y^{-\frac{1}{2}} Y_\eta^2 Y_{\eta\eta}^3 + \frac{27}{4} a Y^{\frac{1}{2}} Y_\eta^2 Y_{\eta\eta\eta}^2 + 13a Y^{\frac{3}{2}} Y_{\eta\eta} Y_{\eta\eta\eta}^2 \right) d\eta.$$

Therefore we have

$$\frac{d}{d\tau} (J_1 + J_2 + J_3) \leq -I_1 + I_2,$$

where $I_1 = \int_\Omega \left(Y^{-\frac{1}{2}} Y_\eta^2 + Y^{\frac{1}{2}} Y_{\eta\eta}^2 + Y^{\frac{3}{2}} Y_{\eta\eta\eta}^2 \right) d\eta \geq 0$ and

$$I_2 = \int_\Omega \left(3 Y^{-\frac{1}{2}} Y_\eta^2 Y_{\eta\eta}^2 + 6 Y^{\frac{1}{2}} Y_{\eta\eta}^3 + \frac{a}{4} Y^{-\frac{3}{2}} Y_\eta^4 \right.$$
$$+ \frac{11}{4} a Y^{-\frac{1}{2}} Y_\eta^2 Y_{\eta\eta}^2 + a Y^{\frac{1}{2}} Y_{\eta\eta}^3 - 2a Y^{-\frac{1}{2}} Y_\eta^2 Y_{\eta\eta}^3$$
$$\left. + \frac{27}{4} a Y^{\frac{1}{2}} Y_\eta^2 Y_{\eta\eta\eta}^2 + 13a Y^{\frac{3}{2}} Y_{\eta\eta} Y_{\eta\eta\eta}^2 \right) d\eta.$$

By estimates (2.11), (2.12), (2.13) we have

$$|I_2| \leq C \left(\|Y_\eta^2 Y^{-1}\|_\infty + \|Y_{\eta\eta}\|_\infty + \|Y_{\eta\eta}\|_\infty^2 + \|Y_{\eta\eta}\|_\infty^3 \right) I_1 \leq C\varepsilon I_1;$$

therefore we get

$$\frac{d}{d\tau} (J_1 + J_2 + J_3) \leq 0$$

if $\varepsilon > 0$ is sufficiently small. Hence estimate (2.10) is true for all $\tau \in [0, T]$. The contradiction obtained proves estimate (2.10) for all $\tau \geq 0$. Now taking a limit $\delta \to 0$, we see that estimate (2.10) is valid for the function Z. Now by estimates (2.10) and (2.13) we get

$$\sup_{\eta \in \Omega} \left| \int_0^\tau A(\tau', \eta) \, d\tau' \right| \leq C\sqrt{\varepsilon} \log(\tau + 1)$$

for all $\tau \geq 0$. Theorem 36 is proved. $\qquad\qquad\qquad\qquad\qquad\qquad \square$

3. Shock wave

We now consider another type of the boundary condition $a_\pm = \mp 1$, which corresponds to the shock wave solutions. We introduce the approximate solution

$$V(t, y) = \sum_{k=0}^{n} e^{-2kt} \varphi_k(y),$$

where $y = xe^t$, $n \geq 2$, and the functions $\varphi_k(y)$, $k \geq 0$ are defined recurrently via equations

$$\varphi_0 \varphi_0' - \varphi_0'' = 0$$

with boundary conditions $\varphi_0(y) \to \mp 1$ as $y \to \pm \infty$. We easily see that $\varphi_0(y) = -\tanh \frac{y}{2}$. Then for all $k \geq 1$ we write

$$(3.1) \qquad -2(k-1)\varphi_{k-1} + y\varphi_{k-1}' + \frac{1}{2}\partial_y \sum_{l=0}^{k} \varphi_{k-l}\varphi_l - \varphi_k'' = 0$$

with boundary conditions $\varphi_k(y) \to 0$ for $y \to \pm \infty$. Integrating this identity with respect to y over $(-\infty, y)$ we obtain

$$\varphi_k' = \varphi_k \varphi_0 + \frac{1}{2}\sum_{l=1}^{k-1} \varphi_{k-l}\varphi_l$$

$$-2(k-1)\int_{-\infty}^{y} \varphi_{k-1}(\tilde\eta)\,d\tilde\eta + \int_{-\infty}^{y} \tilde\eta\varphi_{k-1}'(\tilde\eta)\,d\tilde\eta.$$

Multiplying both sides of the above by $\cosh^2 \frac{y}{2}$ and integrating the resulting equation with respect to y over $(-\infty, y)$ again we have

$$\varphi_k(y) = \int_0^y \frac{\cosh^2 \frac{\eta}{2}}{\cosh^2 \frac{y}{2}} \left(-2(k-1)\int_{-\infty}^{\eta} \varphi_{k-1}(\tilde\eta)\,d\tilde\eta \right.$$

$$\left. + \int_{-\infty}^{\eta} \tilde\eta\varphi_{k-1}'(\tilde\eta)\,d\tilde\eta + \frac{1}{2}\sum_{l=1}^{k-1} \varphi_{k-l}(\eta)\,\varphi_l(\eta) \right) d\eta.$$

We find that $\varphi_k(y)$ is an odd function for any $k \geq 0$. Indeed if φ_{k-1} is an odd function, then

$$\int_{-\eta}^{\eta} \varphi_{k-1}(\tilde\eta)\,d\tilde\eta = \int_{-\eta}^{\eta} \tilde\eta\varphi_{k-1}'(\tilde\eta)\,d\tilde\eta = 0$$

and $\varphi_{k-l}(\eta)\,\varphi_l(\eta) = \varphi_{k-l}(-\eta)\,\varphi_l(-\eta)$ which implies that $\varphi_k(y)$ is an odd function. The function $V(t, y)$ is close to shock wave $\varphi_0(y) = -\tanh \frac{y}{2}$ for large time $t \to \infty$; however, the derivatives with respect to x of $V(t, y)$ do not approximate that of φ_0 as $t \to \infty$. This is why we introduce the higher-order corrections $\varphi_k(y)\,e^{-2kt}$, $k \geq 1$ when we want to consider convergence with derivatives of solution $u(t, x)$ as $t \to \infty$.

By virtue of (1.1) and (3.1) we find for the difference $v\left(t,x\right)=u\left(t,x\right)-V\left(t,y\right)$

$$\partial_t v + e^t v v_x + e^t \partial_x \left(vV\right) - v_{xx} + V_t + e^t V V_x - V_{xx}$$
$$= \partial_t v + e^t v v_x + e^t \partial_x \left(vV\right) - v_{xx}$$
$$+ \sum_{k=0}^{n}\left(-2k\right)e^{-2kt}\varphi_k + + \sum_{k=0}^{n}e^{-2kt}y\varphi_k'$$
$$+ e^{2t}\left(\sum_{k=0}^{n}e^{-2kt}\varphi_k\right)\left(\sum_{k=0}^{n}e^{-2kt}\varphi_k'\right) - e^{2t}\left(\sum_{k=0}^{n}e^{-2kt}\varphi_k''\right)$$
$$= \partial_t v + e^t v v_x + e^t \partial_x \left(vV\right) - v_{xx}$$
$$-2ne^{-2nt}\varphi_n + e^{-2nt}y\varphi_n' + \sum_{k,l=1,k+l>n}^{n}e^{t-2(k+l)t}\varphi_k\partial_x\varphi_l = 0,$$

whence integrating with respect to x we get

(3.2) $$w_t + \frac{1}{2}e^t\left(w_x\right)^2 + e^t V w_x - w_{xx} + R = 0,$$

where $w\left(t,x\right)=\int_{-\infty}^{x}v\left(t,x'\right)dx'$ and

$$R\left(t,y\right) = e^{-2nt-t}\left(y\varphi_n\left(y\right) - \left(2n+1\right)\int_{-\infty}^{y}\varphi_n\left(y'\right)dy'\right)$$
$$+ \frac{1}{2}\sum_{k,l=1,k+l>n}^{n}e^{t-2(k+l)t}\varphi_k\left(y\right)\varphi_l\left(y\right).$$

We suppose that the initial data $u_0\left(x\right)$ for the problem (1.1) are near the approximate solution $V\left(t,y\right)$ so that $w\left(t_0,x\right)\cosh x \in \mathbf{H}^{2,0}$ for some time $t=t_0$, where the initial time $t_0 > 0$ we choose to be sufficiently large. In other words, from the beginning the nonlinear effects dominate the linear ones (we could replace this requirement by considering a large coefficient in the nonlinear term of equation (1.1)).

We now prove the following result.

THEOREM 37. *Let the initial time $t_0 > 0$ be sufficiently large and the initial data $u\left(t_0,x\right)\in \mathbf{H}^{2,0}$ be close to the shock wave $V\left(t_0,xe^{t_0}\right)$, that is*

$$\cosh x\int_{-\infty}^{x}\left(u\left(t_0,x'\right)-V\left(t_0,x'e^{t_0}\right)\right)dx'\in \mathbf{H}^{2,0}.$$

Then there exists a unique solution $u\left(t,x\right)$ to the Cauchy problem (1.1) such that $\cosh x\int_{-\infty}^{x}\left(u\left(t,x'\right)-V\left(t,x'e^{t}\right)\right)dx'\in \mathbf{C}\left(\left[t_0,\infty\right);\mathbf{H}^{2,0}\right)$ and the estimate

$$\left\|\cosh x\int_{-\infty}^{x}\left(u\left(t,x'\right)-V\left(t,x'e^{t}\right)\right)dx'\right\|_{2,0} \le Ce^{3t-2nt}$$

is true for all $t \ge t_0$.

Thus we see that the solution $u\,(t,x)$ of the Cauchy problem (1.1) converges to the shock wave $V\,(t,y)$ as $t \to \infty$ uniformly with respect to $x \in \mathbf{R}$.

PROOF. By virtue of equation (3.2) we have for the function $g\,(t,x) = w\,(t,x)\cosh x$

$$(3.3) \qquad g_t + \frac{e^t}{2\cosh x}\,(g_x - g\tanh x)^2 + \chi g + \psi g_x - g_{xx} + R\cosh x = 0,$$

where

$$\chi = \frac{2}{\cosh^2 x} - 1 - e^t V\,(t,y)\tanh x, \quad \psi = 2\tanh x + e^t V\,(t,y)\,.$$

We denote $g^{(m)} = \partial_x^m g\,(t,x)$. We differentiate equation (3.3) with respect to x, multiply by $g^{(m)}$ and then integrate the result with respect to x over $\mathbf{R}$

$$\frac{d}{dt}\left\|g^{(m)}\right\|^2 + 2\int dx g^{(m)}\left(\partial_x^m\left(\frac{e^t}{2\cosh x}\,(g_x - g\tanh x)^2\right)\right.$$

$$(3.4) \qquad +\partial_x^m\,(\chi g + \psi g_x) + \partial_x^m\,(R\cosh x)\Big) + 2\left\|g^{(m+1)}\right\|^2 = 0.$$

We denote $I\,(t) = \sum_{k=0}^2 e^{4nt - 3kt}\left\|g^{(k)}\,(t)\right\|^2$ and prove that

$$(3.5) \qquad\qquad\qquad I\,(t) < C$$

for all $t > t_0$. By contradiction we suppose that there exists $T > t_0$ such that $I\,(T) = C$ and

$$(3.6) \qquad\qquad\qquad I\,(t) \le C$$

for all $t \in [t_0, T]$. By (3.6) we have $\left\|g^{(k)}\,(t)\right\| \le Ce^{\frac{3}{2}kt - 2nt}$, $k = 0,1,2$ for all $t \in [t_0, T]$. We now estimate different terms in equation (3.4)

$$\left\|g^{(m)}\partial_x^m\,(R\cosh x)\right\|_1 \le \left\|g^{(m)}\right\|\left\|\partial_x^m\,(R\cosh x)\right\|$$

$$\le e^{-\frac{3}{2}t - 2nt + mt}\left\|g^{(m)}\right\|$$

for $m = 0,1,2$. Further we have $\left\|\partial_x^k\chi\right\|_\infty + \left\|\partial_x^k\psi\right\|_\infty \le Ce^{t + kt}$, whence

$$\left|\int g^{(m)}\partial_x^m\,(\chi g + \psi g_x)\,dx\right| \le C\left\|g^{(m)}\right\|\sum_{k=0}^m e^{t + kt}\left\|g^{(m-k)}\right\|$$

$$+C\left\|g^{(m)}\right\|\sum_{k=1}^m e^{t + kt}\left\|g^{(m+1-k)}\right\|$$

$$\le Ce^{2t}\left\|g^{(m)}\right\|^2 + Ce^{2mt}\left\|g^{(1)}\right\|^2 + Ce^{2mt}\left\|g^{(0)}\right\|^2$$

for $m = 1,2$. By (3.6) we estimate the nonlinearity as follows

$$\left|\int dx g^{(m)}\partial_x^m\left(\frac{e^t}{2\cosh x}\,(g_x - g\tanh x)^2\right)\right|$$

$$\le \left\|g^{(m+1)}\right\|^2 + Ce^{-t - 4nt + 3mt}$$

for $m = 0, 1, 2$, since $n \geq 2$. Thus from (3.4) we get

$$\frac{d}{dt} \|g_x\|^2 \leq - \|g_{xx}\|^2 + Ce^{2t} \|g_x\|^2 + Ce^{2t} \|g\|^2 + Ce^{2t-4nt}$$

and

$$\frac{d}{dt} \|g_{xx}\|^2 \leq Ce^{2t} \|g_{xx}\|^2 + Ce^{4t} \|g_x\|^2 + Ce^{4t} \|g\|^2 + Ce^{5t-4nt}.$$

For the case $m = 0$ we apply the estimate

$$
\begin{aligned}
\chi - \frac{1}{2}\psi_x &= \frac{1}{\cosh^2 x} - 1 - e^t V(t, y) \tanh x - \frac{1}{2} e^{2t} V_y(t, y) \\
&= e^t \tanh \frac{y}{2} \tanh x + \frac{e^{2t}}{4 \cosh^2 \frac{y}{2}} + O(1) \\
&\geq C \log t \geq 4n + 1
\end{aligned}
$$

for all $t \geq t_0$, if $t_0 > 0$ is sufficiently large, whence

$$\int g (\chi g + \psi g_x) \, dx = \int \left(\chi - \frac{1}{2}\psi_x \right) g^2 dx \geq (4n + 1) \|g\|^2.$$

Then from (3.4) we get

$$\frac{d}{dt} \|g\|^2 \leq -2 \|g_x\|^2 - (4n + 1) \|g\|^2 + Ce^{-4nt-t}.$$

Therefore we have for $I(t) = \sum_{k=0}^{2} e^{4nt-3kt} \left\| g^{(k)}(t) \right\|^2$

$$
\begin{aligned}
\frac{d}{dt} I &\leq -e^{4nt} \left(1 - Ce^{-t} \right) \left(\|g\|^2 + \|g_x\|^2 + e^{-3t} \|g_{xx}\|^2 \right) + Ce^{-t} \\
&\leq -I + Ce^{-t}
\end{aligned}
$$

for all $t \in [t_0, T]$. Therefore

$$I(t) \leq e^{-(t-t_0)} I(t_0) + C(t - t_0) e^{-t} < C$$

for all $t \in [t_0, T]$. The contradiction obtained proves estimate (3.5) for all $t \geq t_0$; hence the result of the theorem is true. Theorem 37 is proved. $\square$

4. Zero boundary conditions

We now consider the most difficult and intriguing type of the boundary conditions $a_{\pm} = 0$. We know (see [103]) that there exists a unique solution $u(t, x) \in C([0, \infty); L^2) \cap C^\infty ((0, \infty); H^{\infty,0})$ to the Cauchy problem

$$
\begin{cases}
u_t + e^t u u_x - u_{xx} = 0, \ x \in \mathbf{R}, \ t > 0, \\
\quad u(0, x) = u_0(x), \ x \in \mathbf{R}
\end{cases}
$$

if the initial data $u_0 \in L^2$. If the initial datum $u_0(x)$ is an odd function, then the solution $u(t, x)$ remains an odd function for all $t > 0$, and it can be obtained as an odd prolongation of the solution to the following Dirichlet boundary-value problem

(4.1)
$$\begin{cases} u_t + e^t u u_x - u_{xx} = 0, \ x \in (-\infty, 0), \ t > 0, \\ u(t, -\infty) = 0, \ u(t, 0) = 0, \ t > 0, \\ u(0, x) = u_0(x), \ x \in (-\infty, 0). \end{cases}$$

Define $\varphi(t, x)$ as a rarefaction wave constructed in Section 2

$$\begin{cases} \varphi_t + e^t \varphi \varphi_x - \varphi_{xx} = 0, \ x \in \mathbf{R}, \ t > 0, \\ \varphi(0, x) = \varphi_0(x), \ x \in \mathbf{R}, \end{cases}$$

where the initial data $\varphi_0(x)$ are monotonically increasing $\varphi_0'(x) > 0$ for all $x \in \mathbf{R}$ and $\varphi_0(x) \to 0$ as $x \to -\infty$. Now we define $r(t, x)$ as a solution to the Dirichlet boundary value problem

(4.2)
$$\begin{cases} r_t + e^t \varphi r r_x - r_{xx} + e^t \varphi_x r(r - 1) - \frac{2\varphi_x}{\varphi} r_x = 0, \ x \in (-\infty, 0), \ t > 0, \\ r(t, -\infty) = 1, \ r(t, 0) = 0, \ t > 0, \\ r(0, x) = r_0(x), \ x \in (-\infty, 0). \end{cases}$$

Then the function $u = \varphi r$ satisfies problem (4.1).

For example, we suppose that the initial data $\varphi_0(x)$ decay at infinity as $\varphi_0(x) = -\frac{1}{x} + O(e^{-|x|})$ as $x \to -\infty$. The general case $\varphi_0(x) = |x|^{-\alpha} + O(e^{-|x|})$ as $x \to -\infty$, where $\alpha > 0$, can also be considered by our method. We use the method of characteristics applied in Section 2 (see formulas (2.6) and (2.8)) to derive asymptotic expansions for the solution $\varphi(x, t)$. We have $\varphi(t, y(t, \xi)) = \varphi_0(\xi)$, $\varphi_y(t, y(t, \xi)) = \varphi_0'(\xi)(y_\xi(t, \xi))^{-1}$ and

$$y(t, \xi) = \xi + \varphi_0(\xi)(e^t - 1) - \frac{1}{\varphi_0'(\xi)} \partial_\xi \int_0^t \varphi_y(t', y(t', \xi)) \, dt'.$$

Define the curve $\xi_0(t)$ such that $y(t, \xi_0(t)) = 0$. We easily see that $\xi_0(t) \to -\infty$ as $t \to \infty$. Using the estimate of Theorem 36 we obtain

$$\begin{aligned} y(t, \xi) &= \xi - (e^t - 1)\eta \\ &\quad - \partial_\eta \int_0^t \eta^2 \left(1 + \eta^2 \left(e^{t'} + O(1 + t')\right)\right)^{-1} dt' + O\left(e^{-e^t}\right) \end{aligned}$$

for $\xi \to -\infty$, $\eta = \frac{1}{\xi} \to 0$. In the first approximation we write

$$y(t, \xi) = \xi - e^t \eta + O(\eta t)$$

; hence $\xi_0^2(t) = e^t + O(t)$, and

$$\xi_0(t) = -e^{\frac{t}{2}} + O\left(te^{-\frac{t}{2}}\right)$$

as $t \to \infty$. In the second approximation we take

$$\begin{aligned} y(t, \xi) &= \xi - (e^t - 1)\eta \\ &\quad - \partial_\eta \int_0^t \eta^2 \left(1 + \eta^2 e^{t'}\right)^{-1} dt' + O\left(t^2 e^{-2t}\right), \end{aligned}$$

whence
$$\xi_0^2(t) = e^t - 2t + 2\log 2 + O\left(t^2 e^{-t}\right).$$

Therefore the asymptotic expansions
$$
\begin{aligned}
\xi_0(t) &= -e^{\frac{t}{2}} - (t - \log 2)\, e^{-\frac{t}{2}} + O\left(t^2 e^{-\frac{3t}{2}}\right), \\
\varphi(t,0) &= e^{-\frac{t}{2}} - (t - \log 2)\, e^{-\frac{3t}{2}} + O\left(t^2 e^{-\frac{5t}{2}}\right), \\
\varphi_x(t,0) &= e^{-t} - 2(t - \log 2)\, e^{-2t} + O\left(t^2 e^{-3t}\right)
\end{aligned}
$$

are valid for $t \to \infty$. Then by virtue of the Taylor formula
$$\varphi(t,x) = \varphi(t,0) + x\varphi_x(t,0) + \frac{1}{2}x^2 \varphi_{xx}(t,\tilde{x})$$

we get
$$
\begin{aligned}
\varphi(t,x) &= e^{-\frac{t}{2}} - (t - y - \log 2)\, e^{-\frac{3t}{2}} + O\left((t^2 + y^2)\, e^{-\frac{5t}{2}}\right), \\
\varphi_x(t,x) &= e^{-t} - (2t - y - 2\log 2)\, e^{-2t} + O\left((t^2 + y^2)\, e^{-3t}\right)
\end{aligned}
$$

for $t \to \infty$. Continuing this procedure we obtain the asymptotic expansions
$$
\begin{aligned}
e^{\frac{t}{2}}\varphi(t,x) &= \sum_{k=0}^{n} a_k(t,y)\, e^{-tk} + O\left((t+|y|)^{n+1}\, e^{-tn-t}\right), \\
e^t \partial_x \varphi(t,x) &= \sum_{k=0}^{n} b_k(t,y)\, e^{-tk} + O\left((t+|y|)^{n+1}\, e^{-tn-t}\right), \\
e^{\frac{t}{2}}\frac{2\varphi_x(t,x)}{\varphi(t,x)} &= \sum_{k=0}^{n} c_k(t,y)\, e^{-tk} + O\left((t+|y|)^{n+1}\, e^{-tn-t}\right)
\end{aligned}
$$

for $t \to \infty$, where $n \geq 3$, $y = xe^{\frac{t}{2}}$; $a_k(t,y)$, $b_k(t,y)$, and $c_k(t,y)$ are polynomials with respect to t, y of the order less then k.

Now as in Section 3 we construct an approximate solution $\Phi(t,y)$ to problem (4.2) in the form $\Phi(t,y) = \sum_{k=0}^{n} \phi_k(t,y)\, e^{-tk}$. Changing the dependent variable $\tilde{r}(t,y) = r(t,x)$ with $y = xe^{\frac{t}{2}}$ we get
$$\tilde{r}_t + \frac{y}{2}\tilde{r}_y + e^{\frac{3}{2}t}\varphi\tilde{r}\tilde{r}_y - e^t \tilde{r}_{yy} + e^t \varphi_x \tilde{r}(\tilde{r} - 1) - \frac{2\varphi_x}{\varphi}e^{\frac{t}{2}}\tilde{r}_y = 0.$$

Then for the difference $w(t,x) = \tilde{r}(t,y) - \Phi(t,y)$ we obtain
$$
\begin{aligned}
&w_t - w_{xx} + e^t\,(\varphi\Phi w)_x + \frac{1}{2}e^t\,(\varphi w^2)_x + \frac{1}{2}e^t \varphi_x w^2 \\
&\quad + e^t \varphi_x\,(\Phi - 1)\, w - \frac{2\varphi_x}{\varphi}w_x + R = 0,
\end{aligned}
$$

(4.3)

where the remainder term
$$
\begin{aligned}
R \equiv\ & \Phi_t + \frac{y}{2}\Phi_y - e^t \Phi_{yy} + e^{\frac{3}{2}t}\varphi\Phi\Phi_y \\
& + e^t \varphi_x \Phi\,(\Phi - 1) - \frac{2\varphi_x}{\varphi}e^{\frac{t}{2}}\Phi_y.
\end{aligned}
$$

The functions $\phi_k(t, y)$, $0 \leq k \leq n$, are defined recurrently via equations (which are obtained by comparing terms containing e^{t-tk})

(4.4) $$\phi_0'' - a_0\phi_0\phi_0' = 0, \quad \phi_k'' - a_0(\phi_0\phi_k)' = z_k, \quad k \geq 1,$$

where

$$z_k(t, y) = \sum_{j=0}^{k-1}\sum_{l=0}^{j}\left(a_{k-j}\phi_{j-l}\phi_l' + b_{k-1-j}\phi_{j-l}\phi_l\right) + \sum_{l=1}^{k-1}a_0\phi_{k-l}\phi_l'$$

$$- \sum_{j=0}^{k-1}\left(b_{k-1-j}\phi_j + c_{k-1-j}\phi_j'\right) + \partial_t\phi_{k-1} - (k-1)\phi_{k-1} - \frac{y}{2}\phi_{k-1}'$$

for $k \geq 1$. By the boundary conditions we have $\phi_0(y) \to 1$, $\phi_k(y) \to 0$, $k \geq 1$ for $y \to -\infty$ and $\phi_k(0) = 0$, $k \geq 0$. Integrating equations (4.4) with $a_0 = 1$, we get $\phi_0(y) = -\tanh\left(\frac{y}{2}\right)$ and

$$\phi_k(t, y) = \frac{1}{\cosh^2\frac{y}{2}}\int_0^y \cosh^2\frac{\eta}{2}\int_{-\infty}^{\eta} z_k(\eta', t)\,d\eta', \quad \text{for } k \geq 1.$$

We have the estimates

$$|\phi_k(t, y)| \leq C\left(1 + t + y^2\right)^k e^{-|y|}$$

for $k \geq 1$. Therefore for the remainder term R we get

$$R = -\sum_{k=n}^{3n-1}e^{-tk}\sum_{j=k+1-n}^{k+1}\sum_{l=\max(0,j-n)}^{\min(n,j)}a_{k+1-j}\phi_{j-l}\phi_l'$$

$$+ e^{-tn}\left(\partial_t\phi_n - n\phi_n - \frac{y}{2}\phi_n'\right) + \sum_{k=n}^{3n}e^{-tk}\sum_{j=k-n}^{k}\sum_{l=\max(0,j-n)}^{\min(n,j)}b_{k-j}\phi_{j-l}\phi_l$$

$$- \sum_{k=n}^{2n}e^{-tk}\sum_{j=k-n}^{\min(k,n)}\left(b_{k-j}\phi_j + c_{k-j}\phi_j'\right) + e^t\left(e^{\frac{t}{2}}\varphi - \sum_{k=0}^{n}a_ke^{-tk}\right)\Phi\Phi_y$$

$$+ \left(e^t\varphi_x - \sum_{k=0}^{n}b_ke^{-tk}\right)\Phi(\Phi - 1) - \left(\frac{2\varphi_x}{\varphi}e^{\frac{t}{2}} - \sum_{k=0}^{n}c_ke^{-tk}\right)\Phi_y.$$

Hence the estimate

$$\partial_x^j R = O\left(\left(1 + t + y^2\right)^n e^{\frac{tj}{2} - nt}e^{-|y|}\right)$$

is true. Denoting $W(t, x) = \partial_x^{-1}(w(t, x))$, where $\partial_x^{-1} = \int_{-\infty}^{x} dx'$ and integrating equation (4.3) from $-\infty$ to x, we then find

$$W_t - W_{xx} + e^t \varphi \Phi W_x + \frac{1}{2} e^t \varphi w^2$$

$$+ \frac{1}{2} e^t \partial_x^{-1} (\varphi_x w^2) + e^t \varphi_x (\Phi - 1) W - e^t \partial_x^{-1} ((\varphi_x (\Phi - 1))_x W)$$

(4.5) $$- \frac{2\varphi_x}{\varphi} W_x + 2\partial_x^{-1} \left(\left(\frac{\varphi_x}{\varphi} \right)_x w \right) + R_1 = 0$$

with the Neumann boundary condition $W_x(t, 0) = w(t, 0) = 0$, where

$$R_1 = \partial_x^{-1} R = O\left((1 + t + y^2)^n e^{-\frac{t}{2} - nt} e^{-|y|} \right).$$

We suppose that the initial data $r(t_0, x)$ are sufficiently close to $\Phi(t, y)$ so that the function $W(t, x) \cosh 2x \in \mathbf{H}^{2,0}((-\infty, 0))$, and the initial time $t = t_0$ is sufficiently large. The last requirement can be replaced by a sufficiently large coefficient at the nonlinear term in equation (1.1) so that nonlinear effects dominate upon the linear ones from the beginning.

We prove the following result.

THEOREM 38. *Let initial time $t_0 > 0$ be sufficiently large, and the initial data $u_0(x) \in \mathbf{H}^{2,0}$ be an odd function and close to the shock wave $\Phi\left(t_0, xe^{t_0/2}\right)$, that is*

$$\cosh 2x \int_{-\infty}^{x} \left(\frac{u_0(x')}{\varphi_0(x')} - \Phi\left(t_0, e^{t_0/2} x'\right) \right) dx' \in \mathbf{H}^{2,0}((-\infty, 0)),$$

where $\varphi_0(x)$ is such that $\varphi_0'(x) > 0$ for all $x \in (-\infty, 0)$ and $\varphi_0(x) = -\frac{1}{x} + O\left(e^{-|x|}\right)$ as $x \to -\infty$. Then a unique solution $u(t, x)$ to the Cauchy problem (1.1) has asymptotic representation

$$u(t, x) = \varphi(t, -|x|) \Phi\left(t, -e^{t/2} |x|\right) sign(x) + O\left(e^{-t}\right)$$

for $t \to \infty$ uniformly with respect to $x \in \mathbf{R}$.

Since the solution $u(t, x)$ is represented as $u(t, x) = r(t, x) \varphi(t, x)$, then the result of Theorem 38 follows from the next lemma.

LEMMA 38. *Let initial time $t_0 > 0$ be sufficiently large and the initial data $r(t_0, x) \in \mathbf{H}^{2,0}((-\infty, 0))$ be close to the shock wave $\Phi\left(t_0, xe^{t_0/2}\right)$:*

$$\cosh 2x \int_{-\infty}^{x} \left(r(t_0, x') - \Phi\left(t_0, e^{t_0/2} x'\right) \right) dx' \in \mathbf{H}^{2,0}((-\infty, 0)).$$

Then there exists a unique solution $r(t, x)$ to the Cauchy problem (4.2) such that $\cosh x \int_{-\infty}^{x} (u(t, x') - V(t, e^{t/2} x')) dx' \in \mathbf{C}([t_0, \infty); \mathbf{H}^{2,0})$ and the estimate

$$\left\| \cosh x \int_{-\infty}^{x} \left(r(t, x') - \Phi\left(t, e^{t/2} x'\right) \right) dx' \right\|_{2,0} \leq C e^{2t - nt}$$

is true for all $t \geq t_0$, where $n \geq 3$.

Thus we see that the solution $r(t, x)$ to problem (4.2) converges to the shock wave $\Phi(t, y)$ as $t \to \infty$ uniformly with respect to $x \in (-\infty, 0)$.

PROOF. We denote

$$
\begin{aligned}
g(t, x) &= W(t, x) \cosh 2x, \\
h(t, x) &= w(t, x) \cosh 2x, \\
v(t, x) &= W(t, x) \cosh x \text{ and} \\
s(t, x) &= w(t, x) \cosh x.
\end{aligned}
$$

We prove that

$$
(4.6) \qquad I(t) \equiv e^{-21t} \|g\|_\infty^2 + e^{2nt} \left(\|v\|^2 + e^{-t} \|s\|^2 + e^{-2t} \|s_x\|^2 \right) < C
$$

for all $t > t_0$. By the contradiction we suppose that there exists $T > t_0$ such that $I(T) = C$ and

$$
(4.7) \qquad\qquad\qquad\qquad I(t) \leq C
$$

for all $t \in [t_0, T]$. By (4.5) we get

$$
g_t - g_{xx} + \chi_1 g_x - \psi_1 g - e^t \cosh(2x) \partial_x^{-1} \left((\varphi_x (\Phi - 1))_x \frac{g}{\cosh x} \right)
$$

$$
+ e^t \cosh(2x) \partial_x^{-1} \left(\frac{\varphi_x s^2}{2 \cosh^2 x} \right) - \left(\frac{2\varphi_x}{\varphi} - \frac{e^t \varphi s}{2 \cosh x} \right) h
$$

$$
(4.8) \qquad + \cosh(2x) \partial_x^{-1} \left(\left(\frac{\varphi_x}{\varphi} \right)_x \frac{2h}{\cosh 2x} \right) = -R_1 \cosh 2x
$$

with boundary condition $g_x(t, 0) = 0$, where

$$
\chi_1 = 4 \tanh 2x + e^t \varphi \Phi,
$$

$$
\psi_1 = 4 \tanh^2 2x - \frac{4}{\cosh^2 2x} + 2e^t \varphi \Phi \tanh 2x - e^t \varphi_x (\Phi - 1).
$$

Since $\|\varphi\|_\infty \leq Ce^{-\frac{t}{2}}$, $\|\varphi_x\|_\infty \leq Ce^{-t}$, we obtain

$$
\psi_1 + \frac{1}{2} (\chi_1)_x = 4 \tanh^2 2x + 2e^t \varphi \Phi \tanh 2x + \frac{1}{2} e^t \varphi_x \Phi + \frac{1}{2} e^{\frac{3}{2}t} \varphi \Phi_y \leq 5
$$

for $t \geq t_0$ if $t_0 > 0$ is sufficiently large. Also via (4.7) $\|s\|_\infty \leq Ce^{\frac{3t}{4} - nt}$, so by the Young inequality we have the estimates

$$
\left\| e^t \cosh(2x) \partial_x^{-1} \left((\varphi_x (\Phi - 1))_x \frac{g}{\cosh x} \right) \right\|
$$

$$
\leq Ce^t \|\varphi_{xx}\|_\infty \|\Phi - 1\|_1 + Ce^{\frac{3t}{2}} \|\varphi_x\|_\infty \|\Phi_y\|_1 \|g\| \leq Ce^{-nt},
$$

$$
\left\| e^t \cosh(2x) \partial_x^{-1} \left(\frac{\varphi_x s^2}{2 \cosh^2 x} \right) \right\| \leq Ce^t \|\varphi_x\|_\infty \|s\|_\infty \|s\| \leq e^{-nt}
$$

and

$$\left\| \cosh(2x)\, \partial_x^{-1}\left(\left(\frac{\varphi_x}{\varphi}\right)_x \frac{2h}{\cosh 2x}\right)\right\| \;\leq\; C\left\|\left(\frac{\varphi_x}{\varphi}\right)_x h\right\| \leq Ce^{-\frac{t}{2}}\,\|h\|,$$

$$\left\|\left(\frac{2\varphi_x}{\varphi} - \frac{e^t\varphi s}{2\cosh x}\right)h\right\| \;\leq\; Ce^{-\frac{t}{2}}\,\|h\|.$$

Applying the energy method and taking into account the boundary conditions $g_x(t,0) = \chi_1(t,0) = 0$, we get from (4.8)

$$(4.9) \qquad \frac{d}{dt}\,\|g(t)\|^2 + \|h\|^2 \leq 20\,\|g\|^2 + Ce^{-2nt}.$$

Now for the function $h = g_x - 2g\tanh 2x$ we have from equation (4.3)

$$(4.10) \qquad h_t - h_{xx} + \chi_2 h_x - \psi_2 h + \frac{e^t\varphi}{\cosh 2x}\,hh_x + R\cosh 2x = 0$$

with boundary condition $h(t,0) = 0$, where

$$\chi_2 = 4\tanh 2x + e^t\varphi\Phi - \frac{2\varphi_x}{\varphi},$$

$$\psi_2 \;=\; 4\tanh^2 2x - \frac{4}{\cosh^2 2x} - e^t(\varphi\Phi)_x - e^t\varphi_x(\Phi - 1)$$

$$+2e^t\varphi\Phi\tanh 2x - \frac{4\varphi_x}{\varphi}\tanh 2x + 2e^t\frac{\varphi s\tanh 2x}{\cosh x} - \frac{e^t\varphi_x s}{\cosh x}.$$

We have

$$\psi_2 + \frac{1}{2}(\chi_2)_x \leq Ce^t$$

and

$$\left| \int_{-\infty}^0 \frac{e^t\varphi}{\cosh 2x}\,h^2 h_x\,dx\right| \;\leq\; Ce^t\left\|\left(\frac{\varphi}{\cosh 2x}\right)_x h^3\right\|_1$$

$$\leq\; Ce^{\frac{t}{2}}\,\|s\|_\infty\,\|h\|^2 \leq C\,\|h\|^2.$$

Hence by the energy method, taking into account the boundary condition $h(t,0) = 0$, we get from (4.10)

$$(4.11) \qquad \frac{d}{dt}\,\|h(t)\|^2 + 2\,\|h_x\|^2 \leq Ce^t\,\|h\|^2 + Ce^{-2nt}.$$

By virtue of (4.9) and (4.11) we obtain

$$\frac{d}{dt}\left(\|g(t)\|^2 + e^{-2t}\,\|h(t)\|^2\right)$$

$$\leq\; 20\left(\|g(t)\|^2 + e^{-2t}\,\|h(t)\|^2\right) + Ce^{-2nt},$$

whence integrating with respect to t we get

$$\|g(t)\|^2 + e^{-2t}\,\|h(t)\|^2 \leq Ce^{20t};$$

therefore

(4.12)
$$\|g\|_\infty^2 \le C \|g\|^2 + C \|g\| \|h\| \le Ce^{21t}$$

for all $t \in [t_0, T]$.

We now estimate the norms $\|v\|$, $\|s\|$ and $\|s_x\|$. By (4.5) we find for $v(t,x) = W(t,x)\cosh x$

$$v_t - v_{xx} + \chi_3 v_x - \psi_3 v + 2\cosh x \partial_x^{-1} \left(\left(\frac{\varphi_x}{\varphi}\right)_x \frac{s}{\cosh x} \right)$$

$$-e^t \cosh x \partial_x^{-1} \left((\varphi_x(\Phi - 1))_x \frac{v}{\cosh x} \right) + \frac{e^t \varphi s^2}{2\cosh x}$$

(4.13)
$$+e^t \cosh x \partial_x^{-1} \left(\frac{\varphi_x s^2}{2\cosh^2 x} \right) + R_1 \cosh x = 0,$$

where

$$\chi_3 = 2\tanh x + e^t \varphi \Phi - \frac{2\varphi_x}{\varphi},$$

$$\psi_3 = \tanh^2 x - \frac{1}{\cosh^2 x} + e^t \varphi \Phi \tanh x - e^t \varphi_x (\Phi - 1) - \frac{2\varphi_x}{\varphi}\tanh x.$$

By the energy method (we multiply equation (4.13) by v and integrate with respect to x over $(-\infty, 0)$ taking into account the boundary condition $v_x(t, 0) = 0$) we get

(4.14)
$$\frac{d}{dt}\|v\|^2 + \|s\|^2 - \int_{-\infty}^0 ((\chi_3)_x + 2\psi_3) v^2 dx \le C\|v\|^2 + Ce^{-t-2nt},$$

since

$$\left\| \frac{e^t \varphi s^2}{2\cosh x} \right\| \le Ce^{\frac{t}{2}} \|s\|_\infty \|s\| \le \frac{1}{4}\|s\|,$$

and

$$\|R_1 \cosh x\| \le Ce^{-\frac{t}{2}-nt}.$$

Via the Young inequality we also have the estimates

$$\left\| \cosh x \partial_x^{-1} \left(\left(\frac{\varphi_x}{\varphi}\right)_x \frac{s}{\cosh x} \right) \right\| \le Ce^{-t}\|s\| \le \frac{1}{4}\|s\|,$$

$$\left\| e^t \cosh x \partial_x^{-1} \left((\varphi_x(\Phi - 1))_x \frac{v}{\cosh x} \right) \right\|$$
$$\le\quad Ce^t \|\varphi_{xx}\|_\infty \|\Phi - 1\|_1 \|v\| + Ce^{\frac{3t}{2}} \|\varphi_x\|_\infty \|\Phi_y\|_1 \|v\| \le C\|v\|$$

and

$$\left\| e^t \cosh x \partial_x^{-1} \left(\frac{\varphi_x s^2}{2\cosh^2 x} \right) \right\| \le Ce^t \|\varphi_x\|_\infty \|s\|_\infty \|s\| \le \frac{1}{4}\|s\|.$$

We further estimate

$$(\chi_3)_x + 2\psi_3 \quad = \quad -\varphi \left(2e^t \tanh \frac{y}{2} \tanh x + \frac{e^{\frac{3t}{2}}}{2\cosh^2 \frac{y}{2}} \right)$$

$$+ O(1) \quad \le \quad -C \log t$$

for all $|x| \leq e^{\frac{t}{2}}$ since

$$\varphi(t, x) \geq e^{-\frac{t}{2}}$$

in the region $|x| \leq e^{\frac{t}{2}}$. Here we used an important fact that in view of estimate (4.7)

$$\|g\|_\infty^2 \leq Ce^{21t}.$$

Therefore

$$\int_{-\infty}^{-e^{t/2}} v^2(t, x) \, dx = \int_{-\infty}^{-e^{t/2}} e^{-2|x|} e^{2|x|} v^2(t, x) \, dx$$

$$\leq C \|g(t)\|_\infty^2 \int_{-\infty}^{-e^{t/2}} e^{-2|x|} dx \leq Ce^{21t - e^{t/2}} \leq Ce^{-2t - 2nt},$$

whence we get

$$\int_{-\infty}^0 ((\chi_3)_x + 2\psi_3) v^2 dx \leq -C \|v\|^2 \log t + Ce^{-t - 2nt}.$$

In addition from (4.14) we obtain

$$(4.15) \qquad \frac{d}{dt} \|v\|^2 + \|s\|^2 \leq -(2n + 1) \|v\|^2 + Ce^{-t - 2nt}$$

for all $t \in [t_0, T]$.

Let us consider now the estimates for the function $s = v_x - v \tanh x = w \cosh x$. From equation (4.3) we have

$$(4.16) \qquad s_t - s_{xx} + \chi_4 s_x - \psi_4 s + \frac{e^t \varphi s s_x}{\cosh x}$$

$$+ e^t \frac{\varphi_x - \varphi \tanh x}{\cosh x} s^2 + R \cosh x$$

$$= 0,$$

with the boundary condition $s(t, 0) = 0$, where

$$\chi_4 = 2 \tanh x + e^t \varphi \Phi - \frac{2\varphi_x}{\varphi},$$

$$\psi_4 = \frac{1}{\cosh^2 x} - \tanh^2 x + e^t \varphi \Phi \tanh x - e^t (\varphi \Phi)_x$$

$$- \frac{2\varphi_x}{\varphi} \tanh x - e^t \varphi_x (\Phi - 1).$$

Applying the energy method to (4.16) we obtain

$$(4.17) \qquad \frac{d}{dt} \|s\|^2 + \|s_x\|^2 \leq 3e^t \|s\|^2 + Ce^{-2nt}$$

since

$$\|R \cosh x\| \leq Ce^{-nt}, \left\| s^2 ((\chi_4)_x + 2\psi_4) \right\|_1 \leq 2e^t \|s\|^2,$$

$$\left| \int_{-\infty}^{0} \frac{e^t \varphi}{\cosh x} s^2 s_x dx \right| \quad = \quad C e^t \left\| s^3 \left(\frac{\varphi}{\cosh x} \right)_x \right\|_{\mathbf{L}^1}$$

$$\leq \quad C e^{\frac{t}{2}} \|s\|_{\mathbf{L}^\infty} \|s\|_{\mathbf{L}^2}^2 \leq C \|s\|_{\mathbf{L}^2}^2$$

and

$$e^t \left\| \left(\frac{\varphi_x}{\cosh x} - \frac{\varphi}{\cosh x} \tanh x \right) s^2 \right\|_{\mathbf{L}^2} \leq C e^{\frac{t}{2}} \|s\|_{\mathbf{L}^\infty} \|s\|_{\mathbf{L}^2} .$$

In the same way we obtain from (4.16)

$$(4.18) \qquad \frac{d}{dt} \|s_x\|^2 + \|s_{xx}\|^2 \leq e^{2t} \|s\|^2 + 3 e^t \|s_x\|^2 + C e^{-2nt},$$

since

$$\left| \int_{-\infty}^{0} s_x \left(\chi_4 s_x - \psi_4 s \right)_x dx \right| \leq e^{\frac{3t}{2}} \|s\| \|s_x\| + 2 e^t \|s_x\|^2 ,$$

and by the use of the inequality

$$\|s_x\|_{\infty}^2 \leq 2 \|s_x\| \|s_{xx}\| + 2 \|s_x\|^2$$

we have

$$e^t \left| \int_{-\infty}^{0} s_x \left(\frac{\varphi s s_x}{\cosh x} \right)_x dx \right| \leq C e^t \left\| (s_x)^2 \left(\frac{\varphi s}{\cosh x} \right)_x \right\|_1$$

$$\leq \quad C e^{\frac{t}{2}} \|s_x\|_{\infty} \|s_x\|^2 + C \|s\|_{\infty} \|s_x\|^2 \leq C \|s_x\|^2 + \frac{1}{2} \|s_{xx}\|^2 .$$

Also

$$e^t \left\| s_x \left(\left(\frac{\varphi_x}{\cosh x} - \frac{\varphi \tanh x}{\cosh x} \right) s^2 \right)_x \right\|_1$$

$$\leq \quad C e^{\frac{t}{2}} \|s\|_{\infty} \|s\|_{1,0}^2 \leq C \|s\|^2 + C \|s_x\|^2$$

and

$$\|\partial_x \left(R \cosh x \right)\| \leq C e^{-nt}.$$

Thus for

$$J(t) = 15 \|v(t)\|^2 + 4 e^{-t} \|s(t)\|^2 + e^{-2t} \|s_x(t)\|^2$$

we get from (4.15), (4.17) and (4.18)

$$\frac{d}{dt} J \quad \leq \quad 15 \left(- (2n+1) \|v\|^2 - \|s\|^2 + C e^{-t-2nt} \right)$$

$$+ 4 e^{-t} \left(- \|s_x\|^2 + 3 e^t \|s\|^2 + C e^{-2nt} \right)$$

$$+ e^{-2t} \left(e^{2t} \|s\|^2 + 3 e^t \|s_x\|^2 + C e^{-2nt} \right) + C e^{-2nt}$$

$$\leq \quad -30n \|v\|^2 - \|s\|^2 - e^{-t} \|s_x\|^2 + C e^{-2nt}$$

$$\leq \quad - (2n+1) J + C e^{-2nt}.$$

Integrating the last inequality with respect to time we get

$$J(t) \leq e^{-(2n+1)(t-t_0)} J(t_0) + C e^{-(2n+1)t} \left(e^t - e^{t_0} \right) < C$$

for all $t \in [t_0, T]$. Therefore in view of (4.12) we see that estimate (4.6) is valid for all $t \in [t_0, T]$. The contradiction obtained proves estimate (4.6) for all $t \geq t_0$. Thus the result of the lemma is true. Lemma 38 is proved. □

CHAPTER 13

KdVB Equation on a Segment

1. Introduction

We study global existence and large time asymptotic behavior of solutions to the initial-boundary value problem for the Korteweg-de Vries-Burgers equation in the interval $(0, 1)$

(1.1)
$$\begin{cases} u_t + uu_x - u_{xx} + u_{xxx} = 0, \ t > 0, x \in (0, 1) \\ u(x, 0) = u_0(x), \ x \in (0, 1) \\ u(0, t) = u(1, t) = u_x(1, t) = 0, \ t > 0. \end{cases}$$

As far as we know the large time asymptotic behavior for solutions of the initial-boundary value problem for the Korteweg-de Vries-Burgers equation (1.1) on the interval was not studied previously. In this chapter we consider (1.1) in the case of the initial data belonging to $\mathbf{L}^2$. We note here that we do not assume the smallness condition for the data. In the case of large initial data it is more difficult than that of small data to obtain an exact representation of large time asymptotics of solutions, and there are few results (see, for example [76]). Another difficulty in the study of the boundary value problem for the Korteweg-de Vries-Burgers equation (1.1) is that the linear operator $-\partial_x^2 + \partial_x^3$ is not self-adjoint, and we cannot apply the Fourier method when we take the boundary value into account. To avoid this difficulty we apply the Laplace transformation with respect to time to derive the Green function of the resulting equation. For obtaining $\mathbf{L}^p$ - estimates of the Green function we use the method of papers [58] and [60].

To state the results of the present chapter precisely we give some notations. Let us denote

$$\mathbf{H}^1(0, 1) = \left\{ \varphi \in \mathbf{L}^2(0, 1) \, ; \|\varphi\|_{\mathbf{H}^1} = \|\varphi\|_{\mathbf{L}^2} + \|\varphi_x\|_{\mathbf{L}^2} < +\infty \right\}.$$

By the same letter C we denote different positive constants.

We state the main result of this chapter.

THEOREM 39. *Suppose that the initial data $u_0 \in \mathbf{L}^2(0, 1)$. Then there exists a unique solution u of (1.1) such that $u \in \mathbf{C}\left((0, +\infty) \, ; \mathbf{H}^1(0, 1)\right)$. Furthermore the solution has the following asymptotics*

$$u(x, t) = A\Lambda(x)e^{-\xi_0 t} + O\left(e^{-(\xi_0 + \delta)t}\right)$$

for $t \to +\infty$ *uniformly with respect to* $x \in (0, 1)$. *Here* $\xi_0 > 0$ *is the first root of the equation*

(1.2)
$$\sum_{j=1}^{3} e^{-\phi_j} \phi'_j (-\xi) = 0$$

where $\phi_l(\xi)$ *are the roots of the characteristic equation* $-p^2 + p^3 + \xi = 0$, *such that* $\operatorname{Re} \phi_l(\xi) > 0$, $l = 1, 2$, *and* $\operatorname{Re} \phi_3(\xi) < 0$, *for all*

$$\xi \in D_0 = \left\{ \xi \in \mathbf{C} : \operatorname{Re} \xi \geq 0, \xi \notin \left[0, \frac{4}{27} \right] \right\}.$$

A constant $\delta > 0$ *satisfies the condition* $\xi_0 + \delta < \xi_1$, *where* ξ_1 *is the second root of equation (1.2), the constant* A, *and the function* $\Lambda(x) \in \mathbf{L}^\infty$ *are defined by*

$$\Lambda(x) = -\frac{\widetilde{\triangle}(-\xi_0, 1-x)}{\widetilde{\triangle}'(-\xi_0, 1)}, \widetilde{\triangle}(\xi, y) = \sum_{j=1}^{3} e^{-\phi_j y} \phi'_j (\xi),$$

$$\widetilde{\triangle}'(\xi, 1) = \sum_{j=1}^{3} e^{-\phi_j y} \left(\phi''_j (\xi) - (\phi'_j (\xi))^2 \right),$$

$$A = \int_0^1 u_0(y) \widetilde{\triangle}(-\xi_0, y) dy$$

$$+ \int_0^{+\infty} e^{\xi_0 \tau} d\tau \int_0^1 u(y, \tau) u_y(y, \tau) \widetilde{\triangle}(-\xi_0, y) dy.$$

REMARK 16. *By virtue of the numerical computations via Maple program we have* $\xi_0 \approx 70$ *and* $\xi_1 \approx 200$.

We organize this chapter as follows. In Section 2 we solve the linear initial-boundary value problem corresponding to (1.1). In Section 3 we prove the local existence of solutions to (1.1). Section 4 is devoted to the proof of global existence of solutions to (1.1) for the case of small initial data. We prove Theorem 39 in Section 5 by using the time decay estimates of solutions obtained in Section 4.

2. Linear problem

We consider the following linear initial-boundary value problem

(2.1)
$$\begin{cases} u_t - u_{xx} + u_{xxx} = f(x, t), & t > 0, x \in (0, 1), \\ u(x, 0) = u_0(x), & x \in (0, 1), \\ u(0, t) = u(1, t) = u_x(1, t) = 0, t > 0. \end{cases}$$

Taking the Laplace transformation of the problem (2.1) with respect to t we get

(2.2)
$$\begin{cases} \xi \hat{u} - \hat{u}_{xx} + \hat{u}_{xxx} = \hat{f}(x, \xi) + u_0, x \in (0, 1), \\ \hat{u}(0, \xi) = \hat{u}(1, \xi) = \hat{u}_x(1, \xi) = 0 \end{cases}$$

for all $\operatorname{Re}\xi > 0$. We represent the Laplace transform of the solution $\hat{u}$ in the form

$$(2.3) \qquad \hat{u}(x,\xi) = \Psi(x,\xi) + \sum_{i=1}^{3} C_i(\xi)e^{\phi_i(\xi)x},$$

where $\Psi(x,\xi)$ is the particular solution of (2.2). Here $\phi_l(\xi)$ are the roots of the characteristic equation $-p^2 + p^3 + \xi = 0$, such that $\operatorname{Re}\phi_l(\xi) > 0$, $l = 1, 2$, and $\operatorname{Re}\phi_3(\xi) < 0$, for all $\xi \in D_0 = \{\xi \in \mathbf{C} : \operatorname{Re}\xi \geq 0, \xi \notin [0, \frac{4}{27}]\}$. Note that the functions $\phi_l(\xi)$ are analytic in the domain $\{\xi \in \mathbf{C} : \xi \notin (+\infty, \frac{4}{27}]\}$. We represent $p^2 = \frac{\xi}{1-p}$ for $|p| < 1$ and $p^3 = \frac{-\xi}{1-\frac{1}{p}}$ for $|p| > 1$; hence we get the asymptotics

$$(2.4) \qquad \phi_1(\xi) = \begin{cases} \sqrt{\xi} + O(|\xi|), \xi \to 0, \operatorname{Im}\xi > 0, 1 + O(|\xi|), \xi \to 0, \operatorname{Im}\xi < 0, \\ e^{i\frac{\pi}{3}}\sqrt[3]{\xi} + O(1), |\xi| \to +\infty, \end{cases}$$

$$\phi_2(\xi) = \begin{cases} 1 + O(|\xi|), \xi \to 0, \operatorname{Im}\xi > 0, \sqrt{\xi} + O(|\xi|), \xi \to 0, \operatorname{Im}\xi < 0, \\ e^{-i\frac{\pi}{3}}\sqrt[3]{\xi} + O(1), |\xi| \to +\infty, \end{cases}$$

and

$$(2.5) \qquad \phi_3(\xi) = \begin{cases} -\sqrt{\xi} + O(|\xi|), |\xi| \to 0, \\ -\sqrt[3]{\xi} + O(1), |\xi| \to +\infty, \end{cases}$$

for all $\xi \in \mathbf{C} : \xi \notin (+\infty, \frac{4}{27}]$ (by $\sqrt{\xi}$ and $\sqrt[3]{\xi}$ we denote the main value of the analytic function, i.e. $\sqrt{1} = \sqrt[3]{1} = 1$.)

By the method of the variations of constants we look for the particular solution of (2.2) in the form $\Psi(x,\xi) = \sum_{i=1}^{3} A_i(x,\xi)e^{\phi_i(\xi)x}$ and obtain a system for unknown functions $A_i(x,\xi)$, with a parameter $\xi \in D_0$

$$(2.6) \qquad \begin{cases} \sum_{i=1}^{3} A'_i(x,\xi)e^{\phi_i(\xi)x} = 0, \\ \sum_{i=1}^{3} A'_i(x,\xi)\phi_i(\xi)e^{\phi_i(\xi)x} = 0, \\ \sum_{i=1}^{3} A'_i(x,\xi)\phi_i^2(\xi)e^{\phi_i(\xi)x} = \hat{f}(x,\xi) + u_0(x), \end{cases}$$

which is equal to

$$\begin{pmatrix} e^{\phi_1(\xi)x} & e^{\phi_2(\xi)x} & e^{\phi_3(\xi)x} \\ \phi_1(\xi)e^{\phi_1(\xi)x} & \phi_2(\xi)e^{\phi_2(\xi)x} & \phi_3(\xi)e^{\phi_3(\xi)x} \\ \phi_1^2(\xi)e^{\phi_1(\xi)x} & \phi_2^2(\xi)e^{\phi_2(\xi)x} & \phi_3^2(\xi)e^{\phi_3(\xi)x} \end{pmatrix} \begin{pmatrix} A'_1(x,\xi) \\ A'_2(x,\xi) \\ A'_3(x,\xi) \end{pmatrix}$$

$$= \begin{pmatrix} 0 \\ 0 \\ \hat{f}(x,\xi) + u_0(x) \end{pmatrix}.$$

We denote the determinant of the above system by $\triangle$, then we have

$$\triangle = \begin{vmatrix} e^{\phi_1(\xi)x} & e^{\phi_2(\xi)x} & e^{\phi_3(\xi)x} \\ \phi_1(\xi)e^{\phi_1(\xi)x} & \phi_2(\xi)e^{\phi_2(\xi)x} & \phi_3(\xi)e^{\phi_3(\xi)x} \\ \phi_1^2(\xi)e^{\phi_1(\xi)x} & \phi_2^2(\xi)e^{\phi_2(\xi)x} & \phi_3^2(\xi)e^{\phi_3(\xi)x} \end{vmatrix}$$

$$= e^{\phi_1(\xi)+\phi_2(\xi)+\phi_3(\xi)} \begin{vmatrix} 1 & 1 & 1 \\ \phi_1(\xi) & \phi_2(\xi) & \phi_3(\xi) \\ \phi_1^2(\xi) & \phi_2^2(\xi) & \phi_3^2(\xi) \end{vmatrix}$$

$$= e^{\phi_1(\xi)+\phi_2(\xi)+\phi_3(\xi)}(\phi_1(\xi)-\phi_2(\xi))(\phi_2(\xi)-\phi_3(\xi))(\phi_3(\xi)-\phi_1(\xi)) \neq 0$$

in the domain $\xi \in D_0$. For the existence of the inverse Laplace transformation we look for the solutions with the following growth condition $|\Psi(x,\xi)| \leq M(1+|\xi|)^{\beta}$, for all $\mathrm{Re}\,\xi \geq 0$; therefore integrating system (2.6) we obtain

$$\Psi(x,\xi) = -\frac{e^{\phi_1(\xi)x}}{(\phi_1(\xi)-\phi_2(\xi))(\phi_1(\xi)-\phi_3(\xi))}$$

$$\times \int_x^1 e^{-\phi_1(\xi)y}\left(\widehat{f}(y,\xi)+u_0(y)\right)dy$$

$$-\frac{e^{\phi_2(\xi)x}}{(\phi_2(\xi)-\phi_1(\xi))(\phi_2(\xi)-\phi_3(\xi))}$$

$$\times \int_x^1 e^{-\phi_2(\xi)y}\left(\widehat{f}(y,\xi)+u_0(y)\right)dy$$

(2.7)
$$+\frac{e^{\phi_3(\xi)x}}{(\phi_3(\xi)-\phi_1(\xi))(\phi_3(\xi)-\phi_2(\xi))}$$

$$\times \int_0^x e^{-\phi_3(\xi)y}\left(\widehat{f}(y,\xi)+u_0(y)\right)dy.$$

Due to the theory of residues we can rewrite the function $\Psi(x,\xi)$ as

$$\Psi(x,\xi) = \frac{1}{2\pi i}\int_{-i\infty}^{+i\infty}dp\frac{e^{px}}{K(p)+\xi}\int_0^1 e^{-py}\left(\widehat{f}(y,\xi)+u_0(y)\right)dy$$

where $K(p) = -p^2 + p^3, \xi \in D_0$. To satisfy the boundary conditions in system (2.2) we find a system for coefficients C_i in (2.6)

(2.8)
$$\begin{cases} \sum_{i=1}^3 C_i(\xi) = -\Psi(0,\xi) \\ \quad = -\frac{1}{2\pi i}\int_{-i\infty}^{i\infty}\frac{1}{K(p)+\xi}\int_0^1 e^{-py}f(y,\xi)dydp \\ \sum_{i=1}^3 C_i(\xi)e^{\phi_i} = -\Psi(1,\xi) \\ \quad = -\frac{1}{2\pi i}\int_{-i\infty}^{i\infty}\frac{e^p}{K(p)+\xi}\int_0^1 e^{-py}f(y,\xi)dydp \\ \sum_{i=1}^3 C_i(\xi)\phi_i e^{\phi_i} = -\Psi_x(1,\xi) \\ \quad = -\frac{1}{2\pi i}\int_{-i\infty}^{i\infty}\frac{pe^p}{K(p)+\xi}\int_0^1 e^{-py}f(y,\xi)dydp, \end{cases}$$

which is equal to

$$\begin{pmatrix} 1 & 1 & 1 \\ e^{\phi_1(\xi)} & e^{\phi_2(\xi)} & e^{\phi_3(\xi)} \\ \phi_1(\xi)e^{\phi_1(\xi)} & \phi_2(\xi)e^{\phi_2(\xi)} & \phi_3(\xi)e^{\phi_3(\xi)} \end{pmatrix} \begin{pmatrix} C_1(\xi) \\ C_2(\xi) \\ C_3(\xi) \end{pmatrix}$$

$$= \begin{pmatrix} -\Psi(0,\xi) \\ -\Psi(1,\xi) \\ -\Psi_x(1,\xi) \end{pmatrix}.$$

We denote the determinant of this system by $\triangle(\phi_1,\phi_2,\phi_3)$, then it has the form

$$\triangle(\phi_1,\phi_2,\phi_3) = \begin{vmatrix} 1 & 1 & 1 \\ e^{\phi_1(\xi)} & e^{\phi_2(\xi)} & e^{\phi_3(\xi)} \\ \phi_1(\xi)e^{\phi_1(\xi)} & \phi_2(\xi)e^{\phi_2(\xi)} & \phi_3(\xi)e^{\phi_3(\xi)} \end{vmatrix}$$

$$= -\left(e^{\phi_1+\phi_2}(\phi_1-\phi_2) + e^{\phi_2+\phi_3}(\phi_2-\phi_3) + e^{\phi_3+\phi_1}(\phi_3-\phi_1)\right).$$

Since $p^3 - p^2 + \xi = \Pi_{j=1}^3 (p-\phi_j)$ we have $\sum_{j=1}^3 \phi_j = 1$ and

$$\phi_1'(\xi) = -\frac{1}{(\phi_1-\phi_2)(\phi_1-\phi_2)}; \quad \phi_2'(\xi) = -\frac{1}{(\phi_2-\phi_1)(\phi_2-\phi_3)};$$

$$(2.9) \quad \phi_3'(\xi) = -\frac{1}{(\phi_3-\phi_1)(\phi_3-\phi_2)}$$

we can rewrite $\triangle(\phi_1,\phi_2,\phi_3)$ as

$$(2.10) \qquad \triangle(\phi_1,\phi_2,\phi_3) = -e^{\phi_1(\xi)+\phi_2(\xi)+\phi_3(\xi)}V(\xi)\sum_{j=1}^3 e^{-\phi_j(\xi)}\phi_j'(\xi),$$

where

$$V(\xi) = (\phi_1-\phi_2)(\phi_2-\phi_3)(\phi_3-\phi_1).$$

Since $V(\xi) \neq 0$, $\operatorname{Re}\phi_l(\xi) > 0$, $l = 1,2$, and $\operatorname{Re}\phi_3(\xi) < 0$ in domain $\xi \in D_0$ we easily get for $|\xi| \gg 1, \xi \in D_0$

$$\triangle(\phi_1,\phi_2,\phi_3) \neq 0.$$

By numeric computations we can check that $\triangle(\phi_1,\phi_2,\phi_3) \neq 0$ for all $|\xi| \leq C, \xi \in D_0 = \{\xi \in \mathbf{C} : \operatorname{Re}\xi \geq 0, \xi \notin [0,\frac{4}{27}]\}$. Therefore there exists a unique solution of the system (2.8)

$$(2.11) \qquad C_i(\xi) = -\frac{1}{2\pi i \triangle}\int_0^1 f(y,\xi)dy \int_{-i\infty}^{i\infty}\frac{e^{-py}}{K(p)+\xi}\triangle_i(\xi,p)dp,$$

where

$$\triangle_1(\xi,p) = \triangle(p,\phi_2,\phi_3), \triangle_2(\xi,p) = \triangle(\phi_1,p,\phi_3) \text{ and}$$
$$\triangle_3(\xi,p) = \triangle(\phi_1,\phi_2,p).$$

We denote $\widetilde{\triangle}(\xi, y) = \sum_{j=1}^{3} e^{-\phi_j y} \phi_j'(\xi)$. By the method of residues we see that

$$\int_{-i\infty}^{i\infty} \frac{e^{-py}}{K(p) + \xi} \triangle_i(\xi, p) dp$$

$$(2.12) \quad = \quad -2\pi i \widetilde{\triangle}(\xi, y) \begin{cases} e^{\phi_3 + \phi_2} \left(\phi_2 - \phi_3 \right), i = 1 \\ e^{\phi_3 + \phi_1} \left(\phi_3 - \phi_1 \right), i = 2 \\ e^{\phi_2 + \phi_1} \left(\phi_1 - \phi_2 \right) + \frac{\triangle}{\widetilde{\triangle}(\xi, y)} e^{-\phi_3 y} \phi_3', i = 3 \end{cases}$$

for $\xi \in D_0$. Putting (2.12) and (2.11) into (2.6) and using (2.10) we get

$$\hat{u}(x, \xi) \quad = \quad \Psi(x, \xi)$$
$$- \int_0^1 \left(\widehat{f}(y, \xi) + u_0(y) \right) \left(\frac{\widetilde{\triangle}(\xi, 1 - x)\widetilde{\triangle}(\xi, y)}{\widetilde{\triangle}(\xi, 1)} - e^{\phi_3(x-y)} \phi_3' \right) dy.$$

Then taking (2.8) and (2.9) into account we obtain

$$\hat{u}(x, \xi)$$

$$= \quad -\frac{1}{\widetilde{\triangle}(\xi, 1)} \int_0^1 \widehat{f_1}(y, \xi)\widetilde{\triangle}(\xi, 1 - x)\widetilde{\triangle}(\xi, y) dy$$

$$+ \int_x^1 \widehat{f_1}(y, \xi)\widetilde{\triangle}(\xi, y - x) dy$$

$$= \quad -\int_0^x \widehat{f_1}(y, \xi)\frac{\widetilde{\triangle}(\xi, 1 - x)\widetilde{\triangle}(\xi, y)}{\widetilde{\triangle}(\xi, 1)} dy$$

$$+ \int_x^1 \widehat{f_1}(y, \xi) \left(\frac{\widetilde{\triangle}(\xi, y - x)\widetilde{\triangle}(\xi, 1) - \widetilde{\triangle}(\xi, 1 - x)\widetilde{\triangle}(\xi, y)}{\widetilde{\triangle}(\xi, 1)} \right) dy$$

$$(2.13) \quad \equiv \quad \int_0^x \widehat{f_1}(y, \xi)\widehat{F_1}(x, y, \xi) dy + \int_x^1 \widehat{f_1}(y, \xi)\widehat{F_2}(x, y, \xi) dy,$$

where

$$\widehat{f_1}(y, \xi) = \widehat{f}(y, \xi) + u_0(y),$$

$$\widehat{F_1}(x, y, \xi) \quad = \quad -\frac{\widetilde{\triangle}(\xi, 1 - x)\widetilde{\triangle}(\xi, y)}{\widetilde{\triangle}(\xi, 1)},$$

$$\widehat{F_2}(x, y, \xi) \quad = \quad \frac{\widetilde{\triangle}(\xi, y - x)\widetilde{\triangle}(\xi, 1) - \widetilde{\triangle}(\xi, 1 - x)\widetilde{\triangle}(\xi, y)}{\widetilde{\triangle}(\xi, 1)}.$$

Since $\phi_l'(\xi) = O\left(|\xi|^{-\frac{1}{2}}\right)$, $l = 1, 2, 3$ for $|\xi| < 1, \xi \in D_0$, we have

(2.14)
$$\frac{\widetilde{\triangle}(\xi, 1 - x)\widetilde{\triangle}(\xi, y)}{\widetilde{\triangle}(\xi, 1)}$$

$$= \frac{\left(\sum_{j=1}^3 e^{-\phi_j(\xi)(1-x)}\phi_j'\right)\left(\sum_{j=1}^3 e^{-\phi_j(\xi)y}\phi_j'\right)}{\sum_{j=1}^3 e^{-\phi_j(\xi)}\phi_j'} = O\left(|\xi|^{-\frac{1}{2}}\right)$$

and

$$\widetilde{\triangle}(\xi, y - x) = O\left(|\xi|^{-\frac{1}{2}}\right)$$

for $|\xi| < 1, \xi \in D_0$. Because $\mathrm{Re}\,\phi_l(\xi) > 0$, $l = 1, 2$, and $\mathrm{Re}\,\phi_3(\xi) < 0$ for $|\xi| > 1$, $\xi \in D_0$, we obtain for $|\xi| > 1$

$$\frac{1}{\widetilde{\triangle}(\xi, 1)}\widetilde{\triangle}(\xi, 1 - x)\widetilde{\triangle}(\xi, y)$$

$$= \frac{e^{-\phi_3(1-x+y)}(\phi_3')^2\left(1 + \sum_{i=1}^2 O\left(e^{(-\phi_i+\phi_3)(1-x)}\frac{\phi_i'}{\phi_3'}\right)\right)}{e^{-\phi_3}\phi_3'\left(1 + \sum_{j=1}^2 O\left(e^{(-\phi_j+\phi_3)}\frac{\phi_j'}{\phi_3'}\right)\right)}$$

$$\times \left(1 + \sum_{j=1}^2 O\left(e^{(-\phi_j+\phi_3)y}\frac{\phi_j'}{\phi_3'}\right)\right)$$

(2.15)
$$= e^{\phi_3(x-y)}\phi_3'\left(1 + \sum_{i=1}^2 O\left(e^{(-\phi_j+\phi_3)y}\right)\right.$$

$$\left. + \sum_{i=1}^2 O\left(e^{(-\phi_j+\phi_3)(1-x)}\right)\right).$$

Therefore taking the asymptotics (2.4)-(2.5) into account we find that

(2.16)
$$\widehat{F}_1(x, y, \xi) = O\left(\xi^{-\frac{2}{3}}e^{-C\sqrt[3]{|\xi|}(x-y)}\right)$$

for $\xi \in D_0$, $|\xi| > 1$, $x > y$. Using (2.15) we get

$$\widehat{F}_2(x, y, \xi)$$

$$= \widetilde{\triangle}(\xi, y - x) - \frac{\widetilde{\triangle}(\xi, 1 - x)\widetilde{\triangle}(\xi, y)}{\widetilde{\triangle}(\xi, 1)} = \sum_{j=1}^2 \left(\phi_j' e^{-\phi_j(y-x)}\right.$$

$$\left. + O\left(\phi_3' e^{-\mathrm{Re}\,\phi_j y + \mathrm{Re}\,\phi_3 x}\right) + O\left(\phi_3' e^{-\mathrm{Re}\,\phi_j(1-x) + \mathrm{Re}\,\phi_3(1-y)}\right)\right)$$

$$= O\left(\xi^{-\frac{2}{3}}e^{-C\sqrt[3]{|\xi|}(y-x)}\right)$$

for $\xi \in D_0$, $|\xi| > 1$, $y > x$. Therefore the inverse Laplace transforms for the functions $\widehat{F}_1(x, y, \xi)$ and $\widehat{F}_2(x, y, \xi)$ exist. Taking the inverse Laplace transformation of (2.13) we obtain for the solutions of (2.1)

$$u(x,t) = \int_0^x u_0(y)F_1(x,y,t)dy + \int_x^1 u_0(y)F_2(x,y,t)dy$$

$$(2.17)\quad + \int_0^t d\tau \left(\int_0^x f(y,t-\tau)F_1(x,y,\tau)dy + \int_x^1 f(y,t-\tau)F_2(x,y,\tau)dy \right),$$

where

$$F_1(x,y,t) = -\frac{1}{2\pi i}\int_{\Gamma_0} e^{\xi t}\frac{1}{\widetilde{\Delta}(\xi,1)}\widetilde{\Delta}(\xi,1-x)\widetilde{\Delta}(\xi,y)d\xi$$

and

$$F_2(x,y,t)$$
$$= -\frac{1}{2\pi i}\int_{\Gamma_0} d\xi e^{\xi t}\frac{\widetilde{\Delta}(\xi,1-x)\widetilde{\Delta}(\xi,y) - \widetilde{\Delta}(\xi,y-x)\widetilde{\Delta}(\xi,1)}{\widetilde{\Delta}(\xi,1)},$$

where $\Gamma_0 = \partial D_0$, i.e.

$$\Gamma_0 = (-i\infty,-i0) \cup \left[-i0,\frac{4}{27}-i0\right] \cup \left[\frac{4}{27}+i0,i0\right] \cup (i0,i\infty).$$

Since functions $\widehat{F}_l(x,y,\xi)$ are symmetric with respect to $\phi_1,\phi_2,\phi_3,$ using the relations $\overline{\phi_1(\xi)} = \phi_2(\overline{\xi}), \overline{\phi_2(\xi)} = \phi_1(\overline{\xi}), \overline{\phi_3(\xi)} = \phi_3(\overline{\xi})$ for all $\xi \in D_0$ we can change the contour of integration Γ_0 to the imaginary axis $(-i\infty,i\infty)$ to get

$$F_1(x,y,t) = -\frac{1}{2\pi i}\int_{-i\infty}^{i\infty} e^{\xi t}\frac{1}{\widetilde{\Delta}(\xi,1)}\widetilde{\Delta}(\xi,1-x)\widetilde{\Delta}(\xi,y)d\xi$$

and

$$F_2(x,y,t)$$
$$= -\frac{1}{2\pi i}\int_{-i\infty}^{i\infty} d\xi e^{\xi t}\frac{\widetilde{\Delta}(\xi,1-x)\widetilde{\Delta}(\xi,y) - \widetilde{\Delta}(\xi,y-x)\widetilde{\Delta}(\xi,1)}{\widetilde{\Delta}(\xi,1)}.$$

In the next lemma we obtain the estimates of the kernels $F_j(x,y,t)$.

We denote $\widetilde{\Delta}(\xi,y) = \sum_{j=1}^3 e^{-y\phi_j(\xi)}\phi'_j(\xi)$ as above, the roots ϕ_j are defined in Section 2. Define ξ_0 as the first root of equation $\widetilde{\Delta}(-\xi,1) = 0$. By virtue of the numerical computations via Maple program we have $\xi_0 \approx 70$ and the next root $\xi_1 \approx 200$. We choose $\delta > 0$ such that $\xi_0 + \delta < \xi_1$. We denote $\{t\} = \min(1,t)$, $\Lambda(x) = -\frac{\widetilde{\Delta}(-\xi_0,1-x)}{\widetilde{\Delta}'(-\xi_0,1)}$.

LEMMA 39. *We have the asymptotics for large time*

$$(2.18)\qquad F_j(x,y,t) = -e^{-\xi_0 t}\Lambda(x)\widetilde{\Delta}(-\xi_0,y) + O\left(e^{-(\xi_0+\delta)t}\right)$$

and estimates

$$(2.19)\qquad |\partial_x^n F_j(x,y,t)| \le Ce^{-\xi_0 t}\{t\}^{-\alpha}|x-y|^{2\alpha-1-n}$$

for $x,y \in (0,1)$, $x \ne y$, $t > 0$, where $\alpha \in \left[0,\frac{n+1}{2}\right]$, $n = 0,1$, $j = 1,2$.

PROOF. We consider a curve in the complex left-half plane $\operatorname{Re}\xi < 0$ such that $\operatorname{Re}\phi_1(\xi) = 0$, it is defined by the equation $(iy)^2 - (iy)^3 = \xi$ with $y = \operatorname{Im}\phi_1(\xi)$. Therefore there exists a contour

$$C_0 = \left\{\xi \in \mathbf{C}, \operatorname{Re}\xi < 0 : \operatorname{Re}\xi = O\left(|\xi|^{\frac{2}{3}}\right)\right\}$$

such that

$$\operatorname{Re}\phi_l(\xi) > 0, l = 1, 2, \operatorname{Re}\phi_3(\xi) < 0 \text{ for all } \xi \in C_0.$$

We also consider a contour

$$\begin{aligned}
C_1 &= (-\xi_0 - \delta - i\infty, -\xi_0 - \delta - i0) \cup (-\xi_0 - \delta - i0, -i0)\\
&\quad \cup (i0, -\xi_0 - \delta + i0) \cup (-\xi_0 - \delta + i0, -\xi_0 - \delta + i\infty).
\end{aligned}$$

We denote

$$C_0 \cap C_1 = \{z_1, z_2\}, \operatorname{Im} z_1 > 0, \operatorname{Im} z_2 < 0, \operatorname{Re} z_l = -\xi_0 - \delta, l = 1, 2.$$

We now define a contour $C = C_2 \cup C_3$, where

$$C_2 = \{\xi \in C_1 : \operatorname{Im} z_2 \le \operatorname{Im}\xi \le \operatorname{Im} z_1\}$$

and

$$C_3 = \{\xi \in C_0 : \operatorname{Im}\xi > \operatorname{Im} z_1 \text{ or } \operatorname{Im}\xi < \operatorname{Im} z_2\}.$$

Note that the asymptotic formulas (2.4)-(2.5) are valid on the contour C. In view of that we have (2.14)-(2.15) for $\xi \in C$. Therefore changing the contour of integration to C we obtain

$$\begin{aligned}
(2.20) \quad F_1(x, y, \tau) &= -\frac{1}{2\pi i}\int_{\xi \in C_2} e^{\xi t}\frac{1}{\widetilde{\triangle}(\xi, 1)}\widetilde{\triangle}(\xi, 1-x)\widetilde{\triangle}(\xi, y)d\xi\\
&\quad -\frac{1}{2\pi i}\int_{\xi \in C_3} e^{\xi t}\frac{1}{\widetilde{\triangle}(\xi, 1)}\widetilde{\triangle}(\xi, 1-x)\widetilde{\triangle}(\xi, y)d\xi.
\end{aligned}$$

Since $\widetilde{\triangle}(x + i0, q) = \widetilde{\triangle}(x - i0, q)$ we get

$$\begin{aligned}
&-\frac{1}{2\pi i}\int_{-\xi_0-\delta-i0}^{-i0} e^{\xi t}\frac{1}{\widetilde{\triangle}(\xi, 1)}\widetilde{\triangle}(\xi, 1-x)\widetilde{\triangle}(\xi, y)d\xi\\
&-\frac{1}{2\pi i}\int_{+i0}^{-\xi_0-\delta+i0} e^{\xi t}\frac{1}{\widetilde{\triangle}(\xi, 1)}\widetilde{\triangle}(\xi, 1-x)\widetilde{\triangle}(\xi, y)d\xi\\
(2.21) \quad &= -e^{-\xi_0 t}\frac{\widetilde{\triangle}(-\xi_0, 1-x)\widetilde{\triangle}(-\xi_0, y)}{\widetilde{\triangle}'(-\xi_0, 1)}.
\end{aligned}$$

Also by (2.14) we have

$$\begin{aligned}
&\frac{1}{2\pi i}\int_{-\xi_0-\delta+i0}^{z_1} e^{\xi t}\frac{1}{\widetilde{\triangle}(\xi, 1)}\widetilde{\triangle}(\xi, 1-x)\widetilde{\triangle}(\xi, y)d\xi\\
&+\frac{1}{2\pi i}\int_{z_2}^{-\xi_0-\delta-i0} e^{\xi t}\frac{1}{\widetilde{\triangle}(\xi, 1)}\widetilde{\triangle}(\xi, 1-x)\widetilde{\triangle}(\xi, y)d\xi\\
(2.22) \quad &= O\left(e^{-(\xi_0+\delta)t}\right).
\end{aligned}$$

Taking (2.16) into account we get

$$\left| -\frac{1}{2\pi i} \int_{\xi \in C_3} e^{\xi t} \frac{1}{\widetilde{\triangle}(\xi,1)} \widetilde{\triangle}(\xi, 1-x) \widetilde{\triangle}(\xi, y) d\xi \right|$$

$$< \quad Ce^{-(\xi_0+\delta)t} \int_{\xi \in C_3} e^{-Ct|\xi|^{\frac{2}{3}} + t(\xi_0+\delta) - C|x-y||\xi|^{\frac{1}{3}}} |\xi|^{-\frac{2}{3}} d\xi$$

(2.23) $$< \quad Ce^{-(\xi_0+\delta)t} t^{-\alpha} |x-y|^{2\alpha-1}$$

since $C|\xi|^{\frac{2}{3}} - (\xi_0 + \delta) \geq 0$ for $\xi \in C_3$, where $\alpha \in \left[0, \frac{1}{2}\right]$. By (2.21)-(2.23) we have from (2.20)

$$F_1(x, y, t) = -e^{-\xi_0 t} \frac{\widetilde{\triangle}(-\xi_0, 1-x)\,\widetilde{\triangle}(-\xi_0, y)}{\widetilde{\triangle}'(-\xi_0, 1)} + O\left(e^{-(\xi_0+\delta)t}\right)$$

for $x, y > 0, t \geq 1$, and, moreover,

$$|F_1(x, y, t)| \leq Ce^{-\xi_0 t}\left(1 + \{t\}^{-\alpha} |x-y|^{2\alpha-1}\right)$$

for all $x, y \in (0, 1)$, $x \neq y$, $t > 0$, where $\alpha \in \left[0, \frac{1}{2}\right]$. Thus the result of the lemma is true for the case $n = 0$. Consider the case $n = 1$. In view of the asymptotic formulas (2.4)-(2.5) we get

$$\frac{\partial_x \widetilde{\triangle}(\xi, 1-x)\widetilde{\triangle}(\xi, y)}{\widetilde{\triangle}(\xi, 1)}$$

$$= \quad \frac{\left(\sum_{j=1}^{3} e^{-\phi_j(\xi)(1-x)} \left(\phi_j'\phi_j\right)\right)\left(\sum_{j=1}^{3} e^{-\phi_j(\xi)y}\phi_j'\right)}{\sum_{j=1}^{3} e^{-\phi_j(\xi)}\phi_j'} = O(1)$$

and

$$\partial_x \widetilde{\triangle}(\xi, y-x) = O(1)$$

for $|\xi| < 1$, $\xi \in C_2$. As in the proof of the estimate (2.14) we get

$$\frac{\partial_x \widetilde{\triangle}(\xi, 1-x)\widetilde{\triangle}(\xi, y)}{\widetilde{\triangle}(\xi, 1)}$$

$$= \quad e^{-\phi_3(y-x)}\phi_3\phi_3'\left(1 + O\left(e^{-C\sqrt[3]{|\xi|}y}\right) + O\left(e^{-C\sqrt[3]{|\xi|}(1-x)}\right)\right)$$

and

$$\partial_x \widetilde{\triangle}(\xi, y-x) = e^{-\phi_3(y-x)}\phi_3\phi_3'\left(1 + O\left(e^{-C\sqrt[3]{|\xi|}y}\right) + O\left(e^{-C\sqrt[3]{|\xi|}(1-x)}\right)\right)$$

for all $|\xi| > 1$, $\xi \in C_3$. Hence similarly to (2.21)-(2.23) we get

$$|\partial_x F_1(x, y, t)|$$

$$\leq \quad e^{-\xi_0 t} \left| \frac{\widetilde{\triangle}(-\xi_0, y) \partial_x \widetilde{\triangle}(-\xi_0, 1 - x)}{\widetilde{\triangle}'(-\xi_0, 1)} \right| + C e^{-\xi_0 t}$$

$$+ C e^{-(\xi_0 + \delta)t} \int_{\xi \in C_3} e^{-Ct|\xi|^{\frac{2}{3}} + t(\xi_0 + \delta) - C|x - y||\xi|^{\frac{1}{3}}} |\xi|^{-\frac{1}{3}} d\xi$$

$$\leq \quad e^{-\xi_0 t} \left(C + C \{t\}^{-\alpha} |x - y|^{2\alpha - 2} \right)$$

for all $x, y \in (0, 1)$, $x \neq y$, $t > 0$, where $\alpha \in [0, 1]$. The function $F_2(x, y, t)$ is considered in the same manner. Lemma 39 is proved. $\qquad \Box$

Now we prove the local existence for the linear problem (2.1).

THEOREM 40. *Let the initial data $u_0 \in \mathbf{L}^2(0, 1)$ and $f \in \mathbf{L}^2(0, 1)$. Then for any $T > 0$ there exists a unique solution $u \in \mathbf{C}\left((0, T]; \mathbf{H}^1(0, 1)\right)$ of the linear initial-boundary value problem (2.1) such that*

$$\sup_{t \in (0, T]} t^\alpha \|\partial_x^n u(t)\|_{\mathbf{L}^2} \leq C\lambda$$

provided that

$$\lambda = \|u_0\|_{\mathbf{L}^2} + T^{1-\beta} \sup_{t \in (0, T]} t^\beta \|f(t)\|_{\mathbf{L}^2} < +\infty,$$

where $n = 0, 1$, $\alpha \in \left(\frac{n}{2}, 1\right)$, $\beta \in (0, 1)$.

PROOF. From (2.17) and estimates of Lemma 39 we have

$$\|\partial_x^n u(t)\|_{\mathbf{L}^2}$$

$$\leq \quad t^{-\alpha} \left\| \int_0^1 u_0(y) |x - y|^{2\alpha - 1 - n} dy \right\|_{\mathbf{L}^2}$$

$$+ \left\| \int_0^t \tau^{-\alpha} (t - \tau)^{-\beta} d\tau \int_0^1 (t - \tau)^\beta f(y, t - \tau) |x - y|^{2\alpha - 1 - n} dy \right\|_{\mathbf{L}^2}$$

$$\leq \quad C t^{-\alpha} \left(\|u_0(y)\|_{\mathbf{L}^2} + T^{1-\beta} \sup_{t \in (0, T]} t^\beta \|f(t)\|_{\mathbf{L}^2} \right)$$

for $n = 0, 1$. Thus we have the estimate of the theorem and Theorem 40 is proved. $\qquad \Box$

3. Local existence for the nonlinear problem

In this section we prove the following result.

THEOREM 41. *Suppose that the initial data $u_0(x) \in \mathbf{L}^2(0, 1)$. Then there exists a unique solution $u(x, t)$ and a positive constant $T > 0$, which depends on $\|u_0\|_{\mathbf{L}^2}$, such that $u(x, t) \in \mathbf{C}\left((0, T], \mathbf{H}^1(0, 1)\right)$.*

PROOF. We prove the local existence of solutions by the contraction mapping principle. Let $u(x,t)$ be a solution of the following linear problem

(3.1)
$$\begin{cases} u_t + \mathbb{N}(w) - u_{xx} + u_{xxx} = 0, \ t > 0, x \in (0,1), \\ u(x,0) = u_0(x), \ x \in (0,1), \\ u(0,t) = u(1,t) = u_x(1,t) = 0, \ t > 0, \end{cases}$$

where $\mathbb{N}(w) = w w_x$. We take w from the closed ball

$$\mathbf{H}_\rho^1 = \left\{ w \in \mathbf{C}\left((0,T]; \mathbf{H}^1(0,1)\right); \ \sup_{t\in(0,T]} \sum_{n=0}^1 t^{\alpha_n} \|\partial_x^n w\|_{\mathbf{L}^2} \le \rho \right\},$$

where $\alpha_n \in \left(\frac{n}{2}, 1\right)$. We assume that w satisfies the boundary condition $w(0,t) = w(1,t) = w_x(1,t) = 0$. The initial-value problem (3.1) defines a mapping $\mathbb{M}$ by $u = \mathbb{M}(w)$ and we will show that $\mathbb{M}$ is the contraction mapping from $\mathbf{H}_\rho^1$ into itself for a sufficiently small $T > 0$. Since $w \in \mathbf{H}_\rho^1$, using the estimate $\|w\|_{\mathbf{L}^\infty} \le 2 \|w\|_{\mathbf{L}^2} \|w_x\|_{\mathbf{L}^2}$, we have

$$\sup_{t\in(0,T]} t^\beta \|\mathbb{N}(w)(t)\|_{\mathbf{L}^2}$$

$$\le C \sup_{t\in(0,T]} t^\beta \|w(t)\|_{\mathbf{L}^2}^{\frac{1}{2}} \|w_x(t)\|_{\mathbf{L}^2}^{\frac{3}{2}} \le C\rho^2,$$

where $\beta = \frac{\alpha_0 + 3\alpha_1}{2} < 1$. Via Theorem 40 the problem (3.1) has a unique solution $u(x,t) \in \mathbf{C}\left((0,T]; \mathbf{H}^1(0,1)\right)$, such that

$$\sup_{t\in(0,T]} t^{\alpha_n} \|\partial_x^n u\|_{\mathbf{L}^2} \le C\lambda,$$

where

$$\lambda = \|u_0\|_{\mathbf{L}^2} + T^{1-\beta} \sup_{t\in(0,T]} t^\beta \|\mathbb{N}(w)(t)\|_{\mathbf{L}^2}.$$

Therefore we obtain the estimate

(3.2)
$$\sup_{t\in(0,T]} \sum_{n=0}^1 t^{\alpha_n} \|\partial_x^n u\|_{\mathbf{L}^2} \le C\|u_0\|_{\mathbf{L}^2} + CT^{1-\beta}\rho^2 \le \rho$$

if $T = (2C\rho)^{-\frac{1}{1-\beta}} = \left(4C^2 \|u_0\|_{\mathbf{L}^2}\right)^{-\frac{1}{1-\beta}}$ and $C \|u_0\|_{\mathbf{L}^2} \le \frac{\rho}{2}$. Thus the mapping $\mathbb{M}$ transforms the closed ball $\mathbf{H}_\rho^1$ with a center at the origin and a radius ρ into itself. Analogously we can estimate the difference

$$\sup_{t\in(0,T]} \sum_{n=0}^1 t^{\alpha_n} \|\partial_x^n (u - \tilde{u})\|_{\mathbf{L}^2}$$

$$\le \frac{1}{2} \sup_{t\in(0,T]} \sum_{n=0}^1 t^{\alpha_n} \|\partial_x^n (w - \tilde{w})\|_{\mathbf{L}^2}$$

for sufficiently small $T > 0$. Therefore the mapping $\mathbb{M}$ is a contraction mapping in $\mathbf{H}_\rho^1$, and there exists a unique solution $u(x,t) \in \mathbf{C}\left((0,T]; \mathbf{H}^1(0,1)\right)$ of the initial-value problem (1.1). Theorem 41 is proved. □

REMARK 17. *By virtue of estimate (3.2) we see that if the initial data u_0 are small, i.e. the norm $\|u_0\|_{\mathbf{L}^2} \leq \varepsilon$, where $\varepsilon > 0$ is sufficiently small, then there exists $T \geq 1$, such that there exists a unique solution u, which is also small:* $\sup_{t \in (0,T]} \sum_{n=0}^{1} t^{\alpha_n} \|\partial_x^n u\|_{\mathbf{L}^2} < C\varepsilon$.

4. Large time asymptotics

In this section we give sufficient conditions for the global existence of solutions to the initial-boundary value problem (1.1) with small initial data

$$(4.1) \qquad \|u_0\|_{\mathbf{L}^2} < \varepsilon_1,$$

where $\varepsilon_1 > 0$ is sufficiently small.

THEOREM 42. *Suppose that the initial data $u_0 \in \mathbf{L}^2(0,1)$ and satisfy (4.1). Then there exists a unique solution u of (1.1) such that $u \in \mathbf{C}\left((0,+\infty);\mathbf{H}^1(0,1)\right)$. Furthermore the solution has the following asymptotics*

$$(4.2) \qquad u(x,t) = A\Lambda(x)e^{-\xi_0 t} + O\left(e^{-(\xi_0+\delta)t}\right)$$

for $t \to +\infty$ uniformly with respect to $x \in (0,1)$, where $\xi_0 > 0$, $\delta > 0$. The constant A and the function $\Lambda(x) \in \mathbf{L}^\infty(0,1)$ are defined below in the proof.

PROOF. According to Theorem 41 and Remark 17 we see that after some time $T \geq 1$ the solution is small in the norm $\mathbf{H}^1(0,1)$. Therefore we can only consider the initial-boundary problem (1.1) for $t \geq 1$ with small initial data $u(x,1)$ such that $\|u(1)\|_{\mathbf{H}^1} \leq \varepsilon_2$, where $\varepsilon_2 > 0$ is sufficiently small. Let us prove that

$$(4.3) \qquad e^{\xi_0 t}\|u(t)\|_{\mathbf{H}^1} < \varepsilon$$

for all $t \geq 1$, with some small $\varepsilon > 0$. As a contradiction we can find a maximal interval $[1,T_1]$ such that the estimate

$$(4.4) \qquad e^{\xi_0 t}\|u(t)\|_{\mathbf{H}^1} \leq \varepsilon$$

is true for all $t \in [1, T_1]$, and estimate (4.3) is disturbed at time $t = T_1$. From (4.4) and estimates (2.19) of Lemma 39 we obtain for $n = 0, 1$

$$
\left\| u_x^{(n)}(t) \right\|_{\mathbf{L}^2}
$$

$$
\leq \; C e^{-\xi_0 t} \| u(1) \|_{\mathbf{L}^2}
$$

$$
+ \int_1^t d\tau \int_0^x |uu_y(y, \tau)| \, \| \partial_x^n F_1(\cdot, y, t - \tau) \|_{\mathbf{L}+\infty} \, dy
$$

$$
+ \int_1^t d\tau \int_x^1 |uu_y(y, \tau)| \, \| \partial_x^n F_2(\cdot, y, t - \tau) \|_{\mathbf{L}+\infty} \, dy
$$

$$
\leq \; C\varepsilon_2 e^{-\xi_0 t} + C e^{-\xi_0 t} \int_1^t e^{\xi_0 \tau} \| u(\tau) \|_{\mathbf{L}^2}^{\frac{1}{2}} \| u_x(\tau) \|_{\mathbf{L}^2}^{\frac{3}{2}} \{t - \tau\}^{-\frac{3}{4}} \, d\tau
$$

$$
\leq \; C\varepsilon_2 e^{-\xi_0 t} + C\varepsilon^2 e^{-\xi_0 t} \int_1^t e^{-\xi_0 \tau} \{t - \tau\}^{-\frac{3}{4}} \, d\tau
$$

$$
\leq \; C \left(\varepsilon_2 + \varepsilon^2 \right) e^{-\xi_0 t}
$$

for $t \in [1, T_1]$. The contradiction obtained proves (4.3). Now using the estimate (4.3) and Lemma 39 we prove that the solution has asymptotics (4.2) for $t \to +\infty$ uniformly with respect to $x > 0$, where

$$
\Lambda(x) = -\frac{\widetilde{\triangle}(-\xi_0, 1 - x)}{\widetilde{\triangle}'(-\xi_0, 1)},
$$

$$
A \; = \; \int_0^1 u_0(y) \widetilde{\triangle}(-\xi_0, y) dy
$$

$$
+ \int_0^{+\infty} e^{\xi_0 \tau} d\tau \int_0^1 u(y, \tau) u_y(y, \tau) \widetilde{\triangle}(-\xi_0, y) dy.
$$

In fact, due to asymptotics (2.18) of Lemma 39 we have

$$
u(x, t) = A\Lambda(x) e^{-\xi_0 t} + R(x, t),
$$

where in view of (4.3) we attain

$$
|R(x, t)| \; \leq \; e^{-(\xi_0 + \delta)t} \| u(1) \|_{\mathbf{H}^1}
$$

$$
+ C \int_1^t e^{-(\xi_0 + \delta)(t - \tau)} d\tau \int_0^1 |uu_y| \, dy
$$

$$
+ e^{-\xi_0 t} |\Lambda(x)| \int_t^{+\infty} e^{\xi_0 \tau} d\tau \int_0^1 |uu_y| \left| \widetilde{\triangle}(-\xi_0, y) \right| dy
$$

$$
= \; O \left(e^{-(\xi_0 + \delta)t} \right)
$$

for all $t \geq 1$, where $\delta \in (0, \xi_0)$. Theorem 42 is proved. $\qquad \square$

5. Large initial data

We consider the initial-boundary value problem (1.1) with any initial data $\|u_0\|_{\mathbf{L}^2} \leq C$. Multiplying equation (1.1) by u and integrating with respect to $x \in (0, 1)$ we get

$$\frac{d}{dt} \|u(t)\|_{\mathbf{L}^2}^2 + 2 \int_0^1 \left(u^2 u_x - u u_{xx} + u u_{xxx} \right) dx = 0.$$

We have

$$\int_0^1 u^2 u_x dx = \frac{1}{3} u^3 \bigg|_0^1 = 0,$$

$$\int_0^1 u u_{xx} dx = u u_x \big|_0^1 - \int_0^1 u_x^2 dx = - \int_0^1 u_x^2 dx,$$

$$\int_0^1 u u_{xxx} dx = u u_{xx} \big|_0^1 - \frac{1}{2} u_x^2 \big|_0^1 = \frac{1}{2} u_x^2 (0, t)$$

in view of the boundary data; hence

$$\frac{d}{dt} \|u(t)\|_{\mathbf{L}^2}^2 + 2 \int_0^1 u_x^2 dx + u_x^2 (0, t) = 0.$$

Integration with respect to $t > 0$ yields

$$\|u(t)\|_{\mathbf{L}^2} + 2 \int_0^t \|u_x(\tau)\|_{\mathbf{L}^2}^2 d\tau \leq \|u_0\|_{\mathbf{L}^2}$$

for all $t \in (0, +\infty)$. We see that the norm

$$\|u(t)\|_{\mathbf{L}^2} \leq \|u_0\|_{\mathbf{L}^2}$$

for all $t \geq 0$. By the standard continuation process via Theorem 41 we reveal that there exists a unique global solution $u \in \mathbf{C}\left((0, +\infty) ; \mathbf{H}^1(0, 1) \right)$ since the existence time T depends only on $\|u_0\|_{\mathbf{L}^2}$. Moreover we see that for any $\varepsilon > 0$ there exists a time $T > 0$ such that

$$\|u_x(T)\|_{\mathbf{L}^2}^2 < \varepsilon.$$

By the inequality

$$u^2(x, T) = 2 \int_0^x u u_y dy \leq 2 \|u\|_{\mathbf{L}^2} \|u_x\|_{\mathbf{L}^2}$$

we see that the norm $\|u(T)\|_{\mathbf{L}^\infty}$ is small. Hence the norm $\|u(T)\|_{\mathbf{L}^2}$ is also small by the estimate

$$\|u(T)\|_{\mathbf{L}^2} \leq \|u(T)\|_{\mathbf{L}^\infty}.$$

Then we consider the initial-boundary value problem (1.1) for $t \geq T$ and apply Theorem 42, whence the result of Theorem 39 follows and is thus proved.

CHAPTER 14

NLS Equation on Segment

1. Introduction

In this chapter we develop a general theory of the initial-boundary value problem for nonlinear evolution equations with pseudodifferential operators $\mathbb{K}u$ on a segment. There are some results in the case of nonlinear nonlocal equations on a half-line (see [53], [60]). However, to our knowledge there are no results in the case of nonlinear nonlocal equations on a segment. There are many open natural questions which we need to study in this respect. Firstly how many boundary data do we have to pose in the initial-boundary value problem with pseudodifferential operator $\mathbb{K}$ for it's correct solvability? The methods proposed in this chapter can be applied to a wide class of local equations with non self-adjoint differential operators, where because of its incompleteness it is difficult to apply the usual Fourier method if we want to take boundary data into account.

Let us start with the following general case of linear initial-boundary value problem

$$(1.1) \quad \begin{cases} u_t + \mathbb{K}u = f(x,t),\ t > 0,\ x \in (0,a), \\ u(x,0) = u_0(x),\ x \in (0,a), \\ \partial_x^j u(0,t) = h_{0j}(t),\ j = 1, ..., m, \\ \partial_x^l u(a,t) = h_{al}(t),\ l = 1, ..., n, \end{cases}$$

where the pseudodifferential operator $\mathbb{K}u$ on a segment $[0,a]$ we define by the inverse Laplace transformation

$$\mathbb{K}u = \frac{1}{2\pi i} \int_{-i\infty}^{i\infty} e^{px} K(p)$$

$$\times \left(\widehat{u}(p,t) - \sum_{j=1}^{[\alpha]} \frac{\partial_x^{j-1} u(0,t) - e^{-pa} \partial_x^{j-1} u(a,t)}{p^j} \right) dp$$

$$-\theta(x-a) \frac{1}{2\pi i} \int_{\Gamma_1} e^{px} K(p)$$

$$(1.2) \quad \times \left(\widehat{u}(p,t) - \sum_{j=1}^{[\alpha]} \frac{\partial_x^{j-1} u(0,t) - e^{-pa} \partial_x^{j-1} u(a,t)}{p^j} \right) dp.$$

261

Here the contour Γ_1 goes along the boundary of the domain of analyticity of the symbol $K(p)$. We assume that $K(p)$ is always analytic in the domain $\operatorname{Re} p > 0$. Note that in the case of the holomorphic symbol $K(p)$ the last integral in the definition (1.2) is equal to zero; hence we get a usual differential operator. Also we can rewrite the definition (1.2) in the form

$$
\begin{aligned}
\mathbb{K}u \; = \; & (1-\theta(x-a))\frac{1}{2\pi i}\int_{-i\infty}^{i\infty}e^{px}K(p) \\
(1.3) \qquad & \times \left(\widehat{u}(p,t) - \sum_{j=1}^{[\alpha]}\frac{\partial_x^{j-1}u(0,t)-e^{-pa}\partial_x^{j-1}u(a,t)}{p^j}\right)dp,
\end{aligned}
$$

if we take $K(p)=C_\alpha p^\alpha$, $\alpha>0$ for simplicity. We cut along the negative part of the real axis, that is we choose $\arg z \in [-\pi,\pi)$ for any complex $z \in \mathbb{C}$. Here $[\alpha]$ is the integer part of the number α; C_α will be chosen by the dissipation condition $\operatorname{Re} K(p) > 0$ for all $\operatorname{Re} p = 0$. Note that the inverse Laplace transform gives us a function which is equal to 0 for all $x < 0$, so that multiplication by the factor $(1-\theta(x-a))$ forces the operator $\mathbb{K}u$ to vanish outside of the interval $(0,a)$. Thus the solution $u(x,t)$ is considered for all $x \in \mathbf{R}$ prolonged as zero outside of the segment $[0,a]$. We expect that by an analogy with the case of a half-line, the integers n and m are defined by the number of regions, where $\operatorname{Re} K(p) < 0$.

Taking the Laplace transform of the operator $\mathbb{K}u$ we get

$$
\begin{aligned}
\int_0^a e^{-px}\mathbb{K}u\,dx &= \frac{1}{2\pi i}\int_{-i\infty}^{i\infty}\frac{e^{(q-p)a}-1}{q-p}K(q)\widehat{\widetilde{u}}(q,t)dq \\
(1.4) \qquad &= \frac{e^{-pa}}{2\pi i}\int_\Gamma \frac{e^{qa}}{q-p}K(q)\widehat{\widetilde{u}}(q,t)dq + K(p)\widehat{\widetilde{u}}(p,t),
\end{aligned}
$$

where we denote the contour

$$
(1.5) \qquad \Gamma = \left\{q\in\mathbb{C};\; q\in(\infty e^{-i\pi},-i0)\cup\left(+i0,e^{i\pi}\infty\right)\right\}
$$

and

$$
\widehat{\widetilde{u}}(p,t)=\widehat{u}(p,t)-\sum_{j=1}^{[\alpha]}\frac{\partial_x^{j-1}u(0,t)-e^{-pa}\partial_x^{j-1}u(a,t)}{p^j}.
$$

Applying the Laplace transformation with respect to x to problem (1.1) we get

$$
\begin{cases}
\widehat{u}_t+\frac{1}{2\pi i}\int_{-i\infty}^{i\infty}\frac{e^{(q-p)a}-1}{q-p}K(q)\widehat{\widetilde{u}}(q,t)dq=\widehat{f}(p,t), & t>0, \\
\widehat{u}(p,0)=\widehat{u}_0(p), \\
\partial_x^j u(0,t)=h_{0j}(t), j=1,...,n, \\
\partial_x^l u(a,t)=h_{al}(t), l=1,...,m.
\end{cases}
$$

Integrating with respect to time t in view of (1.4) we obtain for the Laplace transform $\widehat{u}(p,t)$

$$
(1.6) \qquad \widehat{u}(p,t)=e^{-K(p)t}\widehat{u}_0(p)+\int_0^t e^{-K(p)(t-\tau)}\widehat{f}_1(p,\tau)d\tau,
$$

where

$$
\begin{aligned}
\widehat{f}_1(p,t) \;=\;\; & \widehat{f}(p,t) + K(p)\sum_{j=1}^{[\alpha]} \frac{\partial_x^{j-1}u(0,t) - e^{-pa}\partial_x^{j-1}u(a,t)}{p^j} \\
& -\frac{1}{2\pi i}\int_\Gamma \frac{e^{(q-p)a}}{q-p}K(q)\widehat{\widetilde{u}}(q,\tau)dq.
\end{aligned}
$$

In order to get the integral formula for the solutions of (1.1), we must know the boundary values $\partial_x^{j-1}u\,(0,t)$, $\partial_x^{j-1}u(a,t)$. Some of the boundary values we include in the problem as given boundary data and the rest we will find from the equation using the growth condition

(1.7) $$|\widehat{u}(p,t)| \le M(1+|p|)^\beta\,\left(1 + \left|e^{-pa}\right|\right) \text{ for all } |p| \ge 1,$$

with some M, $\beta > 0$, which guarantees that the inverse Laplace transform $u(x,t)$ vanishes for all $x < 0$ and $x > a$. It is easy to prove that condition (1.7) is fulfilled in domains $\operatorname{Re} K(p) > 0$. In domains where $\operatorname{Re} K(p) < 0$, we rewrite formula (1.6) as

$$
\begin{aligned}
\widehat{u}(p,t) \;=\;\; & e^{-K(p)t}\left(\widehat{u}_0(p) + \int_0^{+\infty} e^{K(p)\tau}f_1(p,\tau)d\tau\right) \\
& -\int_t^{+\infty} e^{-K(p)(t-\tau)}f_1(p,\tau)d\tau.
\end{aligned}
$$

Clearly the last integral

$$
\int_t^{+\infty} e^{-K(p)(t-\tau)}f_1(p,\tau)d\tau
$$

satisfies condition (1.7) for all $|p| \ge 1$, such that $\operatorname{Re} K(p) < 0$. However the first summand with the exponentially growing factor $e^{-K(p)t}$ does not satisfy condition (1.7); therefore we must satisfy the following conditions

(1.8) $$\widehat{u}_0(p) + \int_0^{+\infty} e^{K(p)\tau}f_1(p,\tau)d\tau = 0$$

for all $|p| > 1$ in the domains, where $\operatorname{Re} K(p) < 0$. We use equations (2.9) to find some of the boundary values $\partial_x^j u(0,t)$, $\partial_x^j u(a,t)$ involved in formula (1.6). Changing the independent variable $K(p) = -\xi$ we transform the domains $\operatorname{Re} K(p) < 0$ to the half-complex plane $\operatorname{Re}\xi > 0$ by $[\alpha]$ different roots $\phi_1(\xi)$, $\phi_2(\xi)$, $\dots$, $\phi_{[\alpha]}(\xi)$, which are analytic functions for all $\operatorname{Re}\xi > 0$ and transform the half-complex plane $\operatorname{Re}\xi > 0$ to domains, where $\operatorname{Re} K(p) < 0$. Then condition (1.8) can be written as a system of $[\alpha]$ equations in the half complex plane $\operatorname{Re} \xi > 0$

$$
\widehat{u}_0(\phi_l) + \widehat{\widehat{f}}(\phi_l, \xi) - \xi \int_0^{+\infty} e^{-\xi\tau} \left(\sum_{j=1}^{[\alpha]} \frac{\partial_x^{j-1} u(0,t) - e^{-\phi_l a}\partial_x^{j-1} u(a,t)}{\phi_l^j} \right) d\tau
$$

$$
(1.9)= \quad -\frac{1}{2\pi i} \int_\Gamma \frac{e^{(q-\phi_l(\xi))a}}{q - \phi_l(\xi)} K(q) \int_0^{+\infty} e^{-\xi\tau} \left(\widehat{u}(q,\tau) \right.
$$

$$
\left. -\sum_{j=1}^{[\alpha]} \frac{\partial_x^{j-1} u(0,t) - e^{-qa}\partial_x^{j-1} u(a,t)}{q^j} \right) d\tau dq,
$$

for $l = 1, 2, ..., [\alpha]$, where $\widehat{u}(p,t)$ is the solution of problem (1.6) and

$$
\widehat{u}_0(\phi_l) = \int_0^a e^{-\phi_l y} u_0(y) dy, \quad \widehat{\widehat{f}}(\phi_l, \xi) = \int_0^{+\infty} \int_0^a e^{-(\phi_l y + \xi t)} f(y,t) dy dt.
$$

We have $[\alpha]$ equations with $2[\alpha]$ unknowns $u_x^{(j-1)}(0,t)$, $u_x^{(j-1)}(a,t)$ so we must add $[\alpha]$ boundary data in the problem (1.1), and the rest $[\alpha]$ boundary values can be found from system (1.9).

In the case $\alpha \in (0,1)$, which is considered in this chapter, we do not need to solve system (1.9), because condition (1.7) is fulfilled automatically for any complex p, due to the estimate $\left| e^{-K(p)t} \right| \le C\left(1 + \left| e^{-pa} \right| \right)$. Our aim in this chapter is to study the global existence and large time asymptotic behavior of solutions to the initial-boundary value problem for the nonlinear Shrödinger equation on a segment $[0,a]$

$$
(1.10) \qquad \begin{cases} u_t + i |u|^2 u + \mathbb{K}u = 0, \ t > 0, x \in (0,a), \\ u(x,0) = u_0(x), \ x \in (0,a), \end{cases}
$$

where the pseudodifferential operator $\mathbb{K}u$ on a segment $[0,a]$ is given by

$$
(1.11) \qquad \mathbb{K}u = (1 - \theta(x-a)) \frac{1}{2\pi i} \int_{-i\infty}^{i\infty} e^{px} K(p) \widehat{u}(p,t) dp,
$$

where $K(p) = p^\alpha$, $\alpha \in (0,1)$.

The nonlinear nonlocal Schrödinger equation (1.10) is a simple model appearing as the first approximation in the description of the dispersive dissipative nonlinear waves; it has many applications in various fields of physics, biology and engineering. In the case of the Cauchy problem the global existence of solutions was proved in papers [70], [49], and the large time asymptotics of solutions was obtained in [76], [103], [54]. In the case of the boundary value problem on a half-line the large time asymptotics of solutions were studied in papers [3], [53], [60], [71]. To our knowledge the global existence and large time asymptotic behavior for solutions of the initial-boundary value problem for the nonlinear nonlocal Schrödinger equation (1.10) on a segment were not studied previously. In this chapter we consider (1.10) in the case of the initial data belonging to space $\mathbf{L}^\infty$. For obtaining

$\mathbf{L}^p$ -estimates of the Green function we use the method of our previous papers [53] and [60].

Let us denote the space $\mathbf{L}^\infty(0, a) = \{\varphi \in \mathbf{L}^\infty(0, a)\,; \|\varphi\|_{\mathbf{L}^\infty} < +\infty\}$. By the same letter C we denote different positive constants. Let $\|\phi\|_{L^p(\mathbf{R}^+)} = \|\phi\|_{L^p}$ and $\|\phi\|_{L^p(0,a)} = \|\phi\|_p$, $1 \le p \le \infty$.

We state the main result of this chapter.

THEOREM 43. *Let the initial data* $u_0 \in \mathbf{L}^\infty(0, a)$ *and the norm* $\|u_0\|_\infty <$ ε, *where* $\varepsilon > 0$ *is sufficiently small. Then there exists a unique solution* $u \in$ $\mathbf{C}\left([0, \infty)\,; \mathbf{L}^\infty(0, a)\right)$ *of problem (1.10). Moreover there exists a constant* A *such that the solution has the following asymptotics*

$$u(x, t) = (1 - \theta(x - a))A(x)t^{-\frac{1}{\alpha}}\Lambda\left(\frac{x}{t^{\frac{1}{\alpha}}}\right) + O\left(t^{-\frac{1+\delta}{\alpha}}\right)$$

for $t \to +\infty$ *uniformly with respect to* $x \in (0, a)$, *where*

$$\Lambda(\xi) = \frac{1}{2\pi i}\int_{-i\infty}^{i\infty} e^{-z^\alpha + z\xi}dz$$

and

$$A = \int_0^x u_0(y)dy + \int_0^{+\infty} d\tau \int_0^x |u(y, \tau)|^2 u(y, \tau)dy < +\infty;$$

here $\xi \in \mathbf{R}^+$, $\delta \in (0, 1 - \alpha)$.

REMARK 18. *Note that the symbols* $K(p)$ *under consideration are not analytic in the left-half complex plane (see definition (1.3)), so the contour of integration in the inverse Laplace transform could not be shifted in order to obtain addition rapid time decay (see formula (2.4) below). As a consequence, the solutions of nonlocal equation (1.1) have a potential decay rate such as* $t^{-\frac{1}{\alpha}}$, *in comparison with the case of the purely differential operator* $\mathbb{K}$. *For example, it is well-known that solutions of the heat equation on a segment decay exponentially with respect to time.*

This chapter is organized as follows. In Section 2 we solve the linear initial-boundary value problem corresponding to (1.10) and prove some preliminary estimates in Lemma 40. Section 3 is devoted to the proof of Theorem 43.

2. Linear problem

We consider the following linear initial-boundary value problem

(2.1)
$$\begin{cases} u_t + \mathbb{K}u = f(x, t), \ t > 0, x \in (0, a)\,, \\ \quad u(x, 0) = u_0(x), \ x \in (0, a)\,, \end{cases}$$

where the pseudodifferential operator $\mathbb{K}u$ on a segment $[0, a]$ is defined in (1.11).

We have for the Laplace transform of operator $\mathbb{K}u$, $p \notin (-\infty, 0)$

$$\int_0^a e^{-px}\mathbb{K}u dx = \frac{1}{2\pi i}\int_{-i\infty}^{i\infty} \frac{e^{(q-p)a}-1}{q-p}K(q)\widehat{u}(q,t)dq$$

$$= \frac{e^{-pa}}{2\pi i}\int_\Gamma \frac{e^{qa}}{q-p}K(q)\widehat{u}(q,t)dq + K(p)\widehat{u}(p,t),$$

where the contour Γ was defined by the formula (1.5).

To derive an integral representation for solutions to the problem (2.1) we suppose that there exists a solution $u(x,t)$ of problem (2.1), which we prolonged by zero outside the interval $(0,a)$:

(2.2) $u(x,t) = 0$ for all $x \notin [0,a]$.

Applying the Laplace transformation with respect to x to the problem (2.1) we get

$$\begin{cases} \widehat{u}_t + \frac{e^{-pa}}{2\pi i}\int_\Gamma \frac{e^{qa}}{q-p}K(q)\widehat{u}(q,t)dq + K(p)\widehat{u}(p,t) = \widehat{f}(p,t), \quad t > 0, \\ \widehat{u}(p,0) = \widehat{u}_0(p). \end{cases}$$

Integrating with respect to time t we obtain for the Laplace transform $\widehat{u}(p,t)$

(2.3) $\widehat{u}(p,t) = e^{-K(p)t}\widehat{u}_0(p) + \int_0^t e^{-K(p)(t-\tau)}\widehat{f}_1(p,\tau)d\tau,$

where

$$\widehat{f}_1(p,t) = \widehat{f}(p,t) + \frac{1}{2\pi i}\int_\Gamma \frac{e^{(q-p)a}}{q-p}K(q)\widehat{u}(q,\tau)dq.$$

Note that by virtue of (2.2) the function $\widehat{u}(p,t)$ is analytic for all complex p, and the condition $0 < \alpha < 1$ implies the condition (1.7) .

Taking the inverse Laplace transform of (2.3) with respect to the space variable we get

$$\begin{aligned} u(x,t) \quad &= \quad \frac{1}{2\pi i}\int_{-i\infty+\varepsilon}^{i\infty+\varepsilon} e^{px-K(p)t}\widehat{u}_0(p)dp \\ &+ \frac{1}{2\pi i}\int_{-i\infty+\varepsilon}^{i\infty+\varepsilon} dp e^{px}\int_0^t e^{-K(p)(t-\tau)}\widehat{f}(p,\tau)d\tau \\ &+ \frac{1}{2\pi i}\int_{-i\infty+\varepsilon}^{i\infty+\varepsilon} dp e^{px}\int_0^t d\tau e^{-K(p)(t-\tau)} \\ &\times \frac{1}{2\pi i}\int_{-i\infty}^{i\infty} \frac{e^{(q-p)a}}{q-p}K(q)\widehat{u}(q,\tau)dq \\ &\equiv \quad I_1 + I_2 + I_3, \end{aligned}$$

(2.4)

where $\varepsilon > 0$.

Now we prove that the last integral in (2.4) is equal to zero for all $x \in [0,a]$. In fact, since Re $K(p) > 0$ for all Re $p > 0$ by the Cauchy Theorem we get for

$\operatorname{Re} q = 0, x \in [0, a], \tau \in (0, t)$

$$\int_{-i\infty+\varepsilon}^{i\infty+\varepsilon} dp e^{p(x-a)} e^{-K(p)(t-\tau)} \frac{1}{q-p} dp = 0.$$

Therefore changing the order of integration we obtain for $x \in [0, a]$ (we can change the order of integration since all integrals converge absolutely)

$$
\begin{aligned}
I_3 &= \frac{1}{2\pi i} \int_0^t d\tau \int_{-i\infty}^{i\infty} e^{qa} K(q) \widehat{u}(q, \tau) dq \\
(2.5) \qquad & \times \frac{1}{2\pi i} \int_{-i\infty+\varepsilon}^{i\infty+\varepsilon} dp e^{p(x-a)} e^{-K(p)(t-\tau)} \frac{1}{q-p} dp = 0.
\end{aligned}
$$

Since $u(x, t) = 0$ for all $x > a$ and for $x < 0$, substituting the Laplace transforms $\widehat{u}_0(p)$ and $\widehat{f}(p, \tau)$ into (2.4) and using (2.5), we obtain the following integral representation for solutions $u(x, t)$ of the problem (2.1)

$$(2.6) \qquad u(x, t) = \int_0^a u_0(y) G(x, y, t) dy + \int_0^t d\tau \int_0^a f(y, \tau) G(x, y, t - \tau) d\tau,$$

where the Green function $G(x, y, t)$ is defined by

$$G(x, y, t) = (1 - \theta(x - a)) \frac{1}{2\pi i} \int_{-i\infty}^{i\infty} e^{p(x-y) - K(p)t} dp.$$

Thus in supposition that there exist solutions of problem (2.1) we get the integral representation (2.6) for these solutions.

Now we prove that the function $u(x, t)$ defined by formula (2.6) gives us a solution to problem (2.1). In fact, taking the Laplace transformation of (2.6) we get for $\operatorname{Re} p = 0$

$$
\begin{aligned}
(2.7) \quad \widehat{u}(p, t) &= \int_0^\infty dx e^{-px} \int_0^a u_0(y) G(x, y, t) dy \\
&+ \int_0^\infty dx e^{-px} \int_0^t d\tau \int_0^a f(y, \tau) G(x, y, t - \tau) dy \\
&= \int_0^a dx e^{-px} \left(\int_0^a u_0(y) \frac{1}{2\pi i} \int_{-i\infty}^{i\infty} e^{q(x-y) - K(q)t} dq dy \right. \\
&\left. + \int_0^t d\tau \int_0^a f(y, \tau) \frac{1}{2\pi i} \int_{-i\infty}^{i\infty} e^{q(x-y) - K(q)(t-\tau)} dq d\tau \right).
\end{aligned}
$$

By analyticity of the symbol $K(p)$ in the complex half-plane $\mathrm{Re}\,p > 0$ and $\alpha < 1$ we have for all $\mathrm{Re}\,p = 0$ and $y \in [0, a)$

$$\frac{1}{2\pi i} \int_{-i\infty}^{i\infty} e^{-K(q)t-qy} \frac{e^{(q-p)a}-1}{q-p} dq$$

$$= \frac{1}{2\pi i} e^{-pa} \int_{-i\infty}^{i\infty} e^{-K(q)t} \frac{e^{q(a-y)}}{q-p} dq - \frac{1}{2\pi i} \int_{-i\infty}^{i\infty} e^{-K(q)t-qy} \frac{1}{q-p} dq$$

$$= e^{-K(p)t-py} + \frac{1}{2\pi i} e^{-pa} \int_{\Gamma} e^{-K(q)t+q(a-y)} \frac{1}{q-p} dq.$$

Thus changing the order of integration in formula (2.7) and calculating the integrals with respect to x we get

$$\widehat{u}(p,t)$$

$$= \frac{1}{2\pi i} \int_0^a u_0(y)dy \int_{-i\infty}^{i\infty} e^{-K(q)t-qy} \frac{e^{(q-p)a}-1}{q-p} dq$$

$$+ \frac{1}{2\pi i} \int_0^t d\tau \int_0^a f(y,\tau)dy \int_{-i\infty}^{i\infty} e^{-qy-K(q)(t-\tau)} \frac{e^{(q-p)a}-1}{q-p} dq$$

$$= e^{-K(p)t} \left(\int_0^a e^{-py} u_0(y)dy + \int_0^t e^{K(p)\tau} d\tau \int_0^a e^{-py} f(y,\tau)dy \right)$$

$$+ \frac{e^{-pa}}{2\pi i} \int_0^a u_0(y)dy \int_{\Gamma} e^{-K(q)t+q(a-y)} \frac{dq}{q-p}$$

$$(2.8) \qquad + \frac{e^{-pa}}{2\pi i} \int_0^t d\tau \int_0^a f(y,\tau)dy \int_{\Gamma} e^{-K(q)(t-\tau)+q(a-y)} \frac{dq}{q-p}.$$

Putting (2.8) into the definition of the pseudodifferential operator $\mathbb{K}u$ (see formula (1.11)) we obtain for all $x \in (0, a)$

$$\mathbb{K}u = \int_0^a u_0(y)dy \frac{1}{2\pi i} \int_{-i\infty}^{i\infty} e^{p(x-y)} e^{-K(p)t} K(p)dp$$

$$+ \int_0^t d\tau \int_0^a f(y,\tau)dy \frac{1}{2\pi i} \int_{-i\infty}^{i\infty} e^{p(x-y)} e^{-K(p)(t-\tau)} K(p)dp$$

$$+ \int_0^a u_0(y)dy \frac{1}{2\pi i} \int_{\Gamma} e^{-K(q)t+q(a-y)} dq \frac{1}{2\pi i} \int_{-i\infty}^{+i\infty} e^{p(x-a)} \frac{K(p)}{q-p} dp$$

$$+ \int_0^t d\tau \int_0^a f(y,\tau)dy \frac{1}{2\pi i} \int_{\Gamma} e^{-K(q)(t-\tau)+q(a-y)} dq$$

$$\times \frac{1}{2\pi i} \int_{-i\infty}^{+i\infty} e^{p(x-a)} \frac{K(p)}{q-p} dp,$$

and by the fact that

$$\int_{-i\infty}^{+i\infty} e^{p(x-a)} \frac{K(p)}{q-p} dp = 0$$

for all $x \in (0, a)$ and $q \in \Gamma$ we obtain via formula (2.6)

$$
\begin{aligned}
\mathbb{K}u \;=\; & \left(-\frac{\partial}{\partial t} \int_0^a u_0(y)dy \frac{1}{2\pi i} \int_{-i\infty}^{i\infty} e^{p(x-y)} e^{-K(p)t} dp \right. \\
& - \frac{\partial}{\partial t} \int_0^t d\tau \int_0^a f(y,\tau)dy \frac{1}{2\pi i} \int_{-i\infty}^{i\infty} e^{p(x-y)} e^{-K(p)(t-\tau)} dp \\
& \left. + \int_0^a f(y,\tau)dy \frac{1}{2\pi i} \int_{-i\infty}^{i\infty} e^{p(x-y)} dp \right) = -u_t(x,t) + f(x,t).
\end{aligned}
$$

Thus the function $u(x,t)$ given by (2.6) satisfies the equation

$$ u_t(x,t) + \mathbb{K}u = f(x,t). $$

Also clearly the initial condition is fulfilled

$$
\begin{aligned}
u(x,0) \;=\;& (1 - \theta(x - a)) \int_0^a u_0(y)G(x - y, 0)dy \\
=\;& (1 - \theta(x - a)) \int_0^{+\infty} u_0(y)\delta(x - y)dy = u_0(x).
\end{aligned}
$$

Thus there exists a solution to the problem (2.1), which is given by formula (2.6). The uniqueness follows from the fact that all solutions have representation (2.6).

Note that by the Cauchy Theorem the Green function $G(x,y,t) = 0$ for all $x < y$ and $t < 0$; therefore formula (2.6) can be written as

$$ (2.9) \qquad u(x,t) = \int_0^x u_0(y)G(x,y,t)dy + \int_0^t d\tau \int_0^x f(y,\tau)G(x,y,t-\tau)d\tau, $$

where

$$ (2.10) \qquad G(x,y,t) = (1 - \theta(x-a)) \frac{1}{2\pi i} \int_{-i\infty}^{i\infty} e^{p(x-y)-K(p)t} dp. $$

Thus we have proved the following result.

THEOREM 44. *Let the initial data $u_0 \in \mathbf{L}^1(0, a)$ and a source*

$$ f(x,t) \in \mathbf{L}_{loc}^1 \left(0, \infty; \mathbf{L}^1(0, a)\right). $$

Then there exists a unique solution $u(x,t)$ of the initial-boundary value problem (2.1), which has representation (2.9).

REMARK 19. *By representation (2.9) we see that $\lim_{x \to +0} u(x,t) = 0$ for all $t > 0$. We emphasize however that we do not need to put the boundary condition $u(0,t) = 0$ into problem (2.1) for its well-posedness, since this is an inherent property of the solutions. For example if we put the boundary condition $u(0,t) = 1$ into problem (2.1), then no solution exists.*

REMARK 20. *Note that the Green function $G(x,y,t)$ is similar to that for the cases of a half-line and the full line. It can be obtained from the full line Green function via multiplication by the step function $(1 - \theta(x - a))$.*

In the next lemma we estimate the kernel $G(x, y, t)$. We denote

$$\Lambda(\xi) = \frac{1}{2\pi i} \int_{-i\infty}^{i\infty} e^{-z^\alpha + z\xi} dz.$$

LEMMA 40. *We have the asymptotics for large time*

$$(2.11) \qquad G(x, y, t) = (1 - \theta(x - a)) t^{-\frac{1}{\alpha}} \Lambda\left(\frac{x}{t^{\frac{1}{\alpha}}}\right) + y^\delta O\left(t^{-\frac{1+\delta}{\alpha}}\right),$$

for $y \in (0, x)$.

PROOF. Changing the variable of integration $p^\alpha t = q^\alpha$ we get

$$
\begin{aligned}
G(x, y, t) &= (1 - \theta(x - a)) \frac{1}{2\pi i} \int_{-i\infty}^{i\infty} e^{p(x-y) - K(p)t} dp \\
&= t^{-\frac{1}{\alpha}} (1 - \theta(x - a)) \frac{1}{2\pi i} \left(\int_{-i\infty}^{i\infty} e^{-q^\alpha + q\widetilde{x}} dq + R(\widetilde{x}, \widetilde{y}) \right),
\end{aligned}
$$

where $\widetilde{x} = x t^{-\frac{1}{\alpha}}, \widetilde{y} = y t^{-\frac{1}{\alpha}}$ and

$$R(\widetilde{x}, \widetilde{y}) = \int_{-i\infty}^{i\infty} e^{-q^\alpha + q\widetilde{x}} (e^{-q\widetilde{y}} - 1) dq.$$

Using estimates $\left|e^{-q\widetilde{y}} - 1\right| < C |q\widetilde{y}|^\delta$ and Re $q^\alpha > 0$ for Re $q = 0$ we easily get

$$t^{-\frac{1}{\alpha}} |R(\widetilde{x}, \widetilde{y})| \le Ct^{-\frac{1}{\alpha}} \int_{-i\infty}^{i\infty} e^{-\text{Re } q^\alpha} |q\widetilde{y}|^\delta dq = y^\delta O\left(t^{-\frac{1+\delta}{\alpha}}\right).$$

Lemma 40 is proved.   □

We denote $\mathcal{G}(t) \phi = \int_0^x G(x, y, t) \phi(y) dy$, where $G(x, t)$ is defined in formula (2.10).

LEMMA 41. *Suppose that the function $\phi \in \mathbf{L}^\infty(0, a)$. Then the estimates are valid*

$$\|\mathcal{G}(t) \phi\|_\infty \le C(1 + t)^{-\frac{1}{\alpha}} \|\phi\|_\infty$$

for all $t > 0$.

Proof. We denote $\widetilde{G}(x) = \mathcal{L}^{-1}(e^{-p^\alpha})$. Note that the function $\widetilde{G}(x)$ is a smooth function $\widetilde{G}(x) \in \mathbf{C}^\infty(\mathbf{R}^+)$ and decays at infinity so that

$$(2.12) \qquad \sup_{x \in \mathbf{R}^+} \langle x \rangle^{1+\gamma} \left|\widetilde{G}(x)\right| \le C,$$

for all $0 < \gamma < 1$. In fact, since Re $p^\alpha > 0$ for Re $p = 0$ we have

$$\left|\widetilde{G}(x)\right| = \left|\frac{1}{2\pi i} \int_{-i\infty}^{i\infty} e^{px - p^\alpha} dp\right| \le C \left\|e^{-p^\alpha}\right\|_{\mathbf{L}^1} \le C.$$

For all $x \geq 1$, integrating by parts and changing the contour of integration we get

$$\left| \tilde{G}\left(x \right) \right| = \left| \frac{1}{2\pi i} \int_{-i\infty}^{i\infty} e^{px - p^\alpha} dp \right| = \left| \frac{\alpha}{2\pi i x} \int_{\infty e^{-\frac{i\pi}{2} + i\varepsilon}}^{\infty e^{\frac{i\pi}{2} - i\varepsilon}} e^{px - p^\alpha} p^{\alpha - 1} dp \right|$$

$$\leq C x^{-1-\gamma} \left| \int_{\infty e^{-\frac{i\pi}{2} + i\varepsilon}}^{\infty e^{\frac{i\pi}{2} - i\varepsilon}} e^{-p^\alpha} p^{-1+\alpha-\gamma} dp \right| \leq C x^{-1-\gamma},$$

where $\varepsilon > 0, 0 < \gamma < 1$. Therefore estimate (2.12) is true. By virtue of (2.12) we find

$$t^{-\frac{1}{\alpha}} \left\| \tilde{G}\left(t^{-\frac{1}{\alpha}} \left(\cdot \right) \right) \right\|_{\mathbf{L}^1} = \left\| \tilde{G}\left(\cdot \right) \right\|_{\mathbf{L}^1} \leq C \left\| \langle x \rangle^{-1-\mu} \right\|_{\mathbf{L}_x^1} \leq C,$$

hence by the Young inequality and using estimate $\|\phi\|_1 < C \|\phi\|_\infty$ we obtain

$$\|\mathcal{G}\left(t \right) \phi\|_\infty \leq C \left\| t^{-\frac{1}{\alpha}} \tilde{G}\left(t^{-\frac{1}{\alpha}} \left(\cdot \right) \right) \right\|_{\mathbf{L}^1} \|\phi\|_\infty \leq C \|\phi\|_\infty$$

and

$$\|\mathcal{G}\left(t \right) \phi\|_\infty \leq C \left\| t^{-\frac{1}{\alpha}} \tilde{G}\left(t^{-\frac{1}{\alpha}} \left(\cdot \right) \right) \right\|_{\mathbf{L}^\infty} \|\phi\|_1$$

$$\leq C t^{-\frac{1}{\alpha}} \|\phi\|_1 \leq C t^{-\frac{1}{\alpha}} \|\phi\|_\infty$$

for all $t > 0$ from which the estimate of the lemma follows. Lemma 41 is proved.

3. Global existence

We prove Theorem 43. We consider the linearized version of problem (1.10)

$$(3.1) \qquad \begin{cases} u_t + \mathbb{K}u = -i \left| v \right|^2 v, \quad t > 0, x \in (0, a), \\ u(x, 0) = u_0(x), x \in (0, a). \end{cases}$$

We suppose that

$$\|u_0\|_\infty < \varepsilon_1$$

and $v \in X_\varepsilon$, where $\varepsilon_1 > 0$ is small enough; $\varepsilon = 100 C \varepsilon_1$ with the constant C from (2.12) and

$$\mathbf{X}_\varepsilon = \left\{ v \in X, \|v\|_{\mathbf{X}} < \varepsilon \right\},$$

$$\mathbf{X} = \left\{ v \in \mathbf{C}([0, +\infty); \mathbf{L}^\infty(0, a)), \|v\|_{\mathbf{X}} = \sup_{t>0} \langle t \rangle^{\frac{1}{\alpha}} \|v\left(t \right)\|_\infty < \varepsilon \right\}.$$

By virtue of (2.9) we have

$$(3.2) \qquad u(x, t) = \mathcal{G}\left(t \right) u_0 - i \int_0^t \mathcal{G}\left(t - \tau \right) \left| v(\tau) \right|^2 v(\tau) d\tau,$$

where

$$\mathcal{G}\left(t \right) \phi(\tau) = \int_0^x G\left(x, y, t \right) \phi\left(y, \tau \right) dy$$

and

$$G(x, y, t) = (1 - \theta\left(x - a \right)) \frac{1}{2\pi i} \int_{-i\infty}^{i\infty} e^{p(x-y) - p^\alpha t} dp.$$

Using the estimate

$$\left\| |v|^2 v(t) \right\|_1 \leq \|v(t)\|_\infty^3 \leq C\varepsilon^3 (1+t)^{-\frac{3}{\alpha}},$$

applying $\mathbf{L}^\infty(0, a)$ norm to formula (3.2) and using the results of Lemma (41) we get

$$\|u(t)\|_\infty \leq C \left\| \mathcal{G}(t) u_0 \right\|_\infty + C \int_0^t \left\| \mathcal{G}(t-\tau) |v(\tau)|^2 v(\tau) \right\|_\infty d\tau$$

$$\leq C(1+t)^{-\frac{1}{\alpha}} \|u_0\|_\infty$$

$$+ \int_0^{\frac{t}{2}} d\tau \left\| |v|^2 v(\tau) \right\|_1 (t-\tau)^{-\frac{1}{\alpha}} d\tau + \int_{\frac{t}{2}}^t d\tau \left\| |v|^2 v(\tau) \right\|_\infty d\tau$$

$$\leq \varepsilon_1 (1+t)^{-\frac{1}{\alpha}} + \varepsilon^3 C \left(\int_0^{\frac{t}{2}} \langle \tau \rangle^{-\frac{3}{\alpha}} (t-\tau)^{-\frac{1}{\alpha}} d\tau + \int_{\frac{t}{2}}^t \langle \tau \rangle^{-\frac{3}{\alpha}} d\tau \right)$$

(3.3) $\leq \varepsilon (1+t)^{-\frac{1}{\alpha}} .$

We introduce the distance in $\mathbf{X}$

$$d(f, g) = \sup_{t>0}(1+t)^{\frac{1}{\alpha}} \|f(t) - g(t)\|_\infty .$$

Then as in the proof of (3.3) we have

(3.4) $d(u_1, u_2) = d(\mathbb{M}v_1, \mathbb{M}v_2) \leq \frac{1}{2} d(v_1, v_2),$

where u_1 and u_2 are solutions of the problems

$$\begin{cases} \partial_t u_j + \mathbb{K} u_j = -i |v_j|^2 v_j, \ t > 0, x \in (0, a), \\ u_j(x, 0) = u_0(x), \ x \in (0, a). \end{cases}$$

Estimates (3.3) and (3.4) show that $\mathbb{M}$ is a contraction mapping from $\mathbf{X}$ into itself. Therefore there exist a unique solution $u(x, t) \in \mathbf{X}$ satisfying estimate $\|u\|_\mathbf{X} < \varepsilon$. This completes the proof of the first part of Theorem 43.

Now using estimate (3.3) we prove that the solution has the following asymptotics

$$u(x, t) = (1 - \theta(x - a))A(x) t^{-\frac{1}{\alpha}} \Lambda \left(\frac{x}{t^{\frac{1}{\alpha}}} \right) + O\left(t^{-\frac{1+\delta}{\alpha}} \right)$$

for $t \longrightarrow +\infty$ uniformly with respect to x, where $\delta \in (0, 1 - \alpha)$

$$\Lambda(x) = \frac{1}{2\pi i} \int_{-i\infty}^{i\infty} e^{-z^\alpha + zx} dz$$

and

$$A(x) = \int_0^x u_0(y)dy - i \int_0^{+\infty} d\tau \int_0^x |u(y, \tau)|^2 u(y, \tau)dy$$

is a bounded function. In fact, in view of asymptotics (2.11) of Lemma 40 we have

(3.5) $u(x, t) = (1 - \theta(x - a))A(x) t^{-\frac{1}{\alpha}} \Lambda \left(\frac{x}{t^{\frac{1}{\alpha}}} \right) + R(x, t),$

where

$$
\begin{aligned}
|R(x,t)| \;\leq\; & Ct^{-\frac{1+\delta}{\alpha}} \left\| (\cdot)^{\delta} u_0(\cdot) \right\|_1 + Ct^{-\frac{1+\delta}{\alpha}} \int_0^t d\tau \int_0^a y^{\delta}\, |u|^3 \, dy \\
& + t^{-\frac{1}{\alpha}} \left| \Lambda \left(\frac{x}{t^{\frac{1}{\alpha}}} \right) \right| \int_t^{+\infty} d\tau \int_0^a |u|^3 \, dy \\
& + \int_0^t d\tau \int_0^a |u(y,\tau)|^3 \, |G(x,y,t-\tau) - G(x,y,t)| \, dy.
\end{aligned}
$$

We have

$$
|G_t(x,y,t)| \leq C \int_{-i\infty}^{i\infty} e^{-\operatorname{Re} |p|^{\alpha} t} \, |p|^{\alpha-1} \, dp \leq Ct^{-\frac{1}{\alpha} - \frac{\alpha-1}{\alpha}} \leq Ct^{-1}.
$$

Therefore we attain

$$
|G(x,y,t-\tau) - G(x,y,t)| \leq Ct^{-1} \tau
$$

and

$$
\begin{aligned}
& \int_0^t d\tau \int_0^a |u(y,\tau)|^3 \, |G(x,y,t-\tau) - G(x,y,t)| \, dy \\
\leq \; & Ct^{-1} \int_0^t \tau (1+\tau)^{-\frac{3}{\alpha}} d\tau \leq Ct^{1-\frac{3}{\alpha}} \leq Ct^{-\frac{1+\delta}{\alpha}}
\end{aligned}
$$

for $0 < \delta < 1 - \alpha$, $t > 1$. Hence by virtue of (3.3) we have

$$
\begin{aligned}
|R(x,t)| \;\leq\; & Ct^{-\frac{1+\delta}{\alpha}} \left\| u_0(\cdot) \right\|_{\infty} + Ct^{-\frac{1+\delta}{\alpha}} \int_0^t (1+\tau)^{-\frac{3}{\alpha}} d\tau \\
& + t^{-\frac{1}{\alpha}} \left| \Lambda \left(\frac{x}{t^{\frac{1}{\alpha}}} \right) \right| \int_t^{+\infty} (1+\tau)^{-\frac{3}{\alpha}} d\tau + Ct^{-\frac{1+\delta}{\alpha}} \\
\leq \; & Ct^{-\frac{1+\delta}{\alpha}}.
\end{aligned}
$$

Theorem 43 is proved.

CHAPTER 15

Periodic Problem

1. Introduction

1.1. Periodic problem and physical examples. In this chapter we study large time asymptotic behavior of solutions to the periodic problem for the following model nonlinear nonlocal evolution equation

(1.1)
$$\begin{cases} u_t + \mathcal{N}(u) + \mathcal{L}u = 0, & x \in \Omega, t > 0, \\ u(0, x) = \phi(x), & x \in \Omega, \end{cases}$$

where $\Omega = [0, 2\pi]$. Here the nonlinearity $\mathcal{N}$ and the linear operator $\mathcal{L}$ are pseudo-differential operators defined by the Fourier series as follows

$$\mathcal{N}(u) = \sum_{n=-\infty}^{\infty} e^{inx} \sum_{j=-\infty}^{\infty} \hat{u}_{n-j} \left(A_{n,j} \hat{u}_j + \sum_{l=-\infty}^{\infty} B_{n,j,l} \hat{u}_{j+l} \overline{\hat{u}_l} \right),$$

and

$$\mathcal{L}u = \sum_{n=-\infty}^{\infty} e^{inx} L_n \hat{u}_n,$$

here and in the rest of the work

$$\hat{u}_n = \frac{1}{2\pi} \int_{\Omega} e^{-inx} u(x) \, dx$$

are the Fourier coefficients of the 2π - periodic function $u(t, x)$. Thus we consider the solutions of equation (1.1), which satisfy the periodic boundary condition $u(t, x) = u(t, 2\pi + x)$ for all $x \in \mathbf{R}$ and $t > 0$, with the 2π - periodic initial data ϕ. We suppose that the coefficients $A_{n,j}$ and $B_{n,j,l}$ are continuous functions with respect to time $t > 0$, and the operators $\mathcal{N}$ and $\mathcal{L}$ have a finite order, i.e. the symbols $A_{n,j}$, $B_{n,j,l}$ and L_n grow with respect to n, j, l no faster than a power of some order κ:

$$|A_{n,j}| \le C (\langle n \rangle + \langle j \rangle)^k, \quad |B_{n,j,l}| \le C (\langle n \rangle + \langle j \rangle + \langle l \rangle)^\kappa, \quad |L_n| \le C \langle n \rangle^\kappa.$$

By C we denote different positive constants. The Japanese brackets $\langle n \rangle = \sqrt{1 + n^2}$. The coefficients L_n are supposed to be independent of time.

Model equation (1.1) describes various wave processes in periodic media. It generalizes many well-known equations of modern mathematical physics. For example, if $u(t, x)$ is a real-valued function and $\mathcal{N}(u) = au^2 + bu^3$ so that $A_{n,j} = a$ and $B_{n,j,l} = b$ with $a, b \in \mathbf{R}$, then equation (1.1) transforms into the generalized

Kolmogorov - Petrovsky - Piskunov equation [88]. In particular, if $L_n = \alpha + \beta n^2$, $\alpha, \beta > 0$ we get the Fisher equation [34]

(1.2) $$u_t + au^2 + bu^3 + \alpha u - \beta u_{xx} = 0.$$

If the solution $u(t, x)$ is a real-valued function and the nonlinearity has the form $\mathcal{N}(u) = auu_x + bu^2 u_x$, that is $A_{n,j} = \frac{ia}{2}n$, $B_{n,j,l} = \frac{ib}{3}n$, $a, b \in \mathbf{R}$, then we obtain the Whitham equation [131]

(1.3) $$u_t + auu_x + bu^2 u_x + \mathcal{L}u = 0,$$

which contains a number of famous nonlinear equations in the theory of water waves such as the following: the Korteweg - de Vries - Burgers (KdV-B) equation

(1.4) $$u_t + auu_x + bu^2 u_x + \alpha u - \beta u_{xx} + u_{xxx} = 0,$$

when $L_n = \alpha + \beta n^2 - in^3$, $\alpha \in \mathbf{R}$ and $\beta > 0$; the Kuramoto-Sivashinsky (KS) equation [89], [121]

(1.5) $$u_t + auu_x + bu^2 u_x + \alpha u - \beta u_{xx} + u_{xxxx} = 0,$$

when $L_n = \alpha + \beta n^2 + n^4$, $\alpha, \beta \in \mathbf{R}$; the Ott - Sudan - Ostrovsky (OSO) equation [108], [109]

(1.6) $$u_t + auu_x + bu^2 u_x + \alpha u + \beta \mathcal{H}u_x + u_{xxx} = 0,$$

when the symbol $L_n = \alpha + \beta |n| - in^3$, $\alpha \in \mathbf{R}$ and $\beta > 0$, here the operator $\mathcal{H}u = \frac{1}{2\pi} \int \cot \frac{x-y}{2} u(t, y) dy$ denotes the Hilbert transformation in the case of the periodic functions. Equation (1.1) contains also the modified Benjamin - Bona - Mahony - Peregrine - Burgers (BBMP-B) equation [6], [111]

(1.7) $$u_t + auu_x + bu^2 u_x + cu_x + \alpha u - \beta u_{xx} - u_{txx} = 0,$$

if we choose $A_{n,j} = \frac{ian}{2(1+n^2)}$, $B_{n,j,l} = \frac{ibn}{3(1+n^2)}$, $L_n = \frac{\alpha + icn + \beta n^2}{1+n^2}$ in (1.1), where $\alpha \in \mathbf{R}$, $\beta > 0$ and $a, b, c \in \mathbf{R}$.

   If we choose the nonlinearity $\mathcal{N}(u) = ib |u|^2 u$ and the linear operator $\mathcal{L}u = \alpha u - \beta u_{xx} - iu_{xx}$, i.e. we let $A_{n,j} = 0$, $B_{n,j,l} = ib$, $L_n = \alpha + (\beta + i)n^2$, then we obtain the nonlinear Schrödinger equation with dissipation, or the generalized Landau - Ginzburg (LG) equation

(1.8) $$u_t + ib |u|^2 u + \alpha u - (\beta + i) u_{xx} = 0,$$

where $\beta > 0$, $\alpha \in \mathbf{R}$ and $b \in \mathbf{C}$. If we take the nonlinearity $\mathcal{N}(u) = a |u|^2 u_x + b \left(|u|^2 u \right)_x$ in equation (1.1), i.e. $A_{n,j} = 0$, $B_{n,j,l} = \frac{ia}{2}(l + n) + ibn$, $a, b \in \mathbf{C}$, then from equation (1.1) we obtain the derivative nonlinear Schrödinger (DNLS) equation with dissipation

(1.9) $$u_t + a |u|^2 u_x + b \left(|u|^2 u \right)_x + \alpha u - (\beta + i) u_{xx} = 0,$$

where $\beta > 0$, $\alpha \in \mathbf{R}$. Finally note that under the condition $\int_\Omega u(t, y) dy = 0$ for all $t \geq 0$ we can introduce a potential $\varphi(t, x) = \int_0^x u(t, y) dy$, which is also a 2π

- periodic function with respect to the space variable x. Then from (1.3) we get the potential Whitham equation

$$(1.10) \qquad \varphi_t + \frac{a}{2}\left(\varphi_x\right)^2 + \frac{b}{3}\left(\varphi_x\right)^3 + \mathcal{L}\varphi = 0,$$

which also follows from (1.1) if we take $A_{n,j} = -\frac{a}{2}(n-j)j$, $B_{n,j,l} = -\frac{ib}{3}(n-j)(j+l)l$, with $a, b \in \mathbf{R}$. In particular, we get potential KdV-B, KS and BBMP-B equations. A number of other interesting nonlinear nonlocal equations important for the theory of waves can be found in book [103]. Thus we see that equation (1.1) is general enough and deserves a comprehensive mathematical study.

Large amount of literature is devoted to the investigation of periodic initial-boundary value problems. The asymptotic stability of stationary periodic solutions for the Fisher equation (1.2) is proved in paper [33]. Periodic problem for the Hopf equation $u_t + uu_x = \alpha u$, where $\alpha > 0$ is studied in paper [61]. It is proved that if the initial data u_0 have mean value zero, the growth from the source term αu is balanced by decay induced by the nonlinear convection term uu_x. This leads to bounded oscillatory profiles for periodic initial data. The asymptotic behavior for large time of solutions to the periodic problem for the Burgers equation with periodic force is found in paper [84]. In papers [7], [10], [62], [116] it is proved the existence of unique global solutions for the periodic problem for KdV-B equation (??) and estimates of the large time decay rates of solutions are found. Also the blow up in finite time is investigated numerically in [10]. The approximate inertial manifolds for the damped forced KdV-B equation are constructed in [95], [126], [135]. In papers [101], [103] the large time asymptotic formulas for the solutions of the periodic problem for the KdV-B equation are obtained. Some time decay estimates of solutions to the KdV-B equation with periodic initial data with zero mean value are found in paper [7], which also reveals that the asymptotic behavior is defined by the higher-order harmonics. In papers [15], [18], [28], [133] the existence of unique global solutions for the periodic problem to GL equation (1.8) is proved. Also time decay estimates in different norms (usually in $\mathbf{L}^2$) of solutions are obtained and the blow up in finite time phenomenon is studied for the case of large initial data. Exact periodic solutions for the LG equation are found in [95]. The approximate inertial manifolds for the LG equation are constructed in [114]. The existence of global solutions of DNLS equation (1.9) is proved in [29]. Global existence and time decay estimates of solutions to the periodic problem for the KS equation (1.5) are obtained in [47], and the attractors for the KS equation are studied in [35]. For the BBMPB equation (1.7), the existence of global solutions and decay estimates of energy are obtained in paper [132], with large time asymptotic formulas for the solutions of periodic problem found in [101], [103]. Large time behavior of solutions to the periodic problem for systems of conservation laws are studied in [23], [120]. Paper [21] proves the existence of global solutions of the periodic problem for the Camassa-Holm equation

$$u_t - u_{xxt} + 3uu_x - 2u_x u_{xx} - uu_{xxx} = 0,$$

and the blow up in finite time is studied.

The aim of this chapter is to find the large time asymptotic representation of solutions to the periodic problem (1.1). We intend to find the main term of the asymptotics and to give an estimate of the remainder in the uniform norm. In the case of the Cauchy problem, the large time asymptotic formulas for solutions of dissipative nonlinear nonlocal equations were obtained in [101] and [103] via the perturbation theory with respect to the parameter, which characterizes the smallness of the initial data. The study of the periodic problem is in many respects easier than the Cauchy problem and, typically, the periodic results are exponential, whereas the problem on the line appears more delicate and often has algebraic decay rates (see [11], [21], [103], [119]). The asymptotic formulas for solutions of the periodic problem to some nonlinear equations were found in [101] and [103] by using the perturbation theory. Here we develop another approach based on the construction of solutions by the contraction principle. We intend to remove the requirement of smallness of the initial data taking into account some additional symmetry of the nonlinearity in equation (1.1). In comparison with the Cauchy problem we can consider not only decaying large time asymptotics, but we can also examine the cases when the solutions exponentially decay with time, oscillate or grow exponentially with time, depending on the linear part of equation (1.1) and the structure of the nonlinearity.

Before stating the results of this chapter we give some notations. The Sobolev space for the case of the periodic functions we define as follows:

$$\mathbf{H}^s = \left\{ \varphi \in \mathbf{D}' : \|\varphi\|_{\mathbf{H}^s}^2 = \sum_{n=-\infty}^{\infty} \langle n \rangle^{2s} |\hat{\varphi}_n|^2 < \infty \right\}$$

for any $s \in \mathbf{R}$. For the functions with zero mean value we introduce the space $\mathbf{H}_0^s = \{ \varphi \in \mathbf{H}^s : \hat{\varphi}_0 = 0 \}$. We also denote

$$\|\varphi\|_{\mathbf{H}_0^s}^2 = \sum_{n=-\infty, n \neq 0}^{\infty} \langle n \rangle^{2s} |\hat{\varphi}_n|^2$$

and $\mathbf{H}^\infty = \cap_{s \geq 0} \mathbf{H}^s$. By $\mathbf{C}\,(\mathbf{I}; \mathbf{B})$ we denote the space of continuous functions from a time interval $\mathbf{I}$ to the Banach space $\mathbf{B}$. Different positive constants might be denoted by the same letter C. The results of this chapter were published in paper [74].

1.2. Small initial data. Suppose that the symbols of the nonlinear operator $\mathcal{N}$ are such that

(1.11) $|A_{n,j}| \leq C \langle n \rangle^{\sigma - \varsigma} \langle n - j \rangle^\varsigma \langle j \rangle^\varsigma$ and $|B_{n,j,l}| \leq C \langle n \rangle^{\sigma - \varsigma} \langle n - j \rangle^\varsigma \langle j + l \rangle^\varsigma \langle l \rangle^\varsigma,$

for all $n, j, l \in \mathbf{Z}$, where $\sigma \geq 0$, $\varsigma \geq 0$. Let the linear operator $\mathcal{L}$ satisfy the dissipation condition

(1.12) $\operatorname{Re} L_n \geq \operatorname{Re} L_0 + \mu |n|^\nu$

for all $n \in \mathbf{Z} \backslash \{0\}$, where $\mu > 0$, $\nu \geq 0$. We denote $\varkappa = 2$ if the nonlinear operator $\mathcal{N}$ is quadratic: $B_{n,j,l} = 0$ for all $n, j, l \in \mathbf{Z}$ and let $\varkappa = 3$ otherwise. Define

$$S = S\left(\sigma, \nu, \zeta, \varkappa\right) = \zeta + \frac{1}{2} - \min\left(\frac{\nu - \sigma}{\varkappa - 1}, \frac{2\nu + 1}{2\varkappa}\right).$$

First we consider the rather general class of nonlinearities in equation (1.1), but we then have to assume the presence of the linear dissipation, which leads to the exponential decay with time of the main term of asymptotics for solutions to the periodic problem (1.1).

THEOREM 45. *Let the linear operator $\mathcal{L}$ satisfy the dissipation condition (1.12) with* $\operatorname{Re} L_0 > 0$, *and the nonlinear operator $\mathcal{N}$ satisfy estimates (1.11) with $\sigma \in [0, \nu]$. Suppose that the initial data $\phi \in \mathbf{H}^s$ have sufficiently small norm $\|\phi\|_{\mathbf{H}^s}$, where $s > S$. Then there exists a unique solution $u(t, x) \in \mathbf{C}^0\left([0, \infty); \mathbf{H}^s\right)$ of the periodic problem (1.1). In the case $\nu > 0$ we also have a smoothing property $u(t, x) \in \mathbf{C}^1\left((0, \infty); \mathbf{H}^\infty\right)$. Moreover, there exists a unique number U, such that the solution u has the following asymptotics for large $t > 0$ uniformly with respect to $x \in \Omega$*

$$(1.13) \qquad\qquad u\left(t, x\right) = U e^{-L_0 t} + O\left(e^{-\chi t}\right),$$

where $\chi > \operatorname{Re} L_0$.

Note that the coefficient U in the asymptotic formula (1.13) can be calculated explicitly via the initial data ϕ by virtue of the recurrence relations as in [103]. Hence we see that the coefficient U in the asymptotics (1.13) does not vanish if the mean value of the initial data ϕ is not zero $\int_\Omega \phi(x)\, dx \neq 0$ and the norm of the initial data is sufficiently small.

Although Theorem 45 is general enough, the application of this theorem to the particular equations mentioned above gives us rather rough results since in Theorem 45 we do not take into account a special character of the nonlinearity. In what follows we intend to remove the following two essential restrictions of Theorem 45: the smallness of the initial data ϕ and the presence of the linear dissipation $\operatorname{Re} L_0 > 0$ in the equation, which leads to the exponential decay of the main term of the asymptotic representation of the solutions.

1.3. Large initial data. As we know, the condition of the strong dissipation (1.12) prevents the effect of blow up for solutions to the Whitham equation and the nonlinear Schrödinger equation (see, for example [103]). Therefore the classical solutions with any large initial data exist globally in time. In particular, this fact is due to some special symmetry of the nonlinearity of these equations, allowing us to estimate easily the $\mathbf{L}^2$ norm of the solution. We now write this symmetry property in the following form

$$(1.14) \qquad\qquad \operatorname{Re} \int_\Omega \overline{\mathcal{J}\varphi} \mathcal{J}\mathcal{N}(\varphi)\, dx = 0$$

for any function $\varphi \in \mathbf{C}^\infty$, where $\mathcal{J}$ is a pseudodifferential operator such that the norm $\|\mathcal{J}u\|_{\mathbf{L}^2}$ is equivalent to the norm of the Sobolev space $\|u\|_{\mathbf{H}^r}$ for some $r \in \mathbf{R}$. The symmetry property (1.14) with $\mathcal{J} = 1$ and $r = 0$ is fulfilled for the Whitham equation (1.3) (and is thus valid for equations (1.4) through (1.6)), and for the nonlinear Schrödinger equations (1.8), (1.9). For equation (1.7) we can take $\mathcal{J} = \sqrt{1 - \partial_x^2}$ and $r = 1$. Condition (1.14) with $\mathcal{J} = \partial_x$, $r = 1$, considered in the space of functions with zero mean value $\mathbf{H}_0^r$, is valid for the case of the potential Whitham equation (1.10).

Now we can state the result analogous to Theorem 45 without any restriction on the size of the initial data.

THEOREM 46. *Let the linear operator $\mathcal{L}$ satisfy dissipation condition (1.12) with $\nu \geq 0$ and $\operatorname{Re} L_0 > 0$, the nonlinear operator $\mathcal{N}$ satisfy estimates (1.11) with $\sigma \in [0, \nu]$ and the symmetry condition (1.14) with $r > S$. The initial data $\phi \in \mathbf{H}^s$ with any $s > S$ if $\nu > 0$ and $\phi \in \mathbf{H}^r$ if $\nu = 0$. Then the result of Theorem 45 holds.*

We can see that the coefficient U in the asymptotic formula (1.13) does not vanish if the mean value of the initial data $\hat{\phi}_0 \neq 0$ and the norm $\|\phi\|_{\mathbf{H}^s}$ is sufficiently small. In the case $\hat{\phi}_0 = 0$ we cannot guarantee that $U \neq 0$; moreover, the main term of the asymptotic formula may be determined by the higher-order harmonics. For example, if the mean value of the nonlinearity $\left(\widehat{\mathcal{N}(\varphi)}\right)_0 = 0$ for any function φ, that is, when

$$(1.15) \qquad\qquad A_{0,j} = 0, \ B_{0,j,l} = 0$$

for all $j, l \in \mathbf{Z}$, and the mean value of the initial data ϕ is equal to zero $\hat{\phi}_0 = 0$, then the mean value of the solution $u(t, x)$ is also equal to zero $\hat{u}_0 = 0$ for all time $t > 0$: that is $u \in \mathbf{H}_0^s$. In addition the asymptotics of the solution of the periodic problem (1.1) is defined by the higher-order harmonics. This property allows us to obtain the following result if we assume that the linear operator $\mathcal{L}$ satisfies the following dissipation condition

$$(1.16) \qquad \operatorname{Re} L_n \geq \min\left(\operatorname{Re} L_1, \operatorname{Re} L_{-1}\right) + \mu |n|^\nu \text{ for all } n \in \mathbf{Z} \backslash \{0, \pm 1\},$$

where $\mu > 0$, $\nu \geq 0$.

THEOREM 47. *Let the linear operator $\mathcal{L}$ satisfy dissipation condition (1.16) with $\nu \geq 0$, the nonlinear operator $\mathcal{N}$ satisfy estimates (1.11) with $\sigma \in [0, \nu]$. Suppose that the equalities (1.15) and the symmetry condition (1.14) in the space $\mathbf{H}_0^r$ with $r > S$ are fulfilled. The initial data $\phi \in \mathbf{H}_0^s$ with any $s > S$ if $\nu > 0$ and $\phi \in \mathbf{H}_0^r$ if $\nu = 0$. Then there exists a unique solution $u(t, x) \in \mathbf{C}^0\left([0, \infty); \mathbf{H}_0^s\right)$ of the periodic problem (1.1). In the case $\nu > 0$ we also have $u(t, x) \in \mathbf{C}^1\left((0, \infty); \mathbf{H}_0^\infty\right)$. Moreover, there exist unique numbers U_n such that the solution $u(t, x)$ has the following asymptotics for large time $t > 0$ uniformly in $x \in \Omega$*

$$(1.17) \qquad\qquad u(t, x) = \sum_{n = \pm 1} U_n e^{inx - L_n t} + O\left(e^{-\chi t}\right),$$

where $\chi > \lambda$.

The coefficients U_n can be calculated via the initial data by virtue of the recurrent relations as in [**103**]. If the initial data have a sufficiently small norm $\|\phi\|_{\mathbf{H}^s}$ and $\hat{\phi}_n \neq 0$, $n = \pm 1$, we can see that the coefficients U_n in the asymptotic formula (1.17) do not vanish. For large initial data we do not have any sufficient condition to ensure that $U_n \neq 0$.

Now we consider the case $\mathrm{Re}\, L_0 \leq 0$ where the main term of the asymptotics can grow with time. As was mentioned in [**103**] the possibility of such behavior of solutions essentially distinguishes the periodic problem from the corresponding Cauchy problem. Note that the nonlinearity of the potential Whitham equation (1.10) does not depend on the mean value $\hat{u}_0(t)$ of the solution $u(t,x)$, i.e. the nonlinearity in equation (1.1) has the following property $\mathcal{N}(u) = \mathcal{N}(u - \hat{u}_0)$. In terms of the symbols $A_{n,j}$ and $B_{n,j,l}$ of the operator $\mathcal{N}$ this property can be expressed as follows

$$(1.18) \qquad A_{n,0} = A_{n,n} = 0 \text{ and } B_{n,j,0} = B_{n,j,-j} = B_{n,n,j} = 0$$

for all $n, j \in \mathbf{Z}$.

THEOREM 48. *Let the linear operator $\mathcal{L}$ satisfy dissipation condition (1.16) with $\nu \geq 0$ and the nonlinear operator $\mathcal{N}$ satisfy estimates (1.11) with $\sigma \in [0, \nu]$. Suppose that the equalities (1.18) and the symmetry condition (1.14) with $r > S$ are valid. The initial data $\phi \in \mathbf{H}^s$ with any $s > S$ if $\nu > 0$ and $\phi \in \mathbf{H}^r$ if $\nu = 0$. Then there exists a unique solution $u(t,x) \in \mathbf{C}^0([0,\infty); \mathbf{H}^s)$ of the periodic problem (1.1) (also we have $u(t,x) \in \mathbf{C}^1((0,\infty); \mathbf{H}^\infty)$ in the case $\nu > 0$). Moreover, there exists unique numbers U_n such that the solution u has the following asymptotics for large $t > 0$ uniformly with respect to $x \in \Omega$*

$$(1.19) \qquad u(t,x) = \sum_{n=0,\pm 1} U_n e^{inx - L_n t} + O\left(e^{-\chi t}\right),$$

where $\chi > \lambda$.

We turn to a very interesting case of the oscillating asymptotics. Suppose that the symbols $A_{n,j}$ and $B_{n,j,l}$ of the nonlinear operator $\mathcal{N}$ in equation (1.1) satisfy the following conditions

$$(1.20) \qquad A_{0,0} = 0 \text{ and } B_{0,0,0} = i\theta(t),$$

where $\theta(t)$ is a real-valued continuous function. Note that equalities (1.20) are fulfilled for equations (1.3) through (1.7) and (1.9) if we choose $\theta = 0$ and, for equation (1.8), if we take $\theta = b$. Now we turn to the case $\mathrm{Re}\, L_0 = 0$ and consider the oscillatory type nonlinearity (as in the Whitham equation (1.3) and in the nonlinear Schrödinger equation (1.8)). If we take $\varphi = \eta = \mathrm{const} \in \mathbf{C}$ in condition (1.14), then we get $\mathrm{Re} \int_\Omega \overline{J_0 \eta} J_0 \left(A_{0,0} \eta^2 + B_{0,0,0} \eta |\eta|^2\right) dx = 0$, whence condition (1.20) follows. Moreover, if we take $\varphi = \eta + w e^{inx}$, with arbitrary complex constants η, w, we obtain

$$
0 = \operatorname{Re} \int_{\Omega} \left((A_{n,0} + A_{n,n}) \, \eta \, |wJ_n|^2 + \left((B_{0,0,n} + B_{0,-n,n}) \, |J_0|^2 \right. \right.
$$
$$
\left. \left. + (B_{n,0,0} + B_{n,n,0}) \, |J_n|^2 \right) |w\eta|^2 + (B_{n,0,n} + B_{n,-n,n}) \, |w^2 J_n|^2 \right) dx,
$$

whence we have the conditions $A_{n,0} + A_{n,n} = 0$,
$\operatorname{Re}(B_{0,0,n} + B_{0,-n,n}) |J_0|^2 + \operatorname{Re}(B_{n,0,0} + B_{n,n,0}) |J_n|^2 = 0$ and
$\operatorname{Re}(B_{n,0,n} + B_{n,-n,n}) = 0$. We denote

$$
\lambda_n^{(j)}(h) = \frac{1}{2} \left(g_n(h) + \overline{g_{-n}(h)} \right)
$$
$$
+ \frac{(-1)^j}{2} \sqrt{\left(g_n(h) - \overline{g_{-n}(h)} \right)^2 + 4 B_{n,n,-n} \overline{B_{-n,-n,n}} \, |h|^4},
$$

where $g_n(h) = (B_{n,n,0} + B_{n,0,0} - i\theta) |h|^2 + L_n - L_0$, $n \in \mathbf{Z} \backslash \{0\}$, $j = 1, 2$, $h \in \mathbf{C}$ (we take the principle value of the square root, so that $\sqrt{1} = 1$). If the coefficients $B_{0,0,0} = i\theta$, $B_{1,1,0}$, $B_{1,0,0}$ and $B_{1,1,-1}$ do not depend on t we can compute the next term of the asymptotic representation of the solution.

THEOREM 49. *Let the linear operator $\mathcal{L}$ be such that $\operatorname{Re} L_0 = 0$. Suppose that the coefficients $B_{n,n,0}$, $B_{n,0,0}$ and $B_{n,n,-n}$ are slowly varying with respect to time $t \geq T$ for some $T > 0$, i.e. i.e.*

$$
(1.21) \qquad \sup_{n \in \mathbf{Z} \backslash \{0\}} \sup_{t \geq T} \left(\left| \dot{B}_{n,n,0}(t) \right| + \left| \dot{B}_{n,0,0}(t) \right| + \left| \dot{B}_{n,n,-n}(t) \right| \right) \leq \varepsilon,
$$

where $\varepsilon > 0$ is sufficiently small and the coefficients $B_{0,0,0} = i\theta$, $B_{1,1,0}$, $B_{1,0,0}$ and $B_{1,1,-1}$ do not depend on t. Assume that $\lambda_n^{(1)}(h) \neq \lambda_n^{(2)}(h)$ for $n = \pm 1, h \in \mathbf{C}$, $\lambda(h) = \min_{n=\pm 1, j=1,2} \operatorname{Re} \lambda_n^{(j)}(h) > 0$ and the eigenvalues $\lambda_n^{(j)}(h)$ satisfy the estimate

$$
\inf_{t \geq T} \operatorname{Re} \lambda_n^{(j)}(h) \geq \lambda(h) + \mu
$$

for all $n \in \mathbf{Z} \backslash \{0, \pm 1\}$. Let the nonlinearity $\mathcal{N}$ satisfy the symmetry condition (1.14) with $r > S$ and estimates (1.11) with $\sigma \in [0, \nu]$. Assume that the initial data $\phi \in \mathbf{H}^s$ with $s > S$ if $\nu > 0$ and $\phi \in \mathbf{H}^r$ if $\nu = 0$. Then there exist a unique solution $u(t,x) \in \mathbf{C}^0 ([0, \infty); \mathbf{H}^s)$ to the periodic problem (1.1) (also $u(t,x) \in \mathbf{C}^1 ((0, \infty); \mathbf{H}^\infty)$ for the case $\nu > 0$). Moreover, there exist unique values U, Φ and $U_n^{(j)}$, $n = \pm 1, j = 1, 2$, such that the following asymptotic formula takes place

$$
(1.22) \quad u(t,x) = e^{-L_0 t - i\Phi - i t\theta |U|^2} \left(U + \sum_{n=\pm 1, j=1,2} U_n^{(j)} e^{inx - \lambda_n^{(j)}(U)t} \right) + O\left(e^{-\chi t} \right),
$$

where $\chi > \lambda(U)$.

For the real-valued solutions of equation (1.1) we have a particular result.

THEOREM 50. *Suppose that* $\Im\mathcal{N}(\varphi) = 0$ *and* $\Im\mathcal{L}\varphi = 0$ *for any real-valued function* φ. *Let the linear operator* $\mathcal{L}$ *be such that* $\operatorname{Re} L_0 = 0$ *and condition (2.8) be true. Assume that the nonlinearity* $\mathcal{N}$ *satisfies conditions (1.14) with* $r > S$ *and estimates (1.11) with* $\sigma \in [0, \nu]$, *and* $\operatorname{Re}(B_{n,0,0} + B_{n,n,0} + B_{n,n,-n}) = 0$ *for all* $n \in \mathbf{Z}$. *Let the initial data* $\phi \in \mathbf{H}^s$ *with* $s > S$ *if* $\nu > 0$ *and* $\phi \in \mathbf{H}^r$ *if* $\nu = 0$. *Then there exists a unique solution* $u(t, x) \in \mathbf{C}^0([0, \infty); \mathbf{H}^s)$ *of the periodic problem (1.1) (also* $u(t, x) \in \mathbf{C}^1((0, \infty); \mathbf{H}^\infty)$ *for the case* $\nu > 0$). *Moreover, there exist unique numbers* U, U_1, *such that the following asymptotic formula takes place for* $t \to \infty$ *uniformly in* $x \in \Omega$

$$u(t, x) = U + 2\operatorname{Re}\left(U_1 e^{ix - L_1 t - U^2 \int_0^t (B_{1,1,0} + B_{1,0,0} + B_{1,1,-1})d\tau}\right)$$

(1.23)
$$+ O\left(e^{-\chi t}\right),$$

where $\chi > \max(0, \operatorname{Re} L_1)$.

1.4. Applications. Now we apply Theorems 45 through 50 to the particular equations (1.2) through (1.10).

For the Fisher equation (1.2) we have $\sigma = \zeta = 0, \nu = 2$; hence $S = -\frac{1}{3}$ if $b \neq 0$ and $S = -\frac{3}{4}$ if $b = 0$. Therefore via Theorem 45 we find that for sufficiently small initial data $\phi \in \mathbf{H}^s$ with $s > -\frac{1}{3}$ if $b \neq 0$ and $s > -\frac{3}{4}$ if $b = 0$, there exists a unique solution $u(t, x) \in \mathbf{C}^0([0, \infty); \mathbf{H}^s) \cap \mathbf{C}^1((0, \infty); \mathbf{H}^\infty)$ which has the asymptotic representation (1.13).

As another example we consider the KdV-B equation (1.4). We have $\sigma = 1$ and $\zeta = 0$ in estimates (1.11) and $\nu = 2$ in the dissipation condition (1.12). Then for the case $\alpha, \beta > 0, a, b \in \mathbf{R}$ by virtue of Theorem 45 we get that if the initial data $\phi \in \mathbf{H}^s$ with $s > 0$ if $b \neq 0$ and $s > -\frac{1}{2}$ if $b = 0$, and the data are sufficiently small in the norm $\mathbf{H}^s$, then there exists a unique solution $u(t, x) \in \mathbf{C}^0([0, \infty); \mathbf{H}^s) \cap \mathbf{C}^1((0, \infty); \mathbf{H}^\infty)$, and the asymptotic representation (1.13) is valid. Due to the zero mean value $\left(\hat{\mathcal{N}}(u)\right)_0 = 0$ of the nonlinearity in the Whitham equation (1.3) the coefficient U of asymptotics (1.13) can be computed explicitly $U = \hat{\phi}_0$ in this case. For the case $b = 0, \alpha = 0$ in the KdV-B equation (1.4) we apply Theorem 50. The symmetry condition (1.14) is fulfilled with $r = 0$. Hence for any (large) initial data $\phi \in \mathbf{H}^s$ with $s > -\frac{1}{3}$, there exists a unique solution $u(t, x) \in \mathbf{C}^0([0, \infty); \mathbf{H}^s) \cap \mathbf{C}^\infty((0, \infty); \mathbf{H}^\infty)$, which has the asymptotics (1.13) with $U = \hat{\phi}_0$. If the initial data ϕ have zero mean value $\int_\Omega \phi(x)\,dx = 0$ we apply Theorem 47 to the KdV-B equation (1.4) with $b = 0, \beta > \max(0, -\alpha), \alpha \in \mathbf{R}$, then for any initial data $\phi \in \mathbf{H}_0^s$, where $s > -\frac{1}{3}$, there exists a unique solution $u(t, x) \in \mathbf{C}^0([0, \infty); \mathbf{H}_0^s) \cap \mathbf{C}^\infty((0, \infty); \mathbf{H}_0^\infty)$ to the periodic problem, and the asymptotics (1.17) is true.

Now consider the case of KS equation (1.5). If $a, b \in \mathbf{R}, \alpha > 0$ and $\beta > -1$, then by using Theorem 53, we find that for any initial data $\phi \in \mathbf{H}^s$, where $s > -1$ if $b \neq 0$ and $s > -\frac{7}{4}$ if $b = 0$, there exists a unique solution $u(t, x) \in \mathbf{C}^0([0, \infty); \mathbf{H}^s) \cap \mathbf{C}^1((0, \infty); \mathbf{H}^\infty)$; also, the asymptotic representation (1.13) is valid with $U = \hat{\phi}_0$.

Theorem 50 gives us the asymptotics (1.23) with $U = \hat{\phi}_0$ of solutions of the periodic problem for the KS equation (1.5) with $a, b \in \mathbf{R}$, $\alpha = 0$ and $\beta > -1$ for any (large) initial data $\phi \in \mathbf{H}^s$, where $s > -1$ if $b \neq 0$ and $s > -\frac{7}{4}$ if $b = 0$. If in addition the initial data have mean value zero then via Theorem 47 we see that there exists a unique solution $u(t, x) \in \mathbf{C}^0\left([0, \infty); \mathbf{H}_0^s\right) \cap \mathbf{C}^\infty\left((0, \infty); \mathbf{H}_0^\infty\right)$ to the periodic problem for the KS equation (1.5) with $a, b \in \mathbf{R}$, $\alpha > -1 - \beta$, and $\beta > -2$ and the asymptotics (1.17) is true.

We apply Theorem 45 to the OSO equation (1.6) with $\alpha, \beta > 0$. Then for small initial data $\phi \in \mathbf{H}^s$ where $s > \frac{1}{2}$ there exists a unique solution $u(t, x) \in \mathbf{C}^0\left([0, \infty); \mathbf{H}^s\right) \cap \mathbf{C}^1\left((0, \infty); \mathbf{H}^\infty\right)$, and the asymptotic representation (1.13) is valid with $U = \hat{\phi}_0$. Note that equation (1.6) is critical for the smoothing effect since $\nu = \sigma$, so the results on smoothing properties from papers [102], [103] are not applicable to equation (1.6).

For the BBMP-B equation (1.7) with $\beta > \alpha > 0$, Theorem 46 yields the existence of a unique solution $u(t, x) \in \mathbf{C}^0\left([0, \infty); \mathbf{H}^1\right)$ for the periodic problem with large initial data $\phi \in \mathbf{H}^1$, and the solution has asymptotics (1.13). Also we can apply Theorem 50 to the BBMP-B equation (1.7) with $\alpha = 0$ for any initial data $\phi \in \mathbf{H}^1$. Thus the main term of asymptotics does not necessarily decay with time.

Theorem 46 can be applied to the LG equation (1.8) with $\alpha > 0$. The symmetry condition (1.14) is fulfilled with $r = 0$. Hence for any initial data $\phi \in \mathbf{H}^s$ with $s > -\frac{1}{3}$, there exists a unique solution $u(t, x) \in \mathbf{C}^0\left([0, \infty); \mathbf{H}^s\right) \cap \mathbf{C}^\infty\left((0, \infty); \mathbf{H}^\infty\right)$ to the periodic problem for the LG equation (1.8) and the asymptotic formula (1.13) takes place. We apply Theorem 49 to equation (1.8) with $\alpha = 0$ if the coefficient $b \geq 0$. Then we have $g_n(U) = i\theta |U|^2 - L_0 + L_n$ and the eigenvalues are equal to $\lambda_n^{(j)}(U) = \beta n^2 + i(-1)^j \sqrt{n^4 + 2b |U|^2}$, so that $\operatorname{Re} \lambda_n^{(j)}(U) \geq \beta n^2$. Therefore for any initial data $\phi \in \mathbf{H}^s$, $s > -\frac{1}{3}$ there exists a unique solution of the periodic problem for equation (1.8) $u(t, x) \in \mathbf{C}^0\left([0, \infty); \mathbf{H}^s\right) \cap \mathbf{C}^\infty\left((0, \infty); \mathbf{H}^\infty\right)$, which has the large time asymptotics

$$u = e^{it\theta |U|^2 + i\Phi}\left(U + \sum_{n=\pm 1, j=1,2} U_n^{(j)} e^{inx - \lambda_n^{(j)}(U)t}\right) + O\left(e^{-\chi t}\right).$$

Via Theorem 45 we find that for any initial data $\phi \in \mathbf{H}^s$ with $s > \zeta$, where $\zeta = 0$ if $a = 0$ and $\zeta = 1$ if $a \neq 0$, there exists a unique solution $u(t, x) \in \mathbf{C}^0\left([0, \infty); \mathbf{H}^s\right) \cap \mathbf{C}^\infty\left((0, \infty); \mathbf{H}^\infty\right)$ to the periodic problem for the derivative nonlinear Schrödinger equation (1.9) with $\alpha, \beta > 0$ and the asymptotic formula (1.13) takes place.

Let us apply Theorem 48 to the potential KdV-B equation

$$(1.24) \qquad u_t + \frac{a}{2} u_x^2 + \frac{b}{2} u_x^3 + \alpha u - \beta u_{xx} + u_{xxx} = 0,$$

with $b = 0$, $\alpha > -\beta$, $\beta > 0$. Then there exists a unique solution $u(t, x) \in \mathbf{C}^0\left([0, \infty); \mathbf{H}_0^s\right) \cap \mathbf{C}^\infty\left((0, \infty); \mathbf{H}^\infty\right)$ to the periodic problem for equation (1.24) with any initial data $\phi \in \mathbf{H}_0^s$, where $s > \frac{1}{2}$ and asymptotic formula (1.19) takes place. We obtain the same asymptotics via Theorem 48 for the potential KS equation

$$(1.25) \qquad u_t + \frac{a}{2}u_x^2 + \frac{b}{2}u_x^3 + \alpha u - \beta u_{xx} + u_{xxxx} = 0,$$

with $\alpha \in \mathbf{R}$, $\beta > -\alpha - 1$.

For the case of the potential BBMP-B equation

$$(1.26) \qquad u_t + \frac{a}{2}u_x^2 + \frac{b}{2}u_x^3 + cu_x + \alpha u - \beta u_{xx} - u_{txx} = 0,$$

with $a, b, c \in \mathbf{R}$, $\beta > 0$, the symmetry condition (1.14) is fulfilled in the space $\mathbf{H}_0^1$ with operator $\mathcal{J} = \partial\sqrt{1 - \partial^2}$ and $r = 2$. Then Theorem 48 yields the existence of a unique solution $u(t, x) \in \mathbf{C}^0\left([0, \infty); \mathbf{H}_0^2\right)$ to the periodic problem for equation (1.26) with any initial data $\phi \in \mathbf{H}_0^2$, and asymptotics (1.19) is true.

The rest of the chapter we organize as follows. In Section 2 we prove Lemmas 42-46, where we obtain some preliminary estimates for the linear and nonlinear operators of equation (1.1). We choose functional spaces $\mathbf{X}_T$ and $\mathbf{Y}_T$ which are suitable for the description of the smoothing property of the linear operator $\mathcal{L}$. In Section 3 we start with Theorem 51 which states the local existence of solutions to the periodic problem (1.1) for the case $\nu = 0$. Then in Theorem 52 we obtain the local smoothing effect for the case $\nu > 0$. Some preliminary results analogous to Theorems 47-49 concerning the case of small initial data are proved in Sections 3.3-3.5. Sections 3.6-3.11 are devoted to the proof of Theorems 45-50. In Section 3.12 we consider the Whitham equation with weak dissipation.

2. Preliminary estimates

Let us define the Green operator $\mathcal{G}$ for the following linear periodic problem

$$(2.1) \qquad \begin{cases} u_t + \mathcal{L}u = f, \ x \in \Omega, t > 0, \\ u(0, x) = \phi, \ x \in \Omega, \end{cases}$$

where the functions f and ϕ are periodic with respect to the spacial variable x : $f(t, x + 2\pi) = f(t, x)$, $\phi(x + 2\pi) = \phi(x)$ for all $x \in \mathbf{R}$, $t \geq 0$. Using the Fourier series we can formally represent the Green operator explicitly

$$\mathcal{G}(t)\psi = \sum_{n=-\infty}^{\infty} \hat{\psi}_n e^{inx - L_n t},$$

or as the integral $\mathcal{G}(t)\psi = \int_\Omega G(t, x - y)\psi(y)\,dy$ where the kernel is $G(t, x) = \sum_{n=-\infty}^{\infty} e^{inx - L_n t}$. Therefore the solution of the problem (2.1) has the form

$$u(t, x) = \mathcal{G}(t)\phi + \int_0^t \mathcal{G}(t - \tau)f(\tau)\,d\tau.$$

We first collect some simple estimates for the solution of the periodic problem (2.1) in the spaces $\mathbf{H}^s$ with $s > \frac{1}{2}$.

LEMMA 42. *Let the linear operator $\mathcal{L}$ satisfy dissipation condition (1.12) with $\nu \geq 0$. Then for any $\phi \in \mathbf{H}^s$ and $f(t) \in \mathbf{H}^s$, where $s > \frac{1}{2}$, we have the following estimates*

$$\|\mathcal{G}(t)\phi\|_{\mathbf{H}^s} \leq e^{-\lambda t}\|\phi\|_{\mathbf{H}^s}$$

and

$$\left\|\int_0^t \mathcal{G}(t-\tau)f(\tau)\,d\tau\right\|_{\mathbf{H}^s} \leq e^{-\lambda t}\int_0^t e^{(\lambda-\Lambda)\tau}d\tau \sup_{0<\tau<t} e^{\Lambda\tau}\|f(\tau)\|_{\mathbf{H}^s}$$

for all $t > 0$, where $\lambda = \operatorname{Re}L_0$, $\Lambda \in \mathbf{R}$. Moreover, if $\phi \in \mathbf{H}^s$ and the function f is such that $e^{\Lambda t}\|f(t)\|_{\mathbf{H}^s}$ is bounded for all $t > 0$ where $\Lambda > \lambda$, $s > \frac{1}{2}$, then the following asymptotics takes place as $t \to \infty$ uniformly with respect to $x \in \Omega$,

$$u(t,x) = e^{-L_0 t}\left(\hat{\phi}_0 + \int_0^\infty e^{L_0\tau}\hat{f}_0(\tau)\,d\tau\right)$$

$$(2.2) \qquad\qquad + O\left(\left(\|\phi\|_{\mathbf{H}^s} + \sup_{t>0} e^{\Lambda t}\|f(t)\|_{\mathbf{H}^s}\right)e^{-\chi t}\right),$$

where $\chi \in (\lambda, \min(\Lambda, \lambda + \mu))$, μ is taken from (1.15).

Proof. By the Parseval equality we find

$$\|\mathcal{G}\phi\|_{\mathbf{H}^s}^2 = \sum_{n=-\infty}^\infty e^{-2\operatorname{Re}L_n t}\langle n\rangle^{2s}\left|\hat{\phi}_n\right|^2 \leq e^{-2\operatorname{Re}L_0 t}\|\phi\|_{\mathbf{H}^s}^2,$$

whence the first estimate of the lemma follows. Furthermore, by the Cauchy inequality we get

$$\left\|\int_0^t \mathcal{G}(t-\tau)f(\tau)\,d\tau\right\|_{\mathbf{H}^s}^2 \leq \sum_{n=-\infty}^\infty \langle n\rangle^{2s}\left(\int_0^t e^{-\operatorname{Re}L_0(t-\tau)}\left|\hat{f}_n(\tau)\right|d\tau\right)^2$$

$$= \int_0^t d\tau\int_0^t d\tau'\,e^{-\lambda(2t-\tau-\tau')}\sum_{n=-\infty}^\infty \langle n\rangle^{2s}\left|\hat{f}_n(\tau)\,\hat{f}_n(\tau')\right|$$

$$\leq e^{-2\lambda t}\left(\int_0^t e^{-(\Lambda-\lambda)\tau}d\tau\right)^2 \sup_{0<\tau<t} e^{2\Lambda\tau}\|f(\tau)\|_{\mathbf{H}^s}^2,$$

whence the second estimate follows. To prove asymptotics (2.2) we extract the mean value in the Fourier series for the Green operator $\mathcal{G}(t)\phi = \hat{\phi}_0 e^{-L_0 t} + \mathcal{G}_0(t)\phi$, where the remainder operator

$$\mathcal{G}_0(t)\phi = \sum_{n=-\infty, n\neq 0}^\infty \hat{\phi}_n e^{inx - L_n t}$$

is estimated as follows

$$\|\mathcal{G}_0(t)\phi\|_{\mathbf{H}^s}^2 = \sum_{n=-\infty, n\neq 0}^\infty \langle n\rangle^{2s}\left|\hat{\phi}_n\right|^2 e^{-2\operatorname{Re}L_n t} \leq Ce^{-2\operatorname{Re}L_0 t - 2\mu t}\|\phi\|_{\mathbf{H}^s}^2,$$

whence by the Sobolev imbedding inequality for $s > \frac{1}{2}$ we obtain the uniform with respect to $x \in \Omega$ asymptotics

$$\mathcal{G}(t)\phi = \hat{\phi}_0 e^{-L_0 t} + O\left(\|\phi\|_{\mathbf{H}^s} e^{-\chi t}\right),$$

where $\chi = \lambda + \mu$. Then we write the identity

$$\int_0^t \mathcal{G}(t-\tau) f(\tau) \, d\tau = e^{-L_0 t} \int_0^\infty e^{L_0 \tau} \hat{f}_0(\tau) \, d\tau - e^{-L_0 t} \int_t^\infty e^{L_0 \tau} \hat{f}_0(\tau) \, d\tau$$

$$+ \int_0^t \mathcal{G}_0(t-\tau) f(\tau) \, d\tau,$$

where the second summand is estimated as follows

$$\left| e^{-L_0 t} \int_t^\infty e^{L_0 \tau} \hat{f}_0(\tau) \, d\tau \right| \leq e^{-\lambda t} \int_t^\infty e^{(\lambda - \Lambda)\tau} d\tau \sup_{t>0} e^{\Lambda t} \|f(t)\|_{\mathbf{H}^s}$$

$$\leq Ce^{-\Lambda t} \sup_{t>0} e^{\Lambda t} \|f(t)\|_{\mathbf{H}^s},$$

and for the third summand we have the estimate

$$\left\| \int_0^t \mathcal{G}_0(t-\tau) f(\tau) \, d\tau \right\|_{\mathbf{H}^s}^2 \leq Ce^{-2\lambda t - 2\mu t} \left(\int_0^t e^{-(\Lambda - \lambda - \mu)\tau} d\tau \right)^2$$

$$\times \sup_{\tau>0} e^{2\Lambda\tau} \sum_{n=-\infty, n\neq 0}^\infty \langle n \rangle^{2s} \left| \hat{f}_n(\tau) \right|^2 \leq Ce^{-2\chi t} \sup_{t>0} e^{2\Lambda t} \|f(t)\|_{\mathbf{H}^s}^2,$$

where $\chi \in (\lambda, \min(\Lambda, \lambda + \mu))$. Thus the estimate of the remainder in asymptotic formula (2.2) is true. Lemma 42 is proved.

In the same manner we have the following result.

LEMMA 43. *Let the linear operator $\mathcal{L}$ satisfy dissipation condition (1.16) with $\nu \geq 0$. Then for any $\phi \in \mathbf{H}_0^s$ and $f(t) \in \mathbf{H}_0^s$, where $s > \frac{1}{2}$, we have the following estimates*

$$\|\mathcal{G}(t)\phi\|_{\mathbf{H}_0^s} \leq e^{-\lambda t} \|\phi\|_{\mathbf{H}_0^s}$$

and

$$\left\| \int_0^t \mathcal{G}(t-\tau) f(\tau) \, d\tau \right\|_{\mathbf{H}_0^s} \leq e^{-\lambda t} \int_0^t e^{(\lambda - \Lambda)\tau} d\tau \sup_{0<\tau<t} e^{\Lambda\tau} \|f(\tau)\|_{\mathbf{H}_0^s}$$

for all $t > 0$, where $\lambda = \min(\operatorname{Re} L_1, \operatorname{Re} L_{-1})$, $\Lambda \in \mathbf{R}$. Moreover, if $\phi \in \mathbf{H}_0^s$ and the function f is such that $e^{\Lambda t} \|f(t)\|_{\mathbf{H}_0^s}$ is bounded for all $t > 0$, where $\Lambda > \lambda$, $s > \frac{1}{2}$, then the following asymptotics takes place as $t \to \infty$ uniformly with respect to $x \in \Omega$,

$$u(t,x) = \sum_{n=\pm 1} e^{inx - L_n t} \left(\hat{\phi}_n + \int_0^\infty e^{L_n \tau} \hat{f}_n(\tau) \, d\tau \right)$$

(2.3) $$+ O\left(\left(\|\phi\|_{\mathbf{H}_0^s} + \sup_{t>0} e^{\Lambda t} \|f(t)\|_{\mathbf{H}_0^s} \right) e^{-\chi t} \right),$$

where $\chi \in (\lambda, \min(\Lambda, \lambda + \mu))$.

In the next lemma we obtain some simple estimates of the nonlinearity in the norms $\mathbf{H}^s$ with $s > \frac{1}{2} + \zeta$, which we will use in the case $\sigma = 0$ and $\nu = 0$.

LEMMA 44. *Let the nonlinear operator $\mathcal{N}$ satisfy estimates (1.11) with $\sigma = 0$. Then for any functions $u, v \in \mathbf{H}^s$, with $s > \frac{1}{2} + \zeta$, the following inequality is valid*

$$\|\mathcal{N}(u) - \mathcal{N}(v)\|_{\mathbf{H}^s} \leq C \left(\|u\|_{\mathbf{H}^s} + \|v\|_{\mathbf{H}^s} + \|u\|_{\mathbf{H}^s}^{\varkappa-1} + \|v\|_{\mathbf{H}^s}^{\varkappa-1} \right) \|u - v\|_{\mathbf{H}^s},$$

where $\varkappa = 2$ if the nonlinear operator $\mathcal{N}$ is quadratic and $\varkappa = 3$ otherwise.

Proof. By the Cauchy inequality we get

$$\sum_{n=-\infty}^{\infty} |\hat{\varphi}_n| \leq \sqrt{\sum_{n=-\infty}^{\infty} \langle n \rangle^{-2\varrho}} \sqrt{\sum_{n=-\infty}^{\infty} \langle n \rangle^{2\varrho} |\hat{\varphi}_n|^2} \leq C \|\varphi\|_{\mathbf{H}^\varrho}$$

for any $\varrho > \frac{1}{2}$; whence by virtue of estimate (1.11) and the Cauchy inequality we obtain

$$\|\mathcal{N}(u) - \mathcal{N}(v)\|_{\mathbf{H}^s}^2$$

$$\leq C \sum_{n=-\infty}^{\infty} \langle n \rangle^{2s} \left| \sum_{j=-\infty}^{\infty} \left(A_{n,j} \left(\hat{u}_{n-j} \hat{u}_j - \hat{v}_{n-j} \hat{v}_j \right) \right. \right.$$

$$\left. \left. + \sum_{l=-\infty}^{\infty} B_{n,j,l} \left(\hat{u}_{n-j} \hat{u}_{j+l} \overline{\hat{u}_l} - \hat{v}_{n-j} \hat{v}_{j+l} \overline{\hat{v}_l} \right) \right) \right|^2$$

$$\leq C \sum_{n=-\infty}^{\infty} \langle n \rangle^{2s-2\zeta} \left(\sum_{j=-\infty}^{\infty} \langle n - j \rangle^\zeta |\hat{u}_{n-j} - \hat{v}_{n-j}| \left(\langle j \rangle^\zeta |\hat{u}_j| + \langle j \rangle^\zeta |\hat{v}_j| \right) \right.$$

$$\left. + (\varkappa - 2) \sum_{l=-\infty}^{\infty} \left(\langle j + l \rangle^\zeta |\hat{u}_{j+l}| \langle l \rangle^\zeta |\hat{u}_l| + \langle j + l \rangle^\zeta |\hat{v}_{j+l}| \langle l \rangle^\zeta |\hat{v}_l| \right) \right)^2$$

$$\leq C \|u - v\|_{\mathbf{H}^s}^2 \left(\|u\|_{\mathbf{H}^s}^2 + \|v\|_{\mathbf{H}^s}^2 + \|u\|_{\mathbf{H}^s}^{2\varkappa-2} + \|v\|_{\mathbf{H}^s}^{2\varkappa-2} \right),$$

from which we have the result of the lemma. Lemma 44 is proved.

When the linear operator $\mathcal{L}$ satisfies the dissipation condition (1.12) with $\nu > 0$ we can obtain the local smoothing effect introducing the following norms

$$\|\varphi\|_{\mathbf{X}_T}^2 = \sum_{n=-\infty, n\neq 0}^{\infty} \left(\int_0^T E_n(t) \langle n \rangle^{s+\nu} |\hat{\varphi}_n(t)| \, dt \right)^2$$

$$+ \sum_{n=-\infty}^{\infty} \sup_{t \in (0,T]} |E_n(t) \langle n \rangle^s \hat{\varphi}_n(t)|^2$$

$$\|\varphi\|_{\mathbf{Y}_T}^2 = \sum_{n=-\infty}^{\infty} \left(\int_0^T E_n(t) \langle n \rangle^s |\hat{\varphi}_n(t)| \, dt \right)^2$$

where $T \in (0,1]$, $E_n(t) = e^{\frac{\mu}{2}t|n|^{\nu_1}}$, $\nu_1 = \min(1,\nu)$, and $\mu > 0$ is taken from the dissipation condition (1.12); $s > S$ (the value $S = S(\sigma,\nu,\zeta,\varkappa) = \zeta + \frac{1}{2} - \min\left(\frac{\nu-\sigma}{\varkappa-1}, \frac{2\nu+1}{2\varkappa}\right)$ is the same as in the introduction of the chapter). We introduce the spaces

$$\mathbf{X}_T = \{\varphi \in \mathbf{C}\left([0,T]\,;\mathbf{H}^s\right) \cap \mathbf{C}\left((0,T]\,;\mathbf{H}^\infty\right) : \|\varphi\|_{\mathbf{X}_T} < \infty\}$$

and $\mathbf{Y}_T = \{\varphi \in \mathbf{C}\left([0,T]\,;\mathbf{H}^s\right) : \|\varphi\|_{\mathbf{Y}_T} < \infty\}$.

LEMMA 45. *Let the linear operator $\mathcal{L}$ satisfy dissipation condition (1.12) with $\nu > 0$. Then for any $\phi \in \mathbf{H}^s$ and $f \in \mathbf{Y}_T$, we have the following estimates*

$$\|\mathcal{G}(t)\phi\|_{\mathbf{X}_T} \le C\|\phi\|_{\mathbf{H}^s} \ \ \text{and} \ \ \left\|\int_0^t \mathcal{G}(t-\tau) f(\tau)\, d\tau\right\|_{\mathbf{X}_T} \le C\|f\|_{\mathbf{Y}_T}.$$

Proof. By virtue of dissipation condition (1.12) with $\nu > 0$ we have

$$(2.4) \qquad \sup_{n \in \mathbf{Z}}\langle n\rangle^\nu \int_0^T e^{-\operatorname{Re} L_n t} E_n(t)\, dt \le \sup_{n \in \mathbf{Z}}\langle n\rangle^\nu \int_0^1 e^{-\frac{\mu}{2}|n|^{\nu}t}\, dt \le C,$$

whence we find

$$\|\mathcal{G}(t)\phi\|_{\mathbf{X}_T} = \sum_{n=-\infty,n\ne 0}^\infty |\hat\phi_n|^2 \langle n\rangle^{2s+2\nu}\left(\int_0^T e^{-\operatorname{Re} L_n t} E_n(t)\, dt\right)^2$$

$$+ \sum_{n=-\infty}^\infty \langle n\rangle^{2s}|\hat\phi_n|^2\left(\sup_{t\in(0,T)} e^{-\operatorname{Re} L_n t} E_n(t)\right)^2 \le C\|\phi\|_{\mathbf{H}^s}^2$$

for all $\phi \in \mathbf{H}^s$. By virtue of (1.12), (2.4), we use the identity $E_n(t) = E_n(t-\tau) E_n(\tau)$, change the order of integration and introduce the variable of integration $t - \tau = t'$

to find

$$\left\| \int_0^t \mathcal{G}(t-\tau) f(\tau) \, d\tau \right\|_{\mathbf{X}_T}$$

$$\leq \sum_{n=-\infty, n\neq 0}^{\infty} \left(\langle n \rangle^{s+\nu} \int_0^T dt E_n(t) \int_0^t d\tau \left| \hat{f}_n(\tau) \right| e^{-\operatorname{Re} L_n(t-\tau)} \right)^2$$

$$+ \sum_{n=-\infty}^{\infty} \left(\langle n \rangle^s \sup_{t\in(0,T)} E_n(t) \int_0^t \left| \hat{f}_n(\tau) \right| e^{-\operatorname{Re} L_n(t-\tau)} d\tau \right)^2$$

$$\leq \sum_{n=-\infty, n\neq 0}^{\infty} \left(\langle n \rangle^s \int_0^T d\tau \left| \hat{f}_n(\tau) \right| E_n(\tau) \int_\tau^T dt \langle n \rangle^\nu E_n(t-\tau) e^{-\operatorname{Re} L_n(t-\tau)} \right)^2$$

$$+ \sum_{n=-\infty}^{\infty} \left(\sup_{t\in(0,T)} \int_0^t \langle n \rangle^s E_n(\tau) \left| \hat{f}_n(\tau) \right| E_n(t-\tau) e^{-\operatorname{Re} L_n(t-\tau)} d\tau \right)^2$$

$$\leq C \sum_{n=-\infty}^{\infty} \left(\int_0^T \langle n \rangle^s E_n(\tau) \left| \hat{f}_n(\tau) \right| d\tau \right)^2 \leq C \|f\|_{\mathbf{Y}_T}^2.$$

Thus the second estimate of the lemma is true. Lemma 45 is proved.

The next lemma is devoted to the estimates of the nonlinearity $\mathcal{N}(u)$ in the space $\mathbf{Y}_T$, when the function $u \in \mathbf{X}_T$.

LEMMA 46. *Let the nonlinear operator $\mathcal{N}$ satisfy condition (1.11) with $\sigma \in [0, \nu]$, $\nu > 0$. Then for any function $u \in \mathbf{X}_T$ the following inequality is valid*

$$\|\mathcal{N}(u) - \mathcal{N}(v)\|_{\mathbf{Y}_T} \leq CT^\gamma \|u - v\|_{\mathbf{X}_T} \left(\|u\|_{\mathbf{X}_T} + \|u\|_{\mathbf{X}_T}^{\varkappa-1} \right.$$

$$\left. + \|v\|_{\mathbf{X}_T} + \|v\|_{\mathbf{X}_T}^{\varkappa-1} \right),$$

where $\varkappa = 2$ if the nonlinear operator $\mathcal{N}$ is quadratic and otherwise $\varkappa = 3$, $\gamma = \min\left(1; \frac{s-S}{2\nu}\right)$.

Proof. We denote $\varrho = \frac{1}{2} + \gamma\nu$, $q = \max\left(0; \frac{\varkappa-1}{\nu}(\varrho + \zeta - s)\right)$, and $p = 1 - \gamma - q \geq 0$, to acquire the following inequalities

$$s - \zeta + p\nu \geq 0, \, s - \zeta - \varrho + \frac{q}{\varkappa-1}\nu \geq 0$$

and

$$s - \zeta + \sigma \leq (s - \zeta + p\nu) + (s - \zeta - \varrho + q\nu).$$

Therefore we get the estimates

$$\langle n \rangle^{s-\zeta+\sigma} \leq C \langle n-j \rangle^{s-\zeta+p\nu} \langle j \rangle^{s-\zeta-\varrho+q\nu} + C \langle n-j \rangle^{s-\zeta-\varrho+q\nu} \langle j \rangle^{s-\zeta+p\nu}$$

and for the cubic part of the nonlinearity we write the estimate (in the case $\varkappa = 3$)

$$
\begin{aligned}
\langle n \rangle^{s-\zeta+\sigma} \quad \leq \quad & C \langle n-j \rangle^{s-\zeta+p\nu} \langle j+l \rangle^{s-\zeta-\varrho+\frac{q}{2}\nu} \langle l \rangle^{s-\zeta-\varrho+\frac{q}{2}\nu} \\
& + C \langle n-j \rangle^{s-\zeta-\varrho+\frac{q}{2}\nu} \langle j+l \rangle^{s-\zeta+p\nu} \langle l \rangle^{s-\zeta-\varrho+\frac{q}{2}\nu} \\
& + C \langle n-j \rangle^{s-\zeta-\varrho+\frac{q}{2}\nu} \langle j+l \rangle^{s-\zeta-\varrho+\frac{q}{2}\nu} \langle l \rangle^{s-\zeta+p\nu}.
\end{aligned}
$$

Then we obtain

$$
\begin{aligned}
\|\mathcal{N}(u) - \mathcal{N}(v)\|_{\mathbf{Y}_T}^2 \leq & \sum_{n=-\infty}^{\infty} \left(\langle n \rangle^s \int_0^T \left| \sum_{j=-\infty}^{\infty} \left(A_{n,j} \left(\hat{u}_{n-j} \hat{u}_j - \hat{v}_{n-j} \hat{v}_j \right) \right. \right. \right. \\
& + \left. \left. \left. \sum_{l=-\infty}^{\infty} B_{n,j,l} \left(\hat{u}_{n-j} \hat{u}_{j+l} \overline{\hat{u}_l} - \hat{v}_{n-j} \hat{v}_{j+l} \overline{\hat{v}_l} \right) \right) \right| E_n(t)\, dt \right)^2 \\
\leq \quad C & \sum_{n=-\infty}^{\infty} \left(\langle n \rangle^{s-\zeta+\sigma} \int_0^T \sum_{j=-\infty}^{\infty} \langle n-j \rangle^{\zeta} |\hat{u}_{n-j} - \hat{v}_{n-j}| \left(\langle j \rangle^{\zeta} (|\hat{u}_j| + |\hat{v}_j|) \right. \right. \\
& + \left. \left. (\varkappa - 2) \sum_{l=-\infty}^{\infty} \langle j+l \rangle^{\zeta} \langle l \rangle^{\zeta} (|\hat{u}_{j+l}| |\hat{u}_l| + |\hat{v}_{j+l}| |\hat{v}_l|) \right) E_n(t)\, dt \right)^2.
\end{aligned}
$$

Therefore

$$
\begin{aligned}
\|\mathcal{N}(u) - \mathcal{N}(v)\|_{\mathbf{Y}_T}^2 \leq C & \sum_{n=-\infty}^{\infty} \left(\int_0^T \sum_{j=-\infty}^{\infty} |\hat{u}_{n-j} - \hat{v}_{n-j}| (|\hat{u}_j| + |\hat{v}_j|) \right. \\
& \times \left. \left(\langle n-j \rangle^{s-\varrho+q\nu} \langle j \rangle^{s+p\nu} + \langle n-j \rangle^{s+p\nu} \langle j \rangle^{s-\varrho+q\nu} \right) E_n(t)\, dt \right)^2 \\
+ (\varkappa - 2) C & \sum_{n=-\infty}^{\infty} \left(\int_0^T \sum_{j=-\infty}^{\infty} \sum_{l=-\infty}^{\infty} |\hat{u}_{n-j} - \hat{v}_{n-j}| (|\hat{u}_{j+l}| |\hat{u}_l| \right. \\
& \times |\hat{v}_{j+l}| |\hat{v}_l|) \left(\langle n-j \rangle^{s+p\nu} \langle j+l \rangle^{s-\varrho+\frac{q}{2}\nu} \langle l \rangle^{s-\varrho+\frac{q}{2}\nu} \right. \\
& + \left. \left. \langle n-j \rangle^{s-\varrho+\frac{q}{2}\nu} \langle j+l \rangle^{s-\varrho+\frac{q}{2}\nu} \langle l \rangle^{s+p\nu} \right) E_n(t)\, dt \right)^2,
\end{aligned}
$$

whence we get

$$\|\mathcal{N}\left(u\right)-\mathcal{N}\left(v\right)\|_{\mathbf{Y}_T}^2 \le C \sum_{n=-\infty}^{\infty} \left(\sum_{j=-\infty}^{\infty} \left(\sup_{t\in(0,T)} E_{n-j}\left(t\right) \langle n-j \rangle^s \phi_{n-j} \right) \right.^{1-p}$$

$$\times \left(\sup_{t\in(0,T)} E_j\left(t\right) \langle j \rangle^s \psi_j \right)^{1-q} \langle j \rangle^{-\varrho}$$

$$\times \int_0^T \left(E_{n-j}\left(t\right) \langle n-j \rangle^{s+\nu} \phi_{n-j} \right)^p \left(E_j\left(t\right) \langle j \rangle^{s+\nu} \psi_j \right)^q dt \right)^2$$

$$+\left(\varkappa-2\right) C \sum_{n=-\infty}^{\infty} \left(\sum_{j=-\infty}^{\infty} \sum_{l=-\infty}^{\infty} \left(\sup_{t\in(0,T)} E_{n-j}\left(t\right) \langle n-j \rangle^s \phi_{n-j} \right)^{1-p} \right.$$

$$\times \left(\sup_{t\in(0,T)} E_{j+l}\left(t\right) \langle j+l \rangle^s \left(|\hat{u}_{j+l}\left(t\right)| + |\hat{v}_{j+l}\left(t\right)| \right) \right)^{1-\frac{q}{2}} \langle j+l \rangle^{-\varrho}$$

$$\times \left(\sup_{t\in(0,T)} E_l\left(t\right) \langle l \rangle^s \psi_l \right)^{1-\frac{q}{2}} \langle l \rangle^{-\varrho} \int_0^T \left(E_{n-j}\left(t\right) \langle n-j \rangle^{s+\nu} \phi_{n-j} \right)^p$$

$$\times \left(E_{j+l}\left(t\right) \langle j+l \rangle^s \left(|\hat{u}_{j+l}\left(t\right)| + |\hat{v}_{j+l}\left(t\right)| \right) \right)^{\frac{q}{2}} \left(E_l\left(t\right) \langle l \rangle^s \psi_l \right)^{\frac{q}{2}} dt \right)^2,$$

where ϕ_j and ψ_j take the values $\phi_j = |\hat{u}_j - \hat{v}_j|$, $\psi_j = |\hat{u}_j| + |\hat{v}_j|$ and vice versa $\phi_j = |\hat{u}_j| + |\hat{v}_j|$, $\psi_j = |\hat{u}_j - \hat{v}_j|$. Then using the Hölder inequality for the integrals with respect to time t and the Cauchy inequality for the sums, we obtain the result of the lemma. Lemma 46 is proved.

3. Proof of theorems

3.1. Local existence. We first prove the local in time existence of the solutions to the periodic problem (1.1).

THEOREM 51. *Let the linear operator $\mathcal{L}$ satisfy the dissipation condition (1.12) with $\nu = 0$ and the nonlinear operator $\mathcal{N}$ satisfy estimates (1.11) with $\sigma = 0$. Suppose that the initial data $\phi \in \mathbf{H}^s$, where $s > \frac{1}{2} + \zeta$. Then for some time $T > 0$ there exists a unique solution $u\left(t,x\right) \in \mathbf{C}^0\left([0,T]; \mathbf{H}^s\right)$ of the periodic problem (1.1).*

Proof. By virtue of the Green operator $\mathcal{G}\left(t\right)$ of the linear periodic problem (2.1) we write the nonlinear periodic problem (1.1) in the form of the integral equation

$$(3.1) \qquad u\left(t\right) = \mathcal{G}\left(t\right)\phi - \int_0^t \mathcal{G}\left(t-\tau\right)\mathcal{N}\left(u\right)\left(\tau\right) d\tau.$$

We solve equation (3.1) by the contraction mapping principle. We define the transformation

$$\mathcal{A}v(t) = \mathcal{G}(t)\phi - \int_0^t \mathcal{G}(t-\tau)\mathcal{N}(v)(\tau)\,d\tau$$

in the space $\mathbf{C}([0,T];\mathbf{H}^s)$, where the value $T > 0$ will be chosen below. By virtue of Lemma 42 and Lemma 44 we get

$$\sup_{t\in[0,T]} \|\mathcal{A}v(t)\|_{\mathbf{H}^s} \leq \sup_{t\in[0,T]} \left(\|\mathcal{G}(t)\phi\|_{\mathbf{H}^s} + \left\|\int_0^t \mathcal{G}(t-\tau)\mathcal{N}(v)(\tau)\,d\tau\right\|_{\mathbf{H}^s}\right)$$

$$\leq \sup_{t\in[0,T]} \left(e^{-\operatorname{Re}L_0 t}\|\phi\|_{\mathbf{H}^s} + CT\|\mathcal{N}(v)(t)\|_{\mathbf{H}^s}\right)$$

$$\leq e^{|\operatorname{Re}L_0|T}\|\phi\|_{\mathbf{H}^s} + CT \sup_{0<\tau<T} \left(\|v(\tau)\|_{\mathbf{H}^s}^2 + \|v(\tau)\|_{\mathbf{H}^s}^{\varkappa}\right),$$

whence we see that there exists a sufficiently small time $T = T(\|\phi\|_{\mathbf{H}^s})$ such that

$$\sup_{t\in[0,T]} \|\mathcal{A}v(t)\|_{\mathbf{H}^s} \leq 2\|\phi\|_{\mathbf{H}^s}.$$

In the same manner we have

$$\sup_{t\in[0,T]} \|\mathcal{A}v(t) - \mathcal{A}\tilde{v}(t)\|_{\mathbf{H}^s}$$

$$\leq \sup_{t\in[0,T]} \left\|\int_0^t \mathcal{G}(t-\tau)(\mathcal{N}(v) - \mathcal{N}(\tilde{v}))(\tau)\,d\tau\right\|_{\mathbf{H}^s}$$

$$\leq CT \sup_{0<\tau<T} \|v(\tau) - \tilde{v}(\tau)\|_{\mathbf{H}^s} \left(\|v(\tau)\|_{\mathbf{H}^s}^2 + \|v(\tau)\|_{\mathbf{H}^s}^{\varkappa}\right)$$

$$+ \|\tilde{v}(\tau)\|_{\mathbf{H}^s}^2 + \|\tilde{v}(\tau)\|_{\mathbf{H}^s}^{\varkappa}\right) \leq \frac{1}{2} \sup_{0<\tau<T} \|v(\tau) - \tilde{v}(\tau)\|_{\mathbf{H}^s}$$

if $T > 0$ is sufficiently small. Therefore the transformation $\mathcal{A}$ is the contraction mapping in the closed ball of the radius $2\|\phi\|_{\mathbf{H}^s}$ in the space $\mathbf{C}([0,T];\mathbf{H}^s)$. Hence there exists a unique solution $u(t,x) \in \mathbf{C}([0,T];\mathbf{H}^s)$ of the periodic problem (1.1). Theorem 51 is proved.

THEOREM 52. *Let the linear operator $\mathcal{L}$ satisfy the dissipation condition (1.12) with $\nu > 0$, and the nonlinear operator $\mathcal{N}$ satisfy estimates (1.11) with $\sigma \in [0,\nu]$. Suppose that the initial data $\phi \in \mathbf{H}^s$, where $s > S$. Then for some time $T > 0$ there exists a unique solution $u(t,x) \in \mathbf{C}^0([0,T];\mathbf{H}^s) \cap \mathbf{C}^1((0,T];\mathbf{H}^\infty)$ of the periodic problem (1.1).*

Proof. We solve equation (3.1) in the space $\mathbf{X}_T$. Using Lemmas 45 and 46 we obtain

$$\|\mathcal{A}v\|_{\mathbf{X}_T} \leq \|\mathcal{G}\phi\|_{\mathbf{X}_T} + \left\|\int_0^t \mathcal{G}(t-\tau)\mathcal{N}(v)(\tau)\,d\tau\right\|_{\mathbf{X}_T}$$

$$\leq C\|\phi\|_{\mathbf{H}^s} + C\|\mathcal{N}(v)\|_{\mathbf{Y}_T}$$

$$\leq C\|\phi\|_{\mathbf{H}^s} + CT^\gamma \left(\|v\|_{\mathbf{X}_T}^2 + \|v\|_{\mathbf{X}_T}^{\varkappa}\right),$$

whence we see that there exists a sufficiently small time $T = T(\|\phi\|_{\mathbf{H}^s})$ such that $\|\mathcal{A}v\|_{\mathbf{X}_T} \leq C\|\phi\|_{\mathbf{H}^s}$. Similarly we have

$$
\begin{aligned}
\|\mathcal{A}v(t) - \mathcal{A}\tilde{v}(t)\|_{\mathbf{X}_T} &\leq \left\|\int_0^t \mathcal{G}(t-\tau)(\mathcal{N}(v) - \mathcal{N}(\tilde{v}))(\tau)\,d\tau\right\|_{\mathbf{X}_T} \\
&\leq C\|(\mathcal{N}(v) - \mathcal{N}(\tilde{v}))\|_{\mathbf{Y}_T} \\
&\leq CT^\gamma \left(\|v\|_{\mathbf{X}_T}^2 + \|v\|_{\mathbf{X}_T}^\varkappa\right)\|v - \tilde{v}\|_{\mathbf{X}_T} \\
&\leq \frac{1}{2}\|v - \tilde{v}\|_{\mathbf{X}_T}
\end{aligned}
$$

if $T > 0$ is sufficiently small. Therefore the transformation $\mathcal{A}$ is the contraction mapping in the closed ball of the radius $C\|\phi\|_{\mathbf{H}^s}$ in the space $\mathbf{X}_T$. Hence there exists a unique solution $u(t,x) \in \mathbf{X}_T$ of the periodic problem (1.1). By the definition of the norm $\mathbf{X}_T$ we have

$$
\begin{aligned}
\sup_{t\in[T_0,T]}\|v(t)\|_{\mathbf{H}^r}^2 &= \sup_{t\in[T_0,T]}\sum_{n=-\infty}^{\infty}\langle n\rangle^{2r}|\hat{v}_n(t)|^2 \\
&\leq C(r,T_0)\sum_{n=-\infty}^{\infty}\left(\sup_{t\in[T_0,T]}E_n(t)\langle n\rangle^s|\hat{v}_n(t)|\right)^2 \\
&\leq C(r,T_0)\|v\|_{\mathbf{X}_T},
\end{aligned}
$$

for all $r > 0$ and $T_0 \in (0,T]$. The derivatives with respect to time $t > 0$ we can estimate directly from equation (1.1) since the symbols $A_{n,j}(t)$, $B_{n,j,l}(t)$ are continuous. Therefore the solution $u(t,x) \in \mathbf{C}^1((0,T];\mathbf{H}^\infty)$. Theorem 52 is proved.

3.2. Proof of Theorem 45. Now we assume that the linear operator $\mathcal{L}$ satisfies the dissipation condition (1.12) with $\lambda = \operatorname{Re} L_0 > 0$, $\nu \geq 0$. Note that if the initial data have a small norm $\|\phi\|_{\mathbf{H}^s} = \varepsilon$ where $s > S$, then the existence time T in Theorem 51 can be chosen such that $T \geq 1$, and the solution at time $t = T$ is also small $\|u(T)\|_{\mathbf{H}^\rho} \leq C\varepsilon$, where $\rho = \frac{1}{2} + \zeta$. In order to apply a standard continuation method to periodic problem (1.1) we estimate the norm $\|u(t)\|_{\mathbf{H}^\rho}$ of the solution. Let us consider the periodic problem starting with the initial time $t = T$. Then we write the integral equation

$$
(3.2) \qquad u(t) = \mathcal{G}(t-T)u(T) - \int_T^t \mathcal{G}(t-\tau)\mathcal{N}(u)(\tau)\,d\tau.
$$

Let us prove the estimate $e^{\lambda t}\|u(t)\|_{\mathbf{H}^\rho} < C\varepsilon$ for all $t \geq T$ with some fixed constant $C > 0$. By the contrary we can find a time $T_1 > T$ such that

$$
\sup_{T\leq t\leq T_1} e^{\lambda t}\|u(t)\|_{\mathbf{H}^\rho} \leq C\varepsilon.
$$

Using Lemma 42 with $\Lambda = 2\lambda$ and Lemma 44, we get

$$\sup_{T \le t \le T_1} e^{\lambda t} \|u(t)\|_{\mathbf{H}^\rho}$$

$$\le \sup_{T \le t \le T_1} e^{\lambda t} \left(\|\mathcal{G}(t - T) u(T)\|_{\mathbf{H}^\rho} + \left\| \int_T^t \mathcal{G}(t - \tau) \mathcal{N}(u)(\tau) d\tau \right\|_{\mathbf{H}^\rho} \right)$$

$$\le C_1 \left(\|u(T)\|_{\mathbf{H}^\rho} + \sup_{T \le t \le T_1} e^{2\lambda t} \|\mathcal{N}(u)(t)\|_{\mathbf{H}^\rho} \right)$$

$$\le C_1 \left(\|\phi\|_{\mathbf{H}^s} + \sup_{T \le t \le T_1} e^{2\lambda t} \|u(t)\|_{\mathbf{H}^\rho}^2 + \sup_{T \le t \le T_1} e^{\varkappa \lambda t} \|u(t)\|_{\mathbf{H}^\rho}^\varkappa \right)$$

$$\le C_1 \left(\varepsilon + (C\varepsilon)^2 + (C\varepsilon)^3 \right) < C\varepsilon,$$

if $C > C_1 > 0$ and since $\varepsilon > 0$ is sufficiently small. We arrive to the contradiction. Now using asymptotics (2.2) we find for large $t \to \infty$ uniformly with respect to $x \in \Omega$

$$u(t, x) = U e^{-L_0 t} + O\left(\varepsilon e^{-\varkappa t} \right)$$

where $\varkappa \in (\lambda, \min(2\lambda, \lambda + \mu))$ and

$$U = \frac{1}{2\pi} e^{L_0 T} \int_\Omega u(T, x) \, dx - \frac{1}{2\pi} \int_T^\infty d\tau e^{L_0 \tau} \int_\Omega \mathcal{N}(u)(\tau, x) \, dx;$$

here we take into account that the initial time is $T > 0$. Note that in view of the estimates for the solution $u(t, x)$, the coefficient U can be calculated approximately with any desired accuracy via the integral equation (3.2). Theorem 45 is proved.

3.3. Initial data with zero mean value.

THEOREM 53. *Let the linear operator $\mathcal{L}$ satisfy the dissipation condition (1.16), where $\min(\mathrm{Re}\, L_1, \mathrm{Re}\, L_{-1}) = \lambda > 0$ and the nonlinear operator $\mathcal{N}$ satisfy (1.11) with $\sigma \in [0, \nu]$ and (1.15). Suppose that the initial data $\phi \in \mathbf{H}_0^s$ with the sufficiently small norm $\|\phi\|_{\mathbf{H}^s}$, where $s > S$. Then there exists a unique solution $u(t, x) \in \mathbf{C}^0([0, \infty); \mathbf{H}_0^s)$ of the periodic problem (1.1). In the case $\nu > 0$ we also have $u(t, x) \in \mathbf{C}^1((0, \infty); \mathbf{H}_0^\infty)$. Moreover, there exist unique numbers U_n such that the solution $u(t, x)$ has the following asymptotics for large time $t > 0$ uniformly in $x \in \Omega$*

$$u(t, x) = \sum_{n = \pm 1} U_n e^{inx - L_n t} + O\left(e^{-\varkappa t} \right),$$

where $\varkappa > \lambda$.

As an example we apply Theorem 53 to the KdV-B equation (1.4) with $\alpha > -\beta$, $\beta > 0$. Then for the small initial data $\phi \in \mathbf{H}_0^s$, where $s > 0$ if $b \ne 0$ and $s > -\frac{1}{2}$ if $b = 0$, there exists a unique solution $u(t, x) \in \mathbf{C}^0([0, \infty); \mathbf{H}_0^s) \cap \mathbf{C}^1((0, \infty); \mathbf{H}^\infty)$, and the asymptotics $u(t, x) = 2\,\mathrm{Re}\left(U_1 e^{ix - \alpha t - \beta t + it} \right) + O\left(e^{-\varkappa t} \right)$ is valid for $t \to \infty$ uniformly with respect to $x \in \Omega$, where $\varkappa > \alpha + \beta$. We have the same asymptotics for the OSO equation (1.6) with $\alpha > -\beta$, $\beta > 0$ and small initial data $\phi \in \mathbf{H}_0^s$, where $s > \frac{1}{2}$.

The proof of Theorem 53 is very similar to that of Theorem 45. We estimate the solution via the integral equation (3.2) in the space $\mathbf{H}_0^\rho$, using Lemma 43 with $\Lambda = 2\lambda$ and Lemma 44

$$\sup_{t \geq T} e^{\lambda t} \|u(t)\|_{\mathbf{H}_0^\rho}$$

$$\leq \sup_{t \geq T} e^{\lambda t} \left(\|\mathcal{G}(t - T) u(T)\|_{\mathbf{H}_0^\rho} + \left\| \int_T^t \mathcal{G}(t - \tau) \mathcal{N}(u)(\tau)\, d\tau \right\|_{\mathbf{H}_0^\rho} \right)$$

$$\leq \|u(T)\|_{\mathbf{H}_0^\rho} + C \sup_{t \geq T} e^{2\lambda t} \|\mathcal{N}(u)(t)\|_{\mathbf{H}_0^\rho}$$

$$\leq C \left(\|\phi\|_{\mathbf{H}_0^s} + \sup_{t \geq T} e^{2\lambda t} \|u(t)\|_{\mathbf{H}_0^\rho}^2 + \sup_{t \geq T} e^{\varkappa \lambda t} \|u(t)\|_{\mathbf{H}^\rho}^{\varkappa} \right),$$

whence we obtain the estimate $\sup_{t \geq T} e^{\lambda t} \|u(t)\|_{\mathbf{H}_0^\rho} \leq C\varepsilon$. Now using asymptotics (2.3) we find (1.17) where $\chi \in (\lambda, \min(2\lambda, \lambda + \mu))$ and

$$U_n = \frac{1}{2\pi} e^{L_0 T} \int_\Omega e^{-inx} u(T, x)\, dx$$
$$- \frac{1}{2\pi} \int_T^\infty d\tau e^{L_0 \tau} \int_\Omega e^{-inx} \mathcal{N}(u)(\tau, x)\, dx.$$

Theorem 53 is proved.

3.4. Potential equations for small initial data. We have the following result.

THEOREM 54. *Let the linear operator $\mathcal{L}$ satisfy the dissipation condition (1.16) with $\min(\operatorname{Re} L_1, \operatorname{Re} L_{-1}) = \lambda > 0$ and $\operatorname{Re} L_0 < 2\lambda$. We assume that the nonlinear operator $\mathcal{N}$ satisfies estimates (1.11) with $\sigma \in [0, \nu]$ and equalities (1.18). Suppose that the initial data $\phi \in \mathbf{H}^s$ have a sufficiently small norm $\|\phi\|_{\mathbf{H}^s}$, where $s > S$. Then there exists a unique solution $u(t, x) \in \mathbf{C}^0([0, \infty); \mathbf{H}^s)$ of the periodic problem (1.1) (also we have $u(t, x) \in \mathbf{C}^1((0, \infty); \mathbf{H}^\infty)$ in the case $\nu > 0$). Moreover, there exist unique numbers U_n such that the solution u has the following asymptotics for large $t > 0$ uniformly with respect to $x \in \Omega$*

$$u(t, x) = \sum_{n=0, \pm 1} U_n e^{inx - L_n t} + O\left(e^{-\chi t}\right),$$

where $\chi > \lambda$.

For example we apply Theorem 54 to the potential KdV-B equation (1.24) with $\alpha > -\beta$, $\beta > 0$. Then for small initial data $\phi \in \mathbf{H}^s$, where $s > 1$ if $b \neq 0$ and $s > \frac{1}{2}$ if $b = 0$, there exists a unique solution $u(t, x) \in \mathbf{C}^0([0, \infty); \mathbf{H}^s)$ $\cap \mathbf{C}^1((0, \infty); \mathbf{H}^\infty)$ which has the asymptotics

$$u = U_0 e^{-\alpha t} + 2 \operatorname{Re}\left(U_1 e^{-\alpha t - \beta t + it + ix}\right) + O\left(e^{-\chi t}\right),$$

where $\chi > \alpha + \beta$.

To prove Theorem 54 we represent the solution $u(t,x)$ of the periodic problem (1.1) in the form $u(t,x) = \eta(t) + w(t,x)$, where $\eta(t) = \hat{u}_0(t)$. By the condition (1.18) we get $\mathcal{N}(u) = \mathcal{N}(w)$. We denote $\mathcal{N}_0(w) = \mathcal{N}(w) - \left(\widehat{\mathcal{N}(w)}\right)_0$. Hence from equation (1.1) we have the following system

$$(3.3) \qquad \begin{cases} \eta_t + \left(\widehat{\mathcal{N}(w)}\right)_0 + L_0\eta = 0, \\ w_t + \mathcal{N}_0(w) + \mathcal{L}w = 0, \end{cases}$$

with the first equation written here for the first harmonic and the second one for the sum of the rest harmonics. The second equation of system (3.3) is independent of the first one and satisfies the conditions of Theorem 53. Hence we get the estimate

$$\sup_{t \geq T} e^{\lambda t} \|w(t)\|_{\mathbf{H}^\rho} \leq C\varepsilon$$

and the asymptotics

$$w(t,x) = \sum_{n=\pm 1} U_n e^{-L_n t + inx} + O\left(e^{-\chi t}\right)$$

for large time t, where $\chi \in (\lambda, \min(2\lambda, \lambda + \mu))$. Using Lemma 44 we obtain the estimate

$$\sup_{t \geq T} e^{2\lambda t} \left|\left(\widehat{\mathcal{N}(w)}(t)\right)_0\right| \leq C\varepsilon^2.$$

Then integrating the first equation of system (3.3) with respect to time we find

$$(3.4) \qquad \eta(t) = e^{-L_0 t}\left(\hat{\phi}_0 - \int_0^t e^{L_0\tau}\left(\widehat{\mathcal{N}(w)}(\tau)\right)_0 d\tau\right).$$

Since $\operatorname{Re} L_0 < 2\lambda$ we get the following asymptotics for large time from equation (3.4)

$$\begin{aligned} \eta(t) &= e^{-L_0 t}\hat{\phi}_0 - e^{-L_0 t}\int_0^\infty e^{L_0\tau}\left(\widehat{\mathcal{N}(w)}(\tau)\right)_0 d\tau \\ &\quad + e^{-L_0 t}\int_t^\infty e^{L_0\tau}\left(\widehat{\mathcal{N}(w)}(\tau)\right)_0 d\tau \\ &= U_0 e^{-L_0 t} + O\left(e^{-\chi t}\right), \end{aligned}$$

where

$$U_0 = \hat{\phi}_0 - \int_0^\infty e^{L_0\tau}\left(\widehat{\mathcal{N}(w)}(\tau)\right)_0 d\tau,$$

$\chi \in (\lambda, 2\lambda)$. From the previous computation we have asymptotics (1.19). The coefficients

$$U_n = \hat{\phi}_n - \int_0^\infty e^{L_n\tau}\left(\widehat{\mathcal{N}(w)}(\tau)\right)_n d\tau,$$

$n = 0, \pm 1$ can be calculated approximately via the initial data by the integral equations associated with system (3.3). Theorem 54 is proved.

3.5. Oscillating nonlinearities for small initial data. We have the following result.

THEOREM 55. *Let the linear operator $\mathcal{L}$ satisfy the dissipation condition (1.12) with $\nu \geq 0$ and also $\mathrm{Re}\, L_0 = 0$. We suppose that the nonlinear operator $\mathcal{N}$ satisfy estimates (1.11) with $\sigma \in [0, \nu]$ and condition (1.20) with function $\theta(t)$, such that $\sup_{t>0} e^{-\mu t} |\theta(t)| \leq C$. Assume that the initial data $\phi \in \mathbf{H}^s$ have a sufficiently small norm $\|\phi\|_{\mathbf{H}^s}$, where $s > S$. Then there exists a unique solution $u(t,x) \in \mathbf{C}^0([0,\infty); \mathbf{H}^s)$ (also $u(t,x) \in \mathbf{C}^1((0,\infty); \mathbf{H}^\infty)$ in the case $\nu > 0$) of the periodic problem (1.1). Moreover, there exist unique numbers U, Φ such that the solution u has the following asymptotics for large $t > 0$ uniformly with respect to $x \in \Omega$*

$$u(t,x) = U \exp\left(-L_0 t - i|U|^2 \int_0^t \theta(\tau)\, d\tau - i\Phi\right) + O\left(e^{-\chi t}\right),$$

where $\chi > 0$.

The amplitude U and phase Φ can be calculated via the initial data by virtue of the recurrent relations. For example, we apply Theorem 55 to the periodic problem for the dissipative nonlinear Schrödinger equation (1.8), where $\alpha = 0$, $\beta > 0$, $a, b \in \mathbf{R}$ with small initial data $\phi \in \mathbf{H}^s$, where $s > -\frac{1}{3}$. Then there exists a unique solution $u(t,x) \in \mathbf{C}^0([0,\infty); \mathbf{H}^s) \cap \mathbf{C}^\infty((0,\infty); \mathbf{H}^\infty)$ which has the large time asymptotics

$$u(t,x) = U \exp\left(i\theta|U|^2 t + i\Phi\right) + O\left(e^{-\chi t}\right)$$

uniformly in $x \in \Omega$, where $\chi > 0$. We see that in contrast to the case of the Cauchy problem the asymptotics (3.11) of the solution to the periodic problem for the nonlinear Schrödinger equation with dissipation (1.8) with $\alpha = 0$ has a non decaying rapidly oscillating character. The effect described in Theorem 55 was considered with less generality in paper [**101**] and in book [**103**].

Proof of Theorem 55. As in the proof of Theorem 54 we represent the solution $u(t,x)$ of the periodic problem (1.1) in the form $u(t,x) = \eta(t) + w(t,x)$, where $\eta(t) = \hat{u}_0(t)$, so that $\hat{w}_0(t) = 0$. Therefore in view of conditions (1.20) we obtain the following system from equation (1.1)

(3.5)
$$\begin{cases} \dot{\eta} + i\theta |\eta|^2 \eta + \mathcal{R}(\eta, w) + L_0 \eta = 0, \\ w_t + \mathcal{N}_0(u) + \mathcal{L}w = 0. \end{cases}$$

where as above $\mathcal{N}_0(u) = \mathcal{N}(u) - \left(\widehat{\mathcal{N}(u)}\right)_0$ is defined by the sum over $n \in \mathbf{Z} \setminus \{0\}$. The nonlinearity $\mathcal{R}(\eta, w)$ is

$$\mathcal{R}(\eta, w) = \sum_{j=-\infty}^{\infty} \left((A_{0,j} + B_{0,j,0}\overline{\eta})\, \hat{w}_{-j}\hat{w}_j \right.$$

$$\left. + (B_{0,0,j} + B_{0,j,j})\, \eta\, |\hat{w}_j|^2 + \sum_{l=-\infty}^{\infty} B_{0,j,l}\hat{w}_{-j}\hat{w}_{j+l}\overline{\hat{w}_l}\right).$$

It can be estimated as follows

$$|\mathcal{R}(\eta, w)| \leq C\, |\eta|\, \|w\|_{\mathbf{H}^\rho}^2 + C\, \|w\|_{\mathbf{H}^\rho}^3,$$

where $\rho > \frac{1}{2} + \zeta$. In order to obtain the global existence of solutions $u(t,x)$ of the periodic problem (1.1) we now show that the norm $\|u(t)\|_{\mathbf{H}^\rho}$ does not grow with time. Integrating equations of system (3.5) with respect to time $t > T$, we get

$$(3.6) \quad \begin{cases} \eta(t) = E(t)\left(\eta(T) - \int_T^t \overline{E(\tau)} \mathcal{R}(\eta, w)(\tau)\, d\tau\right), \\ w(t) = \mathcal{G}(t-T)\, w(T) - \int_T^t \mathcal{G}(t-\tau)\, \mathcal{N}_0(\eta+w)(\tau)\, d\tau, \end{cases}$$

where $E(t) = \exp\left(-i\int_T^t \theta(\tau)\, |\eta(\tau)|^2\, d\tau + L_0 T - L_0 t\right)$.

Define

$$J(t) = \sup_{T \leq \tau \leq t}\left(|\eta(\tau)| + e^{\chi\tau}\, \|w(\tau)\|_{\mathbf{H}_0^\rho}\right),$$

where $\chi \in (0, \mu)$ and note that $J(t)$ is continuous with respect to $t \geq T$. Via the first equation of system (3.6) we get

$$\begin{aligned}
|\eta(t)| &\leq |\eta(T)| + \int_T^t |\mathcal{R}(\eta, w)(\tau)|\, d\tau \\
&\leq C\varepsilon + C\int_T^t \left(|\eta(\tau)| + \|w(\tau)\|_{\mathbf{H}_0^\rho}\right)\|w(\tau)\|_{\mathbf{H}_0^\rho}^2\, d\tau \\
&\leq C\varepsilon + CJ^2 + CJ^3.
\end{aligned}$$

Then we apply Lemma 43 and Lemma 44 to the second equation of system (3.6)

$$\begin{aligned}
\|w(t)\|_{\mathbf{H}_0^\rho} &\leq \|\mathcal{G}(t-T)\, w(T)\|_{\mathbf{H}_0^\rho} + \left\|\int_T^t \mathcal{G}(t-\tau)\, \mathcal{N}_0(\eta+w)(\tau)\, d\tau\right\|_{\mathbf{H}_0^\rho} \\
&\leq C\varepsilon e^{-\chi t} + C\|\mathcal{N}(\eta+w)(t)\|_{\mathbf{H}_0^\rho} \leq Ce^{-\chi t}\left(\varepsilon + J^2 + J^3\right).
\end{aligned}$$

Thus we obtain the inequality

$$J(t) \leq C\varepsilon + J^2(t) + J^3(t)$$

for all $t \geq T$. Since the value $\varepsilon > 0$ is sufficiently small, $J(T) \leq C\varepsilon$ and $J(t)$ is continuous in $t \geq T$, then as in the proof of Theorem 45, we see that $J(t) \leq C\varepsilon$ for all $t \geq T$. Thus we have

$$(3.7) \qquad \sup_{t > T}|\eta(t)| \leq C\varepsilon, \quad \sup_{t > T} e^{\chi t}\|w(t)\|_{\mathbf{H}_0^\rho} \leq C\varepsilon.$$

Via estimates (3.7) it follows that there exist unique solutions $\eta \in \mathbf{C}^0([T, \infty))$ and $w \in \mathbf{C}^0([T, \infty); \mathbf{H}_0^\rho)$ (and for $\nu > 0$ we have $w \in \mathbf{C}^1([T, \infty); \mathbf{H}_0^\infty)$). By virtue of estimate (3.7) we have

$$|\eta(t) - UE(t)| = \left|\int_t^\infty \mathcal{R}(\eta, w)\overline{E(\tau)}d\tau\right| = O\left(e^{-2\chi t}\right),$$

where

$$U = \lim_{t \to \infty} \eta(t)\, \overline{E(t)} = \eta(T) - \int_T^\infty \mathcal{R}(\eta, w)\overline{E(\tau)}d\tau.$$

Since the function $w(t, x)$ decays exponentially with time, it forms a remainder; therefore we obtain the following asymptotics for large time

$$
\begin{aligned}
u(t, x) &= \eta(t) + w(t, x) \\
\text{(3.8)} \qquad &= U \exp\left(-i \int_T^t \theta(\tau) |\eta(\tau)|^2 \, d\tau - L_0 t + L_0 T\right) \\
&\quad + O\left(e^{-\chi t}\right).
\end{aligned}
$$

We denote

$$
\Psi(t) = \int_T^t \theta(\tau) \left(|\eta(\tau)|^2 - |\eta(t)|^2\right) d\tau.
$$

Integrating by parts and taking into account the first equation of system (3.5), we get

$$
\begin{aligned}
\Psi(t) - \Psi(z) &= 2\,\mathrm{Re} \int_t^z \overline{\eta(\tau)} \eta'(\tau) \int_T^\tau \theta(y) \, dy \, d\tau \\
&= -2\,\mathrm{Re} \int_t^z \overline{\eta(\tau)} \mathcal{R}(\eta, w) \int_T^\tau \theta(y) \, dy \, d\tau
\end{aligned}
$$

for all $z > t > T$. By virtue of estimate (3.7) we have $\Psi(t) - \Psi(z) = O(e^{-\chi t})$ for all $z > t > T$; hence there exists a unique limit $\Phi = \lim_{t \to \infty} \Psi(t)$. Using (3.7) we get

$$
\begin{aligned}
\left| |\eta(t)|^2 - |U|^2 \right| &\leq C(|\eta(t)| + |U|) \left| \int_t^\infty \mathcal{R}(\eta, w) \overline{E(\tau)} d\tau \right| \\
&= O\left(e^{-2\chi t}\right).
\end{aligned}
$$

We now write the following identity for the phase of the asymptotic formula (3.8)

$$
\begin{aligned}
\int_T^t \theta(\tau) |\eta(\tau)|^2 \, d\tau &= |U|^2 \int_T^t \theta(\tau) \, d\tau + \Phi \\
&\quad + \Psi(t) - \Phi + \left(|\eta(t)|^2 - |U|^2\right) \int_T^t \theta(\tau) \, d\tau \\
\text{(3.9)} \qquad &= |U|^2 \int_0^t \theta(\tau) \, d\tau + \Phi + O\left(e^{-\chi t}\right).
\end{aligned}
$$

Putting (3.9) into (3.8) yields asymptotics (3.11). Theorem 55 is proved.

3.6. Proof of Theorem 46. For the case $\nu > 0$, due to the smoothing effect, the solution $u(t, x)$ of the periodic problem (1.1) belongs to the space $\mathbf{H}^r$ for all $t > 0$. Therefore let us consider periodic problem (1.1) for $t \geq T > 0$ with initial data from $\mathbf{H}^r$. Multiplying equation (1.1) by $\overline{\mathcal{J} u \mathcal{J}}$, then integrating with respect to $x \in \Omega$, using property (1.14), we get with $\lambda = \operatorname{Re} L_0$

$$\frac{d}{dt} \|\mathcal{J}u\|_{\mathbf{L}^2}^2 = -2\operatorname{Re} \int_\Omega \left(\overline{\mathcal{J}u}\mathcal{J}\mathcal{N}(u) + \overline{\mathcal{J}u}\mathcal{J}\mathcal{L}u\right) dx$$

$$= -2\operatorname{Re} \int_\Omega \overline{\mathcal{J}u}\mathcal{L}\mathcal{J}u\,dx = -2\sum_{n=-\infty}^{\infty} \operatorname{Re} L_n \left|\left(\widehat{\mathcal{J}u}\right)_n\right|^2$$

$$\leq -2\lambda \|\mathcal{J}u\|_{\mathbf{L}^2}^2\,,$$

whence integrating with respect to $t \geq T$, we obtain

$$\|u\|_{\mathbf{H}^r} \leq C\|\mathcal{J}u\|_{\mathbf{L}^2} \leq C\|u(T)\|_{\mathbf{H}^r} e^{-\lambda t}$$

for all $t \geq T$. Therefore there exists a time $T_1 \geq T$ such that the norm $\varepsilon = \|u(T_1)\|_{\mathbf{H}^r}$ is sufficiently small. Now we can consider periodic problem (1.1) for $t \geq T_1 > 0$ with sufficiently small initial data $u(T_1)$ and apply the result of Theorem 45. Theorem 46 is proved.

3.7. Proof of Theorem 47. The proof of Theorem 47 is the same as that of Theorem 46, the only difference being that we consider the functions in the spaces $\mathbf{H}_0^r$ with zero mean value.

3.8. Proof of Theorem 48. Similarly to the proof of Theorem 54 we separate the mean value $\eta(t) = \hat{u}_0(t)$ in the solution $u(t,x) = \eta(t) + w(t,x)$ to get equations (3.3). The symmetry condition (1.14) is fulfilled for the nonlinearity $\mathcal{N}_0$ of the second equation of system (3.3) in the space $w \in \mathbf{H}_0^r$ since from the condition (1.14) for the nonlinearity $\mathcal{N}$ it follows that

$$\operatorname{Re} \int_\Omega \overline{\mathcal{J}w}\mathcal{J}\mathcal{N}_0(w)dx = \operatorname{Re} \int_\Omega \overline{\mathcal{J}w}\mathcal{J}\mathcal{N}(w)dx = 0.$$

Therefore applying Theorem 47 to the second equation of system (3.3) and considering the first equation of system (3.3) as in the proof of Theorem 54, we obtain asymptotics (1.19) in the case of large initial data. Theorem 48 is proved.

3.9. Oscillating asymptotics for large initial data. Now we turn to the case $\operatorname{Re} L_0 = 0$ and consider the oscillatory type nonlinearity. Suppose that the symbols $A_{n,j}$ and $B_{n,j,l}$ of the nonlinear operator $\mathcal{N}$ in equation (1.1) satisfy the following condition $A_{0,0} = 0$ and $B_{0,0,0} = i\theta(t)$, where $\theta(t)$ is a real-valued continuous function. We denote

$$\lambda_n^{(j)}(h) = \frac{1}{2}\left(g_n(h) + \overline{g_{-n}(h)}\right)$$

$$+\frac{(-1)^j}{2}\sqrt{\left(g_n(h) - \overline{g_{-n}(h)}\right)^2 + 4B_{n,n,-n}\overline{B_{-n,-n,n}}|h|^4},$$

where $g_n(h) = (B_{n,n,0} + B_{n,0,0} - i\theta)|h|^2 + L_n - L_0$, $n \in \mathbf{Z}\backslash\{0\}$, $j = 1,2$, $h \in \mathbf{C}$ (we take the principle value of the square root, so that $\sqrt{1} = 1$).

We prove the following result.

THEOREM 56. *Let the linear operator $\mathcal{L}$ be such that* $\operatorname{Re} L_0 = 0, \operatorname{Re} L_n \geq \mu > 0$ *for all* $n \in \mathbf{Z} \backslash \{0\}$. *Suppose that the coefficients* $B_{n,n,0}$, $B_{n,0,0}$ *and* $B_{n,n,-n}$ *are slowly varying with respect to time* $t \geq T$ *for some* $T > 0$, *i.e.*

$$\sup_{n \in \mathbf{Z} \backslash \{0\}} \sup_{t \geq T} \left(\left| \dot{B}_{n,n,0}(t) \right| + \left| \dot{B}_{n,0,0}(t) \right| + \left| \dot{B}_{n,n,-n}(t) \right| \right) \leq \varepsilon,$$

where $\varepsilon > 0$ *is sufficiently small. Assume that the eigenvalues* $\lambda_n^{(j)}(h)$ *satisfy the estimate*

$$(3.10) \qquad\qquad \inf_{t \geq T} \inf_{h \in \mathbf{C}} \operatorname{Re} \lambda_n^{(j)}(h) \geq \mu > 0$$

for all $n \in \mathbf{Z} \backslash \{0\}$. *Let the nonlinearity* $\mathcal{N}$ *satisfy the symmetry condition (1.14) with* $r > S$ *and estimates (1.11) with* $\sigma \in [0, \nu]$. *Assume that the initial data* $\phi \in \mathbf{H}^s$ *with* $s > S$ *if* $\nu > 0$ *and* $\phi \in \mathbf{H}^r$ *if* $\nu = 0$. *Then there exists a unique solution* $u(t, x) \in \mathbf{C}^0([0, \infty); \mathbf{H}^s)$ *to the periodic problem (1.1). Also we have* $u(t, x) \in \mathbf{C}^1((0, \infty); \mathbf{H}^\infty)$ *for the case* $\nu > 0$. *Moreover, there exist unique numbers* U, Φ *such that the solution* u *has the following asymptotics for large* $t > 0$ *uniformly with respect to* $x \in \Omega$

$$(3.11) \qquad u(t, x) = U \exp\left(-L_0 t - i|U|^2 \int_0^t \theta(\tau) \, d\tau - i\Phi \right) + O\left(e^{-\chi t} \right),$$

where $\chi > 0$.

As in the proof of Theorem 55 we write the solution $u(t, x)$ of periodic problem (1.1) in the form $u(t, x) = \eta(t) + w(t, x)$, where $\eta(t) = \hat{u}_0(t)$ so that $\hat{w}_0(t) = 0$. As above for the case $\nu > 0$ the solution $u(t, x)$ of the periodic problem (1.1) belongs to the space $\mathbf{H}^r$ for all $t > 0$ due to the smoothing effect. As in the proof of Theorem 46 we consider periodic problem (1.1) for $t \geq T_1 > 0$ and use property (1.14) to get

$$(3.12) \qquad \frac{d}{dt} \|\mathcal{J}u\|_{\mathbf{L}^2}^2 = -2\operatorname{Re} \int_\Omega \left(\overline{\mathcal{J}u} \mathcal{J} \mathcal{N}(u) + \overline{\mathcal{J}u} \mathcal{J} \mathcal{L}u \right) dx \leq -2\mu \|\mathcal{J}w\|_{\mathbf{L}^2}^2,$$

whence it follows that the solution is bounded $\|u\|_{\mathbf{H}^r} \leq C \|\mathcal{J}u\|_{\mathbf{L}^2} \leq C \|u(T)\|_{\mathbf{H}^r}$ for all $t \geq T_1$. Therefore by a standard continuation argument we get the existence of the unique global solution $u(t, x) \in \mathbf{C}^0([0, \infty); \mathbf{H}^s)$ and also $u(t, x) \in \mathbf{C}^1((0, \infty); \mathbf{H}^\infty)$ in the case $\nu > 0$. Moreover from inequality (3.12) we see that there exists a time $T > 0$ such that the norm $\|w(T)\|_{\mathbf{H}^r}$ is sufficiently small $\|w(T)\|_{\mathbf{H}^r} \leq C \|\mathcal{J}w(T)\|_{\mathbf{L}^2} \leq \varepsilon$. To prove asymptotics (1.23) we now consider system (3.5) for $t \geq T$. We represent the nonlinearity $\mathcal{N}_0(u) = \mathcal{N}(u) - \left(\widehat{\mathcal{N}(u)} \right)_0 = \mathcal{P}(\eta, w) + \mathcal{Q}(\eta, w)$, where the linear part with respect to w is

$$\mathcal{P}(\eta, w) = \sum_{n=-\infty, n \neq 0}^{\infty} e^{inx} \left((B_{n,n,0} + B_{n,0,0}) |\eta|^2 \hat{w}_n + B_{n,n,-n} \eta^2 \overline{\hat{w}_{-n}} \right),$$

and the higher order part is

$$Q(\eta, w) = \sum_{n=-\infty, n\neq 0}^{\infty} e^{inx} \sum_{j=-\infty}^{\infty} \left((A_{n,j} + B_{n,j,0}\overline{\eta}) \, \hat{w}_{n-j}\hat{w}_j \right.$$

$$\left. + (B_{n,-j,j} + B_{n,n,j}) \, \eta\hat{w}_{n+j}\overline{\hat{w}_j} + \sum_{l=-\infty}^{\infty} B_{n,j,l}\hat{w}_{n-j}\hat{w}_{j+l}\overline{\hat{w}_l} \right).$$

We denote $h(t) = \eta(t)\,\overline{E(t)}$ and $v(t) = w(t)\,\overline{E(t)}$, where
$E(t) = \exp\left(-i\int_T^t \theta(\tau)\,|\eta(\tau)|^2\,d\tau + L_0 T - L_0 t\right)$, then from system (3.5) we
have

$$\dot{h} = -\overline{E}R(Eh, Ev), \quad v_t + \mathcal{M}_h v + \overline{E}Q(Eh, Ev) = 0$$

for all $t \geq T$. The linear operator $\mathcal{M}_h$ is defined as follows

$$\mathcal{M}_h v = \sum_{n=-\infty, n\neq 0}^{\infty} e^{inx} \left(g_n(h)\,\hat{v}_n + B_{n,n,-n}h^2\overline{\hat{v}_{-n}} \right),$$

where $g_n(h) = (B_{n,n,0} + B_{n,0,0} - i\theta)\,|h|^2 + L_n - L_0$. Let us prove the estimate

$$(3.13) \qquad \qquad \|v(t)\|_{\mathbf{H}_0^\rho} < C\varepsilon e^{-\chi t}$$

for all $t \geq T$, where $\chi > \mu - C\varepsilon > 0$. We prove (3.13) by the contradiction argument. Since $\|v(T)\|_{\mathbf{H}_0^\rho}$ is small, by continuity we can find a maximal time interval $[T, T_m]$ such that $\|v(t)\|_{\mathbf{H}_0^\rho} \leq C\varepsilon e^{-\chi t}$ for all $t \in [T, T_m]$. Then the nonlinear operators $R(Eh, Ev)$ and $Q(Eh, Ev)$ are exponentially decaying and small for all $t \in [T, T_m]$. Therefore the coefficients of the linear operator $\mathcal{M}_h$ are slowly varying with respect to time $t \geq T$ and we can apply the method of freezing (see book [2], p.115) to the following periodic problem for the linear equation

$$\begin{cases} v_t + \mathcal{M}_h v = 0, \ t \geq T \\ \quad v(T) = w(T), \end{cases}$$

in the space $\mathbf{H}_0^\rho$ where $\rho > \zeta + \frac{1}{2}$, considering the system as a set of systems of two linear differential equations for the Fourier coefficients $\hat{v}_n$ and $\overline{\hat{v}_{-n}}$ for every $n = 1, 2, \ldots$ Therefore we get the estimate $\|v(t)\|_{\mathbf{H}_0^\rho} < C\varepsilon e^{-\chi t}$ for all $t \in [T, T_m]$. The contradiction obtained proves the estimate (3.13) for all $t \geq T$. We find asymptotic formula (3.11) in the same manner as in the proof of Theorem 55. Theorem 56 is proved.

3.10. Proof of Theorem 49. We need to compute in more detail the asymptotics for the function v. We write $v = \psi + f$, where

$$\psi(t, x) = \sum_{n=\pm 1} e^{inx}\hat{v}_n(t)$$

and

$$f(t, x) = \sum_{n\neq 0, \pm 1} e^{inx}\hat{v}_n(t).$$

The evolution of the function ψ is defined by the following two systems

(3.14)
$$\begin{cases} (\hat{v}_n)_t + g_n(U)\hat{v}_n + B_{n,n,-n}U^2\overline{\hat{v}_{-n}} + \hat{r}_n = 0, \\ (\hat{v}_{-n})_t + g_{-n}(U)\hat{v}_{-n} + B_{-n,-n,n}U^2\hat{v}_n + \hat{r}_{-n} = 0, \end{cases}$$

for $n = \pm 1$, where the remainder terms

$$\hat{r}_n = (g_n(h) - g_n(U))\,\hat{v}_n + B_{n,n,-n}\left(h^2 - U^2\right)\overline{\hat{v}_{-n}} + \left(\widehat{EQ\left(Eh, Ev\right)}\right)_n$$

and $U = \lim_{t\to\infty} \eta(t)\,\overline{E(t)}$, as we stated before

$$E(t) = \exp\left(-i\int_T^t \theta(\tau)\,|\eta(\tau)|^2\,d\tau + L_0 T - L_0 t\right).$$

The eigenvalues of the matrix

$$\begin{pmatrix} g_n(U) & B_{n,n,-n}U^2 \\ \overline{g}_{-n}(U) & \overline{B}_{-n,-n,n}\overline{U}^2 \end{pmatrix} \text{ of the system (3.14) are}$$

$$\lambda_n^{(j)}(U) = \frac{1}{2}\left(g_n(U) + \overline{g_{-n}(U)}\right)$$

$$+ \frac{(-1)^j}{2}\sqrt{\left(g_n(U) - \overline{g_{-n}(U)}\right)^2 + 4B_{n,n,-n}\overline{B_{-n,-n,n}}\,|U|^4}.$$

Since the coefficients $g_n(U)$ and $B_{n,n,-n}U^2$ of the linear part of the system (3.14) do not depend on time, the fundamental Cauchy matrix of system (3.14) has the form

$$\begin{pmatrix} \frac{\exp\left(-\lambda_n^{(1)}t\right)}{\lambda_n^{(1)}-\lambda_n^{(2)}}\left(\lambda_n^{(1)} - \overline{g_{-n}}\right) & \frac{\exp\left(-\lambda_n^{(1)}t\right)}{\lambda_n^{(1)}-\lambda_n^{(2)}}B_{n,n,-n}U^2 \\ \frac{\exp\left(-\lambda_{-n}^{(1)}t\right)}{\lambda_{-n}^{(1)}-\lambda_{-n}^{(2)}}\overline{B}_{-n,-n,n}U^2 & \frac{\exp\left(-\lambda_{-n}^{(1)}t\right)}{\lambda_{-n}^{(1)}-\lambda_{-n}^{(2)}}\left(\lambda_{-n}^{(1)} - g_n\right) \end{pmatrix}$$

$$- \begin{pmatrix} \frac{\exp\left(-\lambda_n^{(2)}t\right)}{\lambda_n^{(1)}-\lambda_n^{(2)}}\left(\lambda_n^{(2)} - \overline{g_{-n}}\right) & \frac{\exp\left(-\lambda_n^{(2)}t\right)}{\lambda_n^{(1)}-\lambda_n^{(2)}}B_{n,n,-n}U^2 \\ \frac{\exp\left(-\lambda_{-n}^{(2)}t\right)}{\lambda_{-n}^{(1)}-\lambda_{-n}^{(2)}}\overline{B}_{-n,-n,n}U^2 & \frac{\exp\left(-\lambda_{-n}^{(2)}t\right)}{\lambda_{-n}^{(1)}-\lambda_{-n}^{(2)}}\left(\lambda_{-n}^{(2)} - g_n\right) \end{pmatrix}.$$

We introduce the operator $\mathcal{G}_U(t)$ by the formula

$$\mathcal{G}_U(t)\psi = \sum_{n=\pm 1} e^{inx}\left(\lambda_n^{(1)}(U) - \lambda_n^{(2)}(U)\right)^{-1}\sum_{j=1}^{2}(-1)^j\exp\left(-\lambda_n^{(j)}(U)\,t\right)$$

$$\times\left(B_{n,n,-n}U^2\overline{\hat{\psi}_{-n}} + \hat{\psi}_n\left(\lambda_n^{(j)}(U) - \overline{g_{-n}(U)}\right)\right),$$

so that if ψ is a solution of the linear system (3.14) (with $\hat{r}_n = 0$, $\overline{\hat{r}_{-n}} = 0$), then we can verify by a direct calculation that $\mathcal{G}_U(t)\psi$ is a solution to the Cauchy problem $\frac{d}{dt}\mathcal{G}_U(t)\psi = 0$, $\mathcal{G}_U(0)\psi = \psi$. Applying the operator $\mathcal{G}_U$ to system (3.14) and taking into account that the higher order nonlinear terms $\mathcal{R}(\eta, w)$ and $\mathcal{Q}(\eta, w)$ have a more rapid exponential decay with respect to time we get asymptotics (1.22). Theorem 49 is proved.

3.11. Proof of Theorem 50. Theorem 50 is proved in the same manner as Theorem 49. We only explain the difference in the construction of the Green operator. In the case of the real-valued solution we have the relation for the Fourier coefficients $\hat{w}_n = \overline{\hat{w}_{-n}}$, so that the operator $\mathcal{P}(\eta, w)$ in the second equation of system (3.5) is now equal to $\mathcal{P}(\eta, w) = \sum_{n=-\infty, n\neq 0}^{\infty} e^{inx} \left(g_n(\eta) + \overline{B_{n,n,-n}\eta^2} \right) \hat{w}_n$. The Green operator has the following form

$$\mathcal{G}_U\varphi = \sum_{n=\pm 1}^{\infty} \exp\left(inx - L_n t - \int_0^t \left(g_n(U) + \overline{B_{n,n,-n}U^2} \right) d\tau \right) \hat{\varphi}_n.$$

Using the condition of Theorem 50 we find $\mathrm{Re}\left(g_n(U) + \overline{B_{n,n,-n}U^2} \right) = 0$; therefore estimates of Lemma 42 are valid for the Green operator $\mathcal{G}_h$. All the other considerations are the same as in the proof of Theorem 49. Theorem 50 is proved.

3.12. Whitham equation with weak dissipation. Finally we consider a particular case of the Whitham equation (1.3) with $b = 0$, when the operator $\mathcal{L}$ has the order $\nu > 1$. The symmetry condition (1.14) with $r = 0$ is fulfilled for the Whitham equation (1.3); however we cannot apply Theorems 46 - 50 in the case $\nu \in \left(1, \frac{3}{2}\right]$, since the value $S > 0$ for this region, so that the condition $r > S$ is violated. We can consider the region $\nu \in \left(1, \frac{3}{2}\right]$ if we take into account a special hyperbolic structure of the nonlinear term in the Whitham equation (1.3). We also assume that the linear operator $\mathcal{L}$ can be represented in the integral form

$$(3.15) \qquad \mathcal{L}\varphi = \lambda\varphi + \int_\Omega L(x - y)\varphi_{yy}(y)dy,$$

for any function $\varphi \in \mathbf{H}_0^\infty$, where 2π through periodic kernel $L(x) \in \mathbf{C}^2(0, 2\pi) \cap \mathbf{L}^1(\Omega)$, $\lambda > \mathrm{Re}\, L_0 \geq 0$. Representation (3.15) implies that the order of the operator $\mathcal{L}$ is less than 2. If we assume that the kernel $L(x)$ is convex, that is $L''(x) \leq 0$ for all $x \in \Omega$ (this implies that the operator $\mathcal{L}$ is dissipative), then as in [**103**] we can prove an additional estimate of the solution in the uniform norm $\mathbf{L}^\infty(\Omega)$ and after that in the Sobolev norm $\mathbf{H}^1(\Omega)$. We prove the following result.

THEOREM 57. *Let the linear operator $\mathcal{L}$ satisfy condition (1.12) with $\mathrm{Re}\, L_0 \geq 0$, $\nu > 1$. Also we suppose that $\mathcal{L}$ can be represented in the form (3.15) with kernel $L(x) \in \mathbf{C}^2(0, 2\pi) \cap \mathbf{L}^1(\Omega)$ such that*

$$(3.16) \qquad L''(x) \leq 0 \text{ for all } x \in \Omega \text{ and } x^2(2\pi - x)^2 L''(x) \in \mathbf{L}^1(\Omega).$$

Suppose that the initial data $\phi \in \mathbf{H}^s$ with $s > S$. Then there exists a unique solution $u(t, x) \in \mathbf{C}^0([0, \infty); \mathbf{H}^s) \cap \mathbf{C}^\infty((0, \infty); \mathbf{H}^\infty)$ to the periodic problem for the Whitham equation (1.3) with $b = 0$. Moreover, the asymptotics (1.19) takes place.

As a consequence of Theorem 57 we see in particular that the blow up phenomena is impossible in the case of the order $\nu > 1$ of the linear dissipative operator $\mathcal{L}$. In [**103**], the estimates $\frac{2}{3} \leq \nu_c \leq \frac{3}{2}$ were obtained for the critical value ν_c of the order of the dissipative part of the linear operator $\mathcal{L}$, dividing the regions: $\nu < \nu_c$

where the blow up effect is possible for sufficiently large initial data, and $\nu > \nu_c$ where the blow up phenomena is impossible. Now we have the following estimates $\frac{2}{3} \leq \nu_c \leq 1$ for the critical order. Our conjecture is $\nu_c = 1$.

   Proof of Theorem 57. In view of Theorem 52 we will show that a priory estimate of the solution in the norm $\mathbf{H}^1$ does not grow in time. We denote the mean value $\eta = \hat{u}_0$, then from equation (1.3) we get $\eta_t + L_0 \eta = 0$, whence $\eta(t) = \hat{\phi}_0 e^{-L_0 t}$. For the remainder $w = u - \eta$ we find equation

(3.17) $$w_t + a(w + \eta)w_x + \mathcal{L}w = 0.$$

First let us prove the following estimate

(3.18) $$\|w(t)\|_{\mathbf{L}^\infty} < 2\|w(0)\|_{\mathbf{L}^\infty} e^{-\lambda t}$$

for all $t > 0$. As in paper [103] we define the characteristic function $y(t, \xi)$ as a solution to the following Cauchy problem

$$\frac{dy}{dt} = au(t, y), \quad y(0, \xi) = \xi.$$

Then for the new function $\psi(\xi, t) = w(t, y(\xi, t))$ we get from (3.17)

(3.19) $$\psi_t + \lambda\psi + I(t, \xi) = 0,$$

where

$$I(t, \xi) = \int_\Omega L(x) w_{xx}(y(\xi, t) - x, t) dx.$$

By the contradiction and continuity we assume that there exists the first moment of time $T > 0$, when estimate (3.18) is violated, i.e. for some $\xi = \chi$ we have $\psi(T, \chi) = 2\|w(0)\|_{\mathbf{L}^\infty} e^{-\lambda T}$ or $\psi(T, \chi) = -2\|w(0)\|_{\mathbf{L}^\infty} e^{-\lambda T}$. Consider the case $\psi(T, \chi) = 2\|w(0)\|_{\mathbf{L}^\infty} e^{-\lambda T}$. By virtue of condition (3.15) we see that the integral $I(T, \chi)$ is positive

$$I(T, \chi) = \int_\Omega L''(x)(w(T, y(T, \chi) - x) - w(T, y(T, \chi))) dx > 0,$$

then via continuity we have $I(t, \chi) \geq 0$ for all $t \in [T_1, T]$ where $0 \leq T_1 < T$. Therefore from (3.19) we get $\psi_t \leq -\lambda\psi$ for all $t \in [T_1, T]$, whence integrating with respect to t we find

$$\psi(T, \chi) \leq \psi(T_1, \chi) e^{-\lambda(T - T_1)} < 2\|w(0)\|_{\mathbf{L}^\infty} e^{-\lambda T}.$$

The second case $\psi(T, \chi) = -2\|w(0)\|_{\mathbf{L}^\infty} e^{-\lambda T}$ can be considered in the same manner. Thus we get a contradiction, therefore estimate (3.18) is true for all $t \geq 0$. Now we estimate the solution in the norm $\mathbf{H}^1$. We write equation (3.17) as the integral equation

(3.20) $$w(t) = \mathcal{G}(t) w(0) - a \int_0^t \mathcal{G}(t - \tau) ww_x(\tau) d\tau,$$

where the Green operator $\mathcal{G}$ is defined as follows

$$\mathcal{G}\varphi = \sum_{n=-\infty}^{\infty} \hat{\varphi}_n \exp\left(inx - L_n t - ian \int_0^t \nu(\tau)\, d\tau\right).$$

Using the dissipation condition (1.12), estimate (3.18) and the inequality

$$\langle n \rangle \, e^{-(\mathrm{Re}\, L_0 + \mu |n|^\nu)(t-\tau)} \le C e^{-\lambda(t-\tau)} (t-\tau)^{-\frac{1}{\nu}}$$

for all $n \in \mathbf{Z} \backslash \{0\}$ and $0 < \tau < t$, we get from equation (3.20)

$$\|w(t)\|_{\mathbf{H}^1}$$
$$\le \; e^{-\lambda t} \|w(0)\|_{\mathbf{H}^1}$$
$$+ C \int_0^t \left(\sum_{n=-\infty, n\neq 0}^{\infty} \left(\langle n \rangle \, e^{-(\mathrm{Re}\, L_0 + \mu |n|^\nu)(t-\tau)} \widehat{(w w_x)}_n(\tau) \right)^2 \right)^{\frac{1}{2}} d\tau$$
$$\le \; e^{-\lambda t} \|w(0)\|_{\mathbf{H}^1} + C \int_0^t e^{-\lambda(t-\tau)} \|w w_x(\tau)\|_{\mathbf{L}^2} (t-\tau)^{-\frac{1}{\nu}} d\tau$$
$$\le \; e^{-\lambda t} \|w(0)\|_{\mathbf{H}^1} + C e^{-\lambda t} \int_0^t \|w(\tau)\|_{\mathbf{H}^1} (t-\tau)^{-\frac{1}{\nu}} d\tau,$$

whence by the Gronwall inequality we get the estimate

$$\|w(t)\|_{\mathbf{H}^1} \le C e^{-\lambda t}$$

for all $t \ge 0$. Thus the solution $u(t,x)$ of the periodic problem for the Whitham equation (1.3) exists globally in time. As in Theorem 54 we now obtain asymptotic representation (1.19). Theorem 57 is proved.

Bibliography

[1] M.J. Ablowitz and H. Segur, *Solitons and Inverse Scattering Transform*, SIAM, 1980.

[2] L. Adrianova, *Introduction to Linear Systems of Differential Equations*, Translations of Mathematical Monographs, **146** (1995), Providence, RI: American Mathematical Society.

[3] H. Aikawa and N. Hayashi, *Holomorphic solutions of semilinear heat equations*, Complex Variables, **16** (1991), pp. 115 - 125.

[4] H. Aikawa, N. Hayashi and S. Saitoh, *The Bergman space on a sector and the heat equation*, Complex Variables, **15** (1990), pp. 27 - 36.

[5] C.J. Amick, J.L. Bona and M.E. Schonbek, *Decay of solutions of some nonlinear wave equations*, J. Diff. Eq. **81** (1989), pp. 1-49.

[6] T.B. Benjamin, J.L. Bona and J.J. Mahony, *Model equations for long waves in nonlinear dispersive systems*, Phil. Trans. Roy. Soc. Soc. A, **272** (1972), pp. 47-54.

[7] P. Biller, *Asymptotic behavior in time of solutions to some equations generalizing the Korteweg-de Vries-Burgers equation*, Bull. Polish. Acad. Sc. Math., **32** (1984), pp. 275-282.

[8] P. Biler, G. Karch and W. Woyczynski, *Multifractal and Levy conservation laws*, C.R.Acad. Sci., Paris, Ser.1,Math. **330** (2000), no 5, pp. 343-348.

[9] J.L. Bona and L. Luo, *More results on the decay of the solutions to nonlinear dispersive wave equations*, Discrete and Continuous Dynamical Systems, **1** (1995), pp. 151-193.

[10] J.L. Bona, V.A. Dougalis, O.A. Karakashian and W.R. McKinney, *The effect of dissipation on solutions of the generalized Korteweg-de Vries equation*, J. Comput. Appl. Math. **74** (1996), No.1-2, pp. 127-154.

[11] J.L. Bona, F. Demengel and K.S. Promislow, *Fourier splitting and dissipation nonlinear dispersive waves*, Proc. Roy. Soc. Edinburgh Sect. A **129** (1999), no. 3, pp. 477-502.

[12] J.L. Bona and L. Luo, *Asymptotic decomposition of nonlinear, dispersive wave equations with dissipation*, Advances in Nonlinear Mathematics and Science, Phys. D **152/153** (2001), pp. 363-383.

[13] J. L. Bona, S.M. Sun and B.Y. Zhang, *A non-homogeneous boundary-value problem for the Krteweg-de Vries equation in a quarter plane*, Trans. A.M.S., **354** (2) (2001), pp.427-490.

[14] J. Bourgain, *Fourier transform restriction phenomena for certain lattice subsets and applications to non-linear evolution equations, Part II: the KdV equation*, Geom. Funct. Anal., **3** (1993), pp. 107-156.

[15] C. Bu, R. Shull and K. Zhao, *A periodic boundary value problem for a generalized 2D Ginzburg-Landau equation*, Hokkaido Math. J. **27**(1998), No.1, pp. 197-211.

[16] F. Calodgero and A. Degasperis, *Spectral Transform and Solitons: Tools to Solve and Investigate Nonlinear Evolution Equations*, North-Holland Publishing Co., Amsterdam, 1982.

[17] Th. Cazenave, F. Dickstein, M. Escobedo and F. Weissler, *Self-similar solutions of a non-linear heat equation*, J. Math. Sci. Univ. Tokyo, **8** (2001), no.3, pp. 501-540.

[18] G. Chen and H. Sun, *Cauchy problems for the nonlinear parabolic equations of higer order of generalized Ginzburg-Landau type*, Chin. Ann. Math., Ser. A **15** (1994), No.4, pp. 485-492.

[19] F.M. Christ and M.I. Weinstein, *Dispersion of small amplitude solutions of the generalized Korteweg-de Vries equation,* J. Funct. Anal., **100** (1991), pp. 87-109.

[20] T. Colin and M. Gisclon, *An initial-boundary-value problem that approximate the quarter-plane problem for the Korteweg-de Vries equation,* Nonlinear Anal., **46** (2001), pp. 869-892.

[21] A. Constantin and J. Escher, *Well-posedness, global existence, and blow up phenomena for a periodic quasi-linear hyperbolic equation,* Commun. Pure Appl. Math. **51** (1998), No.5, 475-504.

[22] A. Constantin and J. Escher, *Wave breaking for nonlinear nonlocal shallow water equations,* Acta Mathematica, **181**, No. 2 (1998), pp. 229-243.

[23] C.M. Dafermos, *Large time behavior of periodic solutions of hyperbolic systems of conservation laws,* J. Differ. Equations **121** (1995), No.1, pp. 183-202.

[24] R.K. Dodd, J.C. Eilbeck, J.D. Gibbon and H.C. Morris, *Solitons and Nonlinear Wave Equations,* Academic Press, Inc. N.Y. 1988.

[25] D.B. Dix, *Large Time Behavior of Solutions of Linear Dispersive Equations,* Lecture Notes in Mathematics, Springer-Verlag, Berlin, Heidelberg, **1668,** 1997.

[26] D.B. Dix, *The dissipation of nonlinear dispersive waves; the case of asymptotically weak nonlinearity,* Comm. P. D. E., **17** (1992), pp. 1665-1693.

[27] D.B. Dix, *Temporal asymptotic behavior of solutions of the Benjamin-Ono-Burgers equation,* J. Diff. Eqs. **90** (1991), pp. 238-287.

[28] C.R. Doering, J.D. Gibbon and C.D. Levermore, *Weak and strong solutions of the complex Ginzburg-Landau equation,* Physica D **71** (1994), No.3, 285-318.

[29] J. Duan, P. Holmes and E.S. Titi, *Global existence theory for a generalized Ginzburg-Landau equation,* Nonlinearity **5** (1992), No.6, pp. 1303-1314.

[30] M. Escobedo and O. Kavian, *Asymptotic behavior of positive solutions of a nonlinear heat equation,* Houston J. Math. **13** (1987), no. 4, pp. 39-50.

[31] M. Escobedo, O. Kavian and H. Matano, *Large time behavior of solutions of a dissipative nonlinear heat equation,* Comm. P. D. E., **20** (1995), pp. 1427-1452.

[32] M. Escobedo, J.L. Vazquez and E. Zuazua, *Asymptotic behavior and source-type solutions for a diffusion-convection equations,* Arch. Ration. Mech. Anal. **124** (1993), no. 1 pp. 43-65.

[33] P.C. Fife, *Asymptotic states for equations of reaction and diffusion,* Bull. Am. Math. Soc. **84** (1978), pp. 693-726.

[34] R.A. Fisher, *The Genetical Theory of Natural Selection,* Oxford, Univ. Press, 1930.

[35] C. Foias and I. Kukavica, *Determining nodes for the Kuramoto-Sivashinsky equation,* J. Dyn. Differ. Equations **7** (1995), No.2, pp. 365-373.

[36] A.F. Fokas and B. Polloni, *The solution of certain initial-boundary value problems for the linearized Korteweg-de Vries equation,* Proc. Royal. Soc. London Series A, **454** (1998), pp. 645-657.

[37] G.E. Forsyte, M.A. Malcolm, C.A. Moler, *Computer methods for mathematical computations,* Prentice-Hall, Englewood Clifts, N.J., 1977.

[38] A. Friedman, *Partial Differential Equations,* Krieger, 1983.

[39] H. Fujita, *On the blowing-up of solutions of the Cauchy problem for* $u_t = \Delta u + u^{1+\alpha}$, J. Fac. Sci. Univ. of Tokyo, Sect. I, **13** (1966), pp. 109-124.

[40] V.A. Galaktionov and J.L. Vazquez, *Asymptotic behavior ofnonlinear parabolic equations with critical exponents. A dynamical systems approach,* J. Funct. Anal. **100** (1991), no.2, pp. 435-462.

[41] V.A. Galaktionov, S.P. Kurdyumov and A.A. Samarskii, *On asymptotic eigenfunctions of the Cauchy problem for a nonlinear parabolic equation,* Math. USSR Sbornik, **54** (1986), pp. 421-455.

[42] Y. Giga and T. Kambe, *Large time behavior of the vorticity of two-dimensional viscous flow and its application to vortex formation,* Commun. Math. Phys. **117** (1988), pp. 549-568.

[43] J. Ginibre and G. Velo, *On a class of nonlinear Schrödinger equations. I. The Cauchy problem, general case; II Scattering theory, general case,* J. Funct. Anal., **32** (1979), pp. 1-71.

[44] J. Ginibre and G. Velo, *The Cauchy problem in local spaces for the complex Ginzburg-Landau equation. I. Compactness methods,* Physica D, **95**, (1996), pp. 191-228.

[45] J. Ginibre and G. Velo, *The Cauchy problem in local spaces for the complex Ginzburg-Landau equation. II. Contraction methods,* Commun. Math. Phys., **187** (1997), pp. 45-79.

[46] A. Gmira and L. Veron, *Large time behavior of the solutions of a semilinear parabolic equation in R^N,* J. Diff. Eqs. 53 (1984), pp. 258-276.

[47] B. Guo and X.M. Xiang, *The large time convergence of spectral method for generalized Kuramoto-Sivashinsky equations,* J. Comput. Math. **15** (1997), No.1, pp. 1-13.

[48] K. Hayakawa, *On non-existence of global solutions of some semi-linear parabolic equations,* Proc. Japan Acad., **49** (1973), pp. 503-505.

[49] N. Hayashi and E.I. Kaikina, *Local existence of solutions to the Cauchy problem for nonlinear Schrödinger equations,* SUT J. Math., **34**, No. 2 (1998), pp. 111-137.

[50] N. Hayashi and E.I. Kaikina, *On the local and global existence of solutions to the nonlocal Whitham equation on a half-line,* Day on Diffraction' 99, **1**, (1999), pp. 77-87.

[51] N. Hayashi and E.I. Kaikina, *Initial-boundary value problem for the Korteweg-de Vries-Burgers equation without smallness condition on the data,* submitted.

[52] N. Hayashi, E.I. Kaikina and J.L. Guardado Zavala, *On the boundary-value problem for the Korteweg-de Vries equation,* to appear.

[53] N. Hayashi, E.I. Kaikina and R. Manzo, *Local and global existence of solutions to the nonlocal Whitham equation on a half-line,* Nonlinear Analisys, **48** (2002), pp. 53-75.

[54] N. Hayashi, E.I. Kaikina and P.I. Naumkin, *Large time behavior of solutions to the dissipative nonlinear Schrödinger equation,* Proceedings of the Royal Soc. Edingburgh, **130**-A (2000), pp. 1029-1043.

[55] N. Hayashi, E.I. Kaikina and P.I. Naumkin, *Large time behavior of solutions to the Landau-Ginzburg type equation,* Funcialaj Ekvacioj, **44** (2001), pp. 171-200.

[56] N. Hayashi, E.I. Kaikina and P.I. Naumkin, *Global existence and time decay of small solutions to the Landau-Ginzburg type equations,* to appear in J. Analysé Mathématique.

[57] N. Hayashi, E.I. Kaikina and P.I. Naumkin, *Landau-Ginzburg type equations in the sub critical case,* to appear in Commun. Contemporary Math.

[58] N. Hayashi, E.I. Kaikina and H.F. Ruiz-Paredes, *Boundary-value problem for the Korteweg-de Vries-Burgers type equation,* Nonlinear Differential Equations and Applications, **8** (2002), pp. 439-463.

[59] N. Hayashi, E.I. Kaikina and H.F. Ruiz-Paredes, *Korteweg-de Vries-Burgers equation on a half-line with large initial data,* Journal of Evolution Equations, **2** (2002), pp. 319-347.

[60] N. Hayashi, E.I. Kaikina and I.A. Shishmarev, *Asymptotics of Solutions to the Boundary-Value Problem for the Korteweg-de Vries-Burgers equation on a Half-Line,* J. Math. Anal. Appl., **265** (2002), No. 2, pp. 343-370.

[61] B.T. Hayes, *Stability of solutions to a destabilized Hopf equation,* Commun. Pure Appl. Math. **48** (1995), No.2, pp. 157-166.

[62] G. Huang and D.L. Russell, *Asymptotic properties of solutions of a KdV-Burgers equation with localized dissipation,* J. Math. Syst. Estim. Control **8** (1998), No.4, pp. 467-470.

[63] N. Hayashi and P.I. Naumkin, *Large time asymptotics of solutions to the generalized Korteweg-de Vries equation,* J.Funct.Anal., **156** (1998), pp. 110-136.

[64] N. Hayashi and P.I. Naumkin, *Large time behavior of solutions for the modified Korteweg-de Vries equation,* Inter. Math. Research Notices, (1999), pp. 395-418.

[65] N. Hayashi and P.I. Naumkin, *On the modified Korteweg - de Vries equation,* Mathematical Physics, Analysis and Geometry, **4** (2001), pp. 197-227.

[66] N. Hayashi and P.I. Naumkin, *Asymptotics for the Burgers equation with pumping,* Commun. Math. Phys., to appear.

[67] N. Hayashi, K. Nakamitsu and M. Tsutsumi, *On solutions of the initial value problem for nonlinear Schrödinger equations*, J. Funct. Anal., 71 (1987), pp. 218-245.

[68] A. Hurwitz and R. Courant, *Vorlesungen über allgemeine Funktionentheorie und elliptische Funktionen.* Interscience Publishers, Inc., New York, (1964), 534 pp.

[69] A.M. Il'in and O.A. Oleinik, *Asymptotic behavior of solutions of the Cauchy problem for some quasilinear equations for large values of time*, Mat. Sbornik. *51* (1960), pp. 191-216.

[70] E.I. Kaikina, *The Cauchy problem for the nonlocal Schrödinger equation, I, II*, Mat. Modelirovanie **3** (1991), no. 11, pp. 83-95, 96-108.

[71] E.I. Kaikina, P.I. Naumkin and I.A. Shishmarev, *Asymptotic behavior for large time of solutions to the nonlinear nonlocal Schrödinger equation on a half-line*, SUT Journal of Mathematics, **35**, No.1 (1999), pp. 37-79.

[72] E.I. Kaikina, N. Hayashi and I.A. Shishmarev, *Asymptotics for large time of solutions to the initial-boundary value problem for the Korteweg-de Vries-Burgers equation on a half-line*, Doklady Math. RAN, to appear.

[73] E.I. Kaikina, P.I. Naumkin and I.A. Shishmarev, *Asymptotics of solutions to a periodic problem for a wide class of nonlinear evolution equations*, Doklady Mathematics, **62**, No.2 (2000), pp. 245-248.

[74] E.I. Kaikina, P.I. Naumkin and I.A. Shishmarev, *Periodic problem for a model nonlinear evolution equation*, Advances in Differential Equations, **7**, No. 5 (2002), pp. 581-616.

[75] E.I. Kaikina, P.I. Naumkin and I.A. Shishmarev, *On the initial-boundary value problem for a nonlinear nonlocal Schrödinger equation*, Doklady Mathematics, **61**, No.1 (2000), pp. 120-122.

[76] E.I. Kaikina, I.A. Shishmarev and M. Tsutsumi, *Asymptotics in time for nonlinear Schrödinger equations with a source*, J. Math. Soc. Japan, **51**, No. 2, (1999), pp. 463-483.

[77] E.I. Kaikina, K. Kato, P.I. Naumkin and T. Ogawa, *Wellposedness and analytic smoothing effect for the Benjamin-Ono equations*, Publ. RIMS, Kyoto Univ., **38** (2002), No. 3, pp.

[78] S. Kamin and L.A. Peletier, *Large time behaviour of solutions of the heat equation with absorption*, Ann. Scuola Norm. Sup. Pisa, **12** (1985) pp. 393-408.

[79] S. Kamin and L.A. Peletier, *Large time behavior of solutions of the porous media equation with absorption*, Israel J. Math. **55** (1986), no.2, pp. 129-146.

[80] G. Karch, *Self-similar large time behavior of solutions to the Korteweg-de Vries-Burgers equation*, Nonlinear Anal., T.M.A. **35A** (1999), no.2, pp. 199-219.

[81] O. Kavian, *Remarks on the large time behavior of a nonlinear diffusion equation*, Ann. Inst. Henri Poincaré, Analyse non linéaire, **4** (5) (1987), pp. 423-452.

[82] C.E. Kenig, G. Ponce and L. Vega, *The Cauchy problem for the Korteweg-de Vries equation in Sobolev spaces in negative indices*, Duke Math. J., **71** (1993), pp. 1-21.

[83] C.E. Kenig, G. Ponce and L. Vega, *A bilinear estimate with applications to the KdV equation*, J. Amer. Math. Soc., **9** (1996), pp. 573-603.

[84] W. Kirsch and A. Kutzelnigg, *Time asymptotics for solutions of the Burgers equation with a periodic force*, Math. Z. **232** (1999), No.4, pp. 691-705.

[85] K. Kobayashi, T. Sirao and H. Tanaka, *On the growing up problem for semi-linear heat equations*, J. Math. Soc. Japan, **29** (1977), pp. 407-424.

[86] Yu. A. Kobelev and L.A. Ostrovskii, *Nonlinear acoustics. Theoretical and experimental studies*, Gorky, 1980, pp. 143-160.

[87] A.N. Kolmogorov and S.V. Fomin, *Elements of the Theory of Functions and Functional Analysis*, Graylock Press, N.Y., 1957.

[88] A.N. Kolmogorov, I.G. Petrovskii and N.S. Piskunov, *Étude de l'équation de la chaleur avec croissance de la quantité de matiére et son application á un probléme biologique*, Bull. Moskov. Gos. Univ., Mat. Mech., **1** (1937), No. 6, pp. 1 - 25.

[89] Y. Kuramoto, *Chemical Oscillations, Waves and Turbulence*, Springer-Verlag, Berlin, 1982.

[90] P. Lax, *Hyperbolic Systems of Conservation Laws II*, Comm. Pure Appl. Math. **10** (1957), pp. 537-566.

[91] J. L. Lions, *Quelques méthodes de résolution des problèmes aux limites non-linéaires*, Dunod, Paris, 1969.

[92] J. Lions and E. Magenes, *Non-Homogeneous Boundary Value Problems and Applications*, Springer, N.Y., 1972.

[93] T.P. Liu, *Nonlinear stability of shock waves for viscous conservation laws*, Mem. Amer. Math. Soc. **56** (1985), no. 328, pp. 1-108.

[94] T.-P. Liu, A. Matsumura and K. Nishihara, *Behavior of solutions for the Burgers equation with boundary corresponding to rarefaction waves*, SIAM J. Math. Anal., **29**, No. 2 (1998), pp. 293-308.

[95] D. Lu, L. Tian and Z. Liu, *Wavelet basis analysis in perturbed periodic KdV equation*, Appl. Math. Mech., Engl. Ed. **19** (1998), No.11, pp. 1053-1058.

[96] A. Matsumura and K. Nishihara, *Asymptotics toward the rarefaction waves of the solutions of a one-dimensional model system for compressible viscous gas*, Japan J. Appl. Math., **3**, No. 1 (1986), pp. 1 13.

[97] A. Matsumura and K. Nishihara, *Asymptotics toward the rarefaction wave of the solutions of Burgers' equation with nonlinear degenerate viscosity*, Nonlinear Anal., **23**, No. 5 (1994), pp. 605 614.

[98] A. Matsumura and K. Nishihara, *Asymptotic stability of travelling waves for scalar viscous conservation laws with non-convex nonlinearity*, Comm. Math. Phys., **265**, No. 1 (1994), pp. 83 96.

[99] N. Mizoguchi and E. Yanagida, *Critical exponents for the decay rate of solutions in a semilinear equation*, Arch. Rational Mech. Anal., **145** (1998), pp. 331-342.

[100] N. Mizoguchi and E. Yanagida, *Critical exponents for the blow up of solutions with sign changes in a semilinear parabolic equation*, Math. Ann., **307** (1997), pp. 663-675.

[101] P.I. Naumkin, *Large time asymptotics of solutions to nonlinear nonlocal equations with periodic boundary conditions*, Nonlinear Dispersive Wave Systems, ed. Lokenath Debnath, World Scientific Publ. Co. (1992), pp. 625-635.

[102] P.I. Naumkin and I.A. Shishmarev, *Periodic problem for the Whitham equation*, Math. USSR, Sbornik, **180** (1989), No. 7, pp. 946 - 968.

[103] P.I. Naumkin and I.A. Shishmarev, *Nonlinear Nonlocal Equations in the Theory of Waves*, Translations of Monographs, **133**, A.M.S., Providence, R.I., 1994.

[104] P.I. Naumkin and I.A. Shishmarev, *Asymptotics as time tends to infinity of solutions to nonlinear equations with large initial data*, Matematicheskie Zametki *59* (1996), no.6, pp. 855-864.

[105] P.I. Naumkin and I.A. Shishmarev, *Asymptotic behavior for large time of solutions of Korteweg-de Vries equation with dissipation*, Differential Equations, **29** (1993), pp. 253-263.

[106] P.I. Naumkin and I.A. Shishmarev, *Asymptotic relationship as $t \to +\infty$ between solutions to some nonlinear equations I, II*, Differential Equations **30** (1994), pp. 806-814; 1329-1340.

[107] N. Okazawa and T.Yokota, *Perturbation theory for m-accretive operators and generalized complex Ginzburg-Landau equations*, J. Math. Soc. Japan **54** (2002), pp. 1-19

[108] L.A. Ostrovsky, *Short-wave asymptotics for weak-shock waves and solitons in Mechanics*, Int. J. Non-Linear Mechanics, **11** (1976), pp. 401 - 416.

[109] E. Ott and R.N. Sudan, *Nonlinear theory of ion acoustic waves with Landau damping*, Phys. Fluids, **12**, no. 11 (1969), pp. 2388 - 2394.

[110] R.L. Pego, *Stability in Systems of Conservation laws with Dissipation*, Lect Appl. Math., **23**, (1986), pp. 345 - 357.

[111] D.H. Peregrine, *Calculations of the development of an undular bore*, J. Fluid Mech., **25** (1966), pp. 321-330.

[112] G. Ponce and L. Vega, *Nonlinear small data scattering for the generalized Korteweg - de Vries equation*, J. Funct. Anal., **90** (1990), pp. 445-457.

[113] A.V. Porubov and M.G. Velarde, *Exact periodic solutions of the complex Ginzburg-Landau equation*, J. Math. Phys. **40** (1999), No.2, pp. 884-896.

[114] K. Promislov and R. Temam, *Localization and approximation of attractors for the Ginzburg-Landau equation*, J. Dyn. Differ. Equations **3** (1991), No.4, pp. 491-514.

[115] M. Reed y B. Simon, *Methods of Modern Mathematical Physics, III Scattering Theory*, Academic Press, N.Y. 1970.

[116] D.L. Russell and B.-Y. Zhang, *Smoothing and decay properties of solutions of the Korteweg-de Vries equation on a periodic domain with point dissipation*, J. Math. Anal. Appl. **190** (1995), No.2, pp. 449-488.

[117] M.E. Schonbek, *Uniform decay rates for parabolic conservation laws*, Nonlinear Anal. T.M.A. **10** (1986), no. 9, pp. 943-953.

[118] M.E. Schonbek, *Lower bounds of rates of decay for solutions to the Navier-Stokes equations*, J. Amer. Math. Soc. **4** (1991), no.3, pp. 423-449.

[119] M.E. Schonbek and S.V. Rajopadhye, *Asymptotic behaviour of solutions to the Korteweg-de Vries-Burgers system*, Ann. Inst. Henri Poincare, Anal. Non Lineaire **12** (1995), No.4, pp. 425-457.

[120] C. Sinestrari, *Large time behaviour of solutions of balance laws with periodic initial data*, NoDEA, Nonlinear Differ. Equ. Appl. **2** (1995), No.1, pp. 111-131.

[121] G.I. Sivashinsky, *Instabilities, pattern formation and turbulence formation*, Ann. Rev. Fluid. Mech., **15** (1983), pp. 179 - 199.

[122] W.D. Stevenson, *Power System Analysis*, McGrawHill, 3 edition.

[123] W.A. Strauss, *Nonlinear scattering at low energy*, J. Funct. Anal., **41** (1981), pp. 110-133.

[124] A.G. Sveshnikov and A.N. Tikhonov, *The theory of functions of a complex variable*, "Mir", Moscow, 1982, 333 pp.

[125] S. Takeno, *Dynamical Problems in Soliton Systems*, Springer Ser. Synergetics, **30**, Springer-Verlag, Berlin and New York, 1985.

[126] L. Tian and Z. Xu, *The research of longtime dynamic behavior in weakly damped forced KdV equation*, Appl. Math. Mech., Engl. Ed. **18** (1997), No.10, pp. 1021-1028.

[127] V.V. Varlamov, *On spatially periodic solutions of the damped Boussinesq equation*, Differ. Integral Equ. **10** (1997), No.6, pp. 1197-1211.

[128] M.I. Vishik, *Asymptotic Behavior of Solutions of Evolutionary Equations*, Lezioni Lincee, Cambridge University Press Cambridge, 1992.

[129] V.S. Vladimirov, *Equations of Mathematical Physics*, Moscow, Nauka, 1976.

[130] F.B. Weissler, *Existence and non-existence of global solutions to a nonlinear heat equation*, Israel J. Math., **38** (1988), pp. 29-40.

[131] G.B. Whitham, *Linear and Nonlinear Waves*, Wiley, N.Y., 1974.

[132] J. Xing, *Global strong solution for a class of Burgers-BBM type equation*, Appl. Math., J. Chin. Univ. **6** (1991), No.1, pp. 31-37.

[133] Y. Yang, *Global spatially periodic solutions to the Ginzburg-Landau equation*, Proc. R. Soc. Edinb., Sect. A **110** (1988), No.3/4, pp. 263-273.

[134] K. Yosida, *Functional Analysis*, Springer, N.Y., 1965.

[135] Y. You, *Global dynamics of dissipative generalized Korteweg-de Vries equations*, Chin. Ann. Math., Ser. B **17** (1996), No.4, pp. 389-402.

[136] N.J. Zabusky and C.J. Galvin, *Shallow-water waves, the Korteweg-de Vries equation and solitons*, J.Fluid Mech., **47** (1971), pp. 811-824.

[137] V.E. Zakharov, *Collapse of the Langmuir Waves*, Soviet Phys. JETP **35** (1972), pp. 1745-1759.

[138] L. Zhang, *Decay estimates for the solutions of some nonlinear evolution equations*, J. Diff. Eqs. **116** (1995), pp. 31-58.

[139] E. Zuazua, *A dynamical system approach to the self-similar large time behavior in scalar convection-diffusion equations,* J. Diff. Eqs. **108** (1994), pp. 1-35.

[140] E. Zuazua, *Some recent results on the large time behavior for scalar parabolic conservation laws,* in. Elliptic and Parabolic Problems. Proc. 2nd European Conference, **325** (1995), pp. 251-263.

Index

A

Airy function 150, 154, 172, 180
analytic function 2, 9, 16, 19,
analyticity 262, 268
approximation 173, 215
a priori estimate 47, 126, 222, 225
asymptotic
    behaviour for large time 13, 30
    region xii, 241
asymptotics
    of inverse functions 11, 11
    by boundary data 31
    by initial data 31
    by nonlinearity 57
attractor 277

B

Banach space 13
behavior, asymptotic 13, 30
Benjamin-Ono equation 276
blow up xi, 208, 277
boundary
    data 13
    value 16
bounded
    function 272
    variation 225
Burgers equation 221, 221

C

Cauchy
    inequality 225
    problem 30, 93
    theorem 2, 4
characteristic
    vector 17
    matrix 23
classical solution 13

class of symbols
    K^α 9
    K^α_{diss} 10
    K^α_{disp} 10
closed ball 47, 85, 115
compact
    linear operator 287
    support 123
complete space 6
complex plane 15
conservative operator 149
conservation
    law 222, 277
    of energy 163, 174
contraction
    principle 45
    mapping 28
convergence 224
convolution 112
critical
    operator 171
    value 93

D

decay estimates 33
delta function of Dirac 152
definition of Sobolev space 7
dependence, continuous 4
derivative nonlinear Schrödinger 276
determinant of system 17, 35, 248
differential operator xiii
Dirichlet boundary data 16
dispersive symbol 10
dissipative equation xvii
dissipative symbol 10
domain 15

E